T0213206

Lecture Notes in Computer Science 14234

Founding Editors

Gerhard Goos
Juris Hartmanis

The series Lecture Notes in Computer Science (LNCS), including its subseries Lecture Notes in Artificial Intelligence (LNAI) and Lecture Notes in Bioinformatics (LNBI), has established itself as a medium for the publication of new developments in computer science and information technology research, teaching, and education.

LNCS enjoys close cooperation with the computer science R & D community, the series counts many renowned academics among its volume editors and paper authors, and collaborates with prestigious societies. Its mission is to serve this international community by providing an invaluable service, mainly focused on the publication of conference and workshop proceedings and postproceedings. LNCS commenced publication in 1973.

Gian Luca Foresti · Andrea Fusiello ·
Edwin Hancock
Editors

Image Analysis and Processing – ICIAP 2023

22nd International Conference, ICIAP 2023
Udine, Italy, September 11–15, 2023
Proceedings, Part II

 Springer

Editors
Gian Luca Foresti 🆔
University of Udine
Udine, Italy

Andrea Fusiello 🆔
University of Udine
Udine, Italy

Edwin Hancock 🆔
University of York
York, UK

ISSN 0302-9743   ISSN 1611-3349 (electronic)
Lecture Notes in Computer Science
ISBN 978-3-031-43152-4   ISBN 978-3-031-43153-1 (eBook)
https://doi.org/10.1007/978-3-031-43153-1

This Springer imprint is published by the registered company Springer Nature Switzerland AG
The registered company address is: Gewerbestrasse 11, 6330 Cham, Switzerland

Paper in this product is recyclable.

Preface

The International Conference on Image Analysis and Processing (ICIAP) is a biennial scientific meeting promoted by the Italian Association for Computer Vision, Pattern Recognition and Machine Learning (CVPL - formerly GIRPR), the Italian IAPR Member Society. The 22nd International Conference on Image Analysis and Processing (ICIAP 2023) was held in Udine, Italy, from 11 to 15 September 2023, in the prestigious venue of Palazzo di Toppo – Garzolini – Wasserman. It was co-organised by the Department of Informatics, Mathematics and Physics (DMIF) and the Polytechnic Department of Engineering and Architecture (DPIA) of the University of Udine, and sponsored by ST Microelectronics.

The conference traditionally covers topics related to theoretical and experimental areas of Computer Vision, Image Processing, Pattern Recognition and Machine Learning, with emphasis on theoretical aspects and applications. Keeping with this trend, ICIAP 2023 focused on the following areas: Pattern Recognition, Machine Learning and Deep Learning, 3D Computer Vision and Geometry, Image Analysis: Detection and Recognition, Video Analysis & Understanding, Biomedical and Assistive Technology, Digital Forensics and Biometrics, Multimedia, Cultural Heritage, Robot Vision and Automotive, Shape Representation, Recognition and Analysis, Augmented and Virtual Reality, Geospatial Analysis, and Computer Vision for UAVs.

The ICIAP 2023 main conference received 144 paper submissions from all over the world. The selection process, guided by the three Programme Chairs, resulted in the final selection of 92 high-quality manuscripts, with an overall acceptance rate of 64%.

To ensure the quality of papers ICIAP 2023 implemented a two-round review process. Each submission was managed by two Area Chairs and reviewed by at least three reviewers. Papers were selected through a double-blind peer review process, considering originality, significance, clarity, soundness, relevance and technical content.

The main conference programme included 24 oral presentations, 68 posters and three invited talks by leading experts in computer vision and pattern recognition: Danijel Skočaj (University of Ljubljana), Andrew Fitzgibbon (Graphcore), and Tomas Pajdla (CTU in Prague).

ICIAP 2023 also included 4 tutorials and hosted 15 workshops and 2 competitions, on topics of great relevance with respect to the state of the art. An industrial poster session was organised to bring together papers written by scientists working in industry and with a strong focus on application.

Several awards were presented during the ICIAP 2023 conference. The Eduardo Caianiello award was attributed to the best paper authored or co-authored by at least one young researcher. A Best Paper Award dedicated to Prof. Alfredo Petrosino was also assigned after a careful selection made by an ad hoc appointed committee.

The success of ICIAP 2023 is due to the contribution of many people. Special thanks go to all the reviewers and Area Chairs for their hard work in selecting the papers. Our thanks also go to the organising committee for their tireless efforts, advice and support.

We hope that you will find the papers in this volume interesting and informative, and that they will inspire you to further research in the field of image analysis and processing.

September 2023

Gian Luca Foresti
Andrea Fusiello
Edwin Hancock

Organization

General Chairs

Gian Luca Foresti   University of Udine, Italy
Andrea Fusiello   University of Udine, Italy
Edwin Hancock   University of York, UK

Program Chairs

Michael Bronstein   University of Oxford, UK
Barbara Caputo   Politecnico Torino, Italy
Giuseppe Serra   University of Udine, Italy

Steering Committee

Virginio Cantoni   University of Pavia, Italy
Luigi Pietro Cordella   University of Napoli Federico II, Italy
Rita Cucchiara   University of Modena-Reggio Emilia, Italy
Alberto Del Bimbo   University of Firenze, Italy
Marco Ferretti   University of Pavia, Italy
Gian Luca Foresti   University of Udine, Italy
Fabio Roli   University of Cagliari, Italy
Gabriella Sanniti di Baja   ICAR-CNR, Italy

Workshop Chairs

Federica Arrigoni   Politecnico Milano, Italy
Lauro Snidaro   University of Udine, Italy

Tutorial Chairs

Christian Micheloni   University of Udine, Italy
Francesca Odone   University of Genova, Italy

Publications Chairs

Claudio Piciarelli   University of Udine, Italy
Niki Martinel   University of Udine, Italy

Publicity/Social Chairs

Matteo Dunnhofer   University of Udine, Italy
Beatrice Portelli   University of Udine, Italy

Industrial Liaison Chair

Pasqualina Fragneto   STMicroelectronics, Italy

Local Organization Chairs

Eleonora Maset   University of Udine, Italy
Andrea Toma   University of Udine, Italy
Emanuela Colombi   University of Udine, Italy
Alex Falcon   University of Udine, Italy
Andrea Brunello   University of Udine, Italy

Area Chairs

Pattern Recognition

Raffaella Lanzarotti   University of Milano, Italy
Nicola Strisciuglio   University of Twente, The Netherlands

Machine Learning and Deep Learning

Tatiana Tommasi   Politecnico Torino, Italy
Timothy M. Hospedales   University of Edinburgh, UK

3D Computer Vision and Geometry

Luca Magri   Politecnico Milano, Italy
James Pritts   CTU Prague, Czech Republic

Image Analysis: Detection and Recognition

Giacomo Boracchi   Politecnico Milano, Italy
Mårten Sjöström   Mid Sweden University, Sweden

Video Analysis and Understanding

Elisa Ricci   University of Trento, Italy

Shape Representation, Recognition and Analysis

Efstratios Gavves   University of Amsterdam, The Netherlands

Biomedical and Assistive Technology

Marco Leo   CNR, Italy
Zhigang Zhu   City College of New York, USA

Digital Forensics and Biometrics

Alessandro Ortis   University of Catania, Italy
Christian Riess   Friedrich-Alexander University, Germany

Multimedia

Francesco Isgrò   University of Napoli Federico II, Italy
Oliver Schreer   Fraunhofer HHI, Germany

Cultural Heritage

Lorenzo Baraldi   University of Modena-Reggio Emilia, Italy
Christopher Kermorvant   Teklia, France

Robot Vision and Automotive

Alberto Pretto	University of Padova, Italy
Henrik Andreasson	Örebro University, Sweden
Emanuele Rodolà	Sapienza University of Rome, Italy
Zorah Laehner	University of Siegen, Germany

Augmented and Virtual Reality

Andrea Torsello	University of Venezia Ca' Foscari, Italy
Richard Wilson	University of York, UK

Geospatial Analysis

Enrico Magli	Politecnico Torino, Italy
Mozhdeh Shahbazi	University of Calgary, Canada

Computer Vision for UAVs

Danilo Avola	University of Roma Sapienza, Italy
Parameshachari B. D.	Nitte Meenakshi Institute of Technology, India

Brave New Ideas

Marco Cristani	University of Verona, Italy
Hichem Sahbi	Sorbonne University, France

Endorsing Institutions

International Association for Pattern Recognition (IAPR)
Italian Association for Computer Vision, Pattern Recognition and Machine Learning (CVPL)

Contents – Part II

Contents – Part I

Buffer-MIL: Robust Multi-instance Learning with a Buffer-Based Approach

Gianpaolo Bontempo[1,2], Luca Lumetti[1], Angelo Porrello[1], Federico Bolelli[1(✉)],
Simone Calderara[1], and Elisa Ficarra[1]

[1] University of Modena and Reggio Emilia, Modena, Italy
{gianpaolo.bontempo,luca.lumetti,angelo.porrello,federico.bolelli,
simone.calderara,lisa.ficarra}@unimore.it
[2] University of Pisa, Pisa, Italy
gianpaolo.bontempo@phd.unipi.it

Abstract. Histopathological image analysis is a critical area of research with the potential to aid pathologists in faster and more accurate diagnoses. However, Whole-Slide Images (WSIs) present challenges for deep learning frameworks due to their large size and lack of pixel-level annotations. Multi-Instance Learning (MIL) is a popular approach that can be employed for handling WSIs, treating each slide as a *bag* composed of multiple patches or *instances*. In this work we propose Buffer-MIL, which aims at tackling the covariate shift and class imbalance characterizing most of the existing histopathological datasets. With this goal, a buffer containing the most representative instances of each disease-positive slide of the training set is incorporated into our model. An attention mechanism is then used to compare all the instances against the buffer, to find the most critical ones in a given slide. We evaluate Buffer-MIL on two publicly available WSI datasets, Camelyon16 and TCGA lung cancer, outperforming current state-of-the-art models by 2.2% of accuracy on Camelyon16.

Keywords: Multi-instance Learning · Weakly Supervised Learning · Whole Slide Images

1 Introduction

The histopathological image analysis is a research area with a wide interest as it helps pathologists to carry out accurate diagnosis [12], especially when combined with genomic features [7,14,19]. The most common way to acquire glass slides is by employing Whole-Slide Image (WSI) scanners, which can produce digital high-resolution images [18]. Such resolutions are usually prohibitive for standard deep learning frameworks, and generating pixel-level accurate annotations represent a time-consuming and labor-intensive task. As a consequence, different strategies must be employed to perform automatic WSIs analysis and support clinicians in the daily practice. One of the most common approaches in literature follows the Multi-Instance Learning (MIL) paradigm, where from each slide

G. L. Foresti et al. (Eds.): ICIAP 2023, LNCS 14234, pp. 1–12, 2023.
https://doi.org/10.1007/978-3-031-43153-1_1

(bag) multiple unlabelled patches (instances) are extracted. These patches have a much smaller size w.r.t. the original image and can be directly fed into a deep learning network to obtain a positive or negative prediction (*e.g.,* tumor/not tumor). Once all the patch predictions are obtained, they must be aggregated to provide the final outcome for the entire slide. Indeed, bags can be perceived as a mosaic of interrelated concepts that are comprehensible only when viewed in their entirety [23].

Unfortunately, when dealing with positive bags, we also face the problem of class imbalance, as positive instances usually represent a low percentage of the entire set. Without correct precautions, the model will tend to overfit, and it might misclassify positive instances, leading to a wrong bag-level prediction. A second problem, named covariate shift, occurs when the distribution of instances within positive and negative bags differs between train and test data. This difference can force the model to focus on instances that are not actually related to the correct label [26]. This becomes crucial when dealing with *one-vs-all* cross attention paradigm [13], since the most critical instance drive the attention of all the others. Conversely, the *all-vs-all* attention (*e.g.,* self-attention in transformers) [4,8,11] approach can suffer from high-class imbalance, with instances that are often heterogeneous and noisy, making many comparisons irrelevant and even potentially derailing the final decision.

Motivated by the aforementioned challenges, this work proposes Buffer-MIL to address both class imbalance and covariate shift. To achieve this, our approach incorporates a *buffer-vs-all* strategy that makes use of a buffer to keep track of the most important instances seen during all the training process. This buffer is updated at run-time by selecting the top-k most critical instances of each positive slide in the training set. An attention mechanism is used to compare all the instances against the buffer, enabling the selection of the most critical ones to be incorporated into the learning process. This way, since the morphology of critical instances is more robust to covariate shift, we can leverage their stability to enhance the generalization performance of the model. We evaluate our approach on two publicly available WSI datasets, Camelyon16 and TCGA lung cancer, which demonstrate the effectiveness of the proposed approach. Specifically, Buffer-MIL outperforms the current state-of-the-art models by 2.2% in terms of accuracy and by 2.0% in terms of AUC on a single-scale setting.

Overall, our proposed Buffer-MIL approach provides an effective solution to address both class imbalance and covariate shift in classification tasks by leveraging a buffer containing the most critical instances, which allows for improved model performance. The source-code is available at https://github. com/aimagelab/mil4wsi.

2   Related Work

Multi-instance learning is a popular and well established type of supervised learning, whose application to the classification of WSIs is well known [3,11, 13]. In this section, recent proposals about the application of MIL to WSIs are summarized, and the covariate shift problem is introduced.

2.1   Multi-instance Learning for WSI Analysis

Initially proposed for drug activity prediction [9], the multi-instance learning paradigm gained prominence in the world of histological whole-slide image analysis. Although initially employed as a simple instance classifier, recent studies introduce an attention mechanism to extract bag representations [2,6,13,16,17, 20]. Among them, DS-MIL [13] is based on a dual-stream architecture. Patches are extracted from each considered magnification ($5\times$ and $20\times$ in their study) of the WSIs and used (separately) for self-supervised contrastive learning. Patch embeddings extracted at different resolutions are later concatenated to train the MIL aggregator, which assigns an importance (or criticality) score to each instance. The most critical patch is then selected and compared to all the others (*one-vs-all*). Such comparison is based on a distance measure that recalls an attention mechanism, but it has a substantial difference as two queries are compared instead of using the classical *key and query* approach. All the distances are then aggregated into the final bag-level prediction. Differently, Ilse *et al.* [11] propose a MIL framework (AB-MIL) where the final aggregation function is based on a weighted average. The weights assigned to each instance are computed by a gated attention mechanism. The aim of this method is to find key instances in a fully differentiable and adaptable way, by comparing instances within a bag in an *all-vs-all* fashion.

2.2   Covariate Shift

Covariate shift refers to a marginal training distribution $P_{train}(X)$ that differs from the test one $P_{test}(X)$, maintaining stable the conditional distribution $P(y|X)$ [10,21]. In other words, we have a distribution shift when the training and the test set are not independent and identically distributed. This characteristic lead a neural network to learn features that are not correlated with the correct label. To mitigate these effects a widely used approach is importance weighting, which involves assigning a weight to each training instance x. This weight, denoted as $w(x)$, is calculated as the ratio of the marginal probabilities of the instance in the test and train sets, *i.e.*, $w(x) = P_{test}(X)/P_{train}(X)$. The weight-based approach aims at reducing the discrepancy between the train and test marginals improving the generalization performance of the model [22].

As observed in Stable-MIL [26], in covariate shift settings the meaning and characteristics of noisy instances may change due to the distribution differences between train and test sets. However, critical instances, characterized by their morphology or inherent properties, tend to remain stable and consistent regardless of the covariate shift. In other words, they exhibit robustness to the distribution changes and their predictive behavior remains reliable. Therefore, by focusing on instances that are less affected by the covariate shift, we can improve model stability to also enhance the generalization performance. In our approach, we adopt an attention module to automatically identify these critical instances and store them in a buffer for further analysis and integration into the model. Such buffer is then compared against all the instances of a bag to find patches with the highest contribution.

3 Model

3.1 Notation

Firstly, the notation that will be later used in this paper is introduced to better define the concepts described. With X, X^+, and X^- are denoted generic, positive, and negative bag respectively. Instead, with x we refer to a single instance extracted from a bag.

3.2 Critical Instances

The proposed multi-instance learning framework relies on the concept of critical instances, which play a fundamental role in determining the bag label. Formally, we define x as critical if it satisfies the following two conditions:

- x belongs to a positive bag X^+;
- adding x to a negative bag X^- would change the bag's label from negative to positive, that is, $\phi(X^- \cup \{x\}) = 1$, where ϕ is the function that maps a bag to its label.

The first condition ensures that the critical instance is informative about the positive class, while the second guarantees that the instance is not present in any negative bag that should have a positive label. Thus, critical instances are those that provide evidence for the positive class and cannot be easily explained away as noise. Intuitively, critical instances, x_{crit}, contain the most important information for bag classification. On the other hand, non-critical instances, x_{noisy}, may still contribute to the overall decision but their presence or absence does not have a significant impact on the outcome.

Assumption 1. *Critical instances exhibit similar patterns, unlike x_{noisy}. So, given a feature extractor f pretrained via a self-supervised paradigm, the similarity distance $d(\cdot, \cdot)$ across critical instances is lower than the one with other non-critical instances:*

$$d(f(x_{crit}), f(x_{crit})) < d(f(x_{crit}), f(x_{noisy})) \tag{1}$$

Starting from this assumption, our model builds a buffer containing most critical instances within each positive bag X^+, which is later used to measure how other instances are relevant. Since built over the entire training set, the buffer usage provides a wider knowledge about what is really important w.r.t. using a single instance, as done by DS-MIL.

3.3 Critical Buffer

To rank instances based on their importance within each slide, a standard attention-based DS-MIL [13] is employed. In particular, given a patch x, its embedding is computed as $h = f(x)$, where the function $f(\cdot)$ is obtained from

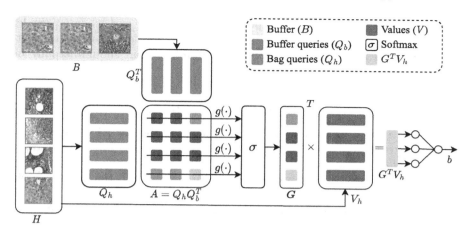

Fig. 1. Visual representation of the proposed model. In particular, given the buffer B and the input slide H, the attention matrix A is computed. The $g(\cdot)$ function is used to select the most informative elements from the matrix into G.

a self-supervised approach. A patch-level classifier $cls_{patch}(\cdot)$ is used to find the index of the most critical patch as:

$$\text{crit} = \text{argmax}(cls_{patch}(f(x))) = \text{argmax}\{W_p f(x_0), ..., W_p f(x_n)\} \qquad (2)$$

where W_p is a weight vector.

The second step is to aggregate instance embeddings into a single bag embedding. This is performed by computing a linear projection of each embedding into a query q_i and a value v_i, using two weight matrices W_q and W_v:

$$q_i = W_q h_i, \quad v_i = W_v h_i \qquad (3)$$

Next, the query relative to the most critical instance, q_{crit}, is obtained and compared to all other queries q_i (including itself) using a distance measure $U(\cdot, \cdot)$ defined as:

$$U(h_i, h_{\text{crit}}) = \frac{\exp(\langle q_i, q_{\text{crit}} \rangle)}{\sum_{k=0}^{N-1} \exp(\langle q_k, q_{\text{crit}} \rangle)} \qquad (4)$$

Finally, the bag score is given by:

$$c_b(B) = W_b \sum_{i=0}^{N-1} U(h_i, h_{\text{crit}}) v_i \qquad (5)$$

where W_b is again a weight vector. The bag score is used to select all the positive bags and extract the top-k instances within each of them. The ranking is given by the score $U(h_i, h_{\text{crit}})$. The buffer is build by training the aforementioned model; at the end of the process it contains the most critical instances of each bag, providing a more stable criticality representation.

The selection of the N most important patches from each slide (N/Slide) is repeated every $freq$ epochs, since the network should learn to assign a better score to bags and instances, better understanding what should actually be considered as critical.

3.4 Bag Embedding Through the Critical Buffer

Figure 1 illustrates how the buffer B is introduced in the attention mechanism. Given the current bag $H = \{h_1, ..., h_i, ..., h_N\}$, composed of N instances, and the buffer $B = \{b_1, ..., b_i, ..., b_M\}$, composed of M critical instances belonging from different slides, a new bag embedding can be computed. First, the weight matrix W_q trained in the previously described steps is used to perform a linear projection of all the instances h_i and all the instances within the buffer b_i, obtaining q_{h_i} and q_{b_i} respectively. An attention matrix A is then built, where $A_{i,j} = \langle q_{h_i}, q_{b_i} \rangle$. This can also be seen as a matrix multiplication, once defined $Q_h \in \mathcal{M}^{N \times K}$ as the row-wise concatenation of every q_{h_i} and $Q_b \in \mathcal{M}^{M \times K}$ as the row-wise concatenation of q_{b_i}, considering K the latent space size where each instance get projected, the attention matrix $A \in \mathcal{M}^{N \times M}$ can be written as follow:

$$A = Q_h Q_b^T \tag{6}$$

As only a single attention score is required for each of the bag instances h_i, an aggregation function $g(\cdot)$ on each row of A must be used to obtain a new matrix $G \in \mathcal{M}^{N \times 1}$ as $G_i = g(\{A_{i,j} : \forall j \in [1, M]\})$.

All the instances h_i are also projected into values v_{h_i} of size L using the W_v weight matrix of the previous step, obtaining $V_h \in \mathcal{M}^{N \times L}$. Finally, the bag embedding is computed as:

$$b = W_b G^T V_h \tag{7}$$

with $W_b \in \mathcal{M}^{1 \times L}$ representing the weight matrix that computes the final bag embedding. In this paper, two different function $g(\cdot)$ are proposed:

- **mean:** the attention scores are computed considering the entire buffer, under the assumption that it is composed of critical instances only. In particular $G_i = \text{mean}\{A_{ij} : \forall j \in [1, M]\}$;
- **max:** considering that the buffer may also contain noisy labels, using a max-pooling operation allows to select only the most representative instances. Specifically, $G_i = \text{max}\{A_{ij} : \forall j \in [1, M]\}$

4   Experimental Settings and Results

4.1 Pre-processing

Each slide has been cropped using the CLAM framework [15], a state-of-the-art tool for selecting tissue patches and removing the WSI background. In particular, each slide has been processed at thumbnail level through a combination of Otsu thresholding [25] and connected components analysis [1], to obtain the tissue contours. After that, each 256×256 patch within the selected contours is extracted without overlapping at $20\times$ scale resolution ($5\times$ and $20\times$ in the multi-scale setting).

Finally, instance embeddings are obtained through a ViT model trained in a self-supervised fashion by means of the DINO paradigm [5]. The training is performed separately on each dataset/resolution. The model has been trained for a week with two NVIDIA GeForce GTX 2080 Ti GPUs using the default parameters proposed by the authors.

4.2 Metrics

The evaluation metrics considered are the Area Under the Curve (AUC) and the accuracy. As the name suggests, the AUC measures the area under the ROC curve, representing the relationship between the true positive rate, $TPR = TP/(TP + FN)$, and the false positive rate, $FPR = FP/(FP + TN)$, for any possible threshold. Once the best threshold for the ROC curve is found, we measure the accuracy as the quantity of TP over the entire test set. Each experiment has been executed with 3 different seeds, reporting the average and the standard deviation.

4.3 Datasets

The proposed method has been extensively tested over two different datasets: Camelyon16 and TCGA Lung. The former has been created with the purpose of automatic detection of metastases in Hematoxylin and Eosin (H&E) stained whole-slide images of lymph node sections, as part of the homonymous challenge held at the International Symposium on Biomedical Imaging (ISBI) in 2016 [2]. The dataset comprises a total of 398 WSIs, out of which 128 are designated as "official test set". The images were acquired through two slide scanners, namely RUMC and UMCU, respectively equipped with $20\times$ and $40\times$ objective lenses. The specimen-level pixel sizes are comparable, *i.e.*, $0.243\,\mu m \times 0.243\,\mu m$ for RUMC and $0.226\,\mu m \times 0.226\,\mu m$ for UMCU. Official training and test set have been employed for our experiments.

The second dataset, publicly available on the GDC Data Transfer Portal, comprises two sub-types of cancer: Lung Adenocarcinoma (LUAD) and Lung Squamous Cell Carcinoma (LUSC), counting 541 and 513 WSIs respectively. In this case, the task is the classification of LUAD vs LUSC. To provide a fair comparison with Li *et al.* [13], we employ the same split between train and test set and remove ten corrupted slides as suggested in the original publication.

4.4 Results

Table 1 compares the proposed Buffer-MIL with state-of-the-art approaches: two MIL models with simple aggregators like mean-pooling and max-pooling, Attention-based MIL (AB-MIL) [11], DS-MIL, and its multi-scale version [13]. We also extend the buffer-based approach to consider multiple resolutions.

Table 1. Performance comparison on Camelyon16 and TCGA Lung dataset. The "†" identifies multi-scale approaches. Buffer aggregation is based on mean in these experiments.

Model	Camelyon16		TCGA Lung	
	Accuracy	AUC	Accuracy	AUC
mean-pooling	0.723 ± 0.004	0.672 ± 0.010	0.823 ± 0.002	0.905 ± 0.001
max-pooling	0.893 ± 0.015	0.899 ± 0.007	0.851 ± 0.008	0.909 ± 0.002
AB-MIL	0.724 ± 0.015	0.744 ± 0.016	0.864 ± 0.009	0.933 ± 0.004
DS-MIL	0.915 ± 0.013	0.952 ± 0.005	0.888 ± 0.005	$\mathbf{0.951 \pm 0.002}$
Buffer-MIL	$\mathbf{0.935 \pm 0.012}$	$\mathbf{0.971 \pm 0.005}$	$\mathbf{0.891 \pm 0.008}$	0.950 ± 0.002
DS-MIL†	0.909 ± 0.020	0.955 ± 0.010	$\mathbf{0.913 \pm 0.005}$	$\mathbf{0.966 \pm 0.002}$
Buffer-MIL†	$\mathbf{0.940 \pm 0.008}$	$\mathbf{0.969 \pm 0.005}$	0.897 ± 0.020	0.956 ± 0.010

Table 2. Comparison between the usage of max and mean aggregation (Agg.) by setting the buffer update frequency to 10.

Agg.	N/slide	Accuracy	AUC
Mean	1	0.934 ± 0.012	0.970 ± 0.006
	2	0.932 ± 0.012	0.968 ± 0.006
	10	$\mathbf{0.935 \pm 0.012}$	$\mathbf{0.971 \pm 0.005}$
Max	1	0.925 ± 0.012	0.966 ± 0.004
	2	0.927 ± 0.020	0.967 ± 0.005
	10	0.930 ± 0.021	0.967 ± 0.003

From a single scale perspective, using the buffer improves the baseline by an average of 2.2% in accuracy and 2.0% in AUC on the Camelyon16 and 0.3% in accuracy for the TCGA Lung dataset. Employing multiple resolutions generally provide better performances: on Camelyon16 the buffer improves the baseline by an average of 3.4% in accuracy and 1.5% in AUC.

4.5   Model Analysis

Our experiments provide evidence that Buffer-MIL is effective at tackling covariate shift, as demonstrated by the higher performance improvement obtained on Camelyon16 compared to TCGA Lung (Table 1). Given the smaller size of Camelyon16, overfitting can become a critical issue, slightly attenuated by the multi-scale approach.

Aggregation Function. Two different aggregation functions have been studied and presented in Table 2. Experimental results reveal that producing the final attention scores by averaging critical representations in the buffer outperforms the use of a max operator.

Table 3. Contribution of buffer update frequency (Freq.) when using mean-based aggregation.

Freq.	N/slide	Accuracy	AUC
1	10	0.919 ± 0.012	0.963 ± 0.004
2	10	0.917 ± 0.009	0.967 ± 0.001
10	10	$\mathbf{0.935 \pm 0.012}$	$\mathbf{0.971 \pm 0.005}$

One possible explanation is that selecting only the most representative disease-positive buffer instances produces a final representation that is not aligned with all the bags. This approach may not capture the diversity of the disease-positive instances and may lead to sub-optimal performance. In contrast, the mean operator takes into account all the critical instances, which allows for a stronger consensus. This approach is better at capturing the diversity of disease-positive instances and is less likely to overfit specific patches. Furthermore, the mean operator is less sensitive to outliers and noise that may be contained in the buffer.

Buffer Update Frequency. This hyperparameter regulates the interval (measured in epochs) between each buffer update. In Table 3, we also investigate the impact of buffer update frequency, which is found to be an important parameter for both max and mean operators.

Our analysis suggests that updating the buffer fewer times generally leads to better performances, as it allows for a better selection of the most representative disease-positive instances across the entire training set. Updating the buffer with an higher frequency prevents its consolidation, and may cause it to be filled with noisy or irrelevant information. Instead, updating the buffer less frequently increases the time interval between buffer creations, causing it to become outdated and failing to capture the most relevant instances. Setting an appropriate interval is required by the model to learn and generalize from the initial training data before incorporating new information into the buffer. In other words, the model can better consolidate the knowledge from the initial training data, and, consequently, perform a better selection of new instances. It is essential to find the right trade-off.

Buffer Size. The buffer is built considering the N most critical instances from each slide. As illustrated in Table 4, our analysis demonstrates that the impact of buffer size is less significant w.r.t. buffer update frequency. Our experiments also suggest that increasing the buffer size does not always lead to improved performance.

One possible explanation is that when the buffer frequency update is low, increasing the buffer size may include more irrelevant or noisy instances, which could negatively impact the model performance. In this scenario, selecting a

Table 4. Buffer size contribution at different update frequencies when using mean-based aggregation.

Freq.	N/slide	Accuracy	AUC
1	1	0.922 ± 0.008	0.962 ± 0.002
	2	$\mathbf{0.925 \pm 0.012}$	$\mathbf{0.961 \pm 0.003}$
	10	0.919 ± 0.012	0.963 ± 0.004
10	1	0.934 ± 0.012	0.970 ± 0.006
	2	0.932 ± 0.012	0.968 ± 0.006
	10	$\mathbf{0.935 \pm 0.012}$	$\mathbf{0.971 \pm 0.005}$

Table 5. Comparison with random sampling when using mean-based aggregation and a frequency update of 10.

N/slide	Our Method		Reservoir Sampling	
	Accuracy	AUC	Accuracy	AUC
1	0.934 ± 0.012	0.970 ± 0.006	0.922 ± 0.014	0.962 ± 0.003
2	0.932 ± 0.012	0.968 ± 0.006	0.922 ± 0.008	0.963 ± 0.004
10	$\mathbf{0.935 \pm 0.012}$	$\mathbf{0.971 \pm 0.005}$	0.925 ± 0.012	0.964 ± 0.004

larger number of instances per slide could cause the buffer to become more "diluted" with irrelevant instances. As a result, the model may not be able to properly consolidate and learn from the most critical instances, leading to a decrease in performance.

On the other hand, when the buffer update frequency is high, the buffer can better capture the most critical disease-positive instances, even if the buffer size is small. In this case, the mean operator typically works better on bigger buffers, but small buffer sizes can still perform comparably well. Selecting the optimal buffer size depends on the specific dataset and task, as well as the buffer update frequency.

Sampling Selection. To provide evidence that selecting proper patches matter, in Table 5 we show a comparison between our proposed method, and the reservoir sampling strategy [24], which is a random-based selection technique. The results demonstrate that our approach outperforms the random selection strategy regardless of the parameters used.

5 Conclusion

In conclusion, our analysis demonstrates that Buffer-MIL is an effective approach for addressing the problem of covariate shift when multi-instance learning is applied to the histopathological context. In particular, the results suggest that performing an appropriate buffer selection approach and identifying the correct interval for updating the buffer are critical to achieve optimal performance.

Further research is needed to investigate how relevant buffers are in more difficult and diverse tasks such as survival prediction. In that case, tissue morphology is not directly connected to the patient outcome and a better storage strategy (*e.g.*, multiple buffers per concept) would be probably needed.

Acknowledgements. This project has received funding from DECIDER, the European Union's Horizon 2020 research and innovation programme under GA No. 965193, and from the Department of Engineering "Enzo Ferrari" of the University of Modena through the FARD-2022 (Fondo di Ateneo per la Ricerca 2022).

References

1. Allegretti, S., Bolelli, F., Cancilla, M., Pollastri, F., Canalini, L., Grana, C.: How does connected components labeling with decision trees perform on GPUs? In: Vento, M., Percannella, G. (eds.) CAIP 2019. LNCS, vol. 11678, pp. 39–51. Springer, Cham (2019). https://doi.org/10.1007/978-3-030-29888-3_4

2. Bejnordi, B.E., et al.: Diagnostic assessment of deep learning algorithms for detection of lymph node metastases in women with breast cancer. JAMA **318**(22), 2199–2210 (2017)

3. Bontempo, G., Porrello, A., Bolelli, F., Calderara, S., Ficarra, E.: DAS-MIL: distilling across scales for MIL classification of histological WSIs. In: Medical Image Computing and Computer Assisted Intervention - MICCAI 2023 (2023)

4. Bruno, P., Amoroso, R., Cornia, M., Cascianelli, S., Baraldi, L., Cucchiara, R.: Investigating bidimensional downsampling in vision transformer models. In: Sclaroff, S., Distante, C., Leo, M., Farinella, G.M., Tombari, F. (eds.) Image Analysis and Processing - ICIAP 2022, pp. 287–299. Springer, Cham (2022). https://doi.org/10.1007/978-3-031-06430-2_24

5. Caron, M., et al.: Emerging properties in self-supervised vision transformers. In: Proceedings of the IEEE/CVF International Conference on Computer Vision (ICCV), pp. 9650–9660 (2021)

6. Chen, R.J., et al.: Scaling vision transformers to gigapixel images via hierarchical self-supervised learning. In: IEEE/CVF Conference on Computer Vision and Pattern Recognition (CVPR), pp. 16144–16155 (2022)

7. Chen, R.J., et al.: Pathomic fusion: an integrated framework for fusing histopathology and genomic features for cancer diagnosis and prognosis. IEEE Trans. Med. Imaging **41**(4), 757–770 (2020)

8. Cornia, M., Baraldi, L., Cucchiara, R.: Explaining transformer-based image captioning models: an empirical analysis. AI Commun. **35**(2), 111–129 (2022)

9. Dietterich, T.G., Lathrop, R.H., Lozano-Pérez, T.: Solving the multiple instance problem with axis-parallel rectangles. Artif. Intell. **89**(1), 31–71 (1997)

10. Huang, J., Gretton, A., Borgwardt, K., Schölkopf, B., Smola, A.: Correcting sample selection bias by unlabeled data. In: Advances in Neural Information Processing Systems, vol. 19 (NIPS) (2006)

11. Ilse, M., Tomczak, J., Welling, M.: Attention-based deep multiple instance learning. In: International Conference on Machine Learning, vol. 80, pp. 2127–2136. PMLR, July 2018

12. Kumar, N., Gupta, R., Gupta, S.: Whole Slide Imaging (WSI) in pathology: current perspectives and future directions. J. Digit. Imaging **33**(4), 1034–1040 (2020)
13. Li, B., Li, Y., Eliceiri, K.W.: Dual-stream multiple instance learning network for whole slide image classification with self-supervised contrastive learning. In: IEEE/CVF Conference on Computer Vision and Pattern Recognition (CVPR), pp. 14318–14328 (2021)
14. Lovino, M., Bontempo, G., Cirrincione, G., Ficarra, E.: Multi-omics classification on kidney samples exploiting uncertainty-aware models. In: Huang, D.-S., Jo, K.-H. (eds.) ICIC 2020. LNCS, vol. 12464, pp. 32–42. Springer, Cham (2020). https://doi.org/10.1007/978-3-030-60802-6_4
15. Lu, M.Y., Williamson, D.F., Chen, T.Y., Chen, R.J., Barbieri, M., Mahmood, F.: Data-efficient and weakly supervised computational pathology on whole-slide images. Nat. Biomed. Eng. **5**(6), 555–570 (2021)
16. Maksoud, S., Zhao, K., Hobson, P., Jennings, A., Lovell, B.C.: SOS: selective objective switch for rapid immunofluorescence whole slide image classification. In: IEEE/CVF Conference on Computer Vision and Pattern Recognition (CVPR), pp. 3862–3871 (2020)
17. Panariello, A., Porrello, A., Calderara, S., Cucchiara, R.: Consistency-based self-supervised learning for temporal anomaly localization. In: Karlinsky, L., Michaeli, T., Nishino, K. (eds.) Computer Vision - ECCV 2022 Workshops, vol. 13805, pp. 338–349. Springer, Cham (2022). https://doi.org/10.1007/978-3-031-25072-9_22
18. Ponzio, F., Urgese, G., Ficarra, E., Di Cataldo, S.: Dealing with lack of training data for convolutional neural networks: the case of digital pathology. Electronics **8**(3) (2019)
19. Roberti, I., Lovino, M., Di Cataldo, S., Ficarra, E., Urgese, G.: Exploiting gene expression profiles for the automated prediction of connectivity between brain regions. Int. J. Mol. Sci. **20**(8), 2035 (2019)
20. Shao, Z., et al.: TransMIL: transformer based correlated multiple instance learning for whole slide image classification. In: Advances in Neural Information Processing Systems (NeurIPS), vol. 34, pp. 2136–2147 (2021)
21. Shimodaira, H.: Improving predictive inference under covariate shift by weighting the log-likelihood function. J. Stat. Plann. Inference **90**(2), 227–244 (2000)
22. Sugiyama, M., Nakajima, S., Kashima, H., Buenau, P., Kawanabe, M.: Direct importance estimation with model selection and its application to covariate shift adaptation. In: Advances in Neural Information Processing Systems (NIPS), vol. 20 (2007)
23. Tu, M., Huang, J., He, X., Zhou, B.: Multiple instance learning with graph neural networks. In: ICML Workshop on Learning and Reasoning with Graph-Structured Representations (2019)
24. Vitter, J.S.: Random sampling with a reservoir. ACM Trans. Math. Softw. **11**(1), 37–57 (1985)
25. Zhang, J., Hu, J.: Image segmentation based on 2D Otsu method with histogram analysis. In: International Conference on Computer Science and Software Engineering, vol. 6, pp. 105–108. IEEE (2008)
26. Zhang, W., Li, J., Liu, L.: Robust multi-instance learning with stable instances. In: ECAI 2020: 24th European Conference on Artificial Intelligence (2019)

Quasi-Online Detection of Take and Release Actions from Egocentric Videos

Rosario Scavo[1,2(✉)], Francesco Ragusa[1,2], Giovanni Maria Farinella[1,2], and Antonino Furnari[1,2]

[1] FPV@IPLAB, DMI - University of Catania, Catania, Italy
rosario.scavo@studium.unict.it
[2] Next Vision s.r.l. - Spinoff of the University of Catania, Catania, Italy
https://iplab.dmi.unict.it/fpv/, https://www.nextvisionlab.it/

Abstract. In this paper, we considered the problem of detecting object take and release actions from untrimmed egocentric videos in an industrial domain. Rather than requiring that actions are recognized as they are observed, in an online fashion, we propose a quasi-online formulation in which take and release actions can be recognized shortly after they are observed, but keeping a low latency. We contribute a problem formulation, an evaluation protocol, and a baseline approach that relies on state-of-the-art components. Experiments on ENIGMA, a newly collected dataset of egocentric untrimmed videos of human-object interactions in an industrial scenario, and on THUMOS'14 show that the proposed approach achieves promising performance on quasi-online take/release action recognition and outperforms methods for online detection of action start on THUMOS'14 by +8.64% when an average latency of $2.19s$ is allowed. Code and supplementary material are available at https://github.com/fpv-iplab/Quasi-Online-Detection-Take-Release.

Keywords: Quasi-Online Action Detection · Low Latency · Take/Release Actions · Egocentric Untrimmed Videos

1 Introduction

Egocentric vision aims to analyze images and videos acquired from the user's point of view through a wearable device equipped with a camera (e.g., smart glasses) to understand how the user interacts with the environment and possibly provide assistance. Understanding users' activities and interactions with objects from egocentric visual signals allows to provide services to support humans in different domains such as homes, kitchens, museums, and industrial workplaces [2,4,13]. Since humans interact with objects using their hands, recognizing actions such as "take an object" and "release an object" can offer crucial insights into the user's intentions, especially in industrial environments.

Supplementary Information The online version contains supplementary material available at https://doi.org/10.1007/978-3-031-43153-1_2.

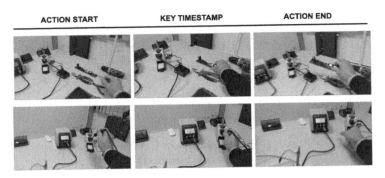

Fig. 1. Two examples of take (top) and release (bottom) actions. For each action, we define a key timestamp that corresponds to the first frame of contact (contact frame) in the take examples and to the last frame of contact (end-of-contact frame) in the release examples. These timestamps are distinct from classic action start and end time which are often ambiguous to annotate [11].

For instance, predicting the user's next action based on the tool they have taken, allows to provide assistance offering feedback to correct erroneous actions (e.g., "Please complete step X before picking up the pliers"). Additionally, this knowledge can be used to suggest ways to improve efficiency and reduce errors, such as recommending the optimal use of the object that was just picked up (e.g., triggering AR contents). Understanding take/release interactions taking place between humans and objects also allows to estimate the usage time of an object, possibly enabling predictive maintenance applications. Critically, such actions should be predicted in a timely fashion in order to provide useful assistance to workers as soon as possible.

In this paper, we consider the problem of detecting two key actions from egocentric videos: "take" and "release". These two actions occur respectively when the user takes an object and when they put it down. We assume a low-latency, quasi-online scenario in which take and release actions can be detected as soon as possible (e.g., in a few seconds) after they are observed from an input streaming video while aiming to keep a low latency to allow making decisions, such as sending an alarm, or notifying that a wrong action occurred in a maintenance procedure. We would like to note that the considered scenario is realistic and of practical relevance in contexts where the system aims to support the user, such as industrial environments. Indeed, in this context, an "after-the-fact" detection of actions with a low latency is useful for the verification of incomplete procedural tasks performed by workers.

To study the considered problem in the industrial domain, we collected and labeled ENIGMA, a dataset of egocentric videos in which several users performed repair and maintenance procedures on electrical boards in an industrial laboratory. Previous literature highlights how labeling start and end times, even for simple take and release actions, can lead to inconsistencies due to the limited agreement between different annotators [11]. In order to avoid bias due to these inconsistencies, each take and release action has been labeled by

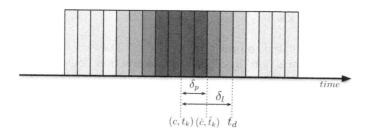

Fig. 2. Given an action of class c executed at the ground truth timestamp t_k, our goal is to estimate its key timestamp $\hat{t}_k$ and its class $\hat{c}$ while keeping a low latency $\delta_l = \max(0, t_d - t_k)$, where t_d is the timestamp in which the prediction is made. For a prediction to be deemed valid, we also expect the distance $\delta_p = |\hat{t}_k - t_k|$ between the estimated key timestamp and the ground truth one to be under a given temporal threshold ϕ.

marking a single timestamp indicating the first frame in which the hand touches an object (contact frame) in the case of take actions, or the first frame in which there is no more contact between the hand and the object (end-of-contact frame) for the case of release actions. The timestamps related to these frames are also referred to as "key timestamps" (see Fig. 1). We hence treat the problem of quasi-online take/release action detection as the one of processing an untrimmed input video and predicting a set of take/release action events with a low latency (see Fig. 2). Based on this problem definition, we designed an evaluation protocol aimed to assess the ability of the models to accurately detect the occurrence of take and release actions, as well as their latency. Therefore, we propose an approach to tackle this task based on a state-of-the-art transformer model for online action detection [19] in conjunction with a post-processing technique that aims to identify action occurrences from the analysis of time series of online prediction scores. Experiments on the collected dataset of egocentric videos show the feasibility of the proposed approach, which achieves promising results despite the task being challenging. Moreover, experimental results on THUMOS'14 demonstrate that our framework outperforms state-of-the-art online detection of action start methods by +8.64% when a quasi-online formulation is considered and an average latency of 2.19s is allowed, which shows the flexibility of the proposed problem formulation and approach in realistic scenarios.

The main contributions of this paper are as follows: 1) We investigate the problem of detecting take and release actions in egocentric videos in a quasi-online manner. 2) We designed an evaluation protocol to assess the accuracy and low-latency performance of the proposed models in predicting take/release actions. 3) We introduced a novel approach to address this task, utilizing cutting-edge transformers and a time series post-processing technique to refine the detection scores. Our code is publicly available to the research community to facilitate future research.

2   Related Works

Our investigation is related to previous works on online action detection, online detection of action starts, and human-object interaction detection.

Online Action Detection. Online action detection aims to detect an action as it happens from video and, ideally, even before it is fully completed [3]. While offline methods assume the entire video is available, online methods process the data up to the current time, predicting the action observed in each frame without looking at future frames. Different approaches over the years have been proposed to solve the online action detection task. Recent works employ Transformer architectures [15], which provide a more effective way to represent and model long data sequences than RNN architectures [10]. OadTR [17] is an encoder-decoder framework based on Transformers that recognize current actions by encoding historical information and predicting future context simultaneously. LSTR [18] explicitly divides the entire history into long- and short-term memories. TeS-Tra [19] is a state-of-the-art approach that uses the same strategy of LSTMs. TeSTra improves the computational efficiency of video transformers by applying temporal smoothing kernels to the cross-attention, resulting in streaming attention that only requires a constant time to update each frame.

The problem formulation proposed in this study is different from online action detection. Indeed, rather than expecting predictions to be made in real-time, we allow models to fully observe actions before determining whether an action should be predicted. While this formulation simplifies the prediction problem, we still assess that models have a low latency to ensure they are practically useful. We build on previous literature on online action detection by incorporating TeSTra into our framework to generate confidence scores for observed actions on a frame-by-frame basis.

Online Detection of Action Start. Some works have explored the problem of Online Detection of Action Start (ODAS) [14] from a third-person perspective. These approaches differ from online action detection in that its primary goal is to detect the start of an action as precisely as possible. Previous authors have employed methodologies such as 3D convolutions [14], a combination of LSTM and reinforcement learning [6], and weakly-supervised learning with video-level labels [7].

The ODAS formulation shares some similarities with the formulation of our problem but it is distinct. Indeed, while ODAS aims to detect the action start by observing the video immediately preceding the action, we aim to detect the key timestamp indicating the execution of a take or release action, after the complete observation of the action, but with a low latency.

Human-Object Interaction Detection. Human-Object Interaction (HOI) detection aims to identify and locate both the human and the object in an image or video and to recognize their interactions. The main HOI detection methods are based on the use of GCN after detecting humans and objects in the scene [12], on human-centric approaches [8], on the detection of the interactions between human-object pairs [16], or on and grasp analysis [1]. In recent years, HOIs have also been studied in Egocentric Vision (Egocentric Human-Object Interaction - EHOI) [13]. However, only few works specifically considered industrial scenarios [13]. Moreover, while the aforementioned works have focused on understanding human-object interactions in still frames, we focus on detecting when take/release actions occur in a streaming video. We expect that our approach could be integrated with HOI analysis to provide a more accurate and grounded output.

3   Problem Definition and Evaluation Protocol

Let (c, t_k) represent a ground truth action, where c is the action class (take or release) and t_k is the related key timestamp (either contact or end-of-contact). Each prediction is represented as a $(\hat{c}, \hat{t}_k, t_d, s)$ tuple, where $\hat{c}$ and $\hat{t}_k$ are respectively the predicted class and key timestamp, t_d is the timestamp in which the prediction is actually made, and s is a confidence score. We define the temporal distance between the correct key timestamp, t_k, and the predicted one, $\hat{t}_k$, as $\delta_p = |\hat{t}_k - t_k|$. We will consider a prediction as correct if δ_p is under a given temporal threshold ϕ. Given a correct prediction, we define its latency as the difference between the timestamp in which the prediction is made, t_d, and the corresponding ground truth key timestamp, t_k: $\delta_l = \max(t_d - t_k, 0)$. Note that the $t_d - t_k$ difference is in general a positive number, but it may assume negative values in rare cases in which the action is predicted a few moments before it happens. In such cases, we will consider a latency equal to zero, hence the max operator in our definition of latency. We do not define latency for incorrect predictions. It is worth noting that, while in an online prediction scenario we impose $t_d = \hat{t}_k$, in the considered quasi-online scenario, we allow the two timestamps to differ, but expect a low latency δ_l. Figure 2 illustrates the considered problem.

Evaluation Protocol. Metrics generally used to evaluate action detection, such as mAP are not suitable in our scenario, as they assume that both action start and end are predicted, while in our case, an action is associated to a single timestamp. Instead, we adopt the point-level detection mAP (**p-mAP**) defined in [14]. Predicted actions are deemed to be correct only when 1) the predicted action class is correct ($c = \hat{c}$), 2) the temporal offset δ_p is smaller than a given evaluation temporal threshold ϕ. Predictions are matched to ground truth actions in a greedy fashion, prioritizing predicted actions with higher confidence scores, checking whether $\delta_p \leq \phi$, and imposing that ground truth actions and predictions are matched at most once. Based on these matches, the point-level Average Precision (AP) for each action class is evaluated and averaged over all action classes to determine the point-level mAP. Consistently with prior works [6,7], we evaluated the p-mAP at temporal offset thresholds ϕ ranging from 1 to 10 s in one-second increments. We defined **mp-mAP** as the average of p-mAP values calculated at different temporal offset thresholds ϕ.

Given our quasi-online problem definition, we complement point-level mAP with an evaluation of the average latency of a given approach. Specifically, we

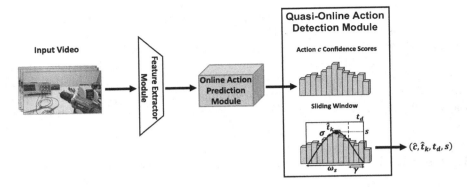

Fig. 3. Overview of the proposed method. Firstly, the input video is processed using a feature extractor. The resulting features are then fed into an online action prediction module, which outputs action confidence scores. Confidence scores are then passed to the quasi-online action detection module which processes them with a sliding window of size ω_s. Confidence scores within the window are smoothed using a Gaussian filter, and a peak detector is employed to temporally detect the action.

compute the average latency as follows: $\overline{\delta}_l = \frac{\sum_{i=1}^{N} tp(i) \cdot \delta_l(i)}{\sum_{i=1}^{N} tp(i)}$, where i is the index of the i-th prediction made by the model, N is the total number of predictions, $tp(i) = 1$ if prediction i is a true positive and $tp(i) = 0$ otherwise, and $\delta_l(i)$ denotes the latency of prediction i.

4   Proposed Approach

Our method comprises three main modules: 1) a feature extraction module that processes the input video and outputs per-frame features, 2) a transformer-based online action prediction module that takes the features as input and predicts actions along with their confidence scores frame-by-frame, in an online fashion, and 3) a quasi-online action detection module which takes as input a window of frame-by-frame action confidence scores and predicts the occurrence of actions in the considered window. Figure 3 shows the overall architecture of the proposed method. The following sections detail each component.

Feature Extraction Module. The feature extraction module takes as input the streaming video and produces per-frame high-level representations. This module may analyze input video clips or single frames. In this work, following [5], we extract per-frame features using a Two-Stream (TS) CNN model. In particular, the TS-features comprise appearance features that focus on the video's visual appearance information and motion features that rely on the user's and objects' movement during the actions. Features are then fused to obtain a unique representation for each video chunk, which is passed to the transformer-based online action prediction module.

Transformer-Based Online Action Prediction Module. We base this module on the state-of-the-art online action prediction approach TeSTra [19]. TeSTra takes pre-extracted features as input and outputs per-frame probability distributions of action classes. In practice, this module allows to predict two confidence scores at each frame: the probability of take actions, and the probability of release actions.

Quasi-Online Action Detection Module. This module takes as input time series of confidence scores produced by the online action prediction module and analyzes them to predict the presence of take/release actions, along with the estimated timestamps in which they take place and associated confidence scores. Rather than requiring the system to predict whether an action is performed in the current frame, we process the time series of predicted confidence scores with a sliding window of size ω_s up to the current timestamp t, and allow the prediction of actions observed within the window. We expect the predicted confidence score of an action to increase before the occurrence of the key timestamp and to decrease after it, as shown in Fig. 4-left. Hence, we aim to predict actions by detecting peaks in the confidence score time series. In practice, confidence score time series tend to be noisy due to the uncertainty of the model and the ambiguity of observations (Fig. 4-middle). To account for this, a Gaussian filter is applied within the current window with a fixed standard deviation σ in order to provide a smooth time series of confidence scores. We hence use a peak detector[1] to find the occurrence of each take/release action. Each peak is considered as a prediction, and the peak height is the associated confidence score (see Fig. 4-right). To avoid predictions due to incomplete observations, we ignore all predictions made in the last γ seconds of the sliding window. We refer to γ as "inhibition time". Note that predictions suppressed at time t because they fall in the inhibition time interval can be recovered later by the model when the sliding window has moved forward and the same prediction does not fall within the inhibition time segment anymore. By iterating over the video with the proposed sliding window approach, the same action may be detected more than once in the video, due to the natural overlap between the considered prediction windows. To avoid multiple detections of the same action, we discard new detections whose difference with previously made predictions is under a given threshold ξ.

5   Experimental Settings and Results

In this section, we report the settings and results of our experimental analysis aimed at evaluating the proposed problem and approach. Please also see the supplementary material for the implementation details and additional analysis of the results.

---

[1] https://docs.scipy.org/doc/scipy/reference/generated/scipy.signal.find_peaks.html

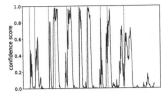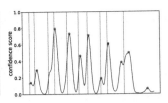

Fig. 4. Left: Ideally, predicted confidence scores of an action should increase before the occurrence of the key timestamp and decreases after it. Middle: In practice, the confidence scores tend to be noisy due to the uncertainty of the model and the ambiguity of observations. Green vertical lines represent ground truth key timestamps. Right: To mitigate the effect of noise, we apply Gaussian smoothing before performing peak detection. Green circles represent the detected peaks. As can be noted, Gaussian smoothing greatly improves the ability to detect key timestamps. (Color figure online)

5.1 Datasets

We perform our experiments on two datasets: ENIGMA, a dataset of egocentric videos collected and labeled for this study, and THUMOS'14 [9].

ENIGMA. In order to study the problem, we collected and labeled a set of egocentric videos of subjects simulating testing and repairing procedures on electrical boards using different laboratory tools. While performing the procedures, the subjects naturally interacted with 13 different objects. We labeled each take and release action with a single key timestamp, denoted with t_k. In the case of take actions, the key timestamp corresponds to the first frame in which the hand touches an object (contact frame), whereas, in the case of release actions, it corresponds to the first frame in which there is no more contact between the hand and the object (end-of-contact frame). See Fig. 1 for two examples. We acquired and labeled a total of 53 videos, which account for 23 hours, 14 minutes, and 8 seconds of video at a resolution of 2272×1278 pixels with a framerate of $30 fps$. We randomly divided the videos into a train, validation, and test set, keeping balanced numbers of take and release actions.[2]

THUMOS'14. We also report results on the THUMOS'14 [9] dataset to assess the ability of the proposed approach to generalize to the problem of online detection of action start, which is closely related to ours. Following prior works [6,7], we evaluated on a test set of 213 untrimmed videos.

5.2 Quasi-Online Detection of Take and Release Actions Results

As previously discussed, the proposed approach relies on a set of parameters, and more precisely: ω_s, the window size, γ the inhibition time, σ the standard deviation of the Gaussian used to smooth the predicted confidence scores, and ξ, the minimum distance at which predictions should be made to avoid multiple

---

[2] For detailed statistics regarding the dataset, please refer to the supplementary material.

Table 1. Evaluations of mp-mAP on ENIGMA for different choices of parameters. Best results per column are reported in **bold**.

mp-mAP (%)	$\overline{\delta}_l$ (s)	w_s (s)	γ (s)	σ (s)
16.46	1.48	2	0.8	0.6
13.72	**1.16**	1	0	0.2
16.08	1.35	2	0	0.6

predictions of the same action. We performed 72 unique experiments considering different combinations of the aforementioned parameters, for w_s varying in the range of [1, 2, 3, 4, 5] seconds, γ varying in the range of [0, 0.2, 0.4, 0.6, 0.8, 1] seconds, and σ varying in the range of [0.2, 0.4, 0.6] seconds. We set $\xi = 2s$. Figure 5-left report boxplots summarizing the distributions of mp-mAP% and average latency $\overline{\delta}_l$ values obtained in the different experiments. As can be noted, while some parameter combinations allow to achieve better results than others, detection performance and average latency values tend to be robust to the choice of parameters. Table 1 reports the performance of the proposed approach for some selected choices of the considered parameters. As can be noted, there is a trade-off between optimizing mp-mAP and reducing latency, but we obtain balanced results setting $w_s = 2s$, $\gamma = 0s$, and $\sigma = 0.6s$, with $mp - mAP = 16.08\%$ and $\overline{\delta}_l = 1.35s$. As can be noted, the proposed approach achieves promising results, but the problem is challenging and there is still room for improvement. Based on these experiments, it is clear that extending the inhibition time leads to better overall mp-mAP outcomes. However, this improvement results in physiologically higher latency. On the other hand, choosing a shorter inhibition time provides a better balance between mp-mAP and latency. This can be explained by observing that the confidence score distribution is almost uniform when far from the start of an action, and the peak detector does not detect a peak due to the Gaussian smoothing.

Average latency gives an indication of the ability of the model to make predictions on time. We further explore how detection performance changes when a given latency threshold is considered. Specifically, given a latency threshold ϵ, we deem as incorrect all predictions with a latency $\delta_l > \epsilon$ and re-compute mp-mAP. Figure 5-right reports mp-mAP% values for different latency thresholds. As can be noted, imposing a practical threshold of about $1s$ reduces mp-mAP% only by about 5% points, leading to a value of about 11%.

5.3 Generalization of the Proposed Approach to the Online Detection of Action Start Problem

The considered problem is closely related to previous investigations on the Online Detection of Action Start problem [6,7,14]. We hence assess how our approach generalizes to such a problem on the THUMOS'14 dataset. Figure 6-left reports the boxplots of the distributions of mp-mAP% and average latency for different parameter choices on THUMOS'14. Also in this case, results are stable for

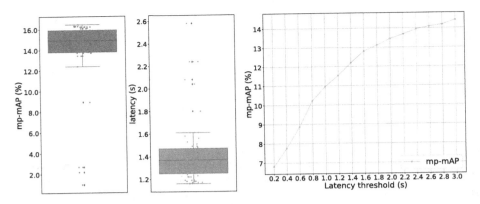

Fig. 5. Left: Boxplots showing the distributions of mp-mAP and average latency. Right: plots showing how mp-mAP% varies using different latency thresholds. Both plots are obtained on ENIGMA.

Table 2. Comparisons of p-mAP under different temporal offset thresholds on THU-MOS'14 for online detection of action start. **Best** and second-best per column are highlighted.

	p-mAP temporal offset threshold ϕ (s)										mp-mAP (%)
	1	2	3	4	5	6	7	8	9	10	
StartNet [6]	21.9	33.5	39.6	42.5	46.2	46.6	47.7	48.3	48.6	49	42.39
WOAD [7]	**28.0**	**40.6**	45.7	48.0	50.1	51.0	51.9	52.4	53.0	53.1	47.38
Ours	17.15	37.82	**48.62**	**55.70**	**60.75**	**64.37**	**67.17**	**68.33**	**69.69**	**70.63**	**56.02**

different parameter choices. We set $\omega_s = 4s$, $\gamma = 0s$ and $\sigma = 0.25s$ for these experiments.[3] Table 2 compares the proposed approach to StartNet [6] and the state-of-the-art WOAD [7], both in terms of p-mAP at different temporal offset thresholds ϕ and mp-mAP. It is worth noting that both competitors aim to perform online detection of action start, while our method focuses on quasi-online detection. As can be noted, the quasi-online relaxation allows our approach to obtain improved performance (+8.64%). The average latency of our approach is 2.19s. Figure 6-right finally compares mp-mAP performance for different latency thresholds. It is worth noting that, when low thresholds are considered, forcing the model to make online predictions, the mp-mAP performance of the model is strongly reduced to about 9%. However, a threshold of 1.5s still allows to achieve an mp-mAP performance of about 23% despite the model not explicitly being designed to tackle this task.

---

[3] See the supplementary material for a study on the influence of the different parameters on THUMOS.

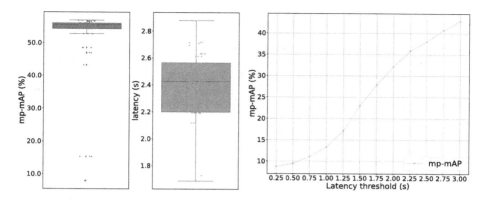

Fig. 6. Left: Boxplots showing the distributions of mp-mAP and average latency. Right: plots showing how mp-mAP% varies using different latency thresholds. Both plots are obtained on THUMOS'14.

6   Conclusion

We study the detection of take and release actions from egocentric videos in a quasi-online fashion. We propose a problem formulation and an initial approach to tackle the task. Experiments show promising results, but the problem is challenging and there is still space for improvement.

Acknowledgements. This research has been supported by Next Vision s.r.l., by the project MISE - PON I&C 2014-2020 - Progetto ENIGMA - Prog n. F/190050/02/X44 - CUP: B61B19000520008, and by Research Program Pia.ce.ri. 2020/2022 Linea 2 - University of Catania.

References

1. Besari, A.R.A., Saputra, A.A., Chin, W.H., Kubota, N., et al.: Feature-based egocentric grasp pose classification for expanding human-object interactions. In: 2021 IEEE 30th International Symposium on Industrial Electronics (ISIE), pp. 1–6. IEEE (2021)
2. Damen, D., et al.: Scaling egocentric vision: the dataset. In: Ferrari, V., Hebert, M., Sminchisescu, C., Weiss, Y. (eds.) ECCV 2018. LNCS, vol. 11208, pp. 753–771. Springer, Cham (2018). https://doi.org/10.1007/978-3-030-01225-0_44
3. De Geest, R., Gavves, E., Ghodrati, A., Li, Z., Snoek, C., Tuytelaars, T.: Online action detection. In: Leibe, B., Matas, J., Sebe, N., Welling, M. (eds.) ECCV 2016. LNCS, vol. 9909, pp. 269–284. Springer, Cham (2016). https://doi.org/10.1007/978-3-319-46454-1_17
4. Farinella, G.M., et al.: VEDI: vision exploitation for data interpretation. In: Ricci, E., Rota Bulò, S., Snoek, C., Lanz, O., Messelodi, S., Sebe, N. (eds.) ICIAP 2019. LNCS, vol. 11752, pp. 753–763. Springer, Cham (2019). https://doi.org/10.1007/978-3-030-30645-8_68

5. Gao, J., Yang, Z., Nevatia, R.: RED: reinforced encoder-decoder networks for action anticipation. arXiv preprint arXiv:1707.04818 (2017)
6. Gao, M., Xu, M., Davis, L.S., Socher, R., Xiong, C.: StartNet: online detection of action start in untrimmed videos. In: Proceedings of the IEEE/CVF International Conference on Computer Vision, pp. 5542–5551 (2019)
7. Gao, M., Zhou, Y., Xu, R., Socher, R., Xiong, C.: WOAD: weakly supervised online action detection in untrimmed videos. In: Proceedings of the IEEE/CVF Conference on Computer Vision and Pattern Recognition, pp. 1915–1923 (2021)
8. Gkioxari, G., Girshick, R., Dollár, P., He, K.: Detecting and recognizing human-object interactions. In: Proceedings of the IEEE Conference on Computer Vision and Pattern Recognition, pp. 8359–8367 (2018)
9. Idrees, H., et al.: The THUMOS challenge on action recognition for videos "in the wild". Comput. Vis. Image Underst. **155**, 1–23 (2017)
10. Karita, S., et al.: A comparative study on transformer vs RNN in speech applications. In: 2019 IEEE Automatic Speech Recognition and Understanding Workshop (ASRU), pp. 449–456. IEEE (2019)
11. Moltisanti, D., Wray, M., Mayol-Cuevas, W., Damen, D.: Trespassing the boundaries: labeling temporal bounds for object interactions in egocentric video. In: Proceedings of the IEEE International Conference on Computer Vision, pp. 2886–2894 (2017)
12. Qi, S., Wang, W., Jia, B., Shen, J., Zhu, S.-C.: Learning human-object interactions by graph parsing neural networks. In: Ferrari, V., Hebert, M., Sminchisescu, C., Weiss, Y. (eds.) ECCV 2018. LNCS, vol. 11213, pp. 407–423. Springer, Cham (2018). https://doi.org/10.1007/978-3-030-01240-3_25
13. Ragusa, F., Furnari, A., Farinella, G.M.: MECCANO: a multimodal egocentric dataset for humans behavior understanding in the industrial-like domain. arXiv preprint arXiv:2209.08691 (2022)
14. Shou, Z., et al.: Online detection of action start in untrimmed, streaming videos. In: Ferrari, V., Hebert, M., Sminchisescu, C., Weiss, Y. (eds.) ECCV 2018. LNCS, vol. 11207, pp. 551–568. Springer, Cham (2018). https://doi.org/10.1007/978-3-030-01219-9_33
15. Vaswani, A., et al.: Attention is all you need. In: Advances in Neural Information Processing Systems, vol. 30 (2017)
16. Wang, T., Yang, T., Danelljan, M., Khan, F.S., Zhang, X., Sun, J.: Learning human-object interaction detection using interaction points. In: Proceedings of the IEEE/CVF Conference on Computer Vision and Pattern Recognition, pp. 4116–4125 (2020)
17. Wang, X., et al.: OadTR: online action detection with transformers. In: Proceedings of the IEEE/CVF International Conference on Computer Vision, pp. 7565–7575 (2021)
18. Xu, M., et al.: Long short-term transformer for online action detection. Adv. Neural. Inf. Process. Syst. **34**, 1086–1099 (2021)
19. Zhao, Y., Krähenbühl, P.: Real-time online video detection with temporal smoothing transformers. In: Avidan, S., Brostow, G., Cissé, M., Farinella, G.M., Hassner, T. (eds.) Computer Vision-ECCV 2022: 17th European Conference, Tel Aviv, Israel, 23–27 October 2022, Proceedings, Part XXXIV, vol. 13694, pp. 485–502. Springer, Cham (2022). https://doi.org/10.1007/978-3-031-19830-4_28

Hashing for Structure-Based Anomaly Detection

Filippo Leveni[1]([☒]) [iD], Luca Magri[1] [iD], Cesare Alippi[1,2] [iD],
and Giacomo Boracchi[1] [iD]

[1] Politecnico di Milano (DEIB), Milan, Italy
{filippo.leveni,luca.magri,cesare.alippi,giacomo.boracchi}@polimi.it
[2] Università della Svizzera italiana, Lugano, Switzerland
cesare.alippi@usi.ch

Abstract. We focus on the problem of identifying samples in a set that do not conform to structured patterns represented by low-dimensional manifolds. An effective way to solve this problem is to embed data in a high dimensional space, called Preference Space, where anomalies can be identified as the most isolated points. In this work, we employ Locality Sensitive Hashing to avoid explicit computation of distances in high dimensions and thus improve Anomaly Detection efficiency. Specifically, we present an isolation-based anomaly detection technique designed to work in the Preference Space which achieves state-of-the-art performance at a lower computational cost. Code is publicly available at https://github.com/ineveLoppiliF/Hashing-for-Structure-based-Anomaly-Detection.

1 Introduction

Anomaly Detection, i.e., the task of identifying anomalous instances, is employed in a wide range of applications such as detection of frauds in financial transactions [1], faults in manufacturing [2], intrusion in computer networks [3], risk analysis in medical data [4] and predictive maintenance [5].

Most anomaly detection approaches identify anomalies as those points that lie in low density regions [6]. However, in many real word scenarios, genuine data lie on low-dimensional manifolds and, in these situations, a density analysis falls short in characterizing anomalies, that are best characterized in terms of their conformity to these low dimensional structures. For example, images of the face of the same subject lie on a low-dimensional subspace [7], while faces of different subjects are far away from that subspace, regardless of data density.

In this work we focus on *Structure-based Anomaly Detection*, i.e., identifying data that do not conform to any structure describing genuine data. Specifically, we introduce RUZHASH-IFOREST, a novel anomaly detection method that is both more accurate and efficient than Preference Isolation Forest (PI-FOREST) [8].

In Fig. 1a we illustrate our method with a toy example. Genuine data G, depicted in green, nearly lies on 1-dimensional structures (lines θ_1 and θ_2), while anomalous data A are far away from them. We embed data in a high-dimensional space (Fig. 1b), called Preference Space and endowed with the Ruziska [10] distance, where anomalies are detected as the most isolated points. The explicit

G. L. Foresti et al. (Eds.): ICIAP 2023, LNCS 14234, pp. 25–36, 2023.
https://doi.org/10.1007/978-3-031-43153-1_3

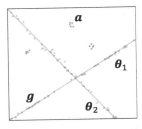 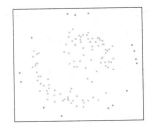 

(a) Input data $X = G \cup A$.   (b) Preference space $\mathcal{P}$.   (c) Anomaly scores α.

Fig. 1. RuzHash-iForest detects anomalies in X that do not conform to structures. (a) Genuine points G, in green, described by two lines of parameters θ_1 and θ_2, and anomalies A in red. (b) Data are mapped to a high-dimensional Preference Space where anomalies result in isolated points (visualized via MDS [9]). (c) Anomaly score $\alpha(\cdot)$ (color coded) is computed via RuzHash-iForest.

computation of distances in the Preference Space, as done in [8], is computational demanding, therefore we propose RuzHash, a Locality Sensitive Hashing (LSH) [11] scheme that efficiently approximates distances. This LSH scheme is embedded in our RuzHash-iForest anomaly detection algorithm, and yields an anomaly score for each point.

We performed experiments on both synthetic and real data to compare RuzHash-iForest and PI-Forest. Results show that our novel LSH-based approach improves PI-Forest both in terms of anomaly detection performance and computational time, with a speed up factor of ×35% to ×70%.

2 Problem Formulation

The problem of structured anomaly detection can be framed as follows. Given a dataset $X = G \cup A \subset \mathcal{X}$, G and A are the set of genuine and anomalous data respectively, and $\mathcal{X}$ is the ambient space. We assume that genuine data $g \in G$ are close to the solutions of a parametric equation $\mathcal{F}(g, \theta) = 0$, that depends on an unknown vector of parameters θ. For example, in Fig. 1a genuine structures are described by lines, thus $\mathcal{F}(g, \theta) = \theta_1 g_1 + \theta_2 g_2 + \theta_3$ and, due to noise, genuine data satisfy the equation only up to a tolerance $\epsilon > 0$, namely $|\mathcal{F}(g, \theta)| < \epsilon$. Moreover, genuine data may be described by multiple models $\{\theta_i\}_{i=1,\dots,k}$ whose number is typically unknown. In contrast, anomalous data $a \in A$ are far from satisfying the parametric equation of any model instance and $\mathcal{F}(a, \theta_i) \gg \epsilon \; \forall \theta_i$.

Structured-based Anomaly Detection aims to produce an anomaly scoring function $\alpha : X \to \mathbb{R}^+$ such that $\alpha(a) \gg \alpha(g)$, as depicted in Fig. 1c where higher scores are in red and lower scores are in blue.

3 Related Work

Among the wide literature on anomaly detection [6], we focus on isolation-based methods, as they reach state-of-the-art performance at low computational and memory requirements. Isolation-based approaches can be traced back to Isolation Forest [12] (iFOREST), where anomalies are separated by building a forest of randomly generated binary trees (iTREE) that recursively partition the data by axis-parallel splits. The number of splits required to isolate any point from the others is inversely related to its probability of being anomalous. In other words, anomalies are more likely to be separated in the early splits of the tree and result in shorter paths. Thus, the average path lengths, computed with respect to a forest of random trees, translates into a reliable anomaly score. Several improvements over the original iFOREST framework have been introduced. Extended Isolation Forest [13] and Generalized Isolation Forest [14] overcome the limitation of axis-parallel splits, while Functional Isolation Forest [15] extends iFOREST beyond the concept of point-anomaly, to identify functional-anomalies. The connection between iFOREST and Locality Sensitive Hashing is investigated in [16]. In particular, the splitting process of an iTREE is interpreted as a Locality Sensitive Hashing (LSH) of the ℓ_1 distance, where points that are nearby according to ℓ_1 are assigned to the same bucket. LSH schemes allow to strike a good trade-off between effectiveness and efficiency, but are limited to identify density-based anomalies.

The literature on structure-based anomaly detection is less explored than its density-based counterpart. The pioneering work was [8], where PI-FOREST, a variant of iFOREST, is presented to detect structured anomalies. PI-FOREST consists to randomly sample a set of low-dimensional structures from data points and to embed data into an high-dimensional space, called Preference Space, where each point is described in terms of its adherence to the sampled structures. Here, PI-FOREST identifies anomalies as the most isolated points according to the Jaccard or Tanimoto distance. Specifically, PI-FOREST leverages on nested Voronoi tessellations built in a recursive way in the Preference Space to instantiate isolation-trees. Although this approach effectively isolates anomalous data, building Voronoi tessellations carries the computational burden of explicitly computing distances in high dimensions, impacting negatively on the computational performance.

In this work, we address this problem by presenting a novel LSH scheme to approximate distances in the Preference Space. The problem of speeding up computation of distances in the Preference Space has been addressed in [17], where MINHASH has been used to cluster points according to the Jaccard distance. However, the focus of [17] was to identify structures rather than anomalies and it is limited to deal with binary preferences.

4   Method

The proposed method, summarized in Algorithm 1, is composed of two main steps. In the first one (lines 1–2), we map input data $X \subset \mathcal{X}$ to $P \subset \mathcal{P}$ via the Preference Embedding $\mathcal{E}(\cdot)$, where $\mathcal{P}$ is an high-dimensional Preference Space. The mapping process $\mathcal{E}(\cdot)$ follows closely the Preference Embedding of [8], and it is reported in Sect. 4.1. The major differences with respect to [8] reside in the

---

Algorithm 1: RUZHASH-IFOREST

Input: X - input data, t - number of trees, ψ - sub-sampling size, b - branching factor

Output: Anomaly scores $\{\alpha(\boldsymbol{x}_j)\}_{j=1,\dots,n}$

    /* Preference Embedding */

1   Sample m models $\{\boldsymbol{\theta}_i\}_{i=1,\dots,m}$ from X

2   $P \leftarrow \{\boldsymbol{p}_j \mid \boldsymbol{p}_j = \mathcal{E}(\boldsymbol{x}_j)\}_{j=1,\dots,n}$

    /* RUZHASH-IFOREST */

3   $F \leftarrow \emptyset$

4   **for** $k = 1$ **to** t **do**

5      $P_\psi \leftarrow$ SUBSAMPLE(P, ψ)

6      $T_k \leftarrow$ RUZHASH-ITREE(P_ψ, b)

7      $F \leftarrow F \cup T_k$

    /* Anomaly score computation */

8   **for** $j = 1$ **to** n **do**

9   **for** $k = 1$ **to** t **do**

10        $\boldsymbol{p}_j \leftarrow j$-th point in P

11        $T_k \leftarrow k$-th RUZHASH-ITREE in F

12        $h_k(\boldsymbol{p}_j) \leftarrow$ HEIGHT$(\boldsymbol{p}_j, T_k)$

13      $\alpha(\boldsymbol{x}_j) \leftarrow$ Eq.(3)

14   **return** $\{\alpha(\boldsymbol{x}_j)\}_{j=1,\dots,n}$

---

second step, and in particular in the computation of distances in the Preference Space. First, we do not use the Tanimoto distance [18], but rather we introduce the Ruzicka distance [10], which demonstrates comparable isolation capabilities. Secondly, we define a novel Locality Sensitive Hashing scheme, called RUZHASH to efficiently approximate Ruzicka distances avoiding their explicit computation. Specifically, we build an ensemble of isolation trees termed RUZHASH-IFOREST (lines 3–7) as described in Sect. 4.2. Each RUZHASH-ITREE of the forest recursively splits the points based on our RUZHASH Local Sensitive Hashing as detailed in Sect. 4.3. This separation mechanism produces an anomaly score $\alpha(\boldsymbol{x})$ for each $\boldsymbol{x} \in X$ (lines 8–14), as discussed in Sect. 4.4.

4.1 Preference Embedding

Preference Embedding has been widely used in the multi-model fitting literature [19–21] and then employed in Structure-based Anomaly Detection [8]. Preference Embedding consists in a mapping $\mathcal{E}: \mathcal{X} \to \mathcal{P}$, from the ambient space $\mathcal{X}$ to the Preference Space $\mathcal{P} = [0, 1]^m$. Such mapping is obtained by sampling a pool $\{\boldsymbol{\theta}_i\}_{i=1,\dots,m}$ of m models from the data X using a RanSaC-like strategy [22] (line 1): the minimal sample sets – containing the minimum number of points necessary to constrain a parametric model – are randomly sampled from the data to determine models parameters. Then (line 2), each sample $\boldsymbol{x} \in \mathcal{X}$ is embedded to a vector $p = \mathcal{E}(x) \in \mathcal{P}$ whose i-th component is defined as:

$$
p_i = \begin{cases} \phi(\delta_i) & \text{if } |\delta_i| \leq \epsilon \\ 0 & \text{otherwise} \end{cases}, \tag{1}
$$

where $\delta_i = \mathcal{F}(x, \theta_i)$, measures the residuals of x with respect to model θ_i and $\epsilon = k\sigma$ defines an inlier threshold proportional to the standard deviation σ of the noise. The preference function ϕ is then defined as:

$$
\phi(\delta_i) = e^{-\frac{1}{2}(\frac{\delta_i}{\sigma})^2}. \tag{2}
$$

We explored also a different definition of ϕ, namely the *binary preference* function that is $\phi(\delta_i) = 1$ when $|\delta_i| \leq \epsilon$ and 0 otherwise. Hereinafter, we will refer to $\mathcal{P} = \{0, 1\}^m$ as the *binary preference space* to distinguish it from the *continuous* one $\mathcal{P} = [0, 1]^m$.

4.2   RuzHash-iForest

We perform anomaly detection in the Preference Space exploiting a forest of isolation trees similarly to [8] (lines 3–14). The fundamental difference of our RuzHash-iForest with respect to [8] is that our ensemble of RuzHash-iTrees bypass the distance computation in the preference space.

We identify two main steps: the *training* of RuzHash-iForest $F = \{T_k\}_{k=1}^t$ (lines 3–7) and the *testing* of vectors P via every RuzHash-iTree $T_k \in F$ (lines 8–14). As regard the training, we build every RuzHash-iTree on a different subset $P_\psi \subset P$ of ψ vectors sampled from P (line 5). During testing, for every point $p_j \in P$ (line 10) and every tree $T_k \in F$ in the forest (line 11), we compute the heights $h_k(p_j)$ reached in T_k by p_j (line 12). The main intuition is that each T_k returns, on average, noticeable smaller heights for anomalies than for genuine points, since isolated points are more likely to be separated early in the recursive splitting process. Thus, the heights are collected in a vector $h(p_j) = [h_1(p_j), \ldots, h_t(p_j)]$ and the anomaly score $\alpha(\cdot)$ is computed (line 13) as:

$$
\alpha(x_j) = 2^{-\frac{E(h(p_j))}{c(\psi)}}, \tag{3}
$$

where $E(h(p_j))$ is the mean over the elements of $h(p_j)$ and $c(\psi) = \log_b \psi$ is an adjustment factor as a function of the tree subsampling size ψ. Our method is agnostic with respect to the specific choice of anomaly score, and other techniques, as discussed in [23], can be employed as well.

4.3   RuzHash-iTree

The construction of each RuzHash-iTree T_k (line 6) is detailed as follow. We are given a subset P_ψ, with cardinality ψ, uniformly sampled from all the input points P embedded in the Preference Space (line 5) and a branching factor $b \in \{1, \ldots, m\}$. At each node, P_ψ is splitted in b branches using RuzHash.

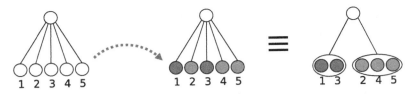

Fig. 2. On the left, split performed by RUZHASH. In the middle, nodes aggregation where the groups has been color coded. On the right, resulting tree with branching factor $b = 2$.

This scheme is recursively executed until either: (i) the current node contains a number of points less than m or (ii) the tree reaches a maximum height, set by default at $\log_m \psi$ (an approximation for the average tree height [24]).

RUZHASH is designed to split the data in m leaves, where m equals the dimension of $\mathcal{P}$. However, we experienced that lower branching factors resulted in slightly better performance. Therefore, we accommodate for a different branching factor b by randomly aggregating in $b < m$ groups the nodes produced by RUZHASH after each split of the tree. Figure 2 shows an example of aggregation process when $m = 5$ and $b = 2$. We can see on the left m leaves performed by RUZHASH, and in the middle the aggregation of the nodes where the groups have been color coded. On the right, we can see the resulting tree with branching factor $b = 2$. The branching factor controls the average tree height. Therefore, the maximum tree height becomes $\log_b \psi$.

4.4 RUZHASH

Instead of leveraging explicitly on distances, the splitting procedure implemented in each node of RUZHASH-ITREE is based on RUZHASH, our novel Locality Sensitive Hashing process, designed to approximate the Ruziska distance. In this way, we greatly reduce the computational burden of the method.

We consider Ruzicka instead of Tanimoto to measure distances in the preference space. This is due to the fact that has not yet been proven whether a hashing scheme for Tanimoto could even exist. Moreover, our experiments demonstrate that Ruzicka and Tanimoto distances achieves comparable performance in isolating anomalous points in the Preference Space.

Given two preferences vectors $p, q \in \mathcal{P}$, their Ruzicka distance is defined as:

$$R(p, q) = 1 - \frac{\sum_{i=1}^{m} \min(p_i, q_i)}{\sum_{i=1}^{m} \max(p_i, q_i)}. \tag{4}$$

In practice, the higher the preferences granted to the same models $\{\theta_i\}_{i=1,\dots,m}$, represented by components p_i and q_i, the closer the vectors p and q are.

Our locality sensitive hashing scheme, RUZHASH, approximates the Ruzicka distance in the Preference Space as follows. First, vectors $p \in \mathcal{P} = [0,1]^m$ are binarized yielding points $p' \in \{0,1\}^m$. The binarization is performed by a

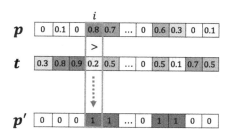

Fig. 3. Example of binarization.

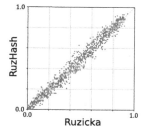

Fig. 4. Correlation between Ruzicka and RuzHash.

component-wise comparison of p with a randomly sampled vector $t \in [0,1]^m$, ad depicted in Fig. 3. Formally, we define a vector $t = [t_1, \ldots, t_m]$ as the realization of a multivariate random vector $T = [T_1, \ldots, T_m]$, where T_i is a uniform random variable $T_i \sim \mathcal{U}_{[0,1)}$, and use it to binarize *all* points $p \in [0,1]^m$. If the i-th component p_i of a point p is greater than t_i, the binarized component equals 1, otherwise it is set to 0. In formulae, the component-wise binarization procedure is defined as:

$$p'_i = \begin{cases} 1 & \text{if } p_i > t_i \\ 0 & \text{otherwise} \end{cases}. \tag{5}$$

Notice that, since t is a realization of random variable, and the binarization procedure (5) depends on t, thus p' is a realization of a random variable. Now, given the i-th binarized components p'_i, q'_i of two preference vectors p, q, the probability that both p'_i and q'_i equals 1, is given by:

$$P(p'_i = 1 \wedge q'_i = 1) = P(p_i > t_i \wedge q_i > t_i) = \min\{p_i, q_i\}. \tag{6}$$

Similarly, the probability that at least one of the binarized component equals 1, is given by $P(p'_i = 1 \vee q'_i = 1) = \max\{p_i, q_i\}$. Note that these two terms are exactly those that appear in (4) in the definition of the Ruziska distance at the numerator and denominator, respectively. Thus, we can rewrite the Ruzicka distance in terms of the previous probabilities between binary vectors:

$$R(p, q) = 1 - \frac{\sum_{i=1}^{m} P(p'_i = 1 \wedge q'_i = 1)}{\sum_{i=1}^{m} P(p'_i = 1 \vee q'_i = 1)}. \tag{7}$$

The Ruzicka distance is estimated by computing the value of (7) for different realizations of vectors p', q', sampling several vectors t. In this way, the problem boils down to counting the number of 1 on the same component i over the total number of non zero entries of p', q', that is their Jaccard distance. This quantity, in turn, can be efficiently estimated using MinHash [25]. The estimated value of Ruzicka converges to its theoretical value (4) as the number of sampled t increases. Figure 4 depicts the nearly exact correlation existing between Ruzicka and RuzHash, computed on pairs of randomly sampled vectors such that the distances between them were uniform with respect to the Ruzicka distance.

Table 1. Differences between iFOREST, PI-FOREST and RUZHASH-IFOREST.

	Splitting scheme	Distance	Computational complexity	
			Training	Testing
PI-FOREST	Voronoi	Tanimoto	$O(\psi\, t\, b\, \log_b \psi)$	$O(n\, t\, b\, \log_b \psi)$
iFOREST	LSH	ℓ_1	$O(\psi\, t\, \log_2 \psi)$	$O(n\, t\, \log_2 \psi)$
RUZHASH-IFOREST	LSH	Ruzicka	$O(\psi\, t\, \log_b \psi)$	$O(n\, t\, \log_b \psi)$

In practice, a different t is sampled at each split of every RUZHASH-ITREE in the forest during the training phase. It follows that, the more splits are performed, the higher the probability that only samples close with respect to the Ruzicka distance fall in the same node and, at the same time, the more isolated points are separated from the others in the early splits.

To conclude, it is worth to notice that Ruzicka distance, as the Tanimoto one, is a generalization of the Jaccard distance when preferences are in $\{0, 1\}^m$. In this case, $\boldsymbol{p} = \boldsymbol{p}'$ and RUZHASH specializes exactly to MINHASH.

4.5   Comparison Between iFOREST, PI-FOREST and RUZHASH-IFOREST

Table 1 summarizes the main differences between iFOREST, PI-FOREST and RUZHASH-IFOREST. As regard the splitting scheme, RUZHASH-IFOREST exploits LSH as done in iFOREST. This can be appreciated from the lower computational complexity compared to the PI-FOREST which builds Voronoi tessellations, and require the explicit computation of distances to the b tessellation centers. As regard the distance employed, RUZHASH-IFOREST exploits the Ruziska distance that is tailored for the Preference Space, rather than relying on the ℓ_1 distance. Moreover, if we compare the computational complexities of iFOREST and RUZHASH-IFOREST, we have a speed up given by the higher branching factor of RUZHASH-IFOREST compared to the fixed branching factor $b = 2$ of iFOREST.

5   Experimental Validation

In this section we evaluate the benefits of our approach for structured anomaly detection on both simulated and real datasets. In particular, we compare the performance of RUZHASH-IFOREST and PI-FOREST both in the *continuous* and *binary* preference space. Results show that RUZHASH-IFOREST performs better in terms of both AUC and execution time.

5.1   Datasets

We consider synthetic datasets and real data employed in [8]. Synthetic datasets consist of 2D points where genuine data G live along parametric structures (lines and circles) and anomalies are uniformly sampled within the range of G such that $\frac{|A|}{|X|} = 0.5$.

We consider a real dataset, the AdelaideRMF dataset [26], that consists stereo images with annotated matching points and anomalies A correspond to mismatches. The first 19 sequences refer to static scenes containing several planes, each giving rise matches described by an homography. The remaining 19 sequences are dynamic with several objects independently moving and give rise to a set of matches described by different fundamental matrices.

5.2 Competing Methods

We compare RuzHash-iForest against PI-Forest [8], both constructed with binary or continuous Preference Space. When preferences are continuous, we use the shorthand RHF and PIF respectively, and we indicate by RHF-B and PIF-B the two competitors when preferences are binary. In order to assess the benefits of Ruziska distance we also considered a version of [8] equipped with Ruziska instead of Tanimoto and denoted this by PIF-R.

Preferences are computed with respect to a pool of $m = 10|X|$ model instances, corresponding to the genuine structures. The inlier threshold ϵ in (2) has been tuned as follows: we first estimate the standard deviation σ of the noise from the data given their ground truth labels, then we fix $\epsilon = k\sigma$ where we choose k to maximize the performance for both RHF and PIF. In particular, we set $k = 3$ for synthetic data, $k = 0.25$ for fundamental matrix and $k = 5$ for homography dataset respectively.

We tested PI-Forest and RuzHash-iForest at the same parameters condition: number of trees in the ensemble $t = 100$, subsampling size to build each tree $\psi = 256$ while the branching factor vary in $b = [2, 4, 8, 16, 32, 64, 128, 256]$.

5.3 Results

Figure 5 shows the aggregated results of our experiments. In particular, in Fig. 5a and 5b we show respectively the average ROC AUC and test time for all the methods at various branching factors. We produced the curves by first averaging the results of 5 executions on synthetic and real datasets separately. We then averaged these results to get the final curves.

The rundown of these experiments is that our RuzHash-iForest achieves higher ROC AUC values in both RHF and RHF-B configurations. More interestingly, our method is the most stable with respect to the choice of the branching factor b. RuzHash-iForest attains accurate results also for small values of b, because the tendency to underestimate the distances of the splitting procedure is compensated by the overestimation due to the greater height of the trees. On the other hand, PIF, PIF-B and PIF-R have a consistent performance loss when the branching factor increases. This can be ascribed to the Voronoi tessellations that enforce each node to contain at least one point. Thus, when $b \geq 32$, trees are constrained to a single level, and the anomaly score does no longer depend on the height of the tree but is fully determined by the adjustment factor alone [8,12], resulting in a degradation of the performances. These trends are confirmed in the average ROC AUC curves computed separately for synthetic and

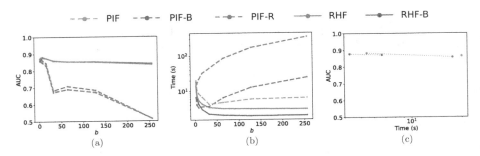

Fig. 5. (a) Average ROC AUCs. (b) Average test times. (c) Relation between best average ROC AUC and corresponding test time.

real datasets, which have not been reported due to space limitations. Furthermore, AUC curves of PIF and PIF-R show that there is not a clear advantage in using the Ruziska distance over the Tanimoto one in the PI-FOREST framework, confirming that the main advantage of our method is due to the LSH scheme and not to the different distance measures involved. The gain in terms of test time is very evident in Fig. 5b and it is consistent with the computational complexities showed in Table 1. In fact, when the branching factor b increases, the test time for RUZHASH-IFOREST decreases according to $\log_b \psi$ in all configurations, while for PI-FOREST increases according to $b \log_b \psi$.

A different visualization of the results is presented in Fig. 5c, where the relation between the best average ROC AUC value and the corresponding test time for each method is shown. We identify for each method the branching factor that maximizes the average ROC AUC (Fig. 5a) and use it to collect the corresponding test time (Fig. 5b). Dashed lines relate RUZHASH-IFOREST and PI-FOREST results that refer to the same underlying distance measure (Ruziska for the continuous case, Jaccard for the binary). It can be appreciated that RUZHASH-IFOREST achieves results that are as accurate as their PI-FOREST counterpart, but with a consistent gain in execution time. Specifically, RHF-B is ×35% faster than PIF-B, while RHF is ×70% faster than PIF-R. Furthermore, RHF is at least ×70% faster than PIF, which employs the Tanimoto distance.

6   Conclusion and Future Directions

We proposed RUZHASH-IFOREST, an efficient algorithm specifically designed to perform Structure-based Anomaly Detection. RUZHASH-IFOREST is an isolation-based anomaly detection algorithm that works in the Preference Space, whose peculiarity resides in the splitting criteria. In particular, a novel Locality Sensitive Hashing, called RUZHASH, has been employed to detect the most isolated points in the Preference Space with respect to the Ruzicka distance. Remarkably, RUZHASH is a generalization of MINHASH to the case where the considered points lie in the continuous space $\mathcal{P} = [0, 1]^m$.

Our empirical evaluation demonstrated that RuzHash-iForest gain efficiency over current state-of-the-art-solutions, and has stable accuracy along different branching factors, both on synthetic and real data. A possible future direction could be to investigate others effective distances for anomaly detection in the Preference Space and to define their corresponding LSH to employ in an isolation-based forest.

References

1. Ahmed, M., Mahmood, A.N., Islam, M.R.: A survey of anomaly detection techniques in financial domain. Future Gener. Comput. Syst. **55**, 278–288 (2016)
2. Miljković, D.: Fault detection methods: a literature survey. In: International Convention on Information. Communication and Electronic Technology, pp. 750–755. IEEE (2011)
3. Lazarevic, A., Ertoz, L., Kumar, V., Ozgur, A., Srivastava, J.: A comparative study of anomaly detection schemes in network intrusion detection. In: International Conference on Data Mining, pp. 25–36. SIAM (2003)
4. Ukil, A., Bandyoapdhyay, S., Puri, C., Pal, A.: IoT healthcare analytics: the importance of anomaly detection. In: Advanced Information Networking and Applications, pp. 994–997. IEEE (2016)
5. De Benedetti, M., Leonardi, F., Messina, F., Santoro, C., Vasilakos, A.: Anomaly detection and predictive maintenance for photovoltaic systems. Neurocomputing **310**, 59–68 (2018)
6. Chandola, V., Banerjee, A., Kumar, V.: Anomaly detection: a survey. ACM Comput. Surv. **41**(3), 1–58 (2009)
7. Ronen Basri and David W Jacobs. Lambertian reflectance and linear subspaces. Trans. Pattern Anal. Mach. Intell. **25**(2), 218–233 (2003)
8. Leveni, F., Magri, L., Boracchi, G., Alippi, C.: PIF: anomaly detection via preference embedding. In: International Conference on Pattern Recognition, pp. 8077–8084. IEEE (2021)
9. Kruskal, J.B.: Nonmetric multidimensional scaling: a numerical method. Psychometrika **29**(2), 115–129 (1964)
10. Ružička, M.: Anwendung mathematisch-statisticher methoden in der geobotanik (synthetische bearbeitung von aufnahmen). Biológia **13**, 647 (1958)
11. Gionis, A., Indyk, P., Motwani, R., et al.: Similarity search in high dimensions via hashing. Conf. Very Large Databases **99**, 518–529 (1999)
12. Liu, F.T., Ting, K.M., Zhou, Z.H.: Isolation-based anomaly detection. Trans. Knowl. DiscData **6**(1), 1–39 (2012)
13. Hariri, S., Kind, M.C., Brunner, R.J.: Extended isolation forest. Trans. Knowl. Data Eng. **33**(4), 1479–1489 (2019)
14. Lesouple, J., Baudoin, C., Spigai, M., Tourneret, J.-Y.: Generalized isolation forest for anomaly detection. Pattern Recogn. Lett. **149**, 109–119 (2021)
15. Staerman, G., Mozharovskyi, P., Clémençon, S., Florence d'Alché, B.: Functional isolation forest. In: Asian Conference on Machine Learning, pp. 332–347. PMLR (2019)
16. Zhang, X., et al.: Lshiforest: a generic framework for fast tree isolation based ensemble anomaly analysis. In: International Conference on Data Engineering, pp. 983–994. IEEE (2017)

17. Magri, L., Fusiello, A.: Reconstruction of interior walls from point cloud data with min-hashed j-linkage. In International Conference on 3D Vision, pp. 131–139. IEEE (2018)
18. Lipkus, A.H.: A proof of the triangle inequality for the tanimoto distance. J. Math. Chem. **26**(1–3), 263–265 (1999)
19. Toldo, R., Fusiello, A.: Robust multiple structures estimation with J-Linkage. In: European Conference on Computer Vision, pp. 537–547 (2008)
20. Magri, L., Fusiello, A.: T-Linkage: a continuous relaxation of J-linkage for multi-model fitting. In: Computer Vision and Pattern Recognition Conference, pp. 3954–3961, June 2014
21. Magri, L., Leveni, F., Boracchi, G.: Multilink: multi-class structure recovery via agglomerative clustering and model selection. In: Computer Vision and Pattern Recognition Conference, pp. 1853–1862 (2021)
22. Fischler, M.A., Bolles, R.C.: Random sample consensus: a paradigm for model fitting with applications to image analysis and automated cartography. Commun. ACM **24**(6), 381–395 (1981)
23. Mensi, A., Bicego, M.: Enhanced anomaly scores for isolation forests. Pattern Recogn. **120**, 108115 (2021)
24. Knuth, D.E.: The Art of Computer Programming: Sorting and Searching, vol. 3. Addison-Wesley Professional, Boston (1998)
25. Broder, A.Z., Charikar, M., Frieze, A.M., Mitzenmacher, M.: Min-wise independent permutations. In: Symposium on Theory of Computing, pp. 327–336 (1998)
26. Wong, H.S., Chin, T.J., Yu, J., Suter, D.: Dynamic and hierarchical multi-structure geometric model fitting. In: International Conference on Computer Vision, pp. 1044–1051. IEEE (2011)

Augmentation Based on Artificial Occlusions for Resilient Instance Segmentation

Nikolaos Kilis[(✉)], Grigorios Tsipouridis, Iason Karakostas, Nikolaos Dimitriou, and Dimitrios Tzovaras

Information Technologies Institute, Centre for Research and Technology Hellas, 57001 Thessaloniki, Greece
{nikolaoskk,tsipurid,iason,nikdim,dimitrios.tzovaras}@iti.gr

Abstract. Real-world instance segmentation applications usually demand real-time identification of objects that are small in size, occluded from other objects, appearing and disappearing in quick succession. For these reasons, instance segmentation requires a lot of representative data to grasp the subtle changes that occur in a scene. In this paper, we propose an augmentation methodology that allows for sufficient training relying on a small number of annotated data. Additionally, we provide two new datasets for instance segmentation including the semantic class firearm, for security applications. By applying the proposed augmentation technique on three datasets, the performance of instance segmentation methods is improved as indicated by the experimental results.

Keywords: data augmentation · image synthesis · synthetic occlusions

1 Introduction

During the past years, significant progress has been made in instance segmentation (IS) mainly due to advances in deep learning [4]. Many efforts were concentrated on improving model accuracy and/or inference speed [7,14]. Methods that employ deep architectures require big training datasets towards achieving a good generalization during training. Such datasets containing common objects are publicly available, however, for specific applications (e.g., security) there is still a big gap in the available annotated training instances and custom datasets should be created. Creating custom datasets usually demands manual annotation of data, which is a painstaking and time-consuming process. This obstacle can be addressed by integrating automated methods for generating such training sets from smaller training sets. For this reason, we were inspired to address the issue in a cost-effective manner by utilizing a data augmentation technique during model training. Our contributions can be summarized in two points. Firstly, we provide two new datasets including annotated images of firearms for IS.

G. L. Foresti et al. (Eds.): ICIAP 2023, LNCS 14234, pp. 37–48, 2023.
https://doi.org/10.1007/978-3-031-43153-1_4

All necessary information on how to obtain the training/validation images and annotations can be provided upon request. Secondly, we introduce a new data augmentation technique that can be used in two ways. In the first case, we can employ our method to generate new training samples by relocating a segmentation mask in a controlled manner, so that it deliberately occludes another object inside a selected image. Additionally, the same method can be used to synthesize new training samples from two images. The first image does not contain any of the training semantic classes and serves as a random background. The second image has to contain at least one object of a minority semantic class and provides a segmentation mask that serves as the foreground in the final image. In this way model training includes both real and more difficult samples with severe or partial occlusions in them. Moreover, a controlled number of additional examples that contain minority semantic classes is added in model training reducing class imbalance.

This method is further analyzed in Sect. 3 followed by the results of the experimental evaluation that showcases the validity of the proposed augmentation technique in Sect. 4. Finally, the conclusions for the proposed methodology are presented in Sect. 5.

2   Data Augmentation Methods

Since data annotation requires significant effort and becomes even more difficult due to data scarcity for certain classes, semi-supervised approaches during model training have been developed [19], where only a part of the dataset needs to be labeled. Although this approach solves the issue of data labeling, there is still an issue with the variety of available data/images that will lead to better-generalized model training. Transfer learning is an interesting approach to tackle the overfitting problem [20]. The main idea is to train a deep neural network on large available datasets of specific tasks (e.g., image classification) and then exploit the learned weights of the network on a new training on a specific task dataset where not enough training data are available. Generative Adversarial Networks (GANs) have also been employed to enhance datasets with additional synthetic samples [22].

Augmented data with artificial occlusions have also been studied in image classification tasks. In [8] the positive effect of varying amounts of occlusion has been explored. In [9] a similar to the proposed augmentation technique is analyzed focusing on generating new data from translating image masks in a random fashion, in contrast with our method in which a deliberate production of partial/severe occlusions between objects occurs, with an additional blurring effect around the relocated mask. In [12] the separation between the content and the style of a pair of images was manipulated in order to create a new image that robustly inherits the content and style of the input images.

In [23] the proposed augmentation strategy exploits patches between pairs of train samples in order to create new samples containing information from both images. Patches between training samples are cut and pasted among them, as well as the labels according to the proportion of the mixture. This augmentation strategy improves the performance of deep neural networks on well-known datasets such as ImageNet [3]. The augmentation technique proposed in [24], selects random patches of an image in the training set and substitutes the pixel values of this area with random values generating images with various levels of occlusions, thus reducing the risk of over-fitting while making the model more robust to occlusions. In [10] a method is proposed where the saliency map of the initial image is exploited in order to detect important regions of the image that are preserved during augmentation leading to better training examples.

In contrast to other augmentation methods, the proposed occlusions are produced via image synthesis, occur deliberately and in a controlled manner, with no randomly generated values or black tiles, solely by exploiting existing objects taken from the initial smaller in size dataset. This is achieved by exploiting a segmentation object mask in one image and translating it so that another object in the same or a different image is partially or severely occluded. The translated object can be scaled, rotated, or flipped, while the ground-truth labels of the image are adjusted accordingly. Our methodology can also be applied together with the baseline augmentation [18], including both photometric (i.e., random contrast/brightness/hue etc.) and geometric distortions (i.e., expand image, random crop/mirror etc.). Moreover, the same technique can be applied to pairs of images, where one image is used as a random background and the other one provides a segmentation mask for the foreground. This process is employed mainly to generate new examples containing minority semantic classes and reduce the overall class imbalance during model training.

3 Object Occlusion-Based Data Augmentation Method

This section describes the proposed data augmentation technique. The core idea is analyzed in Sect. 3.1 and explains how to create additional training examples containing target occlusions by exploiting the existing targets in an image. Apart from the obvious benefit of enriching the training set with more examples, the deep architectures that are trained on such data, are able to learn how to deal with cases where the target to be detected/segmented is occluded by other objects of interest. Figure 1 illustrates some indicative results on a sequence of video frames that include person-to-person occlusions and person entanglement of 3 people. Our findings show that when model training is enhanced with our data augmentation methodology, the model becomes more resilient in locating occluded objects. Moreover, a methodology of reducing class imbalance by generating new examples from the same technique for the minority classes is also analyzed in Sect. 3.2.

3.1 Creating Occlusions

In a given training set $\mathcal{S}$, let $\mathbf{X} \in \mathbb{R}^{H \times W \times C}$, be a training image of $\mathcal{S}$ containing N targets, where H, W corresponds to the height and width of the image

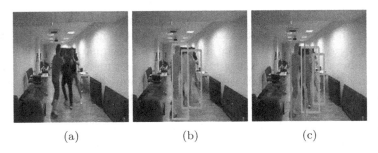

(a)   (b)   (c)

Fig. 1. Prediction masks for the lightweight YOLACT-9 method, (a) Input image, (b) results for training with baseline augmentation [18], (c) results with training with the proposed augmentation method. In (c) the method manages to detect all objects inside the scene.

respectively and C is the number of colour channels. For each target a vector $\mathbf{r}_n$, $n = 0, \ldots, N - 1$, $N \in \mathbb{Z}^+$, is available containing the top left and bottom right pixel coordinates of the groundtruth bounding box, a class label $\mathbf{l}_n$ and a segmentation mask $\mathbf{M}_n \in \mathbb{R}^{H \times W}$ equal to 1 for the pixels where the target lies and 0 otherwise. By exploiting $\mathbf{r}_n$ and $\mathbf{M}_n$, the image $\mathbf{O}_n \subset \mathbf{X}$ can be extracted, being of size equal to the bounding box of the target, $H_n \times W_n$. The extracted image is stored and used for creating synthetic occlusions, given a predetermined probability of occlusion occurrence ($p_o = 0.5$). The proposed augmentation technique can be applied on training examples where $N \geq 1$.

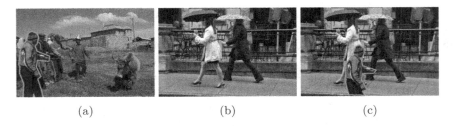

(a)   (b)   (c)

Fig. 2. Process of creating synthetic occlusions, (a) Input image containing the occluder object (highlighted in green for illustration purposes), (b) Input image containing the occluded object, (c) Output image with the synthetic occlusion. (Color figure online)

When our methodology is applied, two objects are randomly selected from the same image $\mathbf{X}$ or from two different images ($\mathbf{X}$ and $\mathbf{Y}$) of the training set (referred as randomPair in Algorithm 1). All available image masks of $\mathbf{X}$ and $\mathbf{Y}$ can be denoted as $\mathbf{M}_n^X$ and $\mathbf{M}_n^Y$ and can be utilized as the foreground and background for the output image respectively. One of the two randomly selected objects is set as the occluder $f \in [0, N - 1]$ and the other as the occluded object $b \in [0, N - 1], b \neq f$. The first step towards the new image, is to compute the scale factor $s = \frac{H_f}{H_b} * d$ by exploiting the dimensions of f and b which is then multiplied by a random number $d \in [-0.3, 0.3]$. The second step, is to calculate the translation vector $\mathbf{v}$, used in translating $\mathbf{O}_f$ on top of $\mathbf{O}_b$. With the aid of $\mathbf{M}_f$ only the pixel values that actually belong to the occluder object are translated. The permutation $\mathbf{P}_f$ matrix, filled with ones on the anti-diagonal and zeros everywhere else can be exploited towards flipping an image in horizontal, vertical or both axes. More specifically, by constructing $\mathbf{P}_v \in \mathbb{R}^{W \times W}$ and $\mathbf{P}_h \in \mathbb{R}^{H \times H}$, the following matrix multiplications, $\mathbf{X}_v = \mathbf{X}\mathbf{P}_v$, $\mathbf{X}_h = \mathbf{P}_h\mathbf{X}$, $\mathbf{X}_b = \mathbf{P}_h\mathbf{X}\mathbf{P}_v$, will result in vertical, horizontal and both axes flipped versions of the initial image.

By pasting image masks on top of other images, boundary artifacts are generated on the edges of the pasted objects. These artifacts have a significant negative impact on methods that strongly depend on local region-based features [6], such as instance segmentation. In order to address this issue, a blending procedure is incorporated in the proposed augmentation method (referred as blurAround-Mask in Algorithm 1). More specifically, the edges of the occluder object are blurred, given a predetermined probability of occurrence ($p_b = \frac{2}{3}$), resulting in a smoothed area 2 pixels wide. As a final step $\mathbf{O}_f$ is utilized to create the new image $\mathbf{X}'_b$, while its corresponding labels and masks are also updated. The entire data augmentation technique for creating synthetic object occlusions is described in Algorithm 1. An example of this process is illustrated in Fig. 2.

Algorithm 1. Data augmentation with object occlusions.

1: **procedure** AUGMOCCLUSION($\mathbf{X}_1, \mathbf{M}_n^X, \mathbf{r}^X, l_1, \mathbf{X}_2, \mathbf{M}_n^Y, \mathbf{r}^Y, l_2, p_o = 0.5$)
2:    **if** length($\mathbf{r}^X$) ≥ 1 **and** length($\mathbf{r}^Y$) ≥ 1 **and** random() $\geq p_o$ **then**
3:        $b, f \leftarrow$ randomPair(l_1, l_2)
4:        $\mathbf{X}_b, \mathbf{O}_b, \mathbf{O}_f, \mathbf{M}_b, \mathbf{M}_f, \mathbf{r}_b, \mathbf{r}_f \leftarrow$ selectIndices($\mathbf{X}_1, \mathbf{O}_1, \mathbf{O}_2, \mathbf{M}_n^X, \mathbf{M}_n^Y, \mathbf{r}^X, \mathbf{r}^Y$)
5:        $s \leftarrow$ scaleFactor($\mathbf{r}_b, \mathbf{r}_f, rand(-0.3, 0.3)$)
6:        $\mathbf{O}_f, \mathbf{M}_f \leftarrow$ translateAndScale($\mathbf{O}_f, \mathbf{M}_f, s$)
7:        $\mathbf{O}_f \leftarrow$ blurAroundMask($\mathbf{O}_f, \mathbf{M}_f, p_b = 0.66$)
8:        $\mathbf{X}'_b \leftarrow$ occludePair($\mathbf{X}_b, \mathbf{O}_f$)
9:        $\mathbf{M}'_b, \mathbf{M}'_f, 1_f \leftarrow$ updateMasksLabels($\mathbf{r}_b, \mathbf{r}_f, \mathbf{M}_b, \mathbf{M}_f, b, f$)
10:        **return** $\mathbf{X}'_b, 1_f, \mathbf{M}'_b, \mathbf{M}'_f$
11:    **end if**
12: **end procedure**

3.2 Generating New Examples

The method described in the previous paragraph can also be used to generate new images for model training. By translating and scaling an image mask of a desired object to another image that contains a random background with irrelevant semantic classes to model training, it is possible to create a controlled number of new images per class. In contrast to the method described in Sect. 3.1, the new image is generated by only exploiting $\mathbf{M}_i, \mathbf{X}_i, i \in [0, N-1]$ of the initial image with the desired label and the random image $\mathbf{Y}$. Some examples can be observed in Fig. 3.

Table 1. DARLENE IS training dataset.

Class:	person	backpack	handbag	suitcase	knife	firearm	TOTAL
Images	64115	5528	6841	2402	4732	1640	85258
Annotations	257253	8714	12342	6112	8291	1795	294507

Particularly for security applications as in [1], it is vital to be able to detect objects that could be related to abnormal events or security crises, namely person, backpack, handbag, suitcase, firearm and knife. To the best of our knowledge, there are no publicly available datasets for IS, regarding class firearm. Moreover, classes backpack, handbag, suitcase and knife are underrepresented in the available datasets in contrast with class person. To this end we have developed the DARLENE IS dataset. The purpose of this new dataset was to, (a) incorporate class firearm into model training, (b) expand the training set by reusing image mask information and (c) reduce class imbalance in model training by augmenting minority classes. The DARLENE IS training dataset consists of annotated images taken from various databases (MS COCO [17], MGD [16], OpenImagesDatasetV6 [13]) and is analyzed in Table 1.

More specifically, regarding class firearm, we manually annotated 1720 examples in total. The majority of firearms (1304 images) were taken from [16], 260 images were from [13] and 76 images were created for DARLENE project, to include a specific type of firearm in model training. From these samples, 1640 were used in model training and 80 for validation. This process was performed with the aid of an online annotation tool, namely VGG Image Annotator [5]. Regarding the DARLENE IS validation dataset, which is summarized in Table 2, we kept the same training-to-validation ratio as in MS COCO dataset for the new class firearm. Secondly, the issue of class imbalance was taken into consideration. This was achieved by creating a controlled number of additional examples for the minority classes and adding them to model training. Thus, by starting from a data distribution that favors heavily class person, we end up with a more balanced distribution per class. The contents of the DARLENE IS balanced training dataset are summarized in Table 2.

4   Experimental Evaluation

The experiments were conducted on three datasets. The first set of experiments employed the validation and test-dev set of MS COCO 2017 dataset [17], while the second one was conducted on DARLENE IS validation dataset. The third dataset, namely DARLENE Security Footage (SF) dataset, consists of 6 security-related scenarios and was created to benchmark the performance of the developed

Fig. 3. Training examples from DARLENE IS dataset. The first 3 images were produced from our augmentation technique, while the last image is a real sample.

Table 2. DARLENE IS balanced training and DARLENE IS validation dataset.

	Class	person	backpack	handbag	suitcase	knife	firearm	TOTAL
Training	Images	64115	7000	7000	7000	7000	7000	99115
	Annotations	257253	10186	12501	10710	10559	7155	308364
Validation	Images	2693	228	292	105	181	80	3579
	Annotations	10772	371	540	299	325	80	12387

methodologies in real-world conditions. The content of these scenarios involved: (a) partial and severe occlusions occurring between objects of interest, (b) person entanglement while overcrowding or fighting, (c) different illumination conditions, (d) recording from different angles in a multi-camera setup and (e) simulation of violence/non-violence storylines with/without the usage of firearms. Some indicative samples of the aforementioned points are presented in Fig. 4. Such a dataset was not publicly available, thus all video recordings and their corresponding ground truth annotations had to be created from the ground up. Each scenario was recorded from 3 cameras for a specific duration (500 frames), thus the total number of frames recorded and annotated for the DARLENE SF dataset reached 9000 video frames.

(a) Partial/severe occlusions.	(b) Person entanglement.	(c) Sufficient illumination.	(d) Poor illumination.

Fig. 4. Special attribute examples from the DARLENE SF dataset. The recordings include (a) partial and severe object occlusions, (b) person entanglement, (c) scenes with sufficient illumination conditions and (d) scenes with poor illumination.

MS-COCO 2017 has 80 semantic classes, while the DARLENE IS dataset includes 6 semantic classes, namely person, backpack, handbag, suitcase, knife and firearm. The instance segmentation models used were YOLACT [2] variants, that differ in the backbone architecture. The backbones employed were

Table 3. Quantitative results of YOLACT variants and YOLO v5 on MS COCO dataset (mAP mask averages for validation and test-dev sets. Comparison between baseline [18] and proposed augmentation technique).

	validation set (baseline/proposed)	test-dev set (baseline/proposed)
YOLACT-18	20.40%/**25.30%**	20.20%/**24.90%**
YOLACT-9	13.31%/**20.65%**	10.20%/**15.70%**
YOLO v5	23.40%/**24.00%**	21.50%/**22.40%**

Table 4. Quantitative results of YOLO V5 and YOLACT variants on DARLENE IS validation dataset. Segmentation mask mean Average Precision per class and average results are reported.

	method	person	backpack	handbag	suitcase	knife	firearm	Avg
Baseline augmentation [18]	YOLACT-18	33.3%	10.3%	8.2%	22.1%	3.9%	30.7%	18.1%
	YOLACT-9	28.9%	7.2%	4.5%	13.1%	2.4%	28.2%	14.1%
	YOLO v5	36.0%	7.8%	6.5%	18.4%	3.8%	27.2%	16.6%
Proposed augmentation	YOLACT-18	34.2%	10.5%	7.7%	20.0%	4.5%	35.6%	**18.7%**
	YOLACT-9	29.7%	8.3%	4.6%	13.4%	2.9%	32.9%	**15.3%**
	YOLO v5	44.2%	8.7%	5.3%	21.8%	10.2%	37.3%	**21.3%**
Copy-Paste	YOLACT-18	33.90%	10.17%	7.51%	22.29%	4.46%	25.90%	17.37%
Augmentation	YOLACT-9	29.66%	8.86%	5.39%	13.56%	2.68%	20.73%	13.48%

ResNet-18 and a self-developed lightweight Fully Convolutional Network as the feature backbone, with 9 convolutional layers (ResNet-9) based on the Residual Networks [11], created for performing the task of instance segmentation in real-time even on devices with limited resources. Both models were pre-trained on ImageNet dataset [3] and their training was extended by utilizing the MS-COCO training set. YOLO v5, having pre-trained weights on MS-COCO, was also employed in this experiment once with and once without the proposed augmentation method, trained for the same number of epochs. This model was modified from its original architecture that performed object detection to be employed for the task of instance segmentation.

The results of the first set of experiments are displayed in Table 3 and show a **0.6%**, a **7.34%** and a **4.9%** increase in performance for YOLO v5, YOLACT with backbone ResNet-9 and ResNet-18 respectively on the validation set. For the test-dev set we also noticed a **0.9%**, a **5.5%** and a **4.7%** increase for YOLO v5, YOLACT with backbone ResNet-9 and ResNet-18 respectively.

The second set of experiments is reported in Table 4 and involves fine-tuning the models for 6 classes on DARLENE IS training dataset. In every case, the models were compared on the DARLENE IS validation dataset, which contains a set of 3579 real samples. The results here show a **1.2%** and a **0.6%** increase in performance for YOLACT with backbone ResNet-9 and ResNet-18 respectively, in comparison with the baseline augmentation. Additionally, the proposed augmentation technique was applied to another segmentation method, namely YOLO

Table 5. Quantitative results of YOLO V5 and YOLACT variants on DARLENE SF dataset. Segmentation mask mean Average Precision per class and average results across all scenarios are reported.

	method	person	backpack	firearm	Avg
Baseline augmentation	YOLACT-18	39.02%	12.64%	11.46%	21.04%
	YOLACT-9	33.51%	10.51%	9.20%	17.74%
	YOLO v5	48.16%	13.50%	14.50%	25.39%
Proposed augmentation	YOLACT-18	39.29%	14.92%	10.65%	**21.62%**
	YOLACT-9	33.46%	12.60%	11.11%	**19.06%**
	YOLO v5	53.25%	16.81%	19.82%	**29.96%**
Copy-Paste	YOLACT-18	40.68%	6.95%	7.44%	18.36%
Augmentation	YOLACT-9	34.41%	10.46%	9.08%	17.98%

v5 [21], where training with the proposed technique, leads to better performance by **+4.7%**. Finally, the proposed augmentation technique is also compared to a different data augmentation technique, namely Copy-Paste Augmentation [9]. As the evaluation results indicate, the proposed technique achieves mean Average Precision metric of **15.3%** and **18.7%** in comparison with Copy-Paste method that achieved **13.48%** and **17.37%** for YOLACT-9 and YOLACT-18 respectively. For the Copy-Paste method we achieved the best results with the following training parameters: Standard Scale Jittering from a scale of 0.8 to 1.25, image padding with a top-left positioning, horizontal image flip and the Copy-Paste augmentation both having a probability of 0.5. We note that this experimental setup could not surpass the baseline augmentation for all semantic classes on the two datasets where it was employed. By examining the per class results in Table 4, we observe that there is an improvement for most classes, however, there is a big drop for class firearm affecting the overall average performance. Due to the random nature of Copy-Paste augmentation during the selection of object classes that are pasted, the representation of minority classes such as firearm can get worse. This results in worse overall performance on this dataset. In some classes we notice a much lower score than others. The reason behind this is that these objects are difficult to detect, especially in cases where their illumination is not good, or they are seen through certain viewing angles, or even because they cover a very small portion of the scene. However, despite the fact that they are hard to detect, for almost all classes in which our augmentation technique was applied, we notice an increase in performance.

The last set of experiments regards evaluating the same models as the second set on the DARLENE SF dataset and is reported in Table 5. The results again show the superiority of our methodology compared to both the baseline augmentation and Copy-paste method. From Table 5 we can extrapolate the best scores on average for all scenarios, achieving **21.62%**, **19.06%** and **29.96%** with YOLACT-18, YOLACT-9 and YOLO v5 respectively. The recorded scenarios included only three out of the six semantic classes the models were trained on, namely person, backpack and firearm.

Fig. 5. Prediction masks for the lightweight YOLACT-18 method, (top) with baseline data augmentations (bottom) with the proposed augmentation technique.

In Fig. 5 we can observe some qualitative results of the YOLACT method trained with the baseline augmentation technique (top) versus the results when the same method is trained exploiting the proposed data augmentation method. These test images were obtained from DARLENE SF and from Crowdpose dataset [15]. In the first image, the detection of a small firearm severely occluded from a person's hand is segmented only for the proposed scheme. In the second image, a difficult detection of a small firearm on top of a background with the same color is correctly identified only when the proposed augmentation technique is employed. The third image includes again a firearm partially occluded by a person and is detected only after training with our augmentation technique. Finally, in the fourth image, a severe person-to-person occlusion (between the child and the biker) was segmented only with the proposed scheme. The rest of the predictions are similar for both methods. Examining these qualitative results, it is observed that the augmentation technique aided the segmentation method to perform better for small and occluded objects.

5   Conclusions

A data augmentation technique is presented, from which new samples from the same dataset can be generated and used for model training. Focus was given on generating examples that deliberately cause occlusions between existing objects so that models can generalize better in unknown objects that are partially occluded. In addition, the same methodology was used to generate additional examples for minority semantic classes, to address the presence of class imbalance. Both of these techniques have shown a performance boost compared to the baseline, ranging from +0.9% up to +5.5% on test-dev set of MS COCO 2017, from +0.7% up to +4.7% on DARLENE IS validation and from +0.58% up to +4.57% on DARLENE SF dataset, for the models tested and serve as a cost-effective way of producing additional training samples to existing datasets. Finally, two new instance segmentation datasets were created which are suitable for model training and testing focused on security applications.

Acknowledgements. This project has received funding from the European Union's Horizon 2020 research and innovation programme under grant agreement No 883297 (project DARLENE). This publication reflects only the authors' views. The European Commission is not responsible for any use that may be made of the information it contains.

References

1. Apostolakis, K.C., Dimitriou, N., Margetis, G., Ntoa, S., Tzovaras, D., Stephanidis, C.: Darlene-improving situational awareness of European law enforcement agents through a combination of augmented reality and artificial intelligence solutions. Open Res. Europe 1(87), 87 (2022)
2. Bolya, D., Zhou, C., Xiao, F., Lee, Y.J.: YOLACT: real-time instance segmentation. In: Proceedings of the IEEE/CVF International Conference on Computer Vision, pp. 9157–9166 (2019)
3. Deng, J., Dong, W., Socher, R., Li, L.J., Li, K., Fei-Fei, L.: ImageNet: a large-scale hierarchical image database. In: Proceedings of the IEEE/CVF Conference on Computer Vision and Pattern Recognition, pp. 248–255. IEEE (2009)
4. Dosovitskiy, A., et al.: An image is worth 16×16 words: transformers for image recognition at scale. In: International Conference on Learning Representations (2020)
5. Dutta, A., Zisserman, A.: The VIA annotation software for images, audio and video. In: Proceedings of the 27th ACM International Conference on Multimedia, MM 2019, ACM, New York, NY, USA (2019)
6. Dwibedi, D., Misra, I., Hebert, M.: Cut, paste and learn: surprisingly easy synthesis for instance detection. In: Proceedings of the IEEE/CVF International Conference on Computer Vision, pp. 1301–1310 (2017)
7. Fang, Y., et al.: EVA: exploring the limits of masked visual representation learning at scale. arXiv preprint arXiv:2211.07636 (2022)
8. Fong, R., Vedaldi, A.: Occlusions for effective data augmentation in image classification. In: 2019 IEEE/CVF International Conference on Computer Vision Workshop, pp. 4158–4166. IEEE (2019)
9. Ghiasi, G., et al.: Simple copy-paste is a strong data augmentation method for instance segmentation. In: Proceedings of the IEEE/CVF Conference on Computer Vision and Pattern Recognition, pp. 2918–2928 (2021)
10. Gong, C., Wang, D., Li, M., Chandra, V., Liu, Q.: KeepAugment: a simple information-preserving data augmentation approach. In: Proceedings of the IEEE/CVF Conference on Computer Vision and Pattern Recognition, pp. 1055–1064 (2021)

11. He, K., Zhang, X., Ren, S., Sun, J.: Deep residual learning for image recognition. In: Proceedings of the IEEE/CVF Conference on Computer Vision and Pattern Recognition, pp. 770–778 (2016)
12. Hong, M., Choi, J., Kim, G.: StyleMix: separating content and style for enhanced data augmentation. In: Proceedings of the IEEE/CVF Conference on Computer Vision and Pattern Recognition, pp. 14862–14870 (2021)
13. Kuznetsova, A., et al.: The open images dataset v4. Int. J. Comput. Vis. **128**(7), 1956–1981 (2020)
14. Li, F., et al.: Mask DINO: towards a unified transformer-based framework for object detection and segmentation. arXiv preprint arXiv:2206.02777 (2022)
15. Li, J., et al.: CrowdPose: efficient crowded scenes pose estimation and a new benchmark. In: Proceedings of the IEEE/CVF Conference on Computer Vision and Pattern Recognition, pp. 10863–10872 (2019)
16. Lim, J., Al Jobayer, M.I., Baskaran, V.M., Lim, J.M., See, J., Wong, K.: Deep multi-level feature pyramids: application for non-canonical firearm detection in video surveillance. Eng. Appl. Artif. Intell. **97**, 104094 (2021)
17. Lin, T.-Y., et al.: Microsoft COCO: common objects in context. In: Fleet, D., Pajdla, T., Schiele, B., Tuytelaars, T. (eds.) ECCV 2014. LNCS, vol. 8693, pp. 740–755. Springer, Cham (2014). https://doi.org/10.1007/978-3-319-10602-1_48
18. Liu, W., et al.: SSD: single shot multibox detector. In: Leibe, B., Matas, J., Sebe, N., Welling, M. (eds.) ECCV 2016. LNCS, vol. 9905, pp. 21–37. Springer, Cham (2016). https://doi.org/10.1007/978-3-319-46448-0_2
19. Ouali, Y., Hudelot, C., Tami, M.: Semi-supervised semantic segmentation with cross-consistency training. In: Proceedings of the IEEE/CVF Conference on Computer Vision and Pattern Recognition, pp. 12674–12684 (2020)
20. Pan, S.J., Yang, Q.: A survey on transfer learning. IEEE Trans. Knowl. Data Eng. **22**(10), 1345–1359 (2009)
21. Redmon, J., Farhadi, A.: YOLOv3: an incremental improvement. CoRR abs/1804.02767 (2018). http://arxiv.org/abs/1804.02767
22. Wang, Y.X., Girshick, R., Hebert, M., Hariharan, B.: Low-shot learning from imaginary data. In: Proceedings of the IEEE/CVF Conference on Computer Vision and Pattern Recognition, pp. 7278–7286 (2018)
23. Yun, S., Han, D., Oh, S.J., Chun, S., Choe, J., Yoo, Y.: CutMix: Regularization strategy to train strong classifiers with localizable features. In: Proceedings of the IEEE/CVF International Conference on Computer Vision (2019)
24. Zhong, Z., Zheng, L., Kang, G., Li, S., Yang, Y.: Random erasing data augmentation. In: Proceedings of the AAAI Conference on Artificial Intelligence, vol. 34, pp. 13001–13008 (2020)

Unsupervised Video Anomaly Detection with Diffusion Models Conditioned on Compact Motion Representations

Anil Osman Tur[1,2], Nicola Dall'Asen[1,3(✉)], Cigdem Beyan[1], and Elisa Ricci[1,2]

[1] University of Trento, Trento, Italy
nicola.dallasen@unitn.it
[2] Fondazione Bruno Kessler, Trento, Italy
[3] University of Pisa, Pisa, Italy

Abstract. This paper aims to address the unsupervised video anomaly detection (VAD) problem, which involves classifying each frame in a video as normal or abnormal, without any access to labels. To accomplish this, the proposed method employs conditional diffusion models, where the input data is the spatiotemporal features extracted from a pre-trained network, and the condition is the features extracted from compact motion representations that summarize a given video segment in terms of its motion and appearance. Our method utilizes a data-driven threshold and considers a high reconstruction error as an indicator of anomalous events. This study is the first to utilize compact motion representations for VAD and the experiments conducted on two large-scale VAD benchmarks demonstrate that they supply relevant information to the diffusion model, and consequently improve VAD performances *w.r.t* the prior art. Importantly, our method exhibits better generalization performance across different datasets, notably outperforming both the state-of-the-art and baseline methods. The code of our method is available HERE.

Keywords: Video anomaly detection · unsupervised learning · video understanding · conditional diffusion models · generative models

1 Introduction

Detecting anomalous events in videos automatically is a crucial task of computer vision that has relevance to numerous applications, including but not limited to intelligent surveillance and activity recognition [2,6,11,14,18,27]. Video anomaly detection (VAD) can be particularly difficult because abnormal events in the real world are infrequent and can belong to an unbounded number of categories. As a result, traditional supervised methods might not be suitable for this task since balanced normal and abnormal samples are typically unavailable for training. Moreover, VAD models are challenged by the contextual and often ambiguous nature of

---

A. Osman Tur and N. Dall'Asen—These authors contributed equally.

G. L. Foresti et al. (Eds.): ICIAP 2023, LNCS 14234, pp. 49–62, 2023.
https://doi.org/10.1007/978-3-031-43153-1_5

abnormal events, despite their sparsity and diversity [23]. As a result, VAD is commonly carried out using a *one-class learning* approach, in which only normal data are provided during training [9,13,22,26,34]. However, given the dynamic nature of real-world applications and the wide range of normal classes, it is not practical to have access to every type of normal training data. Therefore, when using a one-class classifier, there is a high risk of misclassifying an unseen normal event as abnormal because its representation might be significantly different from the representations learned from normal training data [6]. To address the aforementioned challenge of data availability, some researchers have implemented *weakly supervised* VAD that do not require per-frame annotations but instead leverage video-level labels [15,28]. In weakly supervised VAD, unlike its one-class counterpart, a video is considered anomalous if even a single frame within it is labeled as anomalous. Conversely, a video is labeled as normal only when all frames within it are labeled as normal. However, such approaches lack localizing the abnormal portion of the video, which can be impractical when dealing with long videos. Also, it is important to note that labeling a video as normal still requires the inspection of entire frames [32]. A more recent approach to VAD is **unsupervised learning**, in which *unlabelled* videos are used as input and the model learns to classify each frame as normal or anomalous, allowing to localize the abnormal frames. Unlike a one-class classifier, unsupervised VAD does not make any assumptions about the distribution of the training data and does not use any labels during model training. However, it is undoubtedly more challenging to arrive at the performance of other VAD approaches that use labeled training data [32].

This study focuses on performing unsupervised VAD in complex surveillance scenarios by relying solely on the reconstruction capability of the probabilistic generative model called **diffusion models** [12]. The usage of generative models (e.g., autoencoders) is common for one-class VAD [8,19,23]. However, as shown in [32] for unsupervised VAD, the autoencoders might require an additional discriminator to be trained collaboratively to reach a desired level of performance. Instead, our study reveals that diffusion models constitute a more effective category of generative models for unsupervised VAD, displaying superior results when compared to autoencoders, and in some cases, even exceeding the performance of Collaborative Generative and Discriminative Models. Furthermore, we explore the application of **compact motion representations**, namely, **star representation** [7] and **dynamic images** [4] within a conditional diffusion model. This study marks the first attempt at utilizing these motion representations to address the VAD task. The experimental evaluation conducted on two large-scale datasets indicates that using the aforementioned compact motion representations as a condition of diffusion models is more beneficial for VAD. We also explore the transferability of unsupervised VAD methods by assessing their generalization performance when trained on one dataset and tested on another. When performing cross-dataset analysis, it becomes apparent that incorporating compact motion representations as the condition of diffusion models leads to vastly superior performance. This represents a crucial feature of the proposed method in comparison to both the state-of-the-art (SOTA) and baseline models, making it highly valuable for practical applications.

The main contributions can be summarized in three folds. (1) We propose an effective unsupervised VAD method, which uses compact motion representations as the condition of the diffusion models. We show that compact motion representations supply relevant information and further improve VAD performance. (2) Our method leads to enhanced generalization performance across datasets. Its transferability is notably better than the baseline methods and the SOTA. (3) We conduct a hyperparameter analysis for diffusion models, which yields insights into using them for VAD.

2   Related Work

Anomaly Detection. Anomaly refers to an entity that is rare and significantly deviates from normality. Automated anomaly detection models face challenges when detecting abnormal events from images or videos due to their sparsity, diversity, ambiguity, and contextual nature [6,23,33]. Automated anomaly detection is a well-researched subject that encompasses various tasks, *e.g.,* medical diagnosis, defect detection, animal behavior understanding, and fraud detection [2,29,31]. For a review of anomaly detection applications in different domains, interested readers can refer to the survey paper [6]. VAD, the task at hand, deals with complex surveillance scenarios. Zaheer et al. [32] categorized relevant methodologies into four groups: (a) fully supervised approaches requiring normal/abnormal annotations for each video frame in the training data, (b) one-class classification requiring only annotated training data for the normal class, (c) weakly supervised approaches requiring *video-level* normal/abnormal annotations, and (d) unsupervised methods that do not require any annotations.

Labeling data is a costly and time-consuming task, and due to the rarity of abnormal events, it is impractical to gather all possible anomaly samples for fully-supervised learning. Consequently, the most common approach to tackling VAD is to train a **one-class classifier** that learns from the *normal* data [9,13,22,26,34]. Several of these approaches utilize hand-crafted features [16,21], while others rely on deep features that are extracted using pre-trained models [22,26]. Generative models e.g., autoencoders and GANs have also been adapted for VAD [8,19,23]. One-class classifiers often cannot prevent the well-reconstruction of anomalous test inputs, resulting in the misclassification of abnormal instances as normal. Moreover, an unseen normal instance could be misclassified as abnormal because its representation may differ significantly from the representations learned from normal training data. As evident, data collection is still a problem for the one-class approach because it is not practical to have access to every variety of normal training data [6,18]. Therefore, some researchers [15,28] have turned to **weakly supervised VAD**, which does not rely on fine-grained per-frame annotations, but instead use video-level labels. Consequently, a video is labeled as anomalous even if one frame is anomalous, and normal if all frames are normal. This setting is not optimal because labeling a video as normal requires inspecting all frames, and it cannot localize the abnormal portion.

On the other hand, VAD methods that use unlabelled training data are quite rare in the literature. It is important to note that several one-class classifiers [9,13,34] have been referred to as unsupervised, even though they use *labeled normal* data. **Unsupervised VAD** methods analyze unlabelled videos without prior knowledge of normal or abnormal events to classify each frame as normal or anomalous. The only published method addressing this definition is [32], which presents a Generative Cooperative Learning among a generator (an autoencoder) and a discriminator (a multilayer perceptron) with a negative learning paradigm. The autoencoder reconstructs the normal and abnormal instances while the discriminator estimates the probability of being abnormal. Through negative learning, the autoencoder is constrained not to learn the reconstruction of anomalies using the pseudo-labels produced by the discriminator. That approach [32] follows the idea that anomalies occur less frequently than normal events, such that the generator should be able to reconstruct the abundantly available normal representations. Besides, it promotes temporal consistency while extracting relevant spatiotemporal features. Our method differs from [32] in that it relies solely on a generative architecture, specifically a conditional diffusion model. The baseline unconditional diffusion model, in some cases, surpasses the full model of [32] while in all cases it achieves better performance than the autoencoder of [32]. On the other hand, the proposed method improves the achievements of the unconditional diffusion model thanks to using compact motion representations, and importantly, it presents the best generalization results across datasets.

Diffusion Models. They are a family of probabilistic generative models that progressively destruct data by injecting noise, then learn to reverse this process for sample generation. [10,12]. Diffusion models have emerged as a powerful new family of deep generative models with SOTA performance in many applications, including image synthesis, video generation, and discriminative tasks like object detection and semantic segmentation [5,24]. Given that diffusion models have emerged as SOTA generative models for various tasks, we are motivated to explore their potential for VAD through our proposed method.

Star Representation [7]. It aims to represent temporal information existing in a video in a way that the channels of output single RGB image convey the summarized time information by associating the color channels with simplified consecutive moments of the video clip. Such a representation is suitable to be the input of any CNN model and so far in the literature, it was used for dynamic gesture recognition [1,7], while this is the first time it is being used for VAD.

Dynamic Image [4]. It refers to a representation of an input video sequence that summarizes the appearances of objects and their corresponding motions over time by encoding the temporal ordering of the pixels from frame to frame. This can be seen as an early fusion technique since the frames are combined into a single representation before further processing them such as with. It has been used for action and gesture recognition [4,30] and visual activity modeling [3,25], however, it has never been used for VAD.

3   Proposed Method

We design a method to use diffusion models to tackle the unsupervised VAD, *i.e.* to classify each frame in a video as normal or abnormal without using the labels. To

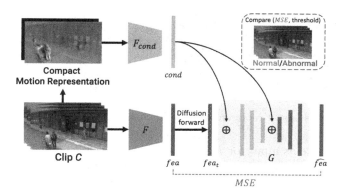

Fig. 1. An illustration of the proposed method. For definitions of the abbreviations used, please refer to the text.

provide a frame-based prediction, we classify a video clip of consecutive N frames and then slide this window along the video. We build our model on top of diffusion models, in particular, *k-diffusion* [12], which has shown better performance *w.r.t* DDPM [10]. To overcome the heavy computational burden of dealing with video clips, we operate in the latent space of a pre-trained network that extracts clip-level features. We then leverage the generative capabilities of diffusion models to reconstruct noised clip features and, based on the reconstruction error, decide whether the clip is normal or abnormal with a data-driven threshold. While this formulation leads to SOTA performance, we further condition the diffusion process with compact motion information coming from the video clip (see Sec. 3.2) to better guide the reverse process and achieve better performance. An overview of our method is provided in Fig. 1.

3.1   Diffusion Model

Diffusion models apply a progressive addition of Gaussian noise ϵ_t of standard deviation σ_t to an input data point x_T sampled from a distribution $p_{data}(x)$ for each timestep $t \in [0, T]$. The noised distribution $p(x, \sigma)$ becomes isotropic Gaussian and allows efficient sampling of new data points $x_0 \sim \mathcal{N}(0, \sigma_{max}^2 \mathbf{I})$. These data are gradually denoised with noise levels $\sigma_0 = \sigma_{max} > \sigma_1 > \cdots > \sigma_{T-1} > \sigma_T = 0$ into new samples. Diffusion models are trained by minimizing the expected L_2 error between predicted and ground truth added noise [10], *i.e.*: $\mathcal{L}_{simple} = \|\epsilon_t - \hat{\epsilon}\|_2$. In this work, we use the diffusion formulation of [12], which allows the network to perform either ϵ or x_0 prediction, or something in between, depending on the noise scale σ_t, nullify the error amplification that happens in DDPM [12]. The denoising network D_θ formulation as follows:

$$D_\theta(x; \sigma_t) = c_{skip}(\sigma_t)\, x + c_{out}(\sigma_t)\, G_\theta\big(c_{in}(\sigma_t)\, x;\ c_{noise}(\sigma - T)\big), \qquad (1)$$

where G_θ becomes the effective network to train, c_{skip} modulates the skip connection, $c_{in}(\cdot)$ and $c_{out}(\cdot)$ scale input and output magnitudes, and $c_{noise}(\cdot)$ scales σ

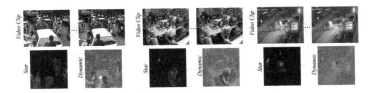

Fig. 2. Examples of star representation and dynamic image for a given video clip.

to become suitable as input for F_θ. Formally, given a video clip C of N frames, i.e. $C \in \mathbb{R}^{N \times 3 \times H \times W}$, we first extract features from a pre-trained 3D-CNN $\mathcal{F}$ to obtain a feature vector $fea \in \mathbb{R}^f$, with f the latent dimension of the network. We then use this latent representation in the diffusion process to reconstruct them without using any label.

We leverage the fact that denoising does not necessarily have to start from noise with variance σ_{max}^2, but it can place at any arbitrary timestep $t \in (0, T]$, as shown in [17]. We can therefore sample $fea_t \sim \mathcal{N}(fea, \sigma_t^2)$ and run the diffusion reverse process on it to reconstruct fea_T. The choice of t allows balancing the amount of information destroyed in the forward process, and we exploit this fact to remove the frequency components associated with anomalies. We then measure the reconstruction goodness in terms of MSE, with a higher reconstruction error *possibly* indicating that the clip is anomalous. When deciding whether a video frame is anomalous or not, we adopt the data-driven thresholding mechanism of [32]. The decision for a single video frame is made by keeping the distribution of the reconstruction loss (MSE) of each clip over a *batch*. The feature vectors resulting in higher reconstruction error refer to anomalous clips and vice versa. This decision is made through the data-driven threshold L_{th}, defined as $L_{th} = \mu_p + k\,\sigma_p$ where k is a constant, μ_p and σ_p are the mean and standard deviation of the reconstruction error for each batch.

3.2   Compact Motion Representations

We further extend the described diffusion model to incorporate compact motion representation in the process to provide rich motion information. We compute this representation using two different approaches: Star representation [7] or Dynamic Image [4]. Visual examples of these two representations are presented in Fig. 2 and a complete description is presented as follows.

Star RGB Images. The objective of using star representation is to depict the time-based data present in an input RGB video [1,7]. The star representation matrix M computation is computed as given in Eq. 2 where $I_k(i,j)$ represents the RGB vectors of a pixel at a given (i, j) position at $k - th$ frame and λ is the cosine similarity of the RGB vectors. By using such a cosine similarity star representation also includes the information change in hue and saturation.

$$M(i,j) = \sum_{k=2}^{N} \left(1 - \frac{\lambda}{2}\right) . |\ \|\ I_{k-1}(i,j)\ \|_2 - \|\ I_{k-1}(i,j)\ \|_2\ |, \qquad (2)$$

where N is the length of the video clip. To create an RGB image as the output, each video segment is divided equally into three sub-videos such that each sub-video is used for generating one of the RGB channels. Thus, the resulting image channels convey the summarized information of consecutive moments.

Dynamic Image Computation. A dynamic image presents a summary of object appearances and their motions throughout an input video sequence by encoding the sequential order of pixels from one frame to another. Dynamic image computation uses RGB images directly by multiplying the video frames by α_t coefficient and summing them to generate the output image with the formula given $d^* = \sum_{k=1}^{N} \alpha_k I_k$, $\alpha_k = 2k - N - 1$, where I_k is the kth image of the video segment and N is the number of frames in the video segment.

Conditioning on Compact Motion Representation. After extracting the compact motion representation of a clip C, we obtain the conditioning feature vector $cond$ through a pre-trained 2D-CNN F_{cond}. We inject this both in the encoder and in the decoder part of our network G by summing with the input features. To deal with the different dimensionality of the two blocks, we use 2 linear projections to obtain vectors of the same size as the input.

4   Experimental Analysis and Results

The **evaluation metric** employed in this study is the Area Under the Receiver Operating Characteristic (ROC) Curve (AUC), which is determined using frame-level annotations of the test videos within the datasets, consistent with established VAD methodologies. In order to evaluate and compare the effectiveness of the proposed approach, the experiments were carried out on two mainstream large-scale unconstrained **datasets**: UCF-Crime [27] and ShanghaiTech [14]. The **UCF-Crime dataset** [27] was obtained from diverse CCTV cameras that possess varying field-of-views. It consists of a total of 128 h of videos, with annotations for 13 distinct anomalous events e.g., road accidents, theft, and explosions. To ensure fair comparisons with the SOTA, we utilized the standardized training and testing splits of the dataset, which consist of 810 abnormal and 800 normal videos for training, and 130 abnormal and 150 normal videos for testing, without utilizing the labels. On the other hand, the **ShanghaiTech dataset** [14] was recorded using 13 distinct camera angles under challenging lighting conditions. For our study, we utilized the training split, which comprises 63 abnormal and 174 normal videos, as well as the testing split, consisting of 44 abnormal and 154 normal videos, in accordance with SOTA conventions.

4.1   Implementation Details

Architecture. In line with [32], we use 16 non-overlapping frames to define a video clip, and we use pre-trained 3D-ResNext101 or 3D-ResNet18 as feature extractor F [6,32]. After computing the compact motion representation, we extract a single conditioning vector with F_{cond} with a pre-trained ResNet50 or ResNet18 due to their widespread use together with such motion representations [3,25]. We use an MLP with an encoder-decoder structure as the denoising network G, and the encoder is comprised of three layers with sizes of {1024, 512, 256}, while the decoder has hidden dimensions of {256, 512, 1024}. The timestep information σ_t is transformed via Fourier embedding and integrated into the network by FiLM layers [20], while the conditioning on compact motion representation is applied after timestep integration by summation to the inputs.

Training and Sampling. The learning rate scheduler and EMA of the model are set to the default values of k-*diffusion*, which include an initial learning rate of 2×10^{-4} and InverseLR scheduling. The weight decay is set at 1×10^{-4}. Training is conducted for 30 epochs with a batch size of 256, while testing is performed on 8192 samples as in previous literature [32]. Several hyperparameters affect the diffusion process in k-*diffusion*, and given the novelty of the task at hand, we do not rely on parameters from prior literature. We, therefore, conduct an extensive exploration of the effects of training and testing noise. Training noise is distributed according to a log-normal distribution with parameters (P_{mean}, P_{std}), while sampling noise is controlled by σ_{min} and σ_{max}, and below, we investigate their role. For the diffusion reverse process, we use LMS sampler with the number of steps T set to 10.

4.2   Results

We first compare our method's results with SOTA and baseline methods. Then, we report the results of the cross-dataset evaluation, where the training and validation sets are from a different domain than the test split. Finally, we analyze how the hyperparameters of the diffusion models affect VAD performance.

Performance Comparisons. The performance of the proposed method together with the SOTA and baseline methods' (i.e., unconditional diffusion model) results are given in Tables 1 and 2 for the ShanghaiTech [14] and UCF-Crime [27] datasets, respectively. These tables also include an ablation study such that the condition of the diffusion models is changed between star representation, dynamic images, and spatiotemporal features, in addition to changing the feature backbone between 3D-ResNext101 and 3D-ResNet18, and the motion representation backbone between ResNet50 and ResNet18.

As seen in Table 1, the proposed method outperforms all others on the ShanghaiTech [14] dataset, achieving the best results by surpassing the SOTA autoencoder [32] by 14.45%, the SOTA collaborative generative and discriminative model [32] by 4.77%, and the SOTA [13] by 20.71%. The proposed method also improves upon the unconditional diffusion models (i.e., baselines) by 1.08%. It is worth noting that other conditional diffusion models, i.e., using spatiotemporal features as the condition and compact motion representation as input, are occasionally less effective than our method, with the proposed method surpassing them by 10.52%. The best performance is achieved by using a 3D-ResNet18 as the feature backbone, star representation as the condition, and ResNet50 as the corresponding backbone. On the other hand, for the UCFC dataset [27] (Table 2),

Table 1. Performance comparisons with the SOTA and the baseline methods on Shang-haiTech [14] dataset. The best results are in bold. The second best results are underlined. The full model of [32] includes generator, negative learning, and discriminator. NA stands for not-applicable. Results with $\diamond$ are taken from [32].

Method	Feature	Condition	AUC (%)
State-of-the-art Methods			
Kim et al. [13]$^\diamond$	3D-ResNext101	NA	56.47
Autoencoder [32]	3D-ResNext101	NA	62.73
Autoencoder [32]	3D-ResNet18	NA	69.02
Full model [32]	3D-ResNext101	NA	72.41
Full model [32]	3D-ResNet18	NA	71.20
Baseline Methods			
Diffusion	3D-ResNext101	-	68.88
Diffusion	3D-ResNet18	-	76.10
Diffusion	Star Rep. [7] w/ ResNet18	-	62.81
Diffusion	Star Rep. [7] w/ ResNet50	-	59.55
Diffusion	Dyn. Img. [4] w/ ResNet18	-	62.88
Diffusion	Dyn. Img. [4] w/ ResNet50	-	64.96
Other Conditional Diffusion Models			
Diffusion	Star Rep. [7] w/ ResNet18	3D-ResNext101	64.87
Diffusion	Star Rep. [7] w/ ResNet50	3D-ResNext101	65.01
Diffusion	Star Rep. [7] w/ ResNet18	3D-ResNext18	64.03
Diffusion	Star Rep. [7] w/ ResNet50	3D-ResNext18	64.15
Diffusion	Dyn. Img. [4] w/ ResNet18	3D-ResNext101	66.66
Diffusion	Dyn. Img. [4] w/ ResNet50	3D-ResNext101	64.24
Diffusion	Dyn. Img. [4] w/ ResNet18	3D-ResNext18	65.02
Diffusion	Dyn. Img. [4] w/ ResNet50	3D-ResNext18	65.26
Proposed Method			
Diffusion	3D-ResNext 101	Star Rep. [7] w/ ResNet18	65.12
Diffusion	3D-ResNext 101	Star Rep. [7] w/ ResNet50	65.17
Diffusion	3D-ResNext 101	Dyn. Img. [4] w/ ResNet18	66.36
Diffusion	3D-ResNext 101	Dyn. Img. [4] w/ ResNet50	65.09
Diffusion	3D-ResNet18	Star Rep. [7] w/ ResNet18	76.36
Diffusion	3D-ResNet18	Star Rep. [7] w/ ResNet50	**77.18**
Diffusion	3D-ResNet18	Dyn. Img. [4] w/ ResNet18	74.61
Diffusion	3D-ResNet18	Dyn. Img. [4] w/ ResNet50	76.16

the proposed method achieves the second-highest score after the more complex model of [32], which employs a generator, discriminator and negative learning. Nonetheless, our method outperforms the SOTA autoencoder [32] by 10.53% and the SOTA [13] by 14.85%. It also demonstrates superior performance compared

Table 2. Performance comparisons with the SOTA and the baseline methods on UCF-Crime [27] dataset. The best results are in bold. The second best results are underlined. The full model of [32] includes generator, negative learning, and discriminator. NA stands for not-applicable. Results with ◇ are taken from [32].

Method	Feature	Condition	AUC (%)
State-of-the-art Methods			
Kim et al. [13]◇	3D-ResNext101	NA	52.00
Autoencoder [32]	3D-ResNext101	NA	56.32
Autoencoder [32]	3D-ResNet18	NA	49.78
Full model [32]	3D-ResNext101	NA	**68.17**
Full model [32]	3D-ResNet18	NA	56.86
Baseline Methods			
Diffusion	3D-ResNext101	-	62.91
Diffusion	3D-ResNet18	-	65.22
Diffusion	Star Rep. [7] w/ ResNet18	-	59.60
Diffusion	Star Rep. [7] w/ ResNet50	-	61.14
Diffusion	Dyn. Img. [4] w/ ResNet18	-	60.14
Diffusion	Dyn. Img. [4] w/ ResNet50	-	62.73
Other Conditional Diffusion Models			
Diffusion	Star Rep. [7] w/ ResNet18	3D-ResNext101	59.26
Diffusion	Star Rep. [7] w/ ResNet50	3D-ResNext101	63.20
Diffusion	Star Rep. [7] w/ ResNet18	3D-ResNext18	61.14
Diffusion	Star Rep. [7] w/ ResNet50	3D-ResNext18	60.78
Diffusion	Dyn. Img. [4] w/ ResNet18	3D-ResNext101	58.23
Diffusion	Dyn. Img. [4] w/ ResNet50	3D-ResNext101	61.04
Diffusion	Dyn. Img. [4] w/ ResNet18	3D-ResNext18	65.06
Diffusion	Dyn. Img. [4] w/ ResNet50	3D-ResNext18	61.27
Proposed Method			
Diffusion	3D-ResNext101	Star Rep. [7] w/ ResNet18	58.82
Diffusion	3D-ResNext101	Star Rep. [7] w/ ResNet50	63.00
Diffusion	3D-ResNext101	Dyn. Img. [4] w/ ResNet18	60.12
Diffusion	3D-ResNext101	Dyn. Img. [4] w/ ResNet50	63.52
Diffusion	3D-ResNet18	Star Rep. [7] w/ ResNet18	63.67
Diffusion	3D-ResNet18	Star Rep. [7] w/ ResNet50	<u>66.85</u>
Diffusion	3D-ResNet18	Dyn. Img. [4] w/ ResNet18	60.69
Diffusion	3D-ResNet18	Dyn. Img. [4] w/ ResNet50	66.11

to the baseline and the other conditional diffusion models by 1.63% and 1.79%, respectively. Furthermore, the optimal performance of the proposed method for this dataset is achieved by utilizing 3D-ResNet18 as the feature backbone, star representation as the condition, and ResNet50 as the condition backbone.

Table 3. Cross-dataset analysis (Training dataset -> Testing dataset). The best results are in bold. The second best results are underlined. The full model of [32] includes generator, negative learning, and discriminator. NA stands for not-applicable.

Method	Feature	Condition	AUC (%)
	UCFC -> ShanghaiTech		
Autoencoder [32]	3D-ResNext101	NA	55.86
Autoencoder [32]	3D-ResNet18	NA	47.48
Full model [32]	3D-ResNext101	NA	55.94
Full model [32]	3D-ResNet18	NA	49.19
Diffusion (Baseline)	3D-ResNet18	-	60.55
Diffusion (Baseline)	Star Rep. [7] w/ ResNet50	-	54.67
Diffusion (Baseline)	Dyn. Img. [4] w/ ResNet50	-	58.14
Diffusion (Proposed)	3D-ResNet18	Star Rep. [7] w/ ResNet50	**64.54**
Diffusion (Proposed)	3D-ResNet18	Dyn. Img. [4] w/ ResNet50	<u>63.58</u>
	ShanghaiTech -> UCFC		
Autoencoder [32]	3D-ResNext101	NA	52.45
Autoencoder [32]	3D-ResNet18	NA	46.53
Full model [32]	3D-ResNext101	NA	52.29
Full model [32]	3D-ResNet18	NA	49.57
Diffusion (Baseline)	3D-ResNet18	-	63.97
Diffusion (Baseline)	Star Rep. [7] w/ ResNet50	-	60.21
Diffusion (Baseline)	Dyn. Img. [4] w/ ResNet50	-	60.75
Diffusion (Proposed)	3D-ResNet18	Star Rep. [7] w/ ResNet50	**65.17**
Diffusion (Proposed)	3D-ResNet18	Dyn. Img. [4] w/ ResNet50	<u>64.97</u>

Cross-dataset Analysis. When performing this analysis, we take into consideration the results presented in Tables 1 and 2 such that we select the combinations of input feature and condition backbone that yield the best results. Table 3 shows that our methods achieve significantly better results in cross-dataset analysis, regardless of which compact motion representation is used as the condition, compared to all other baselines and SOTA methods. Notably, the performance of the proposed method is remarkable (8.6–17.06% better) in comparison to both the generative model and full model proposed by [32]. On the other hand, the baseline unconditional diffusion model that utilizes spatiotemporal features outperforms the baseline unconditional diffusion model that uses compact motion representations. The relative effectiveness of the proposed method is of significant practical importance, as in most cases, the deployment domain differs from the domain on which the model is trained.

Hyperparameter Analysis. We study the effect of the **training noise** on the learning process, and we find that baseline diffusion and our method both achieve higher results with smaller values of noise, meaning a lower P_{mean}. Importantly, our method generally achieves better performance than the baseline, given the same parameters, for a wider choice range of training noise parameters, making it less sensitive to this choice. We explore the effect of $P_{mean} \in [-5, -0.5]$ and $P_{std} \in [0.5, 2.]$. On the other hand, recalling that t closer to zero indicates a point closer to an isotropic Gaussian distribution, we explore the **effect of** different t as **the starting point of the reverse process**. While the baseline unconditional diffusion achieves its best performance with $t = 4$ and $t = 6$, we find that our method achieves better performance in high-noise areas ($t = 1, t = 2$), effectively allowing the removal of more information from the clip vector, and proving the effectiveness of conditioning on motion representation for the task at hand.

5   Conclusions

We have presented a novel approach for unsupervised VAD, which can accurately identify and locate anomalous frames by utilizing only the reconstruction capabilities of diffusion models. Our conditional diffusion model uses features extracted from compact motion representations as the condition while it takes the spatiotemporal features extracted from pre-trained networks as the input. By doing so, we show the contribution of the compact motion representations, *i.e.,* our method succeeded in improving the SOTA VAD results while also demonstrating remarkable transferability across domains. Note that the unsupervised nature of our approach allows for an anomaly detection system to begin identifying abnormalities based solely on observed data, without any human intervention. If no abnormal events have occurred, the system may mistakenly identify rare normal events as abnormal. However, it is expected that such anomaly systems operate for a longer period of time, thus, the likelihood of having no abnormal events decreases significantly. In the future, we aim to modify our method in a way that it can operate on edge devices with near real-time capabilities.

Acknowledgment. The project is partially funded by the European Union (EU) under NextGenerationEU. We acknowledge the support of the MUR PNRR project FAIR - Future AI Research (PE00000013) funded by the NextGenerationEU. E.R. is partially supported by the PRECRISIS, funded by the EU Internal Security Fund (ISFP-2022-TFI-AG-PROTECT-02-101100539). Views and opinions expressed are however those of the author(s) only and do not necessarily reflect those of the EU or The European Research Executive Agency. Neither the EU nor the granting authority can be held responsible for them. The work was carried out in the Vision and Learning joint laboratory of FBK and UNITN.

References

1. Barros, P., Parisi, G.I., Jirak, D.E.A.: Real-time gesture recognition using a humanoid robot with a deep neural architecture. In: IEEE-RAS Humanoids (2014)
2. Beyan, C., Fisher, R.B.: Detecting abnormal fish trajectories using clustered and labeled data. In: ICIP (2013)
3. Beyan, C., Zunino, A., Shahid, M., Murino, V.: Personality traits classification using deep visual activity-based nonverbal features of key-dynamic images. In: IEEE TAC (2019)

4. Bilen, H., Fernando, B., Gavves, E., Vedaldi, A.: Action recognition with dynamic image networks. In: IEEE TPAMI (2017)
5. Blattmann, A., Rombach, R., Ling, H.E.A.: Align your latents: high-resolution video synthesis with latent diffusion models. In: CVPR (2023)
6. Chandola, V., Banerjee, A., Kumar, V.: Anomaly detection: a survey. In: ACM CSUR (2009)
7. Dos Santos, C.C., Samatelo, J.L.A.E.A.: Dynamic gesture recognition by using CNNs and star RGB: a temporal information condensation. Neurocomputing **400**, 238–254 (2020)
8. Gong, D., Liu, L., Le, V., Saha, B., Mansour, M.R., Venkatesh, S., Hengel, A.V.D.: Memorizing normality to detect anomaly: memory-augmented deep autoencoder for unsupervised anomaly detection. In: CVPR (2019)
9. Gutowska, M., Little, S., McCarren, A.: Constructing a meta-learner for unsupervised anomaly detection (2023)
10. Ho, J., Jain, A., Abbeel, P.: Denoising diffusion probabilistic models. In: NeurIPS (2020)
11. Jebur, S.A., Hussein, K.A., Hoomod, H.K.E.A.: Review on deep learning approaches for anomaly event detection in video surveillance. Electronics **12**(1), 29 (2022)
12. Karras, T., Aittala, M., Aila, T., Laine, S.: Elucidating the design space of diffusion-based generative models. In: NeurIPS (2022)
13. Kim, J.H., Kim, D.H., Yi, S., Lee, T.: Semi-orthogonal embedding for efficient unsupervised anomaly segmentation. arXiv preprint:2105.14737 (2021)
14. Liu, W., W. Luo, D.L., Gao, S.: Future frame prediction for anomaly detection - a new baseline. In: CVPR (2018)
15. Majhi, S., Das, S., Brémond, F.: Dam: dissimilarity attention module for weakly-supervised video anomaly detection. In: AVSS (2021)
16. Medioni, G., Cohen, I., Brémond, F., Hongeng, S., Nevatia, R.: Event detection and analysis from video streams. In: IEEE TPAMI (2001)
17. Meng, C., et al.: Sdedit: guided image synthesis and editing with stochastic differential equations. In: ICLR (2021)
18. Mohammadi, B., Fathy, M., Sabokrou, M.: Image/video deep anomaly detection: a survey. arXiv preprint:2103.01739 (2021)
19. Nguyen, T.N., Meunier, J.: Anomaly detection in video sequence with appearance-motion correspondence. In: ICCV (2019)
20. Perez, E., Strub, F., De Vries, H., Dumoulin, V., Courville, A.: Film: visual reasoning with a general conditioning layer. In: AAAI (2018)
21. Piciarelli, C., Micheloni, C., Foresti, G.L.: Trajectory-based anomalous event detection. In: IEEE TCSVT (2008)
22. Ravanbakhsh, M., Nabi, M., Sangineto, E.E.A.: Abnormal event detection in videos using generative adversarial nets. In: ICIP (2017)
23. Ren, J., Xia, F., Liu, Y., Lee, I.: Deep video anomaly detection: opportunities and challenges. In: ICDM workshops (2021)
24. Rombach, R., Blattmann, A., Lorenz, D.E.A.: High-resolution image synthesis with latent diffusion models. In: CVPR (2022)
25. Shahid, M., Beyan, C., Murino, V.: S-VVAD: visual voice activity detection by motion. In: IEEE WACV (2021)
26. Smeureanu, S., Ionescu, R.T., Popescu, M., Alexe, B.: Deep appearance features for abnormal behavior detection in video. In: ICIAP (2017)
27. Sultani, W., Chen, C., Shah, M.: Real-world anomaly detection in surveillance videos. In: CVPR (2018)

28. Tian, Y., Pang, G., Chen, Y., Singh, R., Verjans, J.W., Carneiro, G.: Weakly-supervised video anomaly detection with robust temporal feature magnitude learning. In: ICCV (2021)
29. Wang, D., Lin, J., Cui, P.E.A.: A semi-supervised graph attentive network for financial fraud detection. In: ICDM (2019)
30. Wang, J., Cherian, A., Porikli, F.: Ordered pooling of optical flow sequences for action recognition. In: IEEE WACV (2017)
31. Wolleb, J., Bieder, F., Sandkühler, R., Cattin, P.C.: Diffusion models for medical anomaly detection. In: MICCAI (2022)
32. Zaheer, M.Z., Mahmood, A., Khan, M.H., Segu, M., Yu, F., Lee, S.I.: Generative cooperative learning for unsupervised video anomaly detection. In: CVPR (2022)
33. Zen, G., Ricci, E.: Earth mover's prototypes: a convex learning approach for discovering activity patterns in dynamic scenes. In: CVPR 2011, pp. 3225–3232. IEEE (2011)
34. Zhou, J.T., Du, J., Zhu, H.E.A.: Anomalynet: an anomaly detection network for video surveillance. IEEE Trans. Inf. Forensics Sec. **14**(10), 2537–2550 (2019)

VM-NeRF: Tackling Sparsity in NeRF with View Morphing

Matteo Bortolon[1,2,3]([✉]), Alessio Del Bue[2], and Fabio Poiesi[1,2]

[1] TeV, Fondazione Bruno Kessler, Povo, Italy
{mbortolon,poiesi}@fbk.eu
[2] PAVIS, Istituto Italiano di Tecnologia, Genoa, Italy
{alessio.delbue}@iit.it
[3] DISI, University of Trento, Trento, Italy

Abstract. NeRF aims to learn a continuous neural scene representation by using a finite set of input images taken from various viewpoints. A well-known limitation of NeRF methods is their reliance on data: the fewer the viewpoints, the higher the likelihood of overfitting. This paper addresses this issue by introducing a novel method to generate geometrically consistent image transitions between viewpoints using View Morphing. Our VM-NeRF approach requires no prior knowledge about the scene structure, as View Morphing is based on the fundamental principles of projective geometry. VM-NeRF tightly integrates this geometric view generation process during the training procedure of standard NeRF approaches. Notably, our method significantly improves novel view synthesis, particularly when only a few views are available. Experimental evaluation reveals consistent improvement over current methods that handle sparse viewpoints in NeRF models. We report an increase in PSNR of up to 1.8 dB and 1.0 dB when training uses eight and four views, respectively. Source code: https://github.com/mbortolon97/VM-NeRF.

1 Introduction

Novel View Synthesis (NVS) is the problem of synthesising unseen camera views from a set of known views[1] [8,29]. NVS is a key technology that can enable compelling augmented or virtual reality experiences [10], new entertainment technology [6], and robotics applications [11]. NVS has undergone a significant improvement after the introduction of Neural Radiance Fields (NeRF) [2,17] – a trainable implicit neural representation of a 3D scene that can photorealistically render unseen (novel) views. NeRF is a data-driven model that can synthesise high-quality novel views but in general requiring several multi-view images, e.g. about hundreds of images taken from different and uniformly distributed camera viewpoints around an object of interest [17]. If these viewpoints are few and/or not uniformly distributed, the resulting NeRF model may fail to produce

---

[1] Throughout the paper, we will use the term *viewpoint* to refer to the camera pose, *view* to refer to the scene seen through a certain viewpoint and to *image* to refer to the photometric content captured from a view.

G. L. Foresti et al. (Eds.): ICIAP 2023, LNCS 14234, pp. 63–74, 2023.
https://doi.org/10.1007/978-3-031-43153-1_6

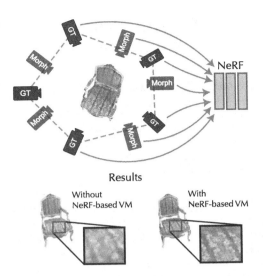

Fig. 1. Given a set of known views (ground truth), View Morphing-NeRF (VM-NeRF) generates image transitions between views (morph) that can be effectively used to train a NeRF model in the case of few-shot view synthesis. Results are of a higher quality when VM-NeRF is used.

satisfactory novel views [12,16]. This detrimental effect is a known drawback of NeRF-based approaches and it is due to the likelihood of overfitting on known viewpoints while decreasing generalisation on novel views that are furthest from the given viewpoints, namely the few-shot view synthesis problem [12].

In this paper, we propose to tackle the problem of training a NeRF model on scenes captured with a sparse set of viewpoints by using a novel geometry-based strategy based on View Morphing [24] (Fig. 1). This purely geometric method can synthesise or morph a new viewpoint that lies in-between two given camera views while ensuring realistic image transitions. Traditionally, view morphing requires a set of accurate point matches between known image pairs in order to successfully perform the morph. As this matching stage is hard to integrate into a NeRF-based learning pipeline, our intuition is to leverage the per-image depth information implicitly estimated by NeRF to obtain dense coordinate matches among views after an image rectification stage (Fig. 2). To this end, we have to relax and modify several steps of the view morphing strategy to be duly integrated in the NeRF learning paradigm. This technique does not require any prior knowledge about the captured 3D scene, and it can synthesise 3D projective transformations (e.g. 3D rotations, translations, shears) of objects by operating entirely on the input images. We evaluate our approach by using the dataset of the original NeRF's paper [17] and we show that PSNR improves up to 1.8dB and 1.0dB when eight and four views are used for training, respectively. We compare our approach with DietNeRF [12], AugNeRF [5] and RegNeRF [19], and show that our approach can produce higher-quality renderings.

To summarise, our contributions are:

- We present a novel and effective method for NeRF to address the problem of few-shot view synthesis;
- We introduce a new view morphing technique based on the NeRF depth output, named VM-NeRF;
- VM-NeRF can achieve higher-quality rendered images than alternative methods in the literature.

2   Related Work

NVS scene synthesis can be solved either by using traditional 3D reconstruction techniques [23] or by adopting methods based on neural rendering [26]. Neural Radiance Fields (NeRF) is a recent neural rendering method that can learn a volumetric representation of an unknown 3D scene approximating its radiance and density fields from a set of known (ground truth) views by using a multilayer perceptron (MLP) [17]. NeRF optimises its parameters on one scene based on a set of known views, thus overfitting can occur when these views are few.

Current approaches addressing few-shot novel view synthesis can be divided into two groups. The first group uses the same trained network to generate novel views of different scenes. This category of methods trains on datasets characterised by similar scenes, such as DTU [1]. Multiple-scene training can introduce datasets biases and may produce low-quality results in contexts outside the training domain [18,27]. SparseNeuS [14] and ShaRF [22] train NVS on multiple scenes by conditioning the MLP with features that encode appearance and geometry of the surface at a 3D location. This can be achieved by using an auxiliary deep network jointly trained with NeRF. The second group uses the original per-scene optimisation procedure of NeRF, so a single network trains and tests only on one scene leading to methods without dataset bias. These methods are more likely to encounter overfit problems on the known views, however they reduce this likelihood by adding either semantic or geometric constraints during training. DietNeRF belongs to this category and exploits the feature representations of known images computed with a CLIP pre-trained image encoder, renders random poses, and processes them by imposing semantic consistency through CLIP features [12]. RegNeRF [19] renders random viewpoints around the known ones, and introduces regularisation constraints between known viewpoints and randomly sampled ones.

Single-scene methods working with few viewpoints may overfit on the known images, producing artefacts when novel views are rendered. In general, we can mitigate overfitting via data augmentation [25], and to the best of our knowledge, the only methods that address data augmentation for NeRF are AugNeRF [5] and GeoAug [4]. AugNeRF aims to improve NeRF generalisation by using adversarial data augmentation to enforce each ray and its augmented version to produce the same result. GeoAug [4] perturbs translation and rotation of the known viewpoints during training. Our proposed approach does not perturb the known input views and rays, instead we create new views (novel 3D projective transformations) using pairs of known views. This allows us to enforce coherence of newly rendered viewpoints between distant viewpoint pairs. At the moment of the acceptance of this paper, we could not replicate the results of GeoAug because the authors have not released their source code.

3 Preliminaries

3.1 NeRF Overview

NeRF's objective is to synthesise novel views of a scene by optimising a volumetric function given a finite set of input views [17]. Let f_θ be the underlying function we aim to optimise. The input to f_θ is a 5D datum that encodes a point on a camera ray, i.e. a 3D spatial location (x, y, z) and a 2D viewing direction (θ, ϕ). Let $c \in \mathbb{R}^3$ be the view-dependent emitted radiance (colour) and σ be the volume density that f_θ predicts at (x, y, z). Novel views are synthesised by querying 5D data along the camera rays. Traditional volume rendering techniques can be used to transform c and σ into an image [13,15]. Because volume rendering is differentiable, f_θ can be implemented as a fully-connected deep network and learned.

Rendering a view from a novel viewpoint consists of estimating the integrals of all 3D rays that originate from the camera optic centre and that pass through each pixel of the camera image plane. Let r be a 3D ray. To make rendering computationally tractable, each ray is represented as a finite set of 3D spatial locations, indexed with i, which are defined between two clipping distances: a near one (t_n) and a far one (t_f). Let Γ be the number of 3D spatial locations sampled between t_n and t_f. Rendering the colour of a pixel is given by

$$\hat{c}(r) = \sum_{i=1}^{\Gamma} s(i) \left(1 - e^{-\hat{\sigma}(r)_i \delta_i}\right) \hat{c}(r)_i, \tag{1}$$

where $\hat{c}(r)_i$ is the colour and $\hat{\sigma}(r)_i$ is the density predicted by the network at i. $\delta_i = t_{i+1} - t_i$ is the distance between adjacent sampled 3D spatial locations, and $s(i)$ is the inverse of the volume density that is accumulated up to the ith spatial location, which is in turn computed as

$$s(i) = e^{-\sum_{j=1}^{i-1} \hat{\sigma}(r)_j \delta_j}, \tag{2}$$

where $(1 - e^{-\hat{\sigma}(r)_i \delta_i})$ is a density-based weight component: the higher the density value σ of a point, the larger the contribution on the final rendered colour. Similarly to Eq. 1, we can render the pixel depth as

$$d(r) = \sum_{i=1}^{\Gamma} s(i) \left(1 - e^{-\hat{\sigma}(r)_i \delta_i}\right) z_i, \tag{3}$$

where z_i is the distance of the ith spatial location with respect to the camera optic centre.

The input required to learn the NeRF parameters is a set of N images and their corresponding camera information. Let $\mathcal{I} = \{I_k\}_{k=1}^N$ be the training images, and $\mathcal{P} = \{P_k\}_{k=1}^N$ and $\mathcal{K} = \{K_k\}_{k=1}^N$ be their corresponding camera poses and intrinsic parameters, respectively. A pose $P = [R, t]$ is composed of rotation R and translation t. We can estimate the depth map of a given view k by rendering the depth of all its pixels, therefore we can define the estimated depth maps as $\mathcal{D} = \{D_k\}_{k=1}^N$.

Learning f_θ is achieved by comparing each ground-truth pixel $c(r)$ with its predicted counterpart $\hat{c}(r)$. The goal is to minimise the following L2-norm objective function

$$\mathcal{L} = \frac{1}{|\mathcal{R}|} \sum_{r \in \mathcal{R}} \left(\|c(r) - \hat{c}_c(r)\|_2^2 + \|c(r) - \hat{c}_f(r)\|_2^2 \right), \tag{4}$$

where $\hat{c}_c(r)$ and $\hat{c}_f(r)$ are the coarse and fine predicted volume colours for ray r, respectively. Please refer to [17] for more details.

3.2 View Morphing Overview

View morphing objective is to synthesise natural 2D transitions between an image pair $\{I_k, I_{k'}\}$ and the approach can be summarised in three steps: *i)* the two images are *prewarped* through rectification, i.e. their image planes are aligned without changing their cameras' optic centres; *ii)* the *morph* is computed between these prewarped images to generate a morphed image whose viewpoint lies on the line connecting the optic centres; *iii)* the image plane of the morphed image is transformed to a desired viewpoint through *postwarping*.

In practice, assuming the two views are prewarped, the morph uses the knowledge of their camera poses $P_k, P_{k'}$, and the pixel correspondences between the images, i.e. $q_k : I_k \Rightarrow I_{k'}, q_{k'} : I_{k'} \Rightarrow I_k$ where q_k is a function that maps a pixel of I_k to the corresponding pixel in $I_{k'}$ [24]. Sparse pixel correspondences can be defined by a user or determined by a keypoint detector, they can then be densified via interpolation to create a dense correspondence map. *This procedure is not viable as is in a learning-based pipeline, hence we have to define a novel view morphing strategy for a NeRF-based network architecture.* A warp function for each image can be computed from the correspondence map through linear interpolation

$$\hat{I}_{k,\alpha} = (1 - \alpha)\hat{I}_k + \alpha q_k(\hat{I}_k)$$
$$\hat{I}_{k',\alpha} = (1 - \alpha)q_{k'}(\hat{I}_{k'}) + \alpha\hat{I}_{k'}$$
$$P_\alpha = (1 - \alpha)P_0 + \alpha P_1, \tag{5}$$

where $\hat{I}_k$ are the coordinates of the image of camera k, $\hat{I}_{k,\alpha}$ are pixel coordinates of the morphed image, and $\alpha \in [0, 1]$ regulates the position of the morphed view along the line connecting the two views. The morphed image can then be computed by averaging the pixel colours of the warped images. Please refer to [24] for more details.

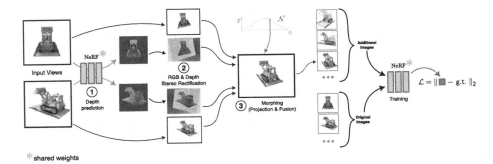

*shared weights

Fig. 2. Block diagram of NeRF-based View Morphing (VM-NeRF). From the left, we (1) predict the depth with NeRF, (2) rectify the input images and predicted depths, and (3) compute the image morphing of a view randomly positioned between the view pair. α determines the new view position and it is sampled from a Gaussian distribution.

4 NeRF-Based View Morphing

The goal of NeRF-based View Morphing (VM-NeRF) is to use the geometrical constraints of the morphing technique to synthesise a set of additional training input views $\mathcal{M} = \{\mathbf{M}_{(k,k'),\alpha}\}$, where $\mathbf{M}_{(k,k'),\alpha}$ is a morphed view generated from the view pair k and k' with a given value of α. Adapting view morphing in a learning-based pipeline is challenging as we need reliable pixel correspondences (q_k and $q_{k'}$) to synthesise morphed views. Our intuition is that it is possible to compute one-to-one correspondences from the disparity information, a function of the depth as in Eq. 3, which we can render with the very same NeRF model. We can then linearly interpolate the photometric content of the view pair to produce the morphed view.

Based on the description in Sect. 3.2, we integrate in NeRF only the steps of prewarping and morphing. We experimentally found that postwarping does not lead to better results. Sect. 4.1 describes how we perform the initial rectification of the two cameras. Section 4.2 describes how the images morphing is computed. Section 4.3 provides detailed information on our practical approach to training NeRF with View Morphing. Figure 2 shows the block diagram of our approach.

4.1 Rectification

Our first step is rectification, which leads to rotating the known camera poses $\boldsymbol{P}_k$ and $\boldsymbol{P}_{k'}$ around their optic centres until their image planes become coplanar. We can then compute the common image plane by using a selection of algorithms such as [7,9]. We represent this plane as the rotation matrix

$$\tilde{\boldsymbol{R}} = [\boldsymbol{a}_x, \boldsymbol{a}_y, \boldsymbol{a}_z], \tag{6}$$

where $\boldsymbol{a}_x, \boldsymbol{a}_y, \boldsymbol{a}_z$ are the axis components of the coplanar plane resulting from the rectification. Stereo rectification is applied to the original images $\{\boldsymbol{I}_k, \boldsymbol{I}_{k'}\}$

and depth maps $\{\boldsymbol{D}_k, \boldsymbol{D}_{k'}\}$ predicted in Eq. 3. The new camera pose of view k is equal to $\tilde{\boldsymbol{P}}_k = [\tilde{\boldsymbol{R}}, \boldsymbol{t}_k]$, where $\boldsymbol{t}_k$ is the translation of the original camera pose $\boldsymbol{P}_k$ (same applies to view k').

Rectification algorithms are typically based on the assumptions that viewpoints are aligned horizontally and that the reference viewpoint is the left-hand side of the camera (from an observer positioned behind the cameras) [7,9]. This is atypical in NeRF, as viewpoints may have arbitrary camera configurations, leading to errors that should be corrected. We mitigate this problem by comparing $\boldsymbol{a}_z$ with the z component of the original view pose. If this angle is greater than $45°$ with respect to both $\boldsymbol{P}_k$ and $\boldsymbol{P}_{k'}$, we rotate the warping matrices and poses by $90°$ or $180°$. The application of this modification to conventional rectification algorithm allows us to correctly generate the following rectified images $\{\tilde{\boldsymbol{I}}_k, \tilde{\boldsymbol{I}}_{k'}\}$ and rectified depth maps $\{\tilde{\boldsymbol{D}}_k, \tilde{\boldsymbol{D}}_{k'}\}$.

4.2 Image Morphing

The second step is image morphing, i.e. fusing the rectified images to obtain the new morphed image. This procedure is divided in three steps: $i)$ finding the pixel correspondences; $ii)$ computing the position of each pixel on the morphed camera; $iii)$ fusing pixels that fall in the same position. To determine the image correspondences, we initially compute the disparity maps as functions of the rectified estimated depths

$$E_k = \frac{f_k}{\tilde{D}_k} \|\boldsymbol{o}_k - \boldsymbol{o}_{k'}\|_2, \quad E_{k'} = \frac{f_{k'}}{\tilde{D}_{k'}} \|\boldsymbol{o}_k - \boldsymbol{o}_{k'}\|_2, \tag{7}$$

where $\{\boldsymbol{o}_k, \boldsymbol{o}_{k'}\}$ are the principal points and $\{f_k, f_{k'}\}$ are the focal lengths of cameras k and k'.

Then, we determine the correspondences of the pixel positions between images defined in Eq. 5 as

$$q_k(\hat{\boldsymbol{I}}_k) = \hat{\boldsymbol{I}}_k + \frac{\tilde{\boldsymbol{b}}_k}{\|\tilde{\boldsymbol{b}}_k\|_2} \mathbf{1}^\top \odot E_k, \tag{8}$$

where $\mathbf{1}$ is a vector of ones, $\odot$ indicates the Hadamard product and $\hat{\boldsymbol{I}}_k$ is the baseline direction with respect to the common plane defined in Eq. 6 that is computed as

$$\tilde{\boldsymbol{b}}_k = \boldsymbol{a}_z \times ((\boldsymbol{o}_k - \boldsymbol{o}_{k'}) \times \boldsymbol{a}_z). \tag{9}$$

The same operation is computed for k'. Then, we apply the warp functions of Eq. 5 to compute the position of each pixel on the morphed view, thus obtaining $\hat{\boldsymbol{I}}_{k,\alpha}$ and $\hat{\boldsymbol{I}}_{k',\alpha}$.

Lastly, a coalescence operation [3] fuses the pixels of the two views k and k'. The coalescence operation concatenates two sets of coordinates and fuse pixels with the same position, preserving only the pixel values of the points that are nearest to the camera. We use $\{\tilde{\boldsymbol{D}}_k, \tilde{\boldsymbol{D}}_{k'}\}$ to determine the distance of the points.

Table 1. Results on the NeRF realistic synthetic 360° dataset.

# views	Method	PSNR ↑	SSIM ↑	LPIPS ↓
100 [17]	NeRF [17]	31.21	0.9513	0.0465
8	NeRF [17]	23.45	0.8673	0.1303
	DietNeRF [12]	22.98	0.8545	0.1258
	AugNeRF [5]	10.04	0.5415	0.3866
	RegNeRF [19]	22.91	0.8756	0.1138
	VM-NeRF	**24.39**	**0.8768**	**0.1146**
8 [12]	NeRF [12]	20.09	0.8220	0.1790
	DietNeRF [12]	23.59	0.8740	0.0970
	VM-NeRF	**24.14**	**0.8729**	**0.1180**
4	NeRF [17]	10.98	0.6550	0.3620
	DietNeRF [12]	12.61	0.6591	0.3302
	AugNeRF [5]	8.14	0.3924	0.4802
	RegNeRF [19]	15.88	**0.7932**	**0.1994**
	VM-NeRF	**16.90**	0.7563	0.2461

Table 2. Ablation study results. Keys: # views: represent the number of views and the relative subset. avg. dist.: average distance between view pairs.

# views	avg. dist.	Method	PSNR ↑	SSIM ↑	LPIPS ↓
8 [12]	5.20	DietNeRF [12]	25.59	0.9120	0.0770
8 [12]	5.20	VM-NeRF	26.90	0.9180	0.0797
8 (s1)	5.18	VM-NeRF	27.87	0.9294	0.0693
8 (s2)	4.38	VM-NeRF	27.13	0.9178	0.0855
8 (s3)	4.73	VM-NeRF	27.48	0.9317	0.0676
8 (s4)	4.69	VM-NeRF	28.39	0.9365	0.0598

4.3 Training with VM-NeRF

VM-NeRF is subject to the same geometric constraints as the original view morphing technique [24]. These constraints impose that singular camera configurations should not exist. These configurations happen whenever the optic centre of a camera is within the field of view of another one [24]. We also discard cameras that are distant from each other more than a threshold γ, as the morphed cameras may be on a transition path that crosses regions where the object of interest is not actually visible (so being rather useless for training a NeRF based model).

Because view morphing allows the synthesis of a new view at any point on the line that connects the known camera pair, we randomly sample new views using a Gaussian distribution centred halfway through the camera pair. Specifically, let us consider a normalised distance between the two cameras. The Gaussian distribution is centred at 0.5 and the standard deviation σ is chosen such that $3\sigma \rightarrow \epsilon$ at the optic centre positions. Therefore, we sample $\alpha \sim \mathcal{N}(0.5, \sigma)$ with $0 \leq \alpha \leq 1$. The depth NeRF can render at the first few iterations is noisy, therefore, we let NeRF warm up on the known views for λ iterations before synthesising and injecting VM-NeRF views in the next training iterations. After the warm-up, for each valid camera pair, we regenerate M new views every η training iterations as the predicted depth improves over time during training.

5 Experiments

5.1 Experimental Setup

We evaluate our method on three training setups using the NeRF realistic synthetic 360° dataset [17], which is composed of eight scenes, i.e. Chair, Drums,

Ficus, Lego, Materials, Ship, Mic, Hot Dog. **First setup:** We select $N = 8$ views out of 100 available for each scene using the Farthest Point Sampling (FPS) [20] (the first view is used for FPS initialisation in each scene). **Second setup:** we use the same $N = 8$ views used in DietNeRF [12]. **Third setup:** we select $N = 4$ views using the previous FPS approach. We test each trained model on all the test views of NeRF realistic synthetic 360°. We quantify the rendering results using the peak signal-to-noise ratio (PSNR) score, the structured similarity index measure (SSIM) [28] and the learned perceptual image patch similarity (LPIPS) [30]. We quantitatively compare our approach against DietNeRF [12] and RegNeRF [19] as the most recent methods for few-shot view synthesis. We also compare against AugNeRF [5] because it is the only data augmentation for NeRF, and data augmentation can be a useful strategy to promote generalisation. We choose to use the Chair scene for our ablation study, which consists of testing VM-NeRF on four different, randomly-chosen, configurations of eight views and on the DietNeRF configuration.

We implement NeRF and our approach in PyTorch Lightning, and run experiments on a single Nvidia A40 with a batch size of 1024 rays. A single scene can be trained in about two days. We use the original implementations of DietNeRF, AugNeRF and RegNeRF to evaluate the different setups. We set the same training parameters as in [17], and set $\gamma = 6$, $\sigma = 0.2$, $M = 1$, $\eta = 5$, $\lambda = 500$.

5.2 Analysis of the Results

Quantitative. Table 1 shows the results averaged over the eight scenes. Our NeRF implementation can achieve nearly the same results reported in [17] on the 100-view setup, i.e. PSNR equal to 31.21 (ours) compared to 31.01 [17].

VM-NeRF can outperform all the other methods in the eight-view setting. Interestingly, the original version of NeRF is the one that performs as second best, followed by DietNeRF and RegNeRF. AugNeRF fails to produce satisfactorily results. We can also observe that VM-NeRF achieves slightly better quality than its version with oracle depth maps, i.e. 24.39 vs. 24.22 PSNR. In fact, we observed that VM-NeRF can effectively leverage the depth information that is estimated during training, although it is noisy. We also evaluate VM-NeRF on the same eight views originally tested by DietNeRF [12]. Also here we can achieve higher quality results on average, i.e. 24.14 vs. 23.59. We also improve in the four-view setup where we obtain an improvement of +1.02 PSNR on average. The results also show that the perturbation of the known input views, done by AugNeRF, has adverse effects in all the tested setups.

Qualitative. Figure 3 shows some qualitative results on Chair, Hot Dog and Lego where we can observe that VM-NeRF produce results with better details than DietNeRF. We speculate that this difference with DietNeRF may be due to its CLIP-based approach that is introduced to leverage a semantic consistency loss for regularisation [21]. The CLIP output is a low-dimensional (global) representation vector of the image, which may hinder the learning of high-definition details. Differently, our approach interpolates the original photometric infor-

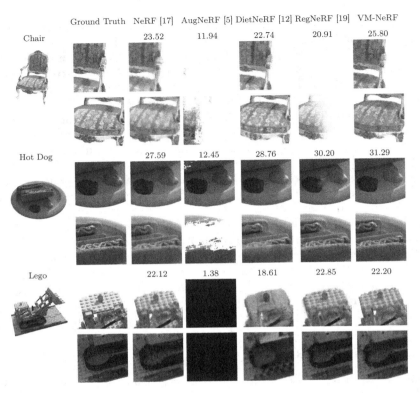

Fig. 3. Comparisons on test-set views of scenes of NeRF realistic synthetic 360°. Unlike AugNeRF [5], VM-NeRF is an effective method that can be used for few-shot view synthesis problems. Unlike DietNeRF [12], VM-NeRF enables NeRF to learn scenes with a higher definition. VM-NeRF produce less artefacts than RegNeRF during rendering [19]. We report the PSNR that we measured for each method and for each rendered image. AugNeRF unsuccessfully learns Chair and Lego (white and black outputs).

mation from two views to produce a new input view, without losing information through the encoding of the low-dimensional representation vector. Figure 3 shows that our approach compared to RegNeRF produces fewer artefacts by correlating the nearby views.

Ablation Study. We assess the stability of VM-NeRF by evaluating the rendering quality when different combinations of views are used to train NeRF. Table 2 shows that the performance is fairly stable throughout different view configurations. We also observed that the algorithm is robust to variations in the distance between view pairs. As long as a view pair is not singular and the distance between cameras is adequate to create acceptable 3D projective transformations of the object, we can successfully synthesise new views with VM-NeRF.

6   Conclusions

We presented a novel method for few-shot view synthesis that blends NeRF and the View Morphing technique [24]. View morphing requires no prior knowledge of the 3D shape and it is based on general principles of projective geometry. We evaluated our approach using the conventional dataset employed by NeRF-based methods, demonstrating that VM-NeRF more effectively learns 3D scenes across various few-shot view synthesis setups. VM-NeRF can interpolate only along the line that connects the optical centres of each camera pair. Therefore, it cannot reconstruct the whole object if only a part of it is viewed during training. Lastly, we designed our approach to be fully differentiable, so an attractive research direction is to integrate our approach into an end-to-end training pipeline.

Acknowledgements. This work was supported by the PNRR project FAIR - Future AI Research (PE00000013), under the NRRP MUR program funded by the NextGenerationEU. This research is partially supported by the project Future Artificial Intelligence Research (FAIR) - PNRR MUR Cod. PE0000013 - CUP: E63C2200194 0006 and the framework "RAISE - Robotics and AI for Socio-economic Empowerment" supported by European Union - NextGenerationEU.

References

1. Aanæs, H., Jensen, R.R., Vogiatzis, G., Tola, E., Dahl, A.B.: Large-scale data for multiple-view stereopsis. Int. J. Comput. Vision **120**(2), 153–168 (2016)
2. Barron, J.T., Mildenhall, B., Tancik, M., Hedman, P., Martin-Brualla, R., Srinivasan, P.P.: Mip-NeRF: a multiscale representation for anti-aliasing neural radiance fields. In: ICCV (2021)
3. Chaitin, G.J.: Register allocation & spilling via graph coloring. ACM Sigplan Not. **17**(6), 98–101 (1982)
4. Chen, D., Liu, Y., Huang, L., Wang, B., Pan, P.: GeoAug: data augmentation for few-shot NeRF with geometry constraints. In: Avidan, S., Brostow, G., Cissé, M., Farinella, G.M., Hassner, T. (eds.) ECCV 2022. LNCS, vol. 13677, pp. 322–337. Springer, Cham (2022). https://doi.org/10.1007/978-3-031-19790-1_20
5. Chen, T., Wang, P., Fan, Z., Wang, Z.: Aug-NeRF: training stronger neural radiance fields with triple-level physically-grounded augmentations. In: CVPR (2022)
6. Devernay, F., Peon, A.R.: Novel view synthesis for stereoscopic cinema: detecting and removing artifacts. In: Workshop on 3D Video Processing (ACMMM) (2010)
7. Fusiello, A., Trucco, E., Verri, A.: A compact algorithm for rectification of stereo pairs. Mach. Vis. Appl. **12**(1), 16–22 (2000)
8. Gallo, O., Troccoli, A., Jampani, V.: Novel View Synthesis: From Depth-Based Warping to Multi-Plane Images and Beyond (2020). https://nvlabs.github.io/nvs-tutorial-cvpr2020/. Conference on Computer Vision and Pattern Recognition
9. Hartley, R.I., Zisserman, A.: Multiple View Geometry in Computer Vision. Cambridge University Press, Cambridge (2004)
10. Hedman, P., Srinivasan, P.P., Mildenhall, B., Barron, J.T., Debevec, P.: Baking neural radiance fields for real-time view synthesis. In: ICCV (2021)
11. Ichnowski, J., Avigal, Y., Kerr, J., Goldberg, K.: Dex-NeRF: using a neural radiance field to grasp transparent objects. In: CRL (2022)

12. Jain, A., Tancik, M., Abbeel, P.: Putting NeRF on a diet: semantically consistent few-shot view synthesis. In: ICCV (2021)
13. Kajiya, J., Herzen, B.: Ray tracing volume densities. In: SIGGRAPH (1984)
14. Long, X., Lin, C., Wang, P., Komura, T., Wang, W.: SparseNeuS: fast generalizable neural surface reconstruction from sparse views. In: Avidan, S., Brostow, G., Cissé, M., Farinella, G.M., Hassner, T. (eds.) ECCV 2022. LNCS, vol. 13692, pp. 210–227. Springer, Cham (2022). https://doi.org/10.1007/978-3-031-19824-3_13
15. Max, N.: Optical models for direct volume rendering. IEEE Trans. Vis. Comput. Graph. 1(2), 99–108 (1995)
16. Mildenhall, B., et al.: Local light field fusion: practical view synthesis with prescriptive sampling guidelines. ACM Trans. Graph. 38(29), 1–14 (2019)
17. Mildenhall, B., Srinivasan, P.P., Tancik, M., Barron, J.T., Ramamoorthi, R., Ng, R.: NeRF: representing scenes as neural radiance fields for view synthesis. In: ECCV (2020)
18. Müller, T., Rousselle, F., Novák, J., Keller, A.: Real-time neural radiance caching for path tracing. ACM Trans. Graph. 40(4), 1–16 (2021)
19. Niemeyer, M., Barron, J.T., Mildenhall, B., Sajjadi, M.S.M., Geiger, A., Radwan, N.: RegNeRF: regularizing neural radiance fields for view synthesis from sparse inputs. In: CVPR (2022)
20. Qi, C.R., Litany, O., He, K., Guibas, L.J.: Deep hough voting for 3D object detection in point clouds. In: ICCV (2019)
21. Radford, A., et al.: Learning transferable visual models from natural language supervision. In: ICML (2021)
22. Rematas, K., Martin-Brualla, R., Ferrari, V.: ShaRF: shape-conditioned radiance fields from a single view. In: ICML (2021)
23. Schönberger, J.L., Frahm, J.M.: Structure-from-motion revisited. In: CVPR (2016)
24. Seitz, S.M., Dyer, C.R.: View morphing. In: Conference on Computer Graphics and Interactive Techniques (1996)
25. Shorten, C., Khoshgoftaar, T.M.: A survey on image data augmentation for deep learning. J. Big Data 6(1), 1–48 (2019)
26. Tewari, A., et al.: Advances in neural rendering. In: Computer Graphics Forum, vol. 41, no. 2, pp. 703–735 (2022)
27. Wang, J., et al.: Generalizing to unseen domains: a survey on domain generalization. IEEE Trans. Knowl. Data Eng. 35(08), 8052–8072 (2023)
28. Wang, Z., Bovik, A., Sheikh, H., Simoncelli, E.: Image quality assessment: from error visibility to structural similarity. IEEE Trans. Image Process. 13(4), 600–612 (2004)
29. Xie, Y., et al.: Neural fields in visual computing and beyond. In: Computer Graphics Forum, vol. 41, no. 2, pp. 641–676 (2022)
30. Zhang, R., Isola, P., Efros, A.A., Shechtman, E., Wang, O.: The unreasonable effectiveness of deep features as a perceptual metric. In: CVPR (2018)

MOVING: A MOdular and Flexible Platform for Embodied VIsual NaviGation

Marco Rosano[1]([✉]), Francesco Ragusa[1,2], Antonino Furnari[1,2], and Giovanni Maria Farinella[1,2,3]

[1] FPV@IPLAB - Department of Mathematics and Computer Science, University of Catania, Catania, Italy
marco.rosano@unict.it
[2] Next Vision s.r.l., Catania, Italy
[3] Cognitive Robotics and Social Sensing Laboratory, ICAR-CNR, Palermo, Italy

Abstract. We present MOVING, a flexible and modular hardware and software platform for visual mapping and navigation in the real world. The platform comprises a flexible sensor configuration consisting of an RGB-D camera, a tracking camera for odometry, and a 2D Lidar, along with a compact processing unit that is equipped with a GPU for running deep learning models. The software is based on ROS, utilizing the RGB-D RTAB-Map SLAM system for mapping and localization and the move base package for path planning and robot movement control. The platform is easily detachable and can be installed on any robot with minimal adaptation required, enabling the reuse of the same robotic software regardless of the robot employed. The effectiveness of the proposed platform was verified through mapping sessions of a large indoor environment, leveraging a Loomo robot. The proposed platform can represent a reasonable solution to speed up the design and testing of new software for autonomous navigation systems, minimizing deployment time in the real world.

Keywords: Robot visual navigation · Visual mapping and navigation · ROS

1 Introduction

Recent years have seen a significant advancement in robotic technology. As a result, increasingly robust and reliable robot platforms have been developed, demonstrating their ability to carry out challenging tasks such as object grasping [11], visual localization [1] and navigation [18], which typically occur in industrial, commercial and even domestic scenarios. Although many hardware components for robotics have been developed for several years now, such as mobile bases with wheels and mechanical arms with gripper, others have been improved only recently. For example, sensors that enable robots to perceive their surroundings, such as cameras and Lidars, have been improved in their performance and

G. L. Foresti et al. (Eds.): ICIAP 2023, LNCS 14234, pp. 75–86, 2023.
https://doi.org/10.1007/978-3-031-43153-1_7

Fig. 1. The proposed MOVING platform, installed on a Loomo robot. The platform consists of a perception module (on the left) formed by an RGB-D camera (bottom-left), a tracking camera (center-left), a 2D Lidar (top-left), and a compact, low-power computational unit (on the right), equipped with a 6-core CPU, a Graphical Processing Unit (GPU) and WiFi connectivity. The sensors are supported by an ad hoc designed 3D printed mount and can be clamped to any robot. The computational unit runs the mapping and navigation software and supports running Deep Learning models on-device.

reliability, and their cost has also decreased. Similarly, the processing units have been improved and compacted to ensure they can be simply installed on any robot and process data directly on-board. On the software side, thanks to the recent progress of Deep Learning algorithms it is now possible to extrapolate richer information from the sensory data, allowing the robot to learn robust control policies and behave more naturally when compared to classic control systems [19]. In Previous works [15,17] several attempts have been made to standardize the software intended for controlling robotic systems. For instance, the ROS project [20] brings together a versatile software stack, developed over the years with the aim of providing the essential tools for the development of robotic control systems. Unfortunately, despite standardization attempts have been made also for robot platforms [3,14], those commercially available do not fully satisfy the different needs of research centers and companies which develop robotic solutions. Additionally, the design of the available robots often lacks modularity and interchangeability of the individual platform components.

In this work we propose a flexible and modular hardware and software platform for SLAM and visual navigation. The platform includes several sensors: an RGB-D camera, a tracking camera which provides odometry and a 2D Lidar. The compact processing unit is equipped with wireless connectivity, including a GPU to run Deep Learning models and can be powered by a simple power bank. The software is based on ROS [20], it uses the RGB-D RTAB-Map SLAM system [12] to map the environment and locate the robot inside it and the move_base package [13] for path planning and robot movement control. The platform, depicted in Fig. 1, can be easily detached and installed on any robot with minimal adaptation required. This allows the reuse of the same SLAM and navigation software regardless of the robot employed, hence reducing deployment time. Moreover, its use encourages to design and test navigation solutions based on the visual platform, rather than on the full stack including sensors, processing units and the robot, which makes the developed solutions more adaptable to different applications. In principle, the same solution could work on a Pepper[1] robot or on a robotic vacuum cleaner. The configuration of the sensors is flexible. It is possible to remove the tracking camera or the Lidar or both according to the precision, size and energy autonomy needed. The sensors' missing information is then estimated at the software level based on the other sensors. The effectiveness of MOVING was verified through several mapping sessions of a large indoor environment, carried out leveraging a Segway Loomo robot[2]. We discuss the flexibility of the proposed platform, which can represent a reasonable solution to speed up the design and test of new software for autonomous navigation systems. Given the remarkable perceptive capacity of the proposed platform and the availability of hardware acceleration (GPU) of the embedded system, MOVING allows to minimize the deployment time of Deep Learning models in the real world.

2   Related Work

Robot Platforms and Applications. Over the years, many robot platforms have been designed with the goal to assist humans in their everyday activities, while working [9], learning [2] or in need of special support [7]. Given the high costs associated with the development of these platforms, the first prototypes were developed by large private companies and large research centers for commercial applications and technological advancement [21]. In the '90s, the work of Hirai et al. [10] represented the first successful attempt to create a working humanoid robot, equipped with a large suite of sensors and able to replicate human activities (e.g. walking, grasping, etc.). The first studies on quadrupedal robots date back to the same period [4]. Regarding wheeled robots, many platforms were developed for research and educational purposes [3]. To foster the development of robotic technologies, many other platforms have been released as

---

[1] https://www.aldebaran.com/en/pepper.
[2] https://www.segway.com/loomo.

open-source [14]. Each of these robot platforms has a heterogeneous set of sensors with different perceptive capabilities and specific software for robot mapping, navigation and control, which support only a limited set of peripherals.

Our platform for SLAM and navigation contrasts with these approaches. In fact, the entire hardware setup can be easily mounted on any robot and, exploiting the ROS software framework [20], it can support a large variety of sensors and leverage open-source software for SLAM and navigation. This gives the flexibility to choose the most suitable set of sensors and to re-use the same software while changing the robot.

Visual Mapping and Navigation. Most navigation systems requires a map of the environment to carry out the operations of localization and path planning. The environment map can be built beforehand starting from a set of RGB images using a Structure From Motion method [22] to return a 3D reconstruction in the form of a point cloud, from which a navigable 2D grid map is extracted. In contrast, Simultaneous Localization And Mapping (SLAM) systems [6] can be used to reconstruct the environment in real time while exploring, starting from RGB-D images [16], from Lidar signal [8], or both [12]. The resulting floor plan is then exploited by the navigation stack to carry out path planning and provide the robot with the actions to execute towards the goal. As an alternative to traditional methods, Deep Learning-based mapping and navigation models have emerged, which are able to learn directly from the data provided, often collected from simulated environments [18]. These methods can replicate the SLAM algorithm pipeline and reduce the complexity of some of its components [5].

In this work we adopt a robust and reliable RGB-D SLAM system called RTAB-Map [12] along with the navigation stack called move_base [13], both integrated in the ROS framework [20]. However, the proposed platform has been designed to supports the use of Deep Learning models at both the software and hardware levels.

3   The MOVING Platform

Figure 2 summarizes the architecture of the proposed platform, which consists of hardware and software components and has been designed to ensure support for a large number of input peripherals, robot platforms and mapping/navigation software. MOVING embraces the concept of "Write once, deploy anywhere", which allows to develop a software application only once and deploying it on as many platforms as desired. To achieve this, we designed a compact hardware system (Fig. 1), easily installable on a large number of robots, with a high-performance and power efficient processing unit to support the computation load while ensuring adequate autonomy.

3.1   Hardware Architecture

The hardware architecture of MOVING consists of the following components:

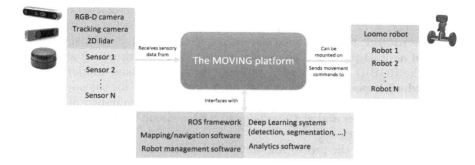

Fig. 2. Architecture of the MOVING platform. In blue we reported the configuration discussed in this paper, whereas in yellow we highlighted the possibility to personalize the platform. With the adoption of the ROS framework [20], MOVING can support a wide range of sensors and can work with classic reliable robotic software as well as custom applications. The platform can be easily installed on a large number of robots, which can be controlled by setting up a simple platform-robot communication interface.

- **RGB-D camera**: the camera acquires RGB images coupled with depth images and sends them to the visual SLAM software to create the environment map. During the navigation, the images can be used to localize the robotic agent within the already acquired map. In the setup discussed in this paper we use a Realsense D455 depth camera[3], which relies on a stereoscopic 3D vision technology to capture a sufficiently accurate depth;
- **Tracking camera**: the tracking camera enabels the robot to use the odometry to track itself in space over time. To avoid pose estimation drift, the system should run at a high frame rate and can integrate visual and inertial information to improve its accuracy. When the device is not available, the odometry can be estimated from RGB or RGB-D images at the expense of less reliability and precision, or from wheel encoders installed on the robot, which however would force to bind the platform to the specific robot. In our setup, we adopt a Realsense T265 tracking camera[4] which combines VO from two fisheye sensors with inertial sensors to compute an highly accurate odometry directly on device;
- **2D Lidar**: the 2D Lidar emits a beam of light (laser) while spinning to measure the distance of obstacles that intersect the scanned virtual horizontal plane. Depending on the model, the 2D Lidar can perceive only the front facing obstacles (180° perception) or all the obstacles around the sensor (360° perception). The perceived range can easily go beyond 10 m even in the case of affordable devices, providing an accurate measurement of the geometry of the environment. In the proposed platform, we adopt the RPLIDAR-A2M12 2D

---

[3] https://www.intelrealsense.com/depth-camera-d455/.
[4] https://www.intelrealsense.com/tracking-camera-t265/.

Lidar[5], which is capable of a 360° scan, has an angular resolution of 0.225°, a max range of 12 m and a frequency of 10 Hz;

- **Sensors mount**: the considered sensors are installed on a mount, specifically designed and 3D printed. The mount allows to position the tracking camera above the RGB-D camera and is extended by an additional support for the Lidar, which is positioned above the cameras and allows for tilt adjustment. The setup can be then hooked to the robot via a joint equipped at one end with a camera screw and at the other end with a clamp for immediate installation on any robot;

- **Processing unit**: the processing unit processes the incoming stream of data from the sensors, runs the mapping and navigation software and manages the communication between the robot and the external services. We have chosen to adopt a ZED Box[6] given its compact form factor and hardware characteristics. It is equipped with a 6-core processor and a Graphics Processing Unit (GPU) to run Deep Learning-based models and comes with Ubuntu OS for software compatibility. The ZED Box has been equipped with WiFi connectivity, a USB hub, and can be powered by a classic power bank equipped with a Power Delivery (PD) output, given its maximum consumption of 20 Wh.

The platform setup, complete with all hardware components, is shown in Fig. 1.

3.2   Software Architecture

The software system is based on the ROS framework [20]. ROS is an open-source project which consists of a set of software libraries and tools for building robotic applications. ROS offers the advantage of: 1) abstracting the robot platform in order to reuse the code; 2) using highly standardized and stable software libraries; 3) integrating custom software easily into the ecosystem; 4) choosing from numerous supported sensors which are easily to integrated in ROS; 5) move the computational load to a third party machine transparently, if the computational power of the embedded system is limited. The software components used are the following:

- **SLAM software**: RTAB-Map [12] is a robust RGB-D, stereo e Lidar graph-based SLAM approach, integrated in the ROS framework as a ROS package. Thanks to a memory management approach, the system supports the scan of large-scale environments while keeping its ability to work in real time. The clear advantage of using visual SLAM over Lidar-based SLAM is the availability of an image-based loop closure detector, which is able to correct the pose estimation drifts during mapping and navigation and provide a first guess on the initial robot's location. RTAB-Map also offers the possibility to estimate the odometry signal from the RGB-D or the Lidar data stream, when an odometry source is not available.

---

[5] https://www.slamtec.ai/home/rplidar_a2/.
[6] https://www.stereolabs.com/zed-box/.

- **Navigation software**: the ROS Navigation Stack offers a collection of software packages that can be used to implement a navigation system. Specifically, the move_base node [13] links together a global planner and a local planner which leverage the environment map to generate an optimal navigable path to the destination. The *move_base* node receives as input the sensors' observations, the map of the environment, the robot's position and the odometry source to output the actions to be performed in the form of linear and angular velocities;
- **Sensors ROS interfaces**: thanks to the broad adoption of the ROS framework in the robotic field, most sensor manufacturers provide the required ROS software packages to interface with the framework. The realsense cameras are supported by the *realsense2_camera* ROS package[7], while the RPLIDAR device is supported by the *rplidar* ROS package[8];
- **ROS-robot communication system**: to forward the actions provided by the *move_base* package, we implemented a socket-based communication system where a sender, integrated in the ROS framework, reads the action commands and sends them to the receiver via network connection.
- **Robot management software**: we developed a web-based, cross-platform robot management software, which allows the user to monitor the robot while it operates in the designated environment, track its movements, and specify new destinations to reach and receive updates on the status of the task. Moreover, given the environment map, an annotation tool can be used to store strategic points of interest inside it and associate them with labels for future use. The backend of the web-based software was developed in Python using the Flask library to build the webserver and the rospy library to interface with ROS, while the frontend was developed in HTML and Javascript.

4   Experimental Settings and Results

We validated the proposed platform while performing the mapping and navigation tasks in an environment, by assessing the quality of the resulting floor plans given the full sensor configuration. To show the contribution of each of the sensors used in our setup, we also performed an ablation study by employing a subset of the sensors (i.e. excluding the 2D lidar, the tracking camera or both of them).

4.1   Mapping of the Environment

To carry out the mapping, the proposed platform was mounted on a Segway Loomo robot[9], which received the action commands through the socket-based communication interface discussed in the Sect. 3.2. Given that Loomo comes equipped with the Android OS, the receiver program was developed in Java and

---

[7] http://wiki.ros.org/realsense2_camera.

[8] http://wiki.ros.org/rplidar.

[9] https://www.segway.com/loomo.

Table 1. List of the performed mapping sessions, considering different sensor configurations. We report if the task has been successfully completed, together with a short comment on the experiment.

Odometry source	Lidar	Sucess	Notes
Tracking camera	2D Lidar	✓	-
Tracking camera	-	✓	Artifacts in the map
Tracking camera	From depth image	✗	Continuous stop and in-place rotations
Lidar - ICP	2D Lidar	✓	-
Lidar - ICP	From depth image	✗	Camera FOV too narrow, unstable lidar signal
RGB-D	-	✓*	*Loss of tracking. Incomplete mapping

used the Loomo SDK to forward the linear and angular velocities to the robot base controller.

To perform experiments in a real scenario, we chose an area of our university department floor[10] that measures approximately 295 square meters. To scan the environment covering the whole area, a path formed by 16 key points has been provided. Points along the track were then sub-sampled at 1.5 m intervals and provided as goals to the autonomous navigation system.

The list of the performed mapping experiments is summarized in Table 1, which reports the sensor configurations used as well as if the mapping was successful or not. Using the complete settings, with both tracking camera and 2D Lidar (first row), the mapping procedure has been complete successfully. When the tracking camera has not been used, the odometry was estimated from the Lidar signal using the Iterative Closest Point (ICP) algorithm (4th and 5th rows) or from visual features extracted from the RGB-D images (last row). In two of the four configurations that do not include the 2D Lidar sensor (3rd and 5th rows), we extracted a Lidar-like signal from the camera's depth images using the *depthimage_to_laserscan* ROS package, which perceives a narrow portion of space and has a shorter perception range (capped to 5 m to avoid noisy measurements), but did not prove to be a useful input signal for the mapping system.

In general, it is possible to observe how four of the six configurations successfully accomplished the mapping task, whereas two experiments, which employed the lidar-like signal, were unsuccessful (3rd and 5th rows). In the first case (3rd

---

[10] Additional details omitted due to the anonymous submission.

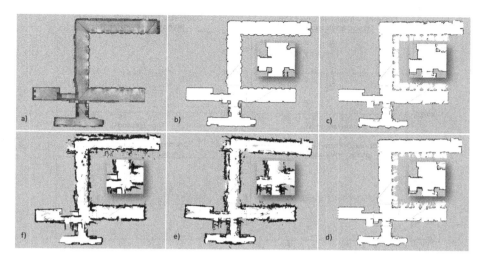

Fig. 3. Floor plans resulting from the mapping sessions performed using different sensor configurations. a) represents the environment acquired with a Matterport 3D scanner, which has been modified (b) in order to be immediately comparable with the maps scanned with the SLAM approach. c) has been acquired using the RGB-D camera+tracking camera+2D Lidar; d) using the RGB-D+ICP odometry+2D Lidar; e) represents the floor plan acquired using the RGB-D+tracking camera; f) represents the floor plan acquired using the RGB-D+RGB-D odometry.

row) the Lidar-like signal was used as a direct replacement of the Lidar for the map construction. In this configuration, the robot was not able to move smoothly, and the navigation was characterized by continuous stops and in-place rotations, probably caused by incorrect measurements. In the second case (5th row) the Lidar-like signal was used only for odometry estimation using the ICP algorithm. Unfortunately, in this configuration, the tracking was immediately lost with continuous odometry resets. In both configurations, it emerged that the Field Of View (FOV) and the perceptive range of the Lidar sensor are essential to benefit from this signal. Concerning the experiment carried out using the odometry estimated from RGB-D images (last row), the system lost tracking of the robot several times after observing textureless walls, resulting in an incomplete map. In fact, it is well known how visual systems based on feature tracking fail in the presence of featureless surfaces.

Figure 3 compares the maps of the environment resulting from the scan sessions performed with the different sensor configurations. For reference, we reported the floor plan of the same environment scanned using a Matterport 3D scanner (Fig. 3a), together with a simplified version (Fig. 3b) which represents a curated optimal mapping. The maps acquired using the RGB-D camera+tracking camera+2D lidar (Fig. 3c) and the RGB-D camera+ICP odometry+2D lidar (Fig. 3d) configurations are both accurate and detailed, presenting a slight distortion in the area of the top corridor. The odometry estimated from

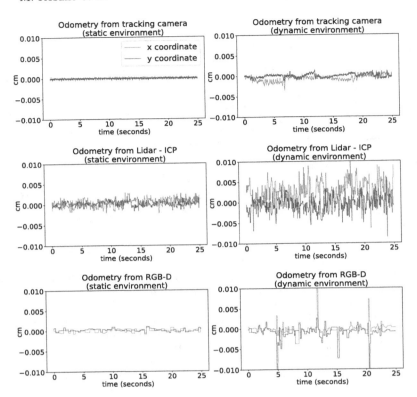

Fig. 4. Influence of the quality of the odometry on the robot tracking system. On the left, we tracked the robot's coordinates when placed in a static environment, for different odometry sources. On the right, the same measurements but in a dynamic environment with people moving around. Since the robot is stationary, the estimated coordinates should be both equal to zero. The chart shows the advantage of using a dedicated odometry device for a reliable localization in challenging scenarios.

the Lidar signal has proven to be reliable in a static scenario as the considered one, but it may exhibit its limitation in a dynamic environment, given the Lidar's perception of motion.

On the contrary, although the maps acquired without Lidar using the RGB-D camera+tracking camera (Fig. 3e) and the RGB-D camera+RGB-D odometry (Fig. 3f) configurations match the geometry of the environment, they have numerous artifacts (irregular and spectral walls), areas incorrectly mapped as obstacles, and unscanned areas. The tracking camera helped limit excessive distortions and misalignments (Fig. 3e), which can instead be observed from the highlighted portion of the map acquired using the odometry estimated from the RGB-D signal (Fig. 3f).

4.2 Odometry and Tracking

To better measure the accuracy and reliability of the odometry signals considered in our experiments, we conducted a further experiment in which the robot was placed in a fixed location of the environment and its position was recorded while observing a static scene (similar to that experienced during mapping) and a dynamic scene (with the presence of people in motion, to replicate a real dynamic environment). As can be observed in Fig. 4, the odometry provided by the tracking camera (first row) is significantly more reliable than the odometry estimated from Lidar (second row) or RGB-D (third row), allowing for a more stable and precise robot tracking even in dynamic environments. While the oscillations of the robot's position derived from the estimated odometry may seem small, it should be noted that, in a scenario where the robot moves around, these inaccuracies can accumulate quickly, resulting in incorrect tracking and localization.

Overall, the results showed how the adoption of different sensor configurations can lead to significantly different results and highlighted the benefits offered by a full sensor configuration.

5   Conclusion

We presented MOVING, a flexible and modular hardware and software platform for SLAM and visual navigation that can be easily installed on any robot requiring a minimal adaptation. The platform allows the reuse of robotic software by encouraging the design and test of navigation solutions based on the visual platform, hence reducing deployment time. The effectiveness of MOVING was verified in a real environment by performing different mapping sessions with several sensor configurations and investigating the reliability of the odometry to track the robot in static and dynamic environments.

Aknowledgment. This research is supported by Next Vision s.r.l. (Next Vision: https://www.nextvisionlab.it/) and by the project Future Artificial Intelligence Research (FAIR) - PNRR MUR Cod. PE0000013 - CUP: E63C22001940006.

References

1. Alkendi, Y., Seneviratne, L., Zweiri, Y.: State of the art in vision-based localization techniques for autonomous navigation systems. IEEE Access **9**, 76847–76874 (2021)
2. Anwar, S., Bascou, N.A., Menekse, M., Kardgar, A.: A systematic review of studies on educational robotics. J. Pre-Coll. Eng. Educ. Res. (J-PEER) **9**(2), 2 (2019)
3. Arvin, F., Espinosa, J., Bird, B., West, A., Watson, S., Lennox, B.: Mona: an affordable open-source mobile robot for education and research. J. Intell. Robot. Syst. **94**, 761–775 (2019)
4. Biswal, P., Mohanty, P.K.: Development of quadruped walking robots: a review. Ain Shams Eng. J. **12**(2), 2017–2031 (2021)

5. Chaplot, D.S., Gandhi, D., Gupta, S., Gupta, A., Salakhutdinov, R.: Learning to explore using active neural slam. In: International Conference on Learning Representations (ICLR) (2020)
6. Durrant-Whyte, H., Bailey, T.: Simultaneous localization and mapping: part I. IEEE Robot. Autom. Mag. **13**(2), 99–110 (2006)
7. Feil-Seifer, D., Matarić, M.J.: Socially assistive robotics. IEEE Robot. Autom. Mag. **18**(1), 24–31 (2011)
8. Grisetti, G., Stachniss, C., Burgard, W.: Improved techniques for grid mapping with rao-blackwellized particle filters. IEEE Trans. Robot. **23**(1), 34–46 (2007)
9. Hägele, M., Nilsson, K., Pires, J.N., Bischoff, R.: Industrial robotics. In: Siciliano, B., Khatib, O. (eds.) Springer Handbook of Robotics, pp. 963–986. Springer, Heidelberg (2016). https://doi.org/10.1007/978-3-540-30301-5_43
10. Hirai, K., Hirose, M., Haikawa, Y., Takenaka, T.: The development of honda humanoid robot. In: Proceedings of 1998 IEEE International Conference on Robotics and Automation (Cat. No. 98CH36146). IEEE (1998)
11. Kleeberger, K., Bormann, R., Kraus, W., Huber, M.F.: A survey on learning-based robotic grasping. Curr. Robot. Rep. **1**, 239–249 (2020)
12. Labbé, M., Michaud, F.: RTAB-map as an open-source lidar and visual simultaneous localization and mapping library for large-scale and long-term online operation. J. Field Robot. **36**(2), 416–446 (2019)
13. Marder-Eppstein, E.: move_base - ros wiki (2016). http://wiki.ros.org/move_base
14. Mondada, F., et al.: Bringing robotics to formal education: the thymio open-source hardware robot. IEEE Robot. Autom. Mag. **24**(1), 77–85 (2017)
15. Montemerlo, M., Roy, N., Thrun, S.: Perspectives on standardization in mobile robot programming: the Carnegie Mellon navigation (carmen) toolkit. In: Proceedings 2003 IEEE/RSJ International Conference on Intelligent Robots and Systems (IROS 2003) (Cat. No. 03CH37453). IEEE (2003)
16. Mur-Artal, R., Tardós, J.D.: ORB-SLAM2: an open-source slam system for monocular, stereo, and RGB-D cameras. IEEE Trans. Robot. **33**(5), 1255–1262 (2017)
17. Muratore, L., Laurenzi, A., Hoffman, E.M., Rocchi, A., Caldwell, D.G., Tsagarakis, N.G.: Xbotcore: a real-time cross-robot software platform. In: 2017 First IEEE International Conference on Robotic Computing (IRC). IEEE (2017)
18. Möller, R., Furnari, A., Battiato, S., Härmä, A., Farinella, G.M.: A survey on human-aware robot navigation. Robot. Auton. Syst. (RAS) **145**, 103837 (2021)
19. Pérez-D'Arpino, C., Liu, C., Goebel, P., Martín-Martín, R., Savarese, S.: Robot navigation in constrained pedestrian environments using reinforcement learning (2020)
20. Quigley, M., et al.: ROS: an open-source robot operating system. In: ICRA Workshop on Open Source Software, Kobe, Japan (2009)
21. Saeedvand, S., Jafari, M., Aghdasi, H.S., Baltes, J.: A comprehensive survey on humanoid robot development. Knowl. Eng. Rev. **34**, e20 (2019)
22. Schönberger, J.L., Frahm, J.M.: Structure-from-motion revisited. In: Conference on Computer Vision and Pattern Recognition (CVPR) (2016)

Evaluation of 3D Reconstruction Pipelines Under Varying Imaging Conditions

Davide Marelli[(✉)][iD], Simone Bianco[iD], and Gianluigi Ciocca[iD]

Department of Informatics, Systems and Communication, University of
Milano-Bicocca, Viale Sarca 336, 20126 Milan, Italy
{davide.marelli,simone.bianco,gianluigi.ciocca}@unimib.it

Abstract. We present IVL-SYNTHSFM-v2, a dataset of 4000 images
depicting five objects from different viewpoints and under different lighting conditions and acquisition setups. The images have been rendered
from 3D scenes with varying camera positions, lights, camera depth of
field, and motion blur. Images depict one of five objects each acquired
using eight different acquisition setups (scenes). 100 images have been
rendered from each scene. The dataset is intended to be used to evaluate 3D reconstruction algorithms. The varying imaging conditions are
introduced to challenge the algorithms to address realistic, non-ideal situations. The dataset provides pixel-precise ground truth to perform accurate evaluations. We demonstrate the usefulness of IVL-SYNTHSFM-v2
by assessing state-of-the-art 3D reconstruction algorithms.

Keywords: Structure from Motion · 3D reconstruction evaluation ·
Synthetic dataset · benchmarking

1 Introduction

Over the years, a variety of techniques and algorithms for 3D reconstruction have
been developed to meet different needs in various fields of application ranging
from active methods that require the use of special equipment to capture geometry information (e.g., laser scanners, structured lights, microwaves, ultrasound)
to passive methods that are based on optical imaging techniques only. The latter
techniques do not require special devices or equipment and are thus easily applicable in different contexts. Among the passive techniques for 3D reconstruction
there is the Structure from Motion (SfM) pipeline [9,15,19,22,25]. Many pipelines
have been proposed in the literature. Here we focus on the most popular ones
with publicly available source code. The most noteworthy pipelines are: COLMAP,
Theia, OpenMVG, VisualSFM, Bundler, MVE, and AliceVision. COLMAP [17]
is an open-source implementation of the incremental SfM and Multi View Stereo
(MVS) pipeline. It is a general-purpose solution usable to reconstruct any scene
introducing enhancements in robustness, accuracy, and scalability. Theia [24] is an

G. L. Foresti et al. (Eds.): ICIAP 2023, LNCS 14234, pp. 87–98, 2023.
https://doi.org/10.1007/978-3-031-43153-1_8

incremental and global SfM open-source library. It includes many algorithms commonly used for feature detection, matching, pose estimation, and 3D reconstruction. OpenMVG [13] is a library created to solve Multiple View Geometry problems. It provides an implementation of the SfM pipeline for both incremental and global cases. Different options are provided for feature detection, matching, pose estimation, and 3D reconstruction. VisualSFM [25] implements the incremental SfM pipeline. Compared to other solutions, only one set of algorithms can be used to make reconstructions. Bundler [22], is one of the first incremental SfM pipeline implementations of success. It reconstructs the scene incrementally using a modified version of the Sparse Bundle Adjustment package of Lourakis and Argyros [10] as the underlying optimization engine. MVE [3], is an incremental SfM implementation. It is designed to allow multi-scale scene reconstruction, comes with a graphical user interface, and also includes an MVS pipeline implementation. Linear SFM [26] decouples the linear and nonlinear components of the pipeline, and performs small reconstructions based on Bundle Adjustment that are afterward joined hierarchically. AliceVision [5] is a complete open source photogrammetry framework implemented in the Meshroom software. The framework's nodal architecture allows the user to create and customize different pipelines adapting them to domain-specific needs.

Driven by the achievements and pending research issues in the 3D reconstruction approaches, many scientific initiatives have been recently proposed to evaluate the current status of available processing methods while boosting further investigations, both in the photogrammetric and computer vision research fields. This encouraged developers and users to deliver comparative performance analyses focusing, in particular, on image-based 3D reconstruction approaches [2,7,18] using a variety of datasets designed for benchmarking them. In the literature there are several datasets that can be used to evaluate reconstruction pipelines. The Middlebury dataset [16] contains 32 + 24 stereo scenes (published in 2014 and 2021, respectively) to evaluate stereo algorithms on 6 and 2-megapixel images. Each scene is composed of a single stereo pair with substantial exposure variations, while scenes and cameras are static. The Middlebury MVS dataset [21] is designed for evaluating multi-view stereo reconstruction algorithms. It consists of undistorted images (640×480 pixels) of a plaster Greek temple and a dinosaur. The DTU Robot Image Data Sets contain two different collections of scenes, one designed to evaluate local features [1], the other for multi-view-stereo (MVS) investigations [6]. The images are at 2 megapixels resolution and contain miniatures with varying illumination conditions. The KITTI dataset [4] is used in the context of autonomous driving for testing different computer vision tasks, such as stereo matching, SLAM, 3D object detection, and depth prediction. The data is derived from several devices mounted on a car. Stretcha [23] is a benchmark dataset designed to compare the reliability of passive 3D reconstruction methods with active stereo systems. Data consists of LiDAR and camera acquisitions of outdoor scenes up to 6 megapixels. Tanks and Temples [8] is a dataset built for image-based 3D reconstruction algorithms. The data consists of real outdoor scenes divided into training and test sets derived from 4K videos (acquired with two rolling shutters and one global-shutter camera). 3DOMCity [14] is a multipurpose and high-resolution $(6,016 \times 4,016$ pixels) benchmark dataset, including 420 nadir and oblique aerial images, to assess the performance of the image-based pipeline for 3D urban reconstruction and 3D data classification. ETH3D [20] is composed of multi-sensors low and high-resolution images and videos for MVS investigations, with scenes acquired both indoor and outdoor, and laser-scanner data as ground truth.

Not all the datasets in the literature comprise the level of detail, scales, and challenges necessary for an accurate and comprehensive evaluation of the reconstruction approaches. Moreover, collecting real data for these benchmarks is often very time-consuming and not trivial. Synthetic datasets can provide realistic data in a less cumbersome way [2]. For instance subjects, camera positions, orientations, lighting conditions, and ambient disturbances can be easily changed. This versatility allows to generate heterogeneous data that can be effectively used to validate reconstruction approaches under varying imaging conditions.

In this paper, we propose the use of the synthetic dataset IVL-SYNTHSFM-v2 to test different SfM reconstruction pipelines. IVL-SYNTHSFM-v2 offers higher-resolution images of different scenes that are rendered with different lighting conditions and camera effects. The dataset has been specifically designed to benchmark 3D reconstruction algorithms under variable and realistic imaging conditions. The pixel-precise Ground Truth (3D geometry, camera parameters and effects) associated with the image data is a unique feature not commonly available in existing datasets and allows accurate evaluation of the algorithms. Moreover, using the tools that we developed for IVL-SYNTHSFM-v2, other datasets can be easily generated on task-specific scenes and image conditions.

The organization of the paper is as follows: Sect. 2 briefly describes the salient characteristics of the IVL-SYNTHSFM-v2 dataset. Section 3, demonstrates how the proposed dataset can be used to evaluate reconstruction algorithms highlighting their strengths and weaknesses under varying imaging conditions. Finally, Sect. 4 reports our conclusions and future directions.

2   The IVL-SYNTHSFM-v2 dataset

The IVL-SYNTHSFM-v2 dataset is an extension of [2]. While maintaining the same goal it also tries to test the robustness of the reconstruction pipelines on different image acquisition setups, such as variations in illumination conditions, depth of field, and motion blur. The dataset is publicly available for download[1]. The dataset was created using Blender for 3D modeling and rendering with the aid of our *SfM Flow* add-on [12]. The rendering of the images uses Cycles, the Blender's path-tracing render engine, that simulates physics-based light interactions. The images in this dataset are generated from five 3D models. Each model is placed in a reference scene and is rendered under different lighting and camera conditions. The list of models/scenes is: Statue, Empire Vase, Hydrant, Bicycle, and Jeep (see Fig. 1).

---

[1] https://doi.org/10.17632/fnxy8z8894.

(q)   (r)   (s)   (t)

Fig. 1. Sample of rendered images of the models. (a) image from the 'fs' set. (b) image from the 'fs-dof-mb' set. (c, d) images from the 'ms' set.

The camera used for the rendering uses a sensor of size 18×32 mm and a 35 mm focal length, all the images are acquired at resolution 1920×1080 pixels. The camera moves in a circle around the vertical axis at the scene center to obtain complete coverage of the object. To simulate realistic manual acquisition the camera position is randomized by 5% of the locations sampled on the circle. For the image sets that make use of a moving sun (ms), a sunlamp is placed at a different position for each image; this is intended to simulate the acquisition during different hours of the day. The sun's movement covers a semicircular path with constant distances between acquisition points; the positions are shuffled for acquisition and kept consistent across different sets of the same target object. The depth of field (DoF) is applied to all images of the sets that make use of it. The DoF effect applied by the render engine that simulates an f-stop ratio of 2.8. Finally, in the sets that make use of motion blur, the effect is introduced randomly (uniform distribution) in 33% of the images. The motion blur effect is simulated considering the camera movement during the time of acquisition of a frame. By changing the acquisition setup, for each scene, eight image sets are created. Each image set is composed of 100 images. The list of the image sets and the corresponding acquisition setup are shown in Table 1. Samples of

Table 1. List of available images sets for each object in the dataset.

Set name	Lighting setup	Depth of field	Motion blur
fs	Sun, fixed position	No	No
fs-dof	Sun, fixed position	Yes, on all images	No
fs-mb	Sun, fixed position	No	Yes, on random images
fs-dof-mb	Sun, fixed position	Yes, on all images	Yes, on random images
ms	Sun, random position	No	No
ms-dof	Sun, random position	Yes, on all images	No
ms-mb	Sun, random position	No	Yes, on random images
ms-dof-mb	Sun, random position	Yes, on all images	Yes, on random images

rendered images are visible in Fig. 1. For the purpose of the evaluation of the reconstruction algorithms, each image set also includes a CSV file ('cameras.csv') containing for each image information about camera position, camera rotation, camera look-at direction, depth of field, motion blur, and sunlighting position. For more details on the IVL-SYNTHSFM-v2 dataset, please refer to [11].

3   Experiments

This section explores the impact of Depth of Field (DoF), Motion Blur (MB), and variations of light intensity and position on the alignment of camera pose in existing SfM pipelines. In our previous work [2] we evaluated the COLMAP, Theia, OpenMVG, and VisualSFM pipelines and found that COLMAP performed best among the four. Here we compare COLMAP against a recent SfM pipeline Meshroom [5] using the new IVL-SYNTHSFM-v2 dataset which provides more challenging data for those effects in both cameras and lighting. Specifically, we want to assess the robustness of the reconstruction pipelines against the combined effects of depth of field, motion blur, and light changes. Table 1 lists all the imaging conditions tested. The two pipelines have been evaluated using standard metrics: the number of registered images, number of keypoints reconstructed, Mean Track Length (MTL), mean observations per image, and Mean Reprojection Error (MRE). Both pipelines have been run using default configurations and forcing only the camera model as a pinhole. In a few cases, other changes were performed: these are reported as table notes.

Table 2. Results on the Bicycle scene using COLMAP.

Set name	Registered images	Total keypoints	MTL	Mean observations per image	MRE [pix]
fs	100	15,016	5.546	832.88	0.619
fs-dof	100	14,820	5.582	827.37	0.623
fs-mb	100	14,427	5.580	805.10	0.640
fs-dof-mb	100	14,383	5.584	803.20	0.642
ms	44	2,940	4.347	290.48	0.939
ms-dof	44	2,947	4.343	290.89	0.944
ms-mb	44	3,438	4.288	335.11	0.971
ms-dof-mb	42	2,879	4.378	300.14	0.945

Table 3. Results on the Empire Vase scene using COLMAP.

Set name	Registered images	Total keypoints	MTL	Mean observations per image	MRE [pix]
fs	100	32,693	5.640	1,843.91	0.391
fs-dof	100	41,789	5.698	2,381.26	0.403
fs-mb	100	26,225	5.534	1,451.45	0.460
fs-dof-mb	100	36,566	5.460	1,996.79	0.432
ms	33	7,239	3.735	819.36	0.394
ms-dof	47	13,659	3.556	1,033.51	0.370
ms-mb	57	3,694	3.231	209.42	0.504
ms-dof-mb	40	9,609	3.589	862.35	0.381

Tables 2, 3, 4, 5 and 6 report the per-object results with the COLMAP pipeline. Across all subjects, except the Statue, the introduction of light intensities and position variations (ms, ms-* sets) drastically reduces the number of registered images as well as the mean observations per image and the MTL. The presence of a non-uniform background in the images simplifies the task of reconstruction for all subjects and allows to correctly register all the images in the fixed light and no camera effect case (fs). However, the Hydrant remains the most critical subject due to its size, filling of images, symmetric geometry, and uniform texture. On this subject, images fail to register in various cases due to the combination of DoF, MB, and characteristics of the subject. Moreover, in the fs-dof-mb it was necessary to perform reconstruction by using sequential matching and forcing the initial pair of images. The fs-dof set required the use of sequential matching instead of the exhaustive one. Furthermore, on the ms* sets almost all images were registered but several of them were aligned in

Table 4. Results on the Hydrant scene using COLMAP.

Set name	Registered images[†]	Total keypoints	MTL	Mean observations per image	MRE [pix]
fs	100	48,029	6.773	3,253.13	0.366
fs-dof [*]	100	56,830	6.733	3,826.87	0.404
fs-mb	100	46,607	6.542	3,049.28	0.406
fs-dof-mb [*,§]	100	53,707	6.467	3,473.62	0.431
ms	96 (17)	28,807	4.035	1,211.04	0.397
ms-dof	96 (27)	33,492	4.001	1,395.99	0.394
ms-mb	94 (40)	24,586	3.972	1,039.10	0.419
ms-dof-mb	96 (40)	26,999	3.876	1,090.21	0.440

† In brackets the images registered with wrong pose;
* Sequential matching used instead of exhaustive matching;
§ Manually fixed initial image pair

Table 5. Results on the Jeep scene using COLMAP.

Set name	Registered images	Total keypoints	MTL	Mean observations per image	MRE [pix]
fs	100	26,327	4.945	1,301.98	0.417
fs-dof	100	25,654	4.944	1,268.49	0.430
fs-mb	100	25,107	4.875	1,224.14	0.443
fs-dof-mb	100	23,679	4.912	1,163.12	0.466
ms	68	5,415	3.202	255.00	0.449
ms-dof	92	7,112	3.285	253.99	0.482
ms-mb	57	3,694	3.231	209.42	0.504
ms-dof-mb	49	3,505	3.151	181.03	0.666

the wrong locations and orientations leading to unusable point clouds. Results are visible in Fig. 2, both pipelines incorrectly aligned the cameras obtaining a wrong multi-part reconstruction of the object.

In a few cases, the *-dof-mb sets provide MREs that are slightly better than the *-mb set but slightly worse than the *-dof set. This behavior is traceable to two different reasons. In some cases, the number of images correctly registered in the *-dof-mb is lower than in the other ones, thus the error is computed on a more robust registered images block. In other cases, the number of registered images is similar, and the lower error is due to the effects of DoF that reduces the mismatch of keypoints located on background elements thus enforcing alignment using more points belonging to the object and leading to a robustly registered image block.

Table 6. Results on the Statue scene using COLMAP.

Set name	Registered images	Total keypoints	MTL	Mean observations per image	MRE [pix]
fs	100	9,417	7.484	704.79	0.580
fs-dof	100	9,349	7.505	701.70	0.583
fs-mb	100	8,763	7.734	677.77	0.614
fs-dof-mb	100	8,826	7.680	677.89	0.614
ms	100	6,111	6.035	368.82	0.621
ms-dof	100	6,112	6.020	368.00	0.619
ms-mb	100	6,162	5.971	367.96	0.634
ms-dof-mb	100	6,066	5.987	363.20	0.634

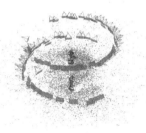   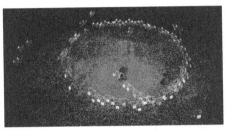

(a) COLMAP   (b) Meshroom

Fig. 2. Results of image registration on the IVL-SYNTHSFM-v2 – Hydrant ms-mb dataset using COLMAP and Meshroom.

Table 7. Results on the Bicycle scene using Meshroom.

Set name	Registered images[†]	Total keypoints	MTL	Mean observations per image	MRE [pix]
fs	55 (55)	8,589	2.421	378.11	1.235
fs-dof	2	47	2.000	47.00	1.353
fs-mb	84 (34)	32,351	2.823	1,087.39	0.783
fs-dof-mb	91 (2)	45,238	3.059	1,520.67	0.589
ms	92 (72)	13,290	2.167	313.08	1.148
ms-dof	92 (73)	18,861	2.424	496.89	0.874
ms-mb	2	33	2.000	33.00	1.243
ms-dof-mb	2	32	2.000	32.00	0.713

† In brackets the images registered with wrong pose

Table 8. Results on the Empire Vase scene using Meshroom.

Set name	Registered images†	Total keypoints	MTL	Mean observations per image	MRE [pix]
fs	100	189,176	4.2119	7,967.89	0.3193
fs-dof	100	186,895	4.3709	8,169.08	0.3787
fs-mb	100	161,652	3.7660	6,087.76	0.3923
fs-dof-mb	100	174,445	4.0093	6,993.97	0.4101
ms	100 (9)	137,732	2.6287	3,620.59	0.2839
ms-dof	100 (17)	137,569	2.7227	3,745.53	0.3423
ms-mb	100 (11)	104,762	2.5741	2,696.69	0.3430
ms-dof-mb	100 (12)	112,589	2.7014	3,041.47	0.3613

† In brackets the images registered with wrong pose

Table 9. Results on the Hydrant scene using Meshroom.

Set name	Registered images†	Total keypoints	MTL	Mean observations per image	MRE [pix]
fs	100 (8)	191,654	5.716	10,955.06	0.332
fs-dof	100 (2)	184,506	6.247	11,525.08	0.407
fs-mb	100 (3)	186,135	5.343	9,945.76	0.387
fs-dof-mb	100 (5)	177,448	5.704	10,121.83	0.448
ms	100 (57)	165,540	2.995	4,958.12	0.313
ms-dof	100 (36)	187,126	3.178	5,946.66	0.346
ms-mb	100 (24)	156,306	2.920	4,564.48	0.350
ms-dof-mb	100 (3)	175,854	3.106	5,462.00	0.391

† In brackets the images registered with wrong pose

Table 10. Results on the Jeep scene using Meshroom.

Set name	Registered images†	Total keypoints	MTL	Mean observations per image	MRE [pix]
fs	100 (15)	101,709	3.808	3,872.60	0.441
fs-dof	100 (10)	102,499	3.761	3,854.75	0.457
fs-mb	99 (9)	102,136	3.742	3,860.86	0.484
fs-dof-mb	97 (7)	100,556	3.731	3,867.31	0.508
ms	96 (74)	62,533	2.582	1,681.73	0.410
ms-dof	99 (72)	61,948	2.619	1,638.53	0.408
ms-mb	90 (47)	47,718	2.585	1,370.69	0.460
ms-dof-mb	99 (70)	55,876	2.567	1,448.48	0.446

† In brackets the images registered with wrong pose

Table 11. Results on the Statue scene using Meshroom.

Set name	Registered images	Total keypoints	MTL	Mean observations per image	MRE [pix]
fs	99	92,230	4.613	4,297.48	0.407
fs-dof	99	91,114	4.630	4,260.94	0.414
fs-mb	100	86,812	4.750	4,123.45	0.428
fs-dof-mb	99	86,960	4.747	4,169.61	0.433
ms	100	67,422	4.061	2,738.01	0.370
ms-dof	100	66,706	4.083	2,723.72	0.365
ms-mb	100	64,295	4.077	2,621.20	0.380
ms-dof-mb	100	61,616	4.100	2,526.31	0.382

Tables 7, 8, 9, 10 and 11 report the per-object results with the Meshroom pipeline. The AliceVision SfM pipeline fails to correctly align the cameras in several cases. In the fs-dof, ms-mb, and ms-dof-mb Bicycle sets, the pipeline was not able to align more than two cameras due to a bad selection of the initial camera pair. In most of the ms-* sets, the pipeline wrongly aligned or discarded numerous images making the result unusable. Thus highlighting the strong impact that changes in the light conditions have on the reconstruction process of the Bicycle, Hydrant, and Jeep models. In the success cases, Alice-Vision was able to align all (or almost all) the cameras and recover a higher number of keypoints than COLMAP despite presenting slightly lower MTLs. The high number of keypoints is directly related to the higher mean observation per image. This also allows the pipeline to estimate strong matches between images and achieve slightly lower MREs than COLMAP. Both pipelines show good results in the reconstruction of the Statue model. Both pipelines achieve low MREs due to the geometry and texture of the object as well as the camera path used for the acquisitions. Interestingly, Meshroom was able to correctly align the cameras for the fs-* sets of the Hydrant where COLMAP required manual steps to succeed. Instead, COLMAP reconstructed the Bicycle model, which is problematic for Meshroom.

4   Conclusion

This paper presented IVL-SYNTHSFM-v2, a dataset of synthetic images aimed to stress 3D reconstruction pipelines in non-ideal conditions using images of various objects acquired with combinations of varying lighting and camera effects. The 4000 images have been used to test the ability of the COLMAP and Meshroom to correctly solve the Structure from Motion task and generate accurate camera alignment and point clouds. The experiments have shown that strong changes in lighting significantly impact the final result in terms of the number of aligned cameras and mean reprojection error. Moreover, motion blur and depth of field have an impact even in the reconstruction of single objects. In the successful reconstruction cases, Meshroom achieved better results than COLMAP presenting higher keypoints count and lower MREs. However, COLMAP provided usable reconstruction in most of the combination of light and camera effects that instead strongly impacted Meshroom. Future work may extend this comparison to other state of the art SfM pipelines and deep learning-based methods. It could also consider the impact of changes to the settings for the used algorithms. The evaluation could also include the subsequent dense reconstruction. Finally, the dataset may be extended with more images to support the training of learning-based methods as well as to include other common camera effects (e.g., vignetting, lens distortions, over/underexposure).

References

1. Aanæs, H., Dahl, A.L., Steenstrup Pedersen, K.: Interesting interest points. Int. J. Comput. Vision **97**(1), 18–35 (2012)
2. Bianco, S., Ciocca, G., Marelli, D.: Evaluating the performance of structure from motion pipelines. J. Imaging **4**(8), 98 (2018)
3. Fuhrmann, S., Langguth, F., Goesele, M.: MVE-a multi-view reconstruction environment. In: GCH, pp. 11–18 (2014)
4. Geiger, A., Lenz, P., Urtasun, R.: Are we ready for autonomous driving? The kitti vision benchmark suite. In: 2012 IEEE Conference on Computer Vision and Pattern Recognition, pp. 3354–3361. IEEE (2012)
5. Griwodz, C., et al.: Alicevision meshroom: an open-source 3D reconstruction pipeline. In: Proceedings of the 12th ACM Multimedia Systems Conference, MMSys 2021, pp. 241–247. Association for Computing Machinery, New York (2021)
6. Jensen, R., Dahl, A., Vogiatzis, G., Tola, E., Aanæs, H.: Large scale multi-view stereopsis evaluation. In: Proceedings of the IEEE Conference on Computer Vision and Pattern Recognition, pp. 406–413 (2014)
7. Jin, Y., et al.: Image matching across wide baselines: from paper to practice. Int. J. Comput. Vision **129**(2), 517–547 (2021)
8. Knapitsch, A., Park, J., Zhou, Q.Y., Koltun, V.: Tanks and temples: benchmarking large-scale scene reconstruction. ACM Trans. Graph. (ToG) **36**(4), 1–13 (2017)
9. Ko, J., Ho, Y.S.: 3D point cloud generation using structure from motion with multiple view images. In: The Korean Institute of Smart Media Fall Conference, pp. 91–92 (2016)
10. Lourakis, M., Argyros, A.: The design and implementation of a generic sparse bundle adjustment software package based on the Levenberg-Marquardt algorithm. Technical report 340, Institute of Computer Science-FORTH, Heraklion, Crete, Greece (2004)
11. Marelli, D., Bianco, S., Ciocca, G.: IVL-SYNTHSFM-v2: a synthetic dataset with exact ground truth for the evaluation of 3D reconstruction pipelines. Data Brief **29**, 105041 (2020)
12. Marelli, D., Bianco, S., Ciocca, G.: SfM flow: a comprehensive toolset for the evaluation of 3D reconstruction pipelines. SoftwareX **17**, 100931 (2022)
13. Moulon, P., Monasse, P., Marlet, R., et al.: OpenMVG an open multiple view geometry library (2013). https://github.com/openMVG/openMVG

14. Özdemir, E., Toschi, I., Remondino, F.: A multi-purpose benchmark for photogrammetric urban 3D reconstruction in a controlled environment. In: Evaluation and Benchmarking Sensors, Systems and Geospatial Data in Photogrammetry and Remote Sensing, vol. 42, pp. 53–60 (2019)
15. Özyeşil, O., Voroninski, V., Basri, R., Singer, A.: A survey of structure from motion. Acta Numer. **26**, 305–364 (2017)
16. Scharstein, D., et al.: High-resolution stereo datasets with subpixel-accurate ground truth. In: Jiang, X., Hornegger, J., Koch, R. (eds.) GCPR 2014. LNCS, vol. 8753, pp. 31–42. Springer, Cham (2014). https://doi.org/10.1007/978-3-319-11752-2_3
17. Schonberger, J.L., Frahm, J.M.: Structure-from-motion revisited. In: Proceedings of the IEEE Conference on Computer Vision and Pattern Recognition, pp. 4104–4113 (2016)
18. Schonberger, J.L., Hardmeier, H., Sattler, T., Pollefeys, M.: Comparative evaluation of hand-crafted and learned local features. In: Proceedings of the IEEE Conference on Computer Vision and Pattern Recognition, pp. 1482–1491 (2017)
19. Schönberger, J.L., Frahm, J.M.: Structure-from-motion revisited. In: Conference on Computer Vision and Pattern Recognition (CVPR), pp. 4104–4113 (2016)
20. Schops, T., et al.: A multi-view stereo benchmark with high-resolution images and multi-camera videos. In: Proceedings of the IEEE Conference on Computer Vision and Pattern Recognition, pp. 3260–3269 (2017)
21. Seitz, S.M., Curless, B., Diebel, J., Scharstein, D., Szeliski, R.: A comparison and evaluation of multi-view stereo reconstruction algorithms. In: 2006 IEEE Computer Society Conference on Computer Vision and Pattern Recognition (CVPR 2006), vol. 1, pp. 519–528. IEEE (2006)
22. Snavely, N., Seitz, S.M., Szeliski, R.: Photo tourism: exploring photo collections in 3D. In: ACM Transactions on Graphics (TOG), vol. 25, pp. 835–846. ACM (2006)
23. Strecha, C., Von Hansen, W., Van Gool, L., Fua, P., Thoennessen, U.: On benchmarking camera calibration and multi-view stereo for high resolution imagery. In: 2008 IEEE Conference on Computer Vision and Pattern Recognition, pp. 1–8. IEEE (2008)
24. Sweeney, C.: Theia multiview geometry library: tutorial & reference (2015). http://theia-sfm.org
25. Wu, C.: Towards linear-time incremental structure from motion. In: 2013 International Conference on 3D Vision-3DV 2013, pp. 127–134. IEEE (2013)
26. Zhao, L., Huang, S., Dissanayake, G.: Linear SFM: a hierarchical approach to solving structure-from-motion problems by decoupling the linear and nonlinear components. ISPRS J. Photogramm. Remote. Sens. **141**, 275–289 (2018)

CarPatch: A Synthetic Benchmark for Radiance Field Evaluation on Vehicle Components

Davide Di Nucci[1]([✉])[ID], Alessandro Simoni[1][ID], Matteo Tomei[2][ID],
Luca Ciuffreda[2], Roberto Vezzani[1][ID], and Rita Cucchiara[1][ID]

[1] Department of Engineering "Enzo Ferrari" (DIEF), University of Modena and
Reggio Emilia, 41125 Modena, Italy
{davide.dinucci,alessandro.simoni,roberto.vezzani,
rita.cucchiara}@unimore.it
[2] Prometeia, 40137 Bologna, Italy
{matteo.tomei,luca.ciuffreda}@prometeia.com

Abstract. Neural Radiance Fields (NeRFs) have gained widespread recognition as a highly effective technique for representing 3D reconstructions of objects and scenes derived from sets of images. Despite their efficiency, NeRF models can pose challenges in certain scenarios such as vehicle inspection, where the lack of sufficient data or the presence of challenging elements (*e.g.* reflections) strongly impact the accuracy of the reconstruction. To this aim, we introduce *CarPatch*, a novel synthetic benchmark of vehicles. In addition to a set of images annotated with their intrinsic and extrinsic camera parameters, the corresponding depth maps and semantic segmentation masks have been generated for each view. Global and part-based metrics have been defined and used to evaluate, compare, and better characterize some state-of-the-art techniques. The dataset is publicly released at https://aimagelab.ing.unimore.it/go/carpatch and can be used as an evaluation guide and as a baseline for future work on this challenging topic.

Keywords: Synthetic vehicle dataset · 3D Reconstruction · Neural radiance fields · Volumetric rendering · RGB-D

1 Introduction

Recent advances in Neural Radiance Fields (NeRFs) [11] strongly improved the fidelity of generated novel views by fitting a neural network to predict the volume density and the emitted radiance of each 3D point in a scene. The differentiable volume rendering step allows having a set of images, with known camera poses, as the only input for model fitting. Moreover, the limited amount of data, *i.e.* (image, camera pose) pairs, needed to train a NeRF model, facilitates its adoption and drives the increasing range of its possible applications. Among these, view synthesis recently emerged for street view reconstruction [12,19] in

ⓒ The Author(s), under exclusive license to Springer Nature Switzerland AG 2023
G. L. Foresti et al. (Eds.): ICIAP 2023, LNCS 14234, pp. 99–110, 2023.
https://doi.org/10.1007/978-3-031-43153-1_9

Fig. 1. A visualization of the *CarPatch* data: RGB images (left), depth images (center), and semantic segmentation of vehicle components (right).

the context of AR/VR applications, robotics, and autonomous driving, with considerable efforts towards vehicle novel view generation. However, these attempts focus on images representing large-scale unbounded scenes, such as those from KITTI [9], and usually fail to achieve high-quality 3D vehicle reconstruction.

In this paper, we introduce an additional use case for neural radiance fields, *i.e. vehicle inspection*, where the goal is to represent an individual high-quality instance of a given car. The availability of a high-fidelity 3D vehicle representation could be beneficial whenever the car body has to be analyzed in detail. For instance, insurance companies or body shops could rely on NeRF-generated views to assess possible external damages after a road accident and estimate their repair cost. Moreover, rental companies could compare two NeRF models trained before and after each rental, respectively, to assign responsibility for any new damages. This would avoid expert on-site inspection or a rough evaluation based on a limited number of captures.

For this purpose, we provide an experimental overview of the state-of-the-art NeRF methods, suitable for vehicle reconstruction. To make the experimental setting reproducible and to provide a basis for new experimentation, we propose *CarPatch*, a new benchmark to assess neural radiance field methods on the *vehicle inspection* task. Specifically, we generate a novel dataset consisting of 8 different synthetic scenes, corresponding to as many high-quality 3D car meshes with realistic details and challenging light conditions. As depicted in Fig. 1, we provide not only RGB images with camera poses, but also binary masks of different car components to validate the reconstruction quality of specific vehicle parts (*e.g.* wheels or windows). Moreover, for each camera position, we generate the ground truth depth map with the double goal of examining the ability of NeRF architectures to correctly predict volume density and, at the same time, enable future works based on RGB-D inputs. We evaluate the novel view generation and depth estimation performance of several methods under diverse settings (both global and component-level). Finally, since the process of image collection for fitting neural radiance fields could be time consuming in real scenarios, we provide the same scenes by varying the number of training images, in order to determine the robustness to the amount of training data.

After an overview of the main related works in Sect. 2, we thoroughly describe the process of 3D mesh gathering, scene setup, and dataset generation in Sect. 3. The evaluation of existing NeRF architectures on *CarPatch* is presented in Sect. 4.

Table 1. Comparison between existing datasets used as benchmarks for neural radiance field evaluation and *CarPatch*. We provide the same scene by varying the amount of training data (40, 60, 80, and 100 images), allowing users to test the robustness of their architectures. We also release depth and segmentation data for all the images.

Dataset	Scenes	Images/scene	Depth	Segmentation
Blender [11]	8	300	✓	✗
Shiny Blender [17]	6	300	✓	✗
BlendedMVG [20]	508	200-4000	✗	✗
CarPatch$_{40}$	8	240	✓	✓
CarPatch$_{60}$	8	260	✓	✓
CarPatch$_{80}$	8	280	✓	✓
CarPatch$_{100}$	8	300	✓	✓

2 Related Work

We provide a brief overview of the latest updates in neural radiance field, including its significant extensions and applications that have influenced our work. NeRF limitations have been tackled by different works, trying to reduce its complexity, increase the reconstruction quality, and develop more challenging benchmarks.

Neural Scene Reconstruction. The handling of aliasing artifacts is a well-known issue in rendering algorithms. Mip-NeRF [1,2] and Zip-NeRF [3] have tackled the aliasing issue by reasoning on volumetric frustums along a cone. These approaches have inspired works such as Able-NeRF [16], which replaces the MLP of the original implementation with a transformer-based architecture. In addition to other sources of aliasing, reflections can pose a challenge for NeRF. Several works have attempted to address the issue of aliasing in reflections by taking into account the reflectance of the scene [4,5,17]. Moreover, computation is a widely recognized concern. Various works in the literature have demonstrated that it is possible to achieve high-fidelity reconstructions while reducing the overall training time. Two notable works in this direction include NSVF [10], which uses a voxel-based representation for more efficient rendering of large scenes, and Instant-NGP [13], which proposes a multi-resolution hash table combined with a light MLP to achieve faster training times. Other approaches such as DVGO [15] and Plenoxels [8] optimize voxel grids of features to enable fast radiance field reconstruction. TensoRF [7] combines the traditional CP decomposition [7] with a new vector-matrix decomposition method [6] leading to faster training and higher-quality reconstruction.

In this work, in order to satisfy real-time performances for vehicle inspection, we select a set of architectures that strike a balance between training time and the quality of the reconstruction.

Scene Representation Benchmarks. One of the most widely used benchmarks for evaluating NeRF is the Nerf Synthetic Blender dataset [11]. This

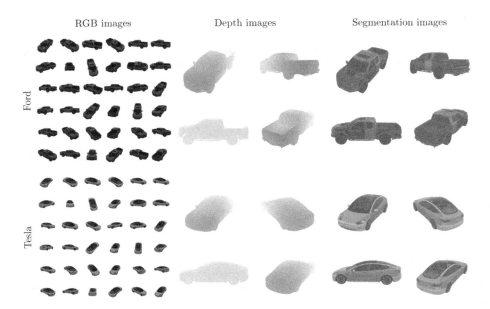

Fig. 2. Sample RGB images (left), depth data (center), and segmentation masks (right) from *CarPatch*, for different car models.

dataset consists of 8 different scenes generated using Blender[1], each with 100 training images and 200 test images. Other synthetic datasets include the Shiny Blender dataset [17], which mostly contains singular objects with simple geometries, and Blend DMVS [20], which provides various scenes to test NeRF implementations at different scales. These works do not provide ground truth information about the semantic meaning of the images. This limitation makes it difficult to study the ability of NeRF to reconstruct certain surfaces compared to others. In our *CarPatch* dataset, we provide ground truth segmentation of vehicle components in the scene, allowing for the evaluation of architectures on specific parts. Table 1 presents a comparison between the most common datasets used as benchmarks and our proposed dataset.

3   The *CarPatch* dataset

In this section, we detail the source data and the procedure exploited for generating our *CarPatch* dataset. In particular, we describe how we gathered 3D models, set up the blender scenes, and designed the image capture process (Fig. 2).

---

[1] http://www.blender.org.

Table 2. Summary of the source 3D models from which our dataset has been generated, including their key features.

Model name	Acronym	#Triangles	#Vertices	#Textures	#Materials
Tesla Model	TESLA	684.3k	364.4k	22	58
Smart	SMART	42.8k	26.4k	0	31
Ford Raptor	FORD	257.1k	156.5k	12	50
BMW M3 E46	BMW	846.9k	442.4k	7	39
Mercedes GLK	MBZ$_1$	1.3M	741.4k	0	15
Mercedes CLS	MBZ$_2$	1.0M	667k	0	18
Volvo S90	VOLVO	3.3M	1.7M	56	44
Jeep Compass	JEEP	334.7k	189.6k	7	39

3.1 Synthetic 3D Models and Scene Setup

All the 3D models included in *CarPatch* scenes have been downloaded from Sketchfab[2], a large collection of free 3D objects for research use. Table 2 provides a detailed list of all the starting models used. Each of them has been edited in Blender to enhance its realism; specifically, we improved the materials, colors, and lighting in each scene to create a more challenging environment.

The scenes have been set up accordingly to the Google Blender dataset [11]. The lighting conditions and rendering settings were customized to create a more realistic environment. The vehicle was placed at the center of the scene at position (0,0,0), with nine lights distributed around the car and varying emission strengths to create shadows and enhance reflections on the materials' surfaces. To improve realism, we resized objects to match their real-world size. The camera and lights were placed in order to provide an accurate representation of the environment, making the scenes similar to real-world scenarios.

3.2 Dataset Building

The dataset was built using the Python interface provided in Blender, allowing us to control objects in the environment. For each rendered image, we captured not only the RGB color values but also the corresponding depth map, as well as the pixel-wise semantic segmentation masks for eight vehicle components: bumpers, lights, mirrors, hoods/trunks, fenders, doors, wheels, and windows. Examples of these segmentation masks can be seen in Fig. 1. Please note that all the pixels belonging to a component (*e.g.* doors) are grouped into the same class, regardless of the specific component location (*e.g.* front/rear/right/left door). The bpycv[3] utility has been used for collecting additional metadata, enabling us to evaluate NeRF models on the RGB reconstruction and depth estimation of the overall vehicle as well as each of its subparts.

---

[2] https://sketchfab.com.
[3] https://github.com/DIYer22/bpycv.

For the rendering of training images, the camera randomly moved on the hemisphere centered in (0,0,0) and above the ground. The camera rotation angle was sampled from a uniform distribution before each new capture. For building the test set, the position of the camera was kept at a fixed distance from the ground and rotated around the Z-axis with a fixed angle equal to $\frac{2\pi}{\#test\_views}$ radians before each new capture.

In order to guarantee the fairness of the current and future comparisons, we explicitly provide four different versions of each scene, by varying the number of training images (40, 60, 80, and 100 images, respectively). Different versions of the same scene have no overlap in training camera poses, while the test set is always the same and contains 200 images for each scene.

We release the code for dataset creation and metrics evaluation at https:// github.com/davidedinuc/carpatch.

4   Benchmark

This section presents the selection and testing of various recent NeRF-based methods [7,13,15] on the presented *CarPatch* dataset, with a detailed description of the experimental setting for each baseline. Additionally, we assess the quality of the reconstructed vehicles in terms of their appearance and 3D surface reconstruction, utilizing depth maps generated during volume rendering.

4.1   Compared Methods

To overcome challenges related to illumination and reflective surfaces during the process of reconstructing vehicles, it is crucial to choose an appropriate neural rendering approach. We tested selected approaches on *CarPatch* without modifying the implementation details available in the original repositories, whenever possible. However, some parameters had to be adjusted in order to fit our models (which are larger compared to reference dataset meshes) to the scene. All tests were performed on a GeForce GTX 1080 Ti. After considering various NeRF systems, we have selected the following baselines:

- **Instant-NGP** [13] Since the original implementation of Instant-NGP is in CUDA, we decided to use an available PyTorch implementation[4] of this approach in order to have a fair comparison with the other approaches. In our experiments, a batch size of 8192 was maintained, with a scene scale of 0.5 and a total of 30,000 iteration steps.
- **TensoRF** [7] In our setting, a batch of 4096 rays was used. Additionally, we increased the overall scale of the scene from 1 to 3.5. These adjustments were made after experimentation and careful consideration of the resulting reconstructions. Training lasts 30,000 iterations.

---

[4] https://github.com/kwea123/ngp_pl.

Table 3. Quantitative results on the *CarPatch* test set for each vehicle model.

Method	Metric	Bmw	Tesla	Smart	Mbz$_1$	Mbz$_2$	Ford	Jeep	Volvo	Avg
iNGP [13]	PSNR↑	39.48	39.46	39.57	36.87	39.15	33.67	35.00	35.93	37.39
DVGO [15]		39.91	39.89	40.34	37.45	39.37	33.82	35.32	36.28	37.80
TensoRF [7]		40.68	39.92	40.38	38.07	40.84	34.33	34.87	36.77	**38.23**
iNGP [13]	SSIM↑	0.985	0.987	0.988	0.985	0.987	0.959	0.978	0.979	0.981
DVGO [15]		0.987	0.988	0.990	0.987	0.988	0.964	0.980	0.981	0.983
TensoRF [7]		0.989	0.987	0.99	0.989	0.991	0.966	0.975	0.982	**0.984**
iNGP [13]	LPIPS↓	0.029	0.029	0.02	0.028	0.024	0.062	0.036	0.032	0.032
DVGO [15]		0.022	0.022	0.014	0.019	0.020	0.051	0.029	0.022	**0.025**
TensoRF [7]		0.023	0.026	0.017	0.02	0.017	0.051	0.039	0.027	0.028
iNGP [13]	D-RMSE↓	0.640	0.369	0.377	0.496	0.500	0.406	0.558	0.674	0.503
DVGO [15]		0.561	0.353	0.305	0.437	0.454	0.339	0.469	0.561	**0.435**
TensoRF [7]		0.590	0.357	0.335	0.467	0.482	0.375	0.536	0.626	0.471
iNGP [13]	SN-RMSE↓	4.24	3.38	3.41	4.26	4.13	5.15	4.60	4.67	4.23
DVGO [15]		4.27	3.48	3.20	4.19	4.24	5.04	4.67	4.71	4.22
TensoRF [7]		3.96	3.24	3.10	4.00	3.91	4.91	4.41	4.48	**4.00**

– **DVGO** [15] In this work, the training process consists of two phases: a coarse training phase of 5,000 iterations, followed by a fine training phase of 20,000 iterations that aims to improve the model's ability to learn intricate details of the scene. In our experiments, we applied a batch size of 8192 while maintaining the default scene size.

4.2 Metrics

The effectiveness of the chosen methods has been assessed thanks to the typical perceptual metrics used in NeRF-based reconstruction tasks, namely PSNR, SSIM [18], and LPIPS [21].

However, the appearance-based metrics are strongly related to the emitted radiance besides the learned volume density. We suggest two supplementary depth-based metrics for the sole purpose of assessing the volume density. Since it is not feasible to obtain ground truth 3D models of the vehicles in real-world scenarios, we utilize the depth map as our knowledge of the 3D surface of the objects. Specifically, we define a depth map as a matrix

$$D = \{d_{ij}\}, d_{ij} \in [0, R] \tag{1}$$

in which each value d_{ij} ranges from 0 to the maximum depth value R. Furthermore, we estimate the surface normals from the depth maps [14]. Initially, we establish the orientation of a surface normal as:

$$\mathbf{d} = \langle d_x, d_y, d_z \rangle = \left(-\frac{\partial d_{ij}}{\partial i}, -\frac{\partial d_{ij}}{\partial j}, 1 \right) \approx \left(d_{(i+1)j} - d_{ij}, d_{i(j+1)} - d_{ij}, 1 \right) \tag{2}$$

Table 4. Quantitative results on the *CarPatch* test set for each vehicle component averaged over the vehicle models.

Method	Component	PSNR↑	SSIM↑	LPIPS↓	D-RMSE↓	SN-RMSE↓
iNGP [13]	*bumper*	33.05	0.986	0.019	0.281	0.79
DVGO [15]		34.41	0.989	0.011	**0.236**	0.72
TensoRF [7]		**35.49**	**0.991**	**0.010**	0.311	**0.68**
iNGP [13]	*light*	28.71	0.993	0.009	0.421	0.48
DVGO [15]		29.10	0.995	**0.006**	**0.384**	0.43
TensoRF [7]		**29.68**	**0.996**	**0.006**	0.438	**0.38**
iNGP [13]	*mirror*	29.60	0.994	0.011	0.427	0.43
DVGO [15]		31.16	**0.996**	**0.007**	**0.345**	**0.38**
TensoRF [7]		**31.68**	**0.996**	0.008	0.372	0.39
iNGP [13]	*hood/trunk*	32.28	0.977	0.052	0.260	1.33
DVGO [15]		32.68	0.981	**0.038**	**0.259**	1.35
TensoRF [7]		**33.75**	**0.983**	0.040	0.302	**1.24**
iNGP [13]	*fender*	32.44	0.990	0.021	0.253	0.87
DVGO [15]		33.55	**0.993**	**0.013**	**0.223**	0.85
TensoRF [7]		**34.36**	**0.993**	0.015	0.267	**0.77**
iNGP [13]	*door*	34.19	0.969	0.079	0.182	0.67
DVGO [15]		35.48	0.977	**0.042**	**0.173**	0.74
TensoRF [7]		**36.25**	**0.979**	0.051	0.191	**0.62**
iNGP [13]	*wheel*	33.12	0.995	0.008	0.391	0.87
DVGO [15]		33.65	0.995	0.006	**0.267**	**0.79**
TensoRF [7]		**34.55**	**0.996**	**0.005**	0.334	**0.79**
iNGP [13]	*window*	26.44	0.897	0.166	0.879	2.52
DVGO [15]		26.54	**0.899**	**0.147**	**0.779**	2.57
TensoRF [7]		**26.74**	0.896	0.160	0.834	**2.38**

where the first two elements represent the depth gradients in the i and j directions, respectively. Afterward, we normalize the normal vector to obtain a unit-length vector $\mathbf{n}(d_{ij}) = \frac{\mathbf{d}}{\|\mathbf{d}\|}$.

We assess the 3D reconstruction's quality through the following metrics:

- **Depth Root Mean Squared Error (D-RMSE)** This metric measures the average difference in meters between the ground truth and predicted depth maps.

$$\text{D-RMSE} = \sqrt{\frac{\sum_{i=0}^{M} \sum_{j=0}^{N} (\hat{d}_{ij} - d_{ij})^2}{M \cdot N}} \tag{3}$$

- **Surface Normal Root Mean Squared Error (SN-RMSE)** This metric measures the average angular error in degrees between the angle direction of

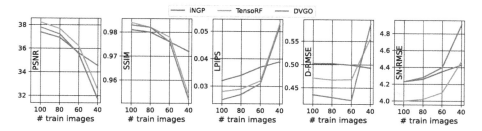

Fig. 3. Performance by varying the number of training images, in terms of PSNR, SSIM, LPIPS, D-RMSE, and SN-RMSE. Despite its lower overall performance, Instant-NGP [13] exhibits low variance with respect to the amount of training data.

the ground truth and predicted surface normals.

$$\text{SN-RMSE} = \sqrt{\frac{\sum_{i=0}^{M} \sum_{j=0}^{N} (\arccos(\mathbf{n}(\hat{d}_{ij})) - \arccos(\mathbf{n}(d_{ij})))^2}{M \cdot N}} \qquad (4)$$

D-RMSE and SN-RMSE are computed only for those pixels with a positive depth value in both GT and predicted depth maps. This avoids computing depth estimation errors on background pixels (which have a fixed depth value of 0).

4.3   Results

The following section presents both quantitative and qualitative results obtained from the selected NeRF baselines. We will discuss their performance on the *CarPatch* dataset, by analyzing the impact of viewing camera angle and the number of training images.

According to Table 3, all the selected NeRF approaches obtain satisfying results. Although the baselines demonstrate similar performances in terms of appearance scores (PSNR, SSIM, and LPIPS), our evaluation using depth-based metrics (D-RMSE and SN-RMSE) reveals significant differences in the 3D reconstruction of the vehicles. DVGO outperforms its competitors by achieving better depth estimation, resulting in a +13.5% improvement compared to iNGP and a +7.6% improvement compared to TensoRF. In contrast, TensoRF predicts a more accurate 3D surface with the lowest angular error on the surface normals.

Since our use case is related to vehicle inspection, in Table 4 we report results computed on each car component. For this purpose, we mask both GT and predictions using a specific component mask before computing the metrics. However, this would lead to an unbalanced ratio between background and foreground pixels, due to the limited components' area, and finally to a biased metric value. By computing D-RMSE and SN-RMSE only on foreground pixels (see Sect. 4.2), depth-based metrics are not affected by this issue. For PSNR, SSIM, and LPIPS, instead, we compute component-level metrics over the image crop delimited by the bounding boxes around each mask. As expected, it is worth noting that NeRF

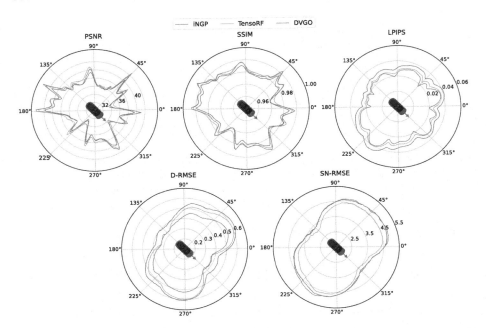

Fig. 4. Performance by camera viewing angle, in terms of PSNR, SSIM, LPIPS, D-RMSE, and SN-RMSE. Depending on the training camera distribution, all the methods struggle wherever the viewpoints are more sparse (*e.g.* between 225° and 270°). The red arrow represents where the front of the vehicle is facing. (Color figure online)

struggles to reconstruct transparent objects (*e.g.* mirrors, lights, and windows) obtaining the highest errors in terms of depth and normal estimation. However, over the single components, TensoRF outperforms the competitors in most of the metrics and in particular on the surface normal estimation. The errors in the reconstruction of specific components' surfaces can also be appreciated in the qualitative results of Fig. 5.

Moreover, we analyze the performances of each method in terms of the number of training images. We trained the baselines on every version of the *CarPatch* dataset and report the results in Fig. 3. It is worth noting that reducing the number of training images has a significant impact on all the metrics independently of the method. However, Instant-NGP demonstrates to be more robust to the number of camera viewpoints having a smoother drop, especially in terms of LPIPS, D-RMSE, and SN-RMSE.

Finally, we discuss how the training camera viewpoints' distribution around the vehicle may affect the performance of each method from certain camera angles. In particular, as depicted in Fig. 4, it is evident how between 180° and 270° and between 0° and 45° there are considerable variations in the metrics. Indeed, in these areas the datasets contain more sparsity in terms of camera viewpoints and, as expected, all the methods are affected.

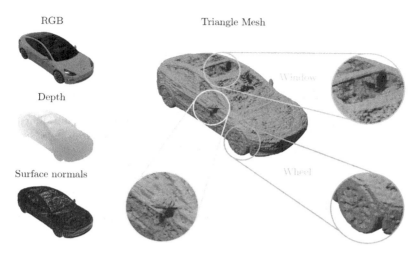

Fig. 5. Sample of 3D reconstruction of the TESLA: (left) the reconstructed RGB, depth, and surface normals, (right) the reconstructed surfaces on the triangle mesh.

5 Conclusion

In this article, we have proposed a new benchmark for the evaluation and comparison of NeRF-based techniques. Focusing on one of the many concrete applications of this recent technology, i.e. *vehicle inspection*, a new synthetic dataset including renderings of 8 vehicles was first created. In addition to the set of RGB views annotated with the camera pose, the dataset is enriched by semantic segmentation masks as well as depth maps to further analyze the results and compare the methods.

The presence of reflective surfaces and transparent parts makes the task of vehicle reconstruction still challenging. Proposed additional metrics, as well as new graphical ways of displaying the results, are proposed to make these limitations more evident. We are confident that *CarPatch* can be of great help as a basis for research on NeRF models in general and, more specifically, in their application to the field of vehicle reconstruction.

Acknowledgements. The work is partially supported by the Department of Engineering Enzo Ferrari, under the project FAR-Dip-DIEF 2022 "AI platform with digital twins of interacting robots and people".

References

1. Barron, J.T., Mildenhall, B., Tancik, M., Hedman, P., Martin-Brualla, R., Srinivasan, P.P.: Mip-NeRF: a multiscale representation for anti-aliasing neural radiance fields. In: Proceedings of the IEEE/CVF ICCV (2021)
2. Barron, J.T., Mildenhall, B., Verbin, D., Srinivasan, P.P., Hedman, P.: Mip-NeRF 360: unbounded anti-aliased neural radiance fields. In: Proceedings of the IEEE/CVF Conference on CVPR (2022)

3. Barron, J.T., Mildenhall, B., Verbin, D., Srinivasan, P.P., Hedman, P.: Zip-NeRF: anti-aliased grid-based neural radiance fields. arXiv:2304.06706 (2023)
4. Bi, S., et al.: Neural reflectance fields for appearance acquisition. arXiv:2008.03824 (2020)
5. Boss, M., Jampani, V., Braun, R., Liu, C., Barron, J., Lensch, H.: Neural-PIL: neural pre-integrated lighting for reflectance decomposition. In: Advances in Neural Information Processing Systems, vol. 34 (2021)
6. Carroll, J.D., Chang, J.J.: Analysis of individual differences in multidimensional scaling via an n-way generalization of "Eckart-Young" decomposition. Psychometrika **35**(3), 283–319 (1970)
7. Chen, A., Xu, Z., Geiger, A., Yu, J., Su, H.: Tensorf: tensorial radiance fields. In: Avidan, S., Brostow, G., Cissé, M., Farinella, G.M., Hassner, T. (eds.) ECCV 2022. LNCS, vol. 13692, pp. 333–350. Springer, Cham (2022). https://doi.org/10.1007/978-3-031-19824-3_20
8. Fridovich-Keil, S., Yu, A., Tancik, M., Chen, Q., Recht, B., Kanazawa, A.: Plenoxels: radiance fields without neural networks. In: Proceedings of the IEEE/CVF Conference on CVPR (2022)
9. Geiger, A., Lenz, P., Urtasun, R.: Are we ready for autonomous driving? The kitti vision benchmark suite. In: Proceedings of the IEEE/CVF Conference on CVPR. IEEE (2012)
10. Liu, L., Gu, J., Zaw Lin, K., Chua, T.S., Theobalt, C.: Neural sparse voxel fields. In: Advances in Neural Information Processing Systems, vol. 33 (2020)
11. Mildenhall, B., Srinivasan, P.P., Tancik, M., Barron, J.T., Ramamoorthi, R., Ng, R.: NeRF: representing scenes as neural radiance fields for view synthesis. Commun. ACM **65**(1), 99–106 (2021)
12. Müller, N., Simonelli, A., Porzi, L., Bulò, S.R., Nießner, M., Kontschieder, P.: AutoRF: learning 3D object radiance fields from single view observations. In: Proceedings of the IEEE/CVF Conference on CVPR (2022)
13. Müller, T., Evans, A., Schied, C., Keller, A.: Instant neural graphics primitives with a multiresolution hash encoding. ACM Trans. Graph. (ToG) **41**(4), 1–15 (2022)
14. Pini, S., Borghi, G., Vezzani, R., Maltoni, D., Cucchiara, R.: A systematic comparison of depth map representations for face recognition. Sensors **21**(3), 944 (2021). https://doi.org/10.3390/s21030944
15. Sun, C., Sun, M., Chen, H.T.: Direct voxel grid optimization: super-fast convergence for radiance fields reconstruction. In: Proceedings of the IEEE/CVF Conference on CVPR (2022)
16. Tang, Z.J., Cham, T.J., Zhao, H.: ABLE-NeRF: attention-based rendering with learnable embeddings for neural radiance field. arXiv:2303.13817 (2023)
17. Verbin, D., Hedman, P., Mildenhall, B., Zickler, T., Barron, J.T., Srinivasan, P.P.: Ref-NeRF: structured view-dependent appearance for neural radiance fields. In: Proceedings of IEEE/CVF Conference on CVPR (2022)
18. Wang, Z., Bovik, A.C., Sheikh, H.R., Simoncelli, E.P.: Image quality assessment: from error visibility to structural similarity. IEEE Trans. Image Process. **13**(4), 600–612 (2004)
19. Xie, Z., Zhang, J., Li, W., Zhang, F., Zhang, L.: S-NeRF: neural radiance fields for street views (2023)
20. Yao, Y., et al.: BlendedMVS: a large-scale dataset for generalized multi-view stereo networks. In: Proceedings of the IEEE/CVF Conference on CVPR (2020)
21. Zhang, R., Isola, P., Efros, A.A., Shechtman, E., Wang, O.: The unreasonable effectiveness of deep features as a perceptual metric. In: Proceedings of the IEEE/CVF Conference on CVPR (2018)

Obstacle Avoidance and Interaction in Extended Reality: An Approach Based on 3D Object Detection

Matteo Martini, Fabio Solari, and Manuela Chessa

Department of Informatics, Bioengineering, Robotics, and Systems Engineering,
University of Genoa, Genoa, Italy
matteo.martini@edu.unige.it, {fabio.solari,manuela.chessa}@unige.it

Abstract. When immersed in Virtual Reality (VR), real objects around us could be dangerous if the users collide with them. On the other hand, such objects could be exploited to improve VR setups and to create Extended Realities where real and virtual elements are coherently mixed up, and interaction is enhanced by passive haptics. This work proposes a system that combines state-of-the-art 3D object detection, depth data from stereo cameras, and VR environment rendering. Our system can detect specific classes of objects, estimate their position and extent, and create a virtual counterpart for each one in an immersive virtual scene. We describe the pipeline of the system and show that our system allows real-time interaction with virtual objects blended with real ones.

Keywords: Virtual Reality · Obstacle avoidance · Human-Computer Interaction · YOLO · stereo camera · passive haptics

1 Introduction

One of the main goals of Virtual Reality (VR) is to immerse the user in a computer-generated 3D environment, excluding the stimuli from the real world. Recent head-mounted displays (HMDs) have robust and efficient tracking systems that allow users to explore VR and act inside it. Moreover, HMDs are generally equipped with some safety system that aims to avoid potential injuries to the users due to the fact that they cannot perceive the real environment while they are wearing the device. Two examples are the *Chaperone* and the *Guardian* safety systems, developed respectively by Valve[1] and Meta[2]. The idea behind them is similar: users, before starting to use any VR application, have to go through a room setup procedure, which lets them define what are the boundaries of the area to be considered walkable and, in general, usable for the simulations, as well as the floor level. From the end of the procedure onwards, the system will constantly track the position of the HMD and the controllers

---

[1] https://www.valvesoftware.com/.
[2] https://www.meta.com/.

G. L. Foresti et al. (Eds.): ICIAP 2023, LNCS 14234, pp. 111–122, 2023.
https://doi.org/10.1007/978-3-031-43153-1_10

in space, checking if they lie in the safe zone, near the border, or even away from it. The two utilities differ in how they visually alert the users that they are going outside the safe area. Nevertheless, none of the two systems offers obstacle detection features, and everything that enters the safe zone after its definition is ignored. For example, if, during the simulation, another person walks around the users, they will never be alerted by either. Moreover, these systems cannot detect obstacles in a realistic home environment. Living rooms or bedrooms may be cluttered with furniture and other objects that are not easily removable or that users will not remove, making it easy for them to hit or trip over something. Other approaches in the literature to perform real-obstacle avoidance while immersed in VR are Hartmann et al.'s RealityCheck [3], Beever and John's *Leveled SR* [1] and Valentini et al.'s system [7]. All of them include the obstacles present in the room in the virtual experience of the users, following different approaches.

In RealityCheck [3], the authors use a Realsense RGB-D camera and a set of Kinectv2s to acquire the real-time depth data of the room, and they use them to augment in different ways, altering the regular rendering pipeline, the virtual world that the user perceives. In the end, they get a significantly more powerful Chaperone-like system that can be tuned to work in different modalities, but that still presents some critical issues. In particular, the system has a high computational cost, and users perceive the additional visual cues related to the real environment without much contextualization in the virtual world, thus potentially losing the sense of presence.

In [1], the authors describe a Mixed Reality (MR) system that allows users to define customized virtual worlds which are aware of objects and obstacles in the room where they are playing, having as a firm point the fact that it should be able to be used by anyone, without needing to invest money in particular equipment beyond the cost of the headset kit. The process starts with defining the obstacles in the room that can be carried out in MR through a smartphone, manually defining the occupation of each one and subsequent automatic mesh segmentation. The generated data is then used to create virtual levels, also substituting the real objects with virtual ones, exploiting their physicality. The system can deliver a higher sense of presence when users play the composed level. However, the object detection procedure has to be carried out manually and does not support dynamic environments.

In [7], the authors aim to enhance users' awareness of the real world around them, procedurally instantiating meshes that are proportional in size to real obstacles which may be present in the room. The first phase of their system consists of the acquisition of a 3D model of the selected room. Then the obtained mesh undergoes a processing phase which, using one of the strategies proposed, generates a voxelization of the room, given a resolution with which to discretize. Voxels are then grouped by closeness into clusters that likely correspond to whole objects. Smaller clusters are discarded, while bigger ones are substituted with individual cuboids of similar volume. Lastly, the cuboids are swapped with the actual meshes that users will see, choosing from a predefined pool and picking that best fits the extent of the cuboid. The system produces good results but, similarly to Leveled SR, works only with static setups and does not react to room changes.

To achieve something more effective, we need to shift to other tools offered by device manufacturers. Specifically, solutions such as Meta's *Scene* and Microsoft's *Spatial Mapping* and *Scene Understanding* can be used. The first allows users to manually define the walls of the room and bounding boxes around specific elements within it. Objects are categorized with a fixed set of nine labels, like "table" or "door", or simply labeled as "other", if none of the proposed types fits them. The information entered, if requested, can be exploited by applications, to instantiate virtual counterparts. The method works, but it is neither automatic nor dynamic. The APIs provided by Microsoft, on the other hand, utilize the sensors of HoloLens to generate a mesh of the environment around the user and analyze its composition. They are primarily used to enable realistic interactions between holograms and real-world elements, considering occlusions and physical aspects such placement and navigation of characters. When activated, the scene understanding component operates automatically. However, it is limited to identifying basic elements such as walls, ceilings, floors and flat surfaces, rather than individual objects.

In this paper, we propose an approach to solve the obstacle avoidance problem and allow interaction with real objects, coherently mixing real and virtual environments to generate an Extended Reality (XR). With respect to the previous works in the literature, our system requires a lower effort for the setup phase and does not rely on a static reconstruction of the room. We exploit state-of-the-art object detection to locate the obstacles we are interested in, and, using depth data coming from a stereo camera, we can obtain their spatial occupation and position. Then, we assign whatever mesh fits well the virtual environment of the in-use application to each class of objects and use them to instantiate virtual obstacles in the scene. Being rendered and not coming from a video stream or a live mesh reconstruction of the environment, they appeared as well as the other virtual elements. This should generate in the users a higher sense of presence with respect to Chaperone-like solutions. Moreover, if the mesh associated with the class of obstacles is different in appearance but similar in shape to the real object, users can also exploit the natural passive haptics provided by the real room, moving pure VR applications to an XR setting.

2   Tools

Before discussing the implementation of the system and its design, we describe here the hardware and software tools we used. It is worth noting that our approach could be extended to similar classes of hardware devices.

In this paper, we use the ZED Mini by Stereolabs Inc.[3], a model characterized by its compact size, meant to be used especially in VR scenarios. Being a stereo camera, it allowed us to acquire not only RGB frames but also the depth data

---

[3] https://www.stereolabs.com.

of the framed scene. Thanks to the ZED SDK, we are then able to compute the position and the size of the objects in world coordinates.

YOLO is the object detection system we relied on to extract semantic information from the frames captured by the ZED camera. It is now a well-known algorithm and it is often chosen for real time and wearable systems (see [2] for a review). The model was originally presented in [4] by Redmon and Farhadi, and then refined first in [5] and then in [6]. In particular, we used YOLOv5, one of the later versions proposed by Ultralytics LLC and suggested by Stereolabs to be used together with their SDK. In order to reduce YOLO's inference time and to optimize the computations performed on each frame, we included into the project also NVIDIA's TensorRT[4]. We used it to recreate YOLO's neural network so that it can be easily and efficiently run on the GPU of the workstation, offloading the CPU and reducing the computation time spent to process each image.

For the management of the virtual environment, we relied on Unreal Engine 5[5], and for its visualization, we used the HTC Vive HMD. The system runs on an AMD Ryzen 9 5900X 12 core/24 thread, with NVIDIA GeForce RTX 3080 Ti (12 GB), and 32 GB of DDR4 RAM, with operating system Ubuntu 20.04 LTS.

3   System Design

Our goal is to create a system that combines the YOLO detector with the capabilities offered by the ZED camera to extract semantic information from the video stream and use it to augment the virtual scene perceived by the user with meaningful models. We managed to do that by defining a four-step pipeline whose blocks have a specific task and can be tuned individually depending on the application's needs.

The first step consists of the creation of a dataset of images to be used to train a YOLO model. This block can be skipped whenever we decide for any reason (saving time, variety of the samples, etc.) to use a pre-built dataset. The second block is the YOLO model setup. The output of this block is a TensorRT engine that can be used to locate objects in images. The third step groups all the operations done from the setup of the ZED camera, to the generation of a three-dimensional bounding box for each one of the detected objects in the current frame. The output data is sent to Unreal Engine through a TCP connection as a string message. The last step consists of all the processing needed to maintain a pool of virtual obstacles that are constantly updated, added, or removed to represent the perception of the ZED camera of the scene as faithfully as possible. The final output is given by rendering the virtual world, augmented with the virtual approximations of the real obstacles that the user can perceive looking inside an HMD.

---

[4] https://developer.nvidia.com/tensorrt.
[5] https://www.unrealengine.com.

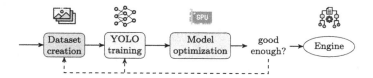

Fig. 1. The offline phase of the system.

The pipeline blocks are neither executed with the same speed, nor in a perfectly sequential way. First of all, the execution is divided into two separate phases, an *offline* and a *real-time* one. Being an offline preparatory step, the offline phase can be executed a priori as many times as necessary until we obtain a model that satisfies our expectations. As soon as this condition is met, it has not to be repeated anymore, unless we want to apply substantial modifications to the model, like for example adding a new class of obstacles. The real-time phase instead has the goal of providing the user frame rate not less than the native HMD one, so all the operations have to be done fast enough in such a way that their results arrive in time for the next frame composition. Moreover, the object detection and the scene management tasks are executed in parallel and as soon as the first produces some data, the second ingests it and computes an according variation to the scene. In the following, we provide details on each block.

3.1 Offline Phase

In Fig. 1 we have a schematic representation of what happens within the offline phase. The first operation is the *dataset creation*. The image collection should be rich enough so that the model can acquire a deeper understanding of the object. Images are then manually annotated and used to train the YOLO model. Since ours is a proof of concept, we have decided to track only specific instances of objects and, therefore, we have built ad hoc datasets. However, for more general applications, it would be sensible to use larger datasets to maximize the number of recognizable classes. We performed training and testing running the Python scripts that Ultralytics provides on its GitHub page relative to YOLOv5. The model has different declensions, whose principal ones are named with sizes from *Nano* to *eXtra large*. After carrying out various tests, we decided to stick with the *Large* size for our cases. We used a capable workstation, so the size and the computational cost of the inference were not a problem, while the stability and reliability of the bounding boxes were pivotal. The selected size turned out to be the best compromise. We trained our models starting from the weights of another one, provided by Ultralytics, which was pre-trained using the COCO dataset. This allowed us to shorten the training times and to start from models that are already trained to recognize objects in general.

Lastly, the conversion from the trained network to TensorRT engine was done by extracting the weights associated with the nodes of the model's network and

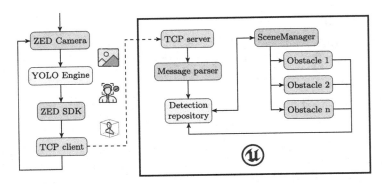

Fig. 2. The real-time phase of the system.

recreating the same exact structure in terms of NVIDIA's SDK. For the network structure, we relied on the YOLOv5 7.0 layout from Wang Xinyu's GitHub repository[6]. This last step returns an engine that can be tested on a real ZED video stream in order to evaluate its performance. If its capabilities are good enough we can keep it and use it in the real-time phase, otherwise, we can loop back and edit either the dataset or the training parameters with the purpose of obtaining a better model.

3.2   Real-Time Phase

As soon as we have a proper model converted into a TensorRT engine we can start the real-time phase. This part is parallelized in many aspects in order to execute all the needed tasks together without creating bottlenecks that can lower the frame rate with a consequent lack of responsiveness. Figure 2 shows all the elements that collaborate when the system is launched.

On the left side, there is the *object detector*: this task is executed in a separate process that, after the initial camera setup and the deserialization of the YOLO engine, for each frame repeats the same sequence of actions. First, it grabs a frame from the ZED camera and performs a little processing needed to obtain an image that is accepted by the TensorRT engine, fixing the size and the color space. Then, the adjusted frame is given to the engine, and the inference procedure is performed asynchronously on the GPU. When it is over, all the bounding box proposals coming from YOLO are acquired moving them from the GPU memory. A non-maximum suppression function filters them to retain just the most meaningful ones. The procedure discards everything with a particularly low confidence score and/or which seems to be a possible duplicate proposal for an already identified object. The survivors are then given to the ZED SDK. This last returns for each of the inputs an extensive set of estimations, from which only some are considered in our case. In particular, we are interested in the unique ID

---

[6] https://github.com/wang-xinyu/tensorrtx.

that allows the tracking of the object across the frames, the label stating what is the object class that YOLO assigned to the content of the detection area, the 3D bounding box, and its centroid. This information is packaged into a message that a TCP client takes care of sending to the corresponding server within the Unreal Engine scene, with which it has previously established a connection.

The last part of the real-time phase is the *scene management*. When the Unreal Engine project is launched, the TCP server starts a separate thread that, asynchronously from the other parts, constantly checks for incoming messages from the client. As soon as one comes in, its content is immediately parsed, and a corresponding detection object is created and filled with the details. The server maintains a detection repository that stores all the latest received updates for each object into a data structure. When a new detection is created, it is inserted there as a new entry or it is used to update the values of an old one, if it is already present.

The scene manager deals with the instantiation and the destruction of the virtual obstacles, while each obstacle is autonomous in terms the position and extent updates. They periodically poll the data structure to check if its current associated values are the same as the ones stored in it and, if the ones in the repository are newer, they update themselves.

4   Results

We preliminary tested our system in two different usage scenarios. The first task was an obstacle avoidance test. However, the pipeline we designed is quite versatile and in principle could be applied to many applications. In order to demonstrate that, we applied it to an object manipulation task and the results obtained were quite promising.

4.1   Obstacle Avoidance

The first scene is set in a 3 by 4.5 m portion of our lab, limited by two walls on one side and two tables on the other. In the middle space, we positioned a few objects with the aim of generating a cluttered environment where it would be very difficult to move while using a VR application and no obstacle detection system enabled. As obstacles, we used a stool, a floor lamp, a school bag, and a couple of nearly identical chairs (see Fig. 3). Every class of objects provided a different challenge to the system. The stool is composed of a simple wooden top and a set of four thin and chromed legs, which being so slim, depending on the framing of the scene, can be difficult to be separated visually from the background and to be registered in the depth map, if the resolution is not high enough. The backpack is quite regular in shape, but it is visually quite different depending on which side it is oriented. The lamp is tall and has a base that is completely different from the lampshade: we used it as a test for objects located further back and to see if YOLO was capable of considering correctly also the base. Lastly, we included the pair of chairs in order to have a case with multiple

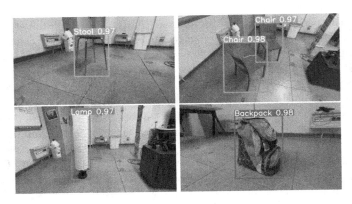

Fig. 3. The obstacles used in the first test and a sample of the YOLO detection.

instances of a class in the scene and to test how well the detection tracking would work. The goal of this test was to check whether the users were able to navigate autonomously in the cluttered space trusting only the elements shown in the virtual scene, having also the possibility of moving them around.

We generated the image dataset of the obstacle collection by shooting different videos using a smartphone. In each one, we focused on a particular object which was positioned in the middle of the area and we circled around it in order to obtain samples from all the points of view. While doing this we kept the other objects near, so that they remained in the background, adding more samples of them in different contexts and in a different scale. The final result is a dataset composed of 927 images and 1449 annotations that is quite balanced across the classes, with the number of chairs a little over the average. The images that compose it have been divided with the proportion 70%-20%-10%, respectively, into the training, validation, and test sets. The detections produced by the model which was trained using them are pretty solid and they are confirmed both by the execution on the test set and on frames coming directly from the ZED camera. On the test set all the object instances were detected (100% of true positives, with an accuracy score equal or greater than 90% for 97% of them) and only two false positives were generated. Figure 4 contains the plots of the losses over the training procedure, that was iterated for 60 epochs and using a batch size of 7 images, as well as the confusion matrix obtained on the test set with the final model.

Figure 5 shows some snapshots of the virtual environment created from the detection of the four classes of obstacles (first column). In the middle column, simpler meshes have been used to replace the obstacles, in order to highlight their spatial occupation. However, chairs for example have no physical part in the upper half, except for the backrest and this can be confusing when touching objects. In the right column, to give users more coherent feedback we also defined the second model association map using more realistic meshes (in this case,

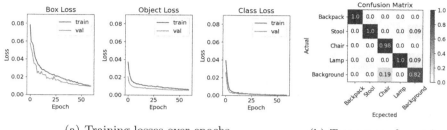

(a) Training losses over epochs

(b) Test set conf. matrix

Fig. 4. Training and test data of the YOLO model for obstacles.

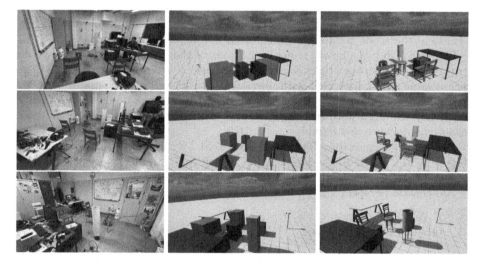

Fig. 5. Example of obstacle detection and mapping in the VR environment.

meshes similar to the corresponding real objects). Also, here[7] and here[8] you can observe two corresponding videos showing the system in action.

The users were able to navigate through the cluttered area without stumbling upon objects, having the possibility of moving the real objects while visually perceiving the virtual ones shifting coherently.

The main limitation is that the ZED SDK does not recover the full 6-DOF pose of the targets and, as for the 2D bounding boxes, if the object is rotated the ZED SDK will just enlarge the detection volume to continue to contain it and this generates a general overestimation of the obstacle's size. Aside from that, the system provided qualitative good results, generating reliable virtual twins of the objects for which it was trained. The most challenging class to be correctly

---

[7] https://youtu.be/cKD5Pj3w_Ps.

[8] https://youtu.be/1NYZric_NlA.

Fig. 6. The objects used in the second test and a sample of the YOLO detection.

reconstructed by the system was the stool. All the items having legs and an empty space between them got reasonably into trouble the ZED SDK, which sometimes struggled to find the correct distance of the objects from the camera. This manifested with some jittering of the virtual obstacles and, particularly in the case of the stool, with a constant, often non-negligible, error in depth.

The test also brought out some limitations of the system overall. The whole setup relies on just one static camera, so occlusions are probably the most significant issue. Having to deal with obstacles with legs, the overlapping between objects was rarely complete and, accomplice the good work done by the YOLO model, objects often were still detected and instantiated in the scene, but their measurements were less reliable.

The user occlusion problem could be solved by attaching the ZED camera directly to the HMD instead of keeping it still and framing the world from a fixed point of view, though such a solution would arise other problems due to the ego-centric perspective.

4.2   Object Manipulation

In the second test, the defined task was quite simple. Three different soda cans, shown in Fig. 6, were positioned on a desk and the user was requested to move them around and swap their position trusting only the virtual counterparts, generated by our detector and visualized in an Unreal Engine scene. In this case, the camera was directly attached in front of the HTC Vive through our custom 3D printed mount.

We proceeded in the same way as described before, starting from the dataset generation and we ended up with a balanced dataset composed of 657 images and 1181 annotations, subdivided again with the 70%-20%-10% scheme. Also this time we trained a YOLOv5 network, starting the training from a COCO pre-trained model. We obtained again pretty solid detections, and in Fig. 7 you can find the training losses and the confusion matrix of the final model on the test set. The cans were correctly labeled and the confidence scores are consistent.

The system behaved correctly and as expected it instantiated the correct meshes in the position of the real cans. The manipulation of the objects is definitely possible, and the cans can be picked up and moved on the table by the user. The major issue, in this case, was given by the strong occlusions produced by the arms of the users that can get in the way between the camera and the cans

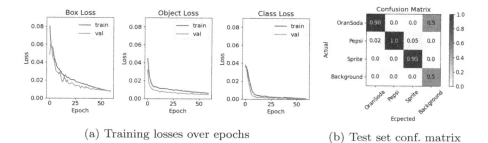

(a) Training losses over epochs   (b) Test set conf. matrix

Fig. 7. Training and test data of the YOLO model for soda cans.

Fig. 8. Examples of object manipulation and mapping in the VR environment.

when people try to reach items located on the opposite side of the field of view. Reasonably, this cannot be prevented because it is part of the natural way people grasp objects. The occlusions generated by the hands and the fingers instead have generally not been a problem. In Fig. 8 you can find a comparison of the real and the reconstructed scene, while here[9] you can watch the corresponding video demonstration.

5   Conclusions and Future Work

In this paper, we presented an XR system for obstacle avoidance and interaction, based on the state-of-the-art object detector YOLO and the stereopsis-based capabilities of the ZED camera. We preliminary tested the software in two different scenarios. In the first one, users have to move around in a cluttered space and they need to be opportunely informed on what and where the obstacles are with respect to their position, in order to move accordingly and avoid injuries. The second one is a proof of concept about the versatility of the developed system, to show that it can be used to allow interaction with objects in XR. In

---

[9] https://youtu.be/10g3zlHGqDw.

particular, the system allows natural object manipulation with a virtual counterpart to exploit the benefits of passive haptics. In both cases, we obtained interesting results: the detection and mapping of the objects is coherent (100% of true positives in the test set, with an accuracy score equal or greater than 90% for 97% of them), and the XR environment is enjoyable. This opens further discussions and future improvements.

The main upgrade from which the system would benefit most would surely be stepping from the basic extended 2D to 3D bounding boxes to a full 6-DOF pose tracking. The benefits provided by this variation can be numerous, of which as an example we mention only greater reliability in estimating the spatial occupation and a more realistic positioning of the virtual meshes, especially in the case of object manipulation. Moving more on the technical side, a few objectives are worth mentioning, e.g., the porting of the Linux project to the Windows platform. This would solve several issues about the use of the HTC Vive HMD, and would also expand the basin of integrable devices.

After that, a quantitative evaluation of the system would allow us to compute the error in the 6-DOF localization of the objects, that would affect interaction capabilities. Moreover, user studies to quantify both obstacle avoidance in dynamic situations and interaction will be our main future goals.

References

1. Beever, L., John, N.W.: Leveled SR: a substitutional reality level design workflow. In: IEEE Conference on Virtual Reality and 3D User Interfaces, VR 2022, Christchurch, New Zealand, 12–16 March 2022, pp. 130–138. IEEE (2022)
2. Ghasemi, Y., Jeong, H., Choi, S.H., Park, K.B., Lee, J.Y.: Deep learning-based object detection in augmented reality: a systematic review. Comput. Ind. **139**, 103661 (2022)
3. Hartmann, J., Holz, C., Ofek, E., Wilson, A.D.: Realitycheck: blending virtual environments with situated physical reality. In: Brewster, S.A., Fitzpatrick, G., Cox, A.L., Kostakos, V. (eds.) Proceedings of the 2019 CHI Conference on Human Factors in Computing Systems, CHI 2019, Glasgow, Scotland, UK, 04–09 May 2019, p. 347. ACM (2019)
4. Redmon, J., Divvala, S.K., Girshick, R.B., Farhadi, A.: You only look once: unified, real-time object detection. CoRR (2015)
5. Redmon, J., Farhadi, A.: YOLO9000: better, faster, stronger. In: 2017 IEEE Conference on Computer Vision and Pattern Recognition, CVPR 2017, Honolulu, HI, USA, 21–26 July 2017, pp. 6517–6525. IEEE Computer Society (2017)
6. Redmon, J., Farhadi, A.: Yolov3: an incremental improvement. CoRR (2018)
7. Valentini, I., Ballestin, G., Bassano, C., Solari, F., Chessa, M.: Improving obstacle awareness to enhance interaction in virtual reality. In: IEEE Conference on Virtual Reality and 3D User Interfaces, VR 2010, Atlanta, GA, USA, 22–26 March 2020, pp. 44–52. IEEE (2020)

HMPD: A Novel Dataset for Microplastics Classification with Digital Holography

Teresa Cacace[1], Marco Del-Coco[2(✉)], Pierluigi Carcagnì[2],
Mariacristina Cocca[3], Melania Paturzo[1], and Cosimo Distante[2]

[1] Institute of Applied Science and Intelligent System ISASI, National Research
Council of Italy, Pozzuoli, Italy
[2] Institute of Applied Science and Intelligent System ISASI, National Research
Council of Italy, Lecce, Italy
`marco.delcoco@cnr.it`
[3] Institute for Polymers Composites and Biomaterials, National Research Council of
Italy, Pozzuoli, Italy

Abstract. Microfiber shedding caused by the washing of synthetic textiles is widely recognized as a major source of environmental microplastic pollution. Such a phenomenon pushed the scientific community to develop ever better methodologies for the detection and the identification of synthetic microfibers dispersed in water involving, mainly because of the microscopic nature of the elements under analysis, a wide range of competencies ranging from instrumental acquisition to data analysis. This work deploys a large dataset of holographic images of microplastic and non-microplastic samples carefully labeled by experts. Data acquisition starts from video sequences collected by means of a holographic microscope recording the flow of water samples; then, holography postprocessing and object detection are performed to retrieve the patches of interest. Finally, classification tests exploiting deep learning strategies have been performed to validate the proposed dataset. The dataset and the benchmarking code are available at: https://github.com/beppe2hd/HMPD.

Keywords: dataset · microplastic · holography · deep learning

1 Introduction

Among the wide range of pollutant elements affecting the environment, microplastics (MPs) have been attracting a growing interest in recent years [1], mainly due to the high concentration found in animals or humans. In other words, other than hurting the ecosystem, microplastic affects our food chain and health.

Microfiber shedding caused by the washing of synthetic textiles has been identified as one of the primary MPs sources in the environment [2]. Other than that, many studies highlighted a strong correlation between the number of MPs released during a wash and the influence of different washing treatments or textile characteristics [7,9,13].

G. L. Foresti et al. (Eds.): ICIAP 2023, LNCS 14234, pp. 123–133, 2023.
https://doi.org/10.1007/978-3-031-43153-1_11

Different methodologies have been employed to detect and identify synthetic MPs dispersed in water [9]. A preparation phase and successive visual inspection by expert users are usually mandatory [6,7,13]. Unfortunately, such procedures are not applicable for continuous monitoring of the environment task due to their intrinsically time-consuming modality. Indeed, expert operators are demanded to perform the sample preparation procedure, implemented in a strictly controlled environment, and then to proceed with a visual inspection devoted to disentangling MPs from other pollutant factors or elements naturally present in the sample.

Therefore, a portable system, capable to operate "out of the box" a secure sample handling procedure and an automated microparticle identification and counting is highly desirable.

In this work, we introduce a large sized dataset of *microplastic/non-microplastic* images retrieved using a low-cost holographic microscope. More precisely, the acquisition has been performed by exploiting a direct pipeline, including an initial water sample acquisition using the holographic microscope and the subsequent detection processing devoted to retrieving the image patches of the particles in the sample under test.

An overview of the pipeline devoted to the dataset generation is reported in Fig. 1. The holographic microscope records MPs samples dispersed in water and flowing through a microchannel. The device implements the architecture for off-axis Digital Holography (DH), a non-destructive, label-free, and single-shot imaging technique that provides full-field information on the samples. The outcomes are the flowing samples' raw Amplitude and Phase video sequences. The *Object Detector* leverage on image processing techniques that take care of recognizing the flowing object and isolating it as a stand-alone image. Finally, careful labeling has been performed by experts in the environmental and optical fields that classified each patch as *microplastic* or *non-microplastic*. The employed water samples come from wastewater from a washing test of polyester textiles in a commercial washing machine. Nevertheless, 3 robust classifiers, usually asking for large annotated datasets, have been trained and tested to provide a classification baseline for the proposed dataset.

Such a dataset could play a pivotal role in enhancing specific classification procedures capable of enabling additional processing (i.e., MPs density estimation) if assembled with the described patches retrieving process. In addition, the availability of a large-sized labeled dataset can leverage the use of SOTA detection and classification algorithms such as YOLO [14].

The remainder of the paper is organized as follows: Sect. 2 describes the experimental setup devoted to the dataset creation, annotation procedure, and validation, while Sect. 3 goes through a detailed description of the annotated dataset highlighting the characteristics of the employed water samples and Sect. 4 reports results of classification performed with some mainstream deep-learning networks. Finally, Sect. 5 concludes the paper.

2   Experimental Set-Up

The construction of the dataset is performed using a straightforward experimental pipeline (Fig. 1), including holography analysis, object detection and storing, and an annotation step enabling the field experts to conduct a fast and reliable labeling process.

The pipelines start with the acquisition step performed by the *Holographic Microscope*; the water sample is pumped through the microscope microchannel, and the raw video sequence is acquired and processed by returning the *amplitude, phase*, and *raw* video outputs. The phase component is then provided to the *Object Detection* module to detect flowing objects. Detected objects are extracted as stand-alone patches (amplitude, phase, and raw components) and stored in a database with the relative metadata. At this point, the stored data can be labeled to create a dataset suitable for supervised training procedures.

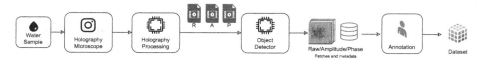

Fig. 1. Experimental set-up for dataset generation and annotation.

2.1   Digital Holography Microscope

We have employed a custom-made compact and portable DH design for image acquisitions, which implements off-axis holography. This imaging technique records the interference between the light scattered by the sample (object wave) and a clean reference wave, which is then numerically processed to yield amplitude images and quantitative phase maps of the sample. It allows fast, dynamic, non-destructive, and label-free specimen measurement. Our setup implements the off-axis configuration through a wavefront division design [3–5] - the final device, described in detail in ref. [5], is structured to enable fast substitution of the commercial PMMA microfluidic chip where the samples flow when necessary.

A pressure-driven pump is connected to the microfluidic chip, keeping the samples flowing while recording holographic videos. The flowing velocity is regulated to observe every object on several frames, while the width of the channel (1 mm) is recorded entirely by using a low magnification of about 4×. The collected videos are numerically processed frame-by-frame: every frame (hologram) undergoes a standard procedure for hologram demodulation based on linear filtering in the Fourier domain. An amplitude image and the corresponding quantitative phase map [4] are reconstructed from each frame. In addition, a reference hologram, obtained as the mean over the whole video, is used to remove the effect of optical aberrations by subtraction. Finally, the processed frames are recombined to produce two separate video sequences, one of amplitude (equivalent to bright-field) images and another of phase maps.

2.2   Object Detection and Data Storage

Object detection is based on a Gaussian Mixture Model [15] (GMM) foreground detector. It is performed on the phase video sequence due to its high contrast appearance, which maximizes the detection capabilities of the algorithm. The algorithm discriminates foreground from background pixels using a continuously updated model based on a mixture of Gaussian's representative of a non-stationary background. The pixels that do not fit the model are considered to be foreground. This approach is particularly suitable for the proposed application, where the background is subject to slight fluctuations due to residual vibrations during the recordings. Once the foreground pixels have been detected, connected components are found, grouping them into objects which are then defined by a bounding box. Overlapping bounding boxes are merged together to avoid object fragmentation caused by miss-detection. More precisely, the GMM available in *opencv* python library, with a threshold on the variance set to 6 and a history length of 100 frames, has been employed. Additionally, a morphological close (available on *opencv* with a kernel size of 5) is applied to foreground pixels. Finally, the connected components are retrieved and labeled through the special functions of the scikit-image Python library. Other than that, all the patches below the threshold area of 100 pixels are discarded to avoid small noisy elements. A phase video frame, with detected objects, is depicted in Fig. 2.

Finally, raw, amplitude, and Phase data are saved in a database that stores the reference frame and bounding box coordinates for each bounding box. The data storage has been then exploited to facilitate the annotation and the generation of multiple datasets. The labeling procedure has been done by two experts in the environmental and optical fields.

Fig. 2. Example of a frame (phase channel) showing the detected flowing objects.

2.3   Classification

The ability to discern between MPs and non-MPs objects is a classification problem. In the past years, handcrafted features represented the standard de facto to provide visual patches description to classification algorithms. Such approaches are still employed nowadays when the complexity of the problem (problem size and variability in the patterns) does not ask for frontier processing like the ones using Deep Learning paradigm that rapidly invested most of the computer vision field thanks to their excellent flexibility and performances. Anyway, previous analysis shows that straightforward approaches are not sufficient to retrieve satisfying classification results.

Against this background, 3 well-known convolutional neural network (CNN) architectures, in their lighter version, have been considered to produce a baseline benchmark for classifying *microplastic* or *non-microplastic* objects. In such a way, we exploited the ability of CNNs to retrieve the most significant features while keeping low both the architectural complexity and the over-fitting issues. Going into detail, the employed networks are:

- **AlexNet**: comes out with the work presented in [11], and it is among the first architectures introducing the use of *Relu* instead of *Tanh* to add non-linearity and make use of dropout instead of regularisation to deal with over-fitting. It can be considered one of the most important baselines in classification problems. In this work, the final output layer has been modified to provide a binary output.
- **VGG**: is a typical CNN exhibiting numerous layers and introduced in [12] (the abbreviation VGG stands for Visual Geometry Group). The original architecture consists of 16 layers, with the input layer sized at a resolution of 224 by 224 and 1000 category output layer. In this work, a reduced version with 11 layers and a binary output has been employed.
- **ResNet**: architectures [10] try to overcome the well known degradation problem, i.e., the impossibility of increasing the network performances simply by stacking layers and the resulting vanishing gradient problem. ResNet architectures introduce the *residual learning framework*, which exploits shortcut identity connections between convolutional layers (*Conv Block*) to maximize information flow and reinvigorate gradient flow for weights adaptation. In this work, the lightest version, named ResNet-18, has been employed with binary output.

3   Dataset

One of the main scopes of this work is to provide the scientific community with a new dataset devoted to the MPs classification and made publicly available at: https://github.com/beppe2hd/HMPD along with the benchmark code employed in Sect. 4 The sample of wastewater employed for the dataset was obtained from the washing process of synthetic fabric performed at a real scale. The washing test was performed using a *Bosch washing machine Serie 4 Varioperfect*

WLG24225i, at 40 °C, for 107 min and 1200 rpm, using a commercial liquid detergent in the recommended doses, and a washing load of 2–2.5 kg made of 100% polyester t-shirts. The wastewater coming directly from the drainpipe of the washing machine is subjected to a multi-step filtration procedure. For this study, we have selected an aliquot of wastewater collected after passing through a nylon net filter with a 20 μm pore size (Merck millipore). This selection prevents any clogging of the microchannel, which has a rectangular section of 1.0×0.2 mm. Finally, to increase the density of MPs and speed up the measurement process, the sample volume has been halved by evaporation under reduced pressure in a rotary evaporator.

It is worth noting as changing washing parameters such as washing machine, temperature, and detergent can affect the number of microfibers released [7]. Other than that, comparative studies between different textiles show a similar morphology in the released microfibers, except for natural fibers and blends, which present higher levels of entanglement [8].

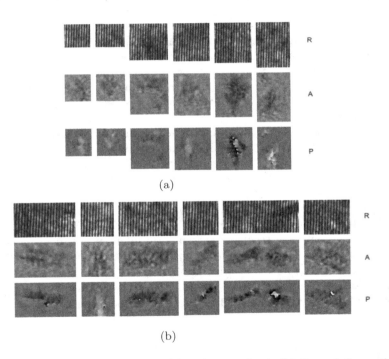

(a)

(b)

Fig. 3. Examples of *non-microplastic* (a) and *microplastic* (b). In each figure the rows are respectively about the raw (R), amplitude (A), and phase (P) components of the objects.

Two different 3 ml aliquots of the described wastewater sample have been provided to the Digital Holography Microscope, producing two videos for 5422 frames.

The object detection process extracted 11285 objects, and for each, raw holo-gram, amplitude, and Phase were saved as separated patches. The annotation process has been done by two experts that independently labeled each patch as *microplastic* (3312) and *non-microplastic* (8140); patches provided with un-matching label categories have been considered as *possible-microplastics* (980). In case of difficulties in identifying the objects, the experts employed numeri-cal propagation. This procedure, commonly used in Digital Holography, calcu-lates the optical field at used-imposed distances from the acquisition plane, thus enabling numerical, a-posteriori refocusing of the objects. In dubious cases, the spot containing the uncertain object has been propagated, providing additional information for the labeling procedure.

The final dataset, named Holography Micro-Plastic Dataset (HMPD), has been made considering all the 3293 positive and 3293 negative samples, randomly selected among the *non-microplastic*, to obtain a well-balanced dataset. A sec-ond version of the dataset, named HMPD-P, including the *possible-microplastics* occurrences as positive samples, has been considered with a cardinality of 4260 positive and 4260 random negative samples.

Additionally, a folding procedure that shuffles and randomly splits the ele-ments producing 5 different training/testing (80% 20% respectively) configura-tions, has been applied to favor comparison among future proposals. Each con-figuration has been stored and provided with the dataset allowing at the same time a standard k-fold validation procedure and the use of fixed training/testing file lists.

Figure 3 shows some examples (by columns) of *non-microplastic* and *microplastic* patches where rows report raw hologram, amplitude, and phase channels, respectively. A careful examination of the figure reveals that it is pretty complicated to discern some samples, if not for an expert operator provided with broader context information (i.e., the full video sequence), which category they belong to. Therefore it becomes evident that an automatic classification is a mandatory, as well as challenging, task.

In Fig. 4 are depicted the patches area distributions for both the *non-microplastic* and *microplastic* classes, whereas in Table 1 are reported the max-imum, minimum, and average area and maximum and minimum values of the height and width of the patches. All data is referred to the HMPD version of the dataset. Both patch examples and area distributions show as, although most of the *non-microplastic* samples are characterized by a small-sized section, the area of a non-negligible part of them is comparable with the area of most of the *microplastic* elements making this features a non-discriminating factor.

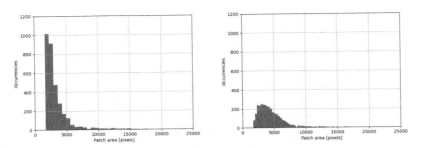

Fig. 4. Statistical distribution of the patches area: *non-microplastic* samples in the left histogram and *microplastic* samples in the right histogram.

Table 1. HMPD dataset statistics: count represents the number of available samples; successively minimum, maximum and average patches areas in pixels are reported, whereas the last four columns report the minimum and maximum values for height and width, respectively.

label	count	min-area	max-area	avg-area	min h	max h	min w	max w
non-microplastic	3293	1681	66276	3833.50	41	263	41	299
microplastic	3293	1681	31980	4705.02	41	214	41	211

4   Baseline Benchmarks

To provide a baseline for the proposed dataset and prove its significance, the architectures listed in Subsect. 2.3 have been tested on the HMPD version of the dataset. More precisely, the train/test configuration mentioned in sub-section 3 has been employed, and the Receiving Operator Curves (ROC), average values of accuracy, precision, recall, and F1-score have been computed. All the networks have been providexxx d with patch images resized at 80 × 80 pixels, and trained for 25 epochs with a learning rate = 0.00001 and a batch size of 32. Moreover, to minimize the over-fitting issues, data augmentation has been employed: patches have been resized at a size of 100 × 100 pixels, randomly cropped at the desired 80 × 80 final shape and successively randomly rotated in a range −90 to 90 degrees; finally, a random horizontal flip has been performed. All the tests have been performed on the three considered channels: raw hologram, amplitude, and phase.

Results have been resumed in Table 2 and by means of ROCs depicted in Fig. 5. It is clear that the raw channel cannot provide useful information, leading all the tested networks to return unsatisfying performances. On the other hand, amplitude and phase channels led to an accuracy exceeding 0.809 with a non-negligible advantage of the phase channel on the amplitude one. Similar behaviour can be observed in the ROC plot, where the curves related to the phase channel are slightly better than the ones computed by exploiting the amplitude channel. This is consistent with the examples provided in Fig. 3 where the phase channel exhibits the higher contrast, whereas the raw channel appears as the noisiest one. Regarding networks, VGG and AlexNet get the best performances gathering a non-negligible advantage on the classification capabilities of ResNet.

Table 2. Values of the tested networks' average accuracy, recall, precision, and F1-score. A, P, and R stand respectively for Amplitude, Phase, and Raw.

network	channel	accuracy	recall	precision	F1
AlexNet	A	0.824	0.875	0.793	0.831
AlexNet	P	0.837	0.842	**0.834**	0.838
AlexNet	R	0.705	0.682	0.708	0.696
ResNet18	A	0.809	0.857	0.782	0.817
ResNet18	P	0.835	0.850	0.826	0.838
ResNet18	R	0.701	0.713	0.697	0.704
VGG11	A	0.840	0.868	0.821	0.844
VGG11	P	**0.855**	**0.893**	0.829	**0.867**
VGG11	R	0.704	0.642	0.733	0.685

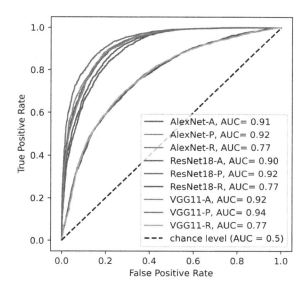

Fig. 5. ROC curves for HMPD dataset on raw, amplitude and phase channels.

To evaluate the influence of *possible-microplastic* samples on the proposed dataset, the experiments have been repeated on HMPD-P; for brevity, only the phase channel has been considered. Once again, the results are provided in the form of ROCs, accuracy, precision, recall, and F1 score reported respectively in

Fig. 6 and Table 3. A quick comparison of the results highlights as most of the metrics decrease when compared with their correspondent of the HMPD version of the dataset. Such a behaviour could be due to the intrinsic uncertainty of the introduced samples that do not provide additional information. A similar trend can be found by observing the ROC results of Fig. 6, where both the results for HMPD and HMPD-P are reported.

Table 3. HMPD-P values of average accuracy, recall, precision and F1-score of the tested networks. P stands for Phase.

network	channel	accuracy	recall	precision	F1
AlexNet	P	0.837	**0.870**	0.816	**0.843**
ResNet18	P	0.820	0.838	0.81	0.824
VGG11	P	**0.839**	0.821	**0.853**	0.836

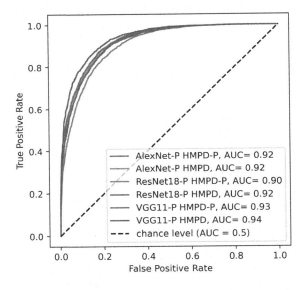

Fig. 6. ROC curves for HMPD and HMPD-P on phase channel.

5   Conclusions

In this work, a new dataset devoted to the classification of *microplastic* vs. *non-microplastic* objects acquired using a holographic microscope has been deployed and made publicly available (https://github.com/beppe2hd/HMPD). Moreover, some streamlined CNNs have been employed to provide a baseline for the classification problem. Experiments showed that the phase channel is the most discriminating one, highlighting the key contribution of holography analysis. Other than that, it has been proven that the problem can be satisfyingly treated with small sized CNN networks and that some architecture strategies, such as the residual block ones, bring marginal advantage to more straightforward solutions. It is worth noting that the dataset enables the use of more recent approaches, such as YOLO [14] that perform the detection and the classification in a unique step, providing more reliable results, usually with real time processing capabilities. In light of this, future works will be devoted to testing SOTA detection and classification solutions, eventually increasing the dataset dimensionality and verifying the possible advantage of using multiple focal planes.

References

1. Agache, I., et al.: Climate change and global health: a call to more research and more action. Allergy **77**(5), 1389–1407 (2022)
2. Boucher, J., Friot, D.: Primary Microplastics in the Oceans: A Global Evaluation of Sources, vol. 10. Iucn Gland, Switzerland (2017)
3. Cacace, T., et al.: Compact off-axis holographic slide microscope: design guidelines. Biomed. Opt. Express **11**(5), 2511–2532 (2020)
4. Cacace, T., et al.: Compact holographic microscope for imaging flowing microplastics. In: 2021 International Workshop on Metrology for the Sea; Learning to Measure Sea Health Parameters (MetroSea), pp. 229–233. IEEE (2021)
5. Cacace, T., et al.: Compact holographic imaging and machine learning for microfibers quantification in laundry wastewater. In: Proceedings of the International Conference on Microplastic Pollution in the Mediterranean Sea, Springer (2023). (in press)
6. Corami, F., Rosso, B., Bravo, B., Gambaro, A., Barbante, C.: A novel method for purification, quantitative analysis and characterization of microplastic fibers using micro-FTIR. Chemosphere **238**, 124564 (2020)
7. De Falco, F., Di Pace, E., Cocca, M., Avella, M.: The contribution of washing processes of synthetic clothes to microplastic pollution. Sci. Rep. **9**(1), 6633 (2019)
8. Dreillard, M., et al.: Quantification and morphological characterization of microfibers emitted from textile washing. Sci. Total Environ. **832**, 154973 (2022)
9. Gaylarde, C., Baptista-Neto, J.A., da Fonseca, E.M.: Plastic microfibre pollution: how important is clothes' laundering? Heliyon **7**(5), e07105 (2021)
10. He, K., Zhang, X., Ren, S., Sun, J.: Deep residual learning for image recognition. In: Proceedings of the IEEE Conference on Computer Vision and Pattern Recognition, pp. 770–778 (2016)
11. Krizhevsky, A., Sutskever, I., Hinton, G.E.: Imagenet classification with deep convolutional neural networks. Commun. ACM **60**(6), 84–90 (2017)
12. Simonyan, K., Zisserman, A.: Very deep convolutional networks for large-scale image recognition. arXiv preprint arXiv:1409.1556 (2014)
13. Volgare, M., et al.: Washing load influences the microplastic release from polyester fabrics by affecting wettability and mechanical stress. Sci. Rep. **11**(1), 19479 (2021)
14. Wang, C.Y., Bochkovskiy, A., Liao, H.Y.M.: Yolov7: trainable bag-of-freebies sets new state-of-the-art for real-time object detectors. arXiv preprint arXiv:2207.02696 (2022)
15. Zivkovic, Z.: Improved adaptive gaussian mixture model for background subtraction. In: Proceedings of the 17th International Conference on Pattern Recognition, 2004. ICPR 2004, vol. 2, pp. 28–31. IEEE (2004)

Early Detection of Hip Periprosthetic Joint Infections Through CNN on Computed Tomography Images

Francesco Guarnera[1]([✉]) [iD], Alessia Rondinella[1] [iD], Oliver Giudice[1] [iD],
Alessandro Ortis[1] [iD], Sebastiano Battiato[1] [iD], Francesco Rundo[2] [iD],
Giorgio Fallica[3], Francesco Traina[4], and Sabrina Conoci[5] [iD]

[1] Department of Mathematics and Computer Science, University of Catania,
Catania, Italy
francesco.guarnera@unict.it, alessia.rondinella@phd.unict.it,
{giudice,ortis,battiato}@dmi.unict.it
[2] STMicroelectronics, ADG R&D Power and Discretes Division, Catania, Italy
francesco.rundo@st.com
[3] IBMTech s.r.l., via Napoli 116, Catania, Italy
giorgio.fallica@ibmtech.it
[4] University of Bologna, Rizzoli Orthopedic Institute of Bologna, Bologna, Italy
francesco.traina@ior.it
[5] Department ChBioFarAm, University of Messina, Messina, Italy
sabrina.conoci@unime.it

Abstract. Early detection of an infection prior to prosthesis removal (e.g., hips, knees or other areas) would provide significant benefits to patients. Currently, the detection task is carried out only retrospectively with a limited number of methods relying on biometric or other medical data. The automatic detection of a periprosthetic joint infection from tomography imaging is a task never addressed before. This study introduces a novel method for early detection of the hip prosthesis infections analyzing Computed Tomography images. The proposed solution is based on a novel ResNeSt Convolutional Neural Network architecture trained on samples from more than 100 patients. The solution showed exceptional performance in detecting infections with an experimental high level of accuracy and F-score.

Keywords: Periprosthetic Joint Infection detection · Hip Arthoplasthy · Medical Imaging · Artificial Intelligence

1 Introduction

The surgical replacement of human joints has become increasingly common in recent years as a treatment to pathologies like osteoarthritis or rheumatoid arthritis. However, the Periprosthetic Joint Infection (PJI) that unfortunately occurs around a joint implant, still represents a serious concern for patients

G. L. Foresti et al. (Eds.): ICIAP 2023, LNCS 14234, pp. 134–143, 2023.
https://doi.org/10.1007/978-3-031-43153-1_12

and physicians [1]. A PJI can lead to pain, joint dysfunction, and the need for revision surgery, which typically pose an increased potential for further complications for the patients, as well as additional costs [2]. To address this challenge, medical imaging plays a critical role in the early and accurate detection of PJI. Traditionally, PJI detection is carried out retrospectively by means of a combination of methods relying on biometric or other medical data [3] which could be time-consuming and limited in term of detection accuracy. To counter this, Machine Learning (ML) techniques could train on past data and be applied to PJI detection, with the aim of improving the accuracy of diagnosis [4,5]. Certainly, the use of Computed Tomography (CT) images provided several advantages in the diagnosis of infection [6], giving that the periosteal response should be a strong prior when the infection is present [7]. With this hypothesis, the information obtained from CT scans could be employed as input to supervised ML approaches to develop models able to accurately predict PJIs, reducing the risk of recurrent infections and improving patient outcomes. However, the PJI is particularly challenging to be diagnosed, even for physicians, but recent evolution in Convolutional Neural Network (CNN) architecture have proven great results in medical imaging [9] giving inspiration to this work. In this paper, a Convolutional Neural Network (CNN) solution for the classification of infected and aseptic patients with hip replacements by analyzing CT scans is presented [8].

The proposed solution exploits the attention mechanism of recent ResNeSt [10] architecture to proper extract features from CT scans and build a predictive model that can accurately differentiate between infected and non-infected (aseptic) patients. A private dataset acquired at Rizzoli Orthopedic Institute (IOR) was employed for training and final evaluation of the solution (approved by ethical committee). The dataset was manually labeled by expert radiologists. The obtained results showed the effectiveness of the proposed solution in the PJI detection task, demonstrating its potential to improve overall medical outcomes. This highlights the importance of medical imaging in the diagnosis of PJI and underscores the potential of machine learning techniques in this area.

The remainder of this paper is organized as follows: Sect. 2 briefly presents related works; Sect. 3 details the proposed solution and Sect. 4 shows the experimental results with discussion. Conclusions are summarized in Sect. 5.

2   Related Works

The state of the art (SOTA) methods for the detection of PJIs employ mainly statistical methods, such as regression and Fisher test. In [11] a risk prediction model is proposed for PJI detection within 90 days after surgery using Least Absolute Shrinkage and Selection Operator (LASSO) regression analysis [12]. Other approaches propose the use of the Fisher's test, to detect the infection from CT images at the site of hip prosthesis before surgery [13,14]. While these methods demonstrated to be effective, they are limited in capturing complex relationships between imaging features thus better understanding the PJI status. For instance, many patients' problems, such as mechanical loosening, can

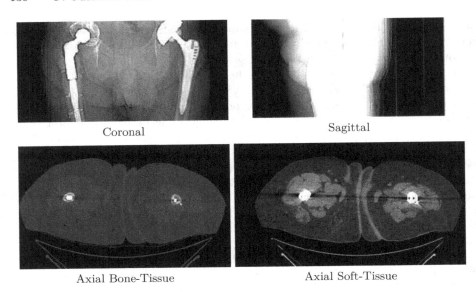

Coronal   Sagittal

Axial Bone-Tissue   Axial Soft-Tissue

Fig. 1. Examples of acquisitions in CT.

be detected only retrospectively by studying the bone-prosthesis system and its evolution [15]. Only a limited number of studies exploited machine learning (ML) techniques. This is likely due to issue concerning datasets: availability of reliable labeling, variability in image quality and in patients samples, etc. However, ML approaches have the potential to significantly enhance the accuracy of PJI detection. Several studies have demonstrated the potential of deep learning approaches for PJI prediction, with promising results in terms of accuracy, sensitivity, and specificity. The authors of [16] proposed a ResNet architecture to classify pathological sections of patients with PJI achieving high accuracy. In [5], a ML approach was proposed to anticipate recurrent infections in patients who have undergone revision total knee arthroplasty due to a PJI. In some cases, the performance of these models has been shown to be superior to traditional diagnostic methods that rely solely on clinical inspection and laboratory tests. To the best of the authors' knowledge, as far as hip CT images are concerned, there are no SOTA methods able to early detect PJIs. This specific medical imaging problem is still in its early stages and there are several obstacles that need to be overcome: the most important one is the scarcity of large datasets. Given this, a specific dataset of hip CT scan images was collected and labeled by experts. Thus, a novel CNN solution was introduced in order to advance the state of the art in early PJI detection task, achieving the first automatic approach with promising high level of accuracy.

3   Methodology

The CT scans are a widely used medical imaging data providing detailed cross-sectional images of internal structures of a human body. In the context of hip replacements, they play an important role in detecting a PJI. The axial plane imaging mode has the ability to highlight both bone and soft tissue (see Fig. 1), thus facilitating the identification of a orthopedic joint and bone infections.

3.1   Hip CT Scans Image Dataset

At first, patients who underwent hip replacement surgery and subsequent CT scans for various reasons, including postoperative follow-up, suspected PJI, or routine monitoring were identified at Rizzoli Orthopedic Institute of Bologna. As infection is not easily detectable, and there is no standard protocol for medical experts to identify it. However, physicians are able to detect infection clues in proximity to the bone borders, with particular attention to the acetabular and femoral area. Finally, they are able to confirm the presence of infection by means of cultures: this is a retrospective method. In order to have a first early detection solution, a dataset consisting of 102 CT scans of patients with hip replacements was collected (52 samples of an infected patient and 50 aseptic ones). Given the before mentioned insight, only images regarding the axial bone tissue were employed. The labeling of this dataset posed a further challenge: the labels had to be assigned at patient level. In other words, each patient is labeled as either being infected or aseptic, rather than each individual CT scan image. Thus, all images from a specific patient will be labeled as infected or aseptic. Also each CT scan contains a different number of images per patient. These are big issues for ML solutions as it means that no explicit information is available about the presence or absence of a PJI on each individual images. All collected images are in Digital Imaging and Communication in Medicine (DICOM) format[1].

3.2   Pre-processing Pipeline

In order to identify what images contain the prosthesis, every pixel of the images have been converted through the Hounsfield scale, as follows:

$$h_p = p * s + i \tag{1}$$

where p is a pixel value, s and i are the slope and the intercept respectively contained in the DICOM metadata, and h_p is the value of the pixel in the Hounsfield scale. The Hounsfield scale is a quantitative measure for describing radio-density which is the property of relative transparency of a material to the passage of the X-ray portion of the electromagnetic spectrum. Images with $h_p > 3000$ were selected as being images presenting non-human material like a metallic prosthesis [17]. This allowed the selection of images coming from the upper part of the acetabulum to the lower part of the stem only.

---

[1] https://www.dicomstandard.org/.

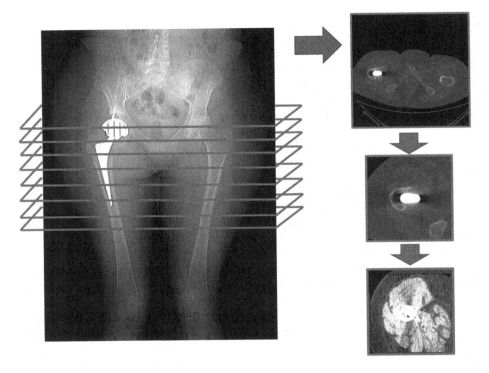

Fig. 2. Pre-processing pipeline: Slice selection based on Hounsfield selection, contour-based ROI detection and final histogram equalization.

A further preprocessing operation was carried out extracting the contours of the bone relative to the prothesis [18] on images having $h_p > 1000$. Once the contours were extracted, the corresponding centroid was computed and a sub-image of 188×188 pixels was extracted. Finally, histogram equalization was carried out on these patches to adjust the contrast and to normalize different image histograms. The overall pre-processing operations are graphically summarized in Fig. 2.

3.3   The Proposed Approach for PJI Detection

In this section the proposed CNN-based solution for the early detection of hip PJI detection is described. A ResNeSt architecture [10] was employed to classify each data sample (CT scan images from a patient with hip replacements) between infected or aseptic classes. The ResNeSt architecture is an evolution of the ResNet neural network [19], incorporating split-attention blocks instead of residual blocks, thus emphasizing the correlation of informative feature maps. ResNeSt was selected for being the best on the specific task, after testing other similar architectures. More specifically, performances obtained by the ResNet and the ResNeXt architectures [20] were compared, However, the ResNeSt network architecture, incorporating split-attention mechanisms, demonstrated superior results in both the validation set and the test set. The other evaluated architectures, although achieving similar levels of accuracy, demonstrated poor generalization performances, as indicated by their "randomic behaviour" on heatmap analysis. The proposed approach takes input images of size 188×188 that have undergone pre-processing, as described in Sect. 3.2. The images are then processed within the architecture and the loss function is computed based on the comparison between the output predictive values and truth labels of the data. The loss function employed was the cross entropy and moreover, in order to improve the stability of the model, the Jacobian regularization was applied [21]. Jacobian regularization permitted to reduce the impact of input perturbations caused by metallic artifacts in the prostheses. The use of ResNeSt with the addition of regularization techniques allowed to exploit the strengths of this state-of-the-art network architecture to extract features from CT scans and accurately classify infected and aseptic patients with hip replacements (with explainable non-randomic heatmaps).

4   Experiments and Results

As outlined in Sect. 3.1, the dataset collected and employed for experiments is composed of 50 aseptic and 52 infected patients. To evaluate the generalizing properties of the proposed solution, a balanced test set (D_x) of 12 patients (6 aseptic and 6 infected) was extracted from the full dataset. Additionally, to effectively control overfitting during the training process, we employed four distinct balanced validation sets (D_{v1}, D_{v2}, D_{v3}, and D_{v4}), each consisting of 12 different patients. The remaining 42 patients were considered as D_t. The following configurations were considered for the training phase:

- C_1: train $\{D_t \cup D_{v2} \cup D_{v3} \cup D_{v4}\}$ and valid D_{v1};
- C_2: train $\{D_t \cup D_{v1} \cup D_{v3} \cup D_{v4}\}$ and valid D_{v2};
- C_3: train $\{D_t \cup D_{v1} \cup D_{v2} \cup D_{v4}\}$ and valid D_{v3};
- C_4: train $\{D_t \cup D_{v1} \cup D_{v2} \cup D_{v3}\}$ and valid D_{v4}.

Training was performed for 100 epochs, employing ADAM as optimizer with a starting learning rate of $1e-4$, a weight decay equal to $1e-4$ and a batch size fixed at 4. The proposed approach was implemented in Python language (version 3.9.7) using the Pytorch package. All experiments were done on a NVIDIA Quadro RTX 6000 GPU. The overall training procedure took 50 h.

Tables 1 and 2 shows the results obtained applying the best model obtained for each configuration and tested on D_x. As explained in Sect. 3.3 the proposed technique makes a classification (positive or infected vs. negative or aseptic) for each image of a patient; this implies that the model is independent to the number of images per patient, as each patient has a different number of images based on their length of the prosthesis (third column of Tables 1) and 2. It has to be noted that, in order to accurately classify a patient as infected or aseptic through the images predicted by the model, a threshold strategy must be employed. The

Table 1. Accuracy and F-score obtained on C_1, C_2 configurations employing the same test set D_x.

Patient number	Patient type	Number images	Accuracy / F-score configuration C_1	Accuracy / F-score configuration C_2
1	aseptic	67	0.88/0.94	0.88/0.93
2	aseptic	39	0.66/0.8	0.41/0.58
3	aseptic	69	0.85/0.85	0.76/0.87
4	aseptic	68	0.52/0.69	0.85/0.92
5	aseptic	77	0.79/0.88	0.79/0.88
6	aseptic	53	0.96/0.98	1/1
7	infected	80	0.46/0.63	0.51/0.68
8	infected	115	1/1	1/1
9	infected	191	0.98/0.99	0.91/0.96
10	infected	71	0.18/0.31	0.13/0.20
11	infected	77	0.57/0.72	0.89/0.94
12	infected	117	0.95/0.98	0.96/0.98

Table 2. Accuracy and F-score obtained on C_3, C_4 configurations employing the same test set D_x.

Patient number	Patient type	Number images	Accuracy / F-score configuration C_3	Accuracy / F-score configuration C_4
1	aseptic	67	0.91/0.95	0.89/0.94
2	aseptic	39	0.92/0.96	0.56/0.72
3	aseptic	69	0.76/0.87	0.67/0.8
4	aseptic	68	0.70/0.83	0.57/0.73
5	aseptic	77	0.96/0.98	0.82/0.9
6	aseptic	53	1/1	0.92/0.96
7	infected	80	0.39/0.55	0.65/0.79
8	infected	115	0.98/0.99	1/1
9	infected	191	0.82/0.90	0.92/0.96
10	infected	71	0.03/0.05	0.13/0.22
11	infected	77	0.52/0.68	0.91/0.95
12	infected	117	0.99/0.99	0.91/0.95

choice of the threshold could be done considering the right trade-off between false positive and false negative.

Accuracy (Eq. 2) and F-score (Eq. 3) were employed as performance metrics as shown in Tables 1 and 2.

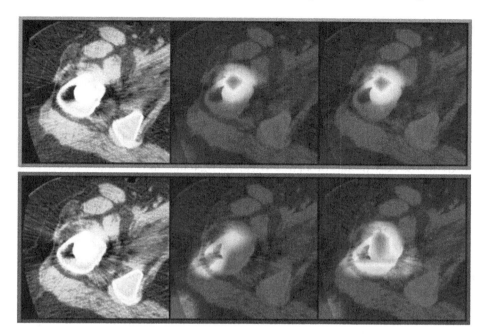

Fig. 3. Input images and corresponding heatmaps of trained model attention produced by means of GradCam and GradCam++ libraries. Green border indicates a correct prediction while the red border shows a wrong one. (Color figure online)

Accuracy is defined as follows:

$$ACC_y = Y_{rp}/PNI \tag{2}$$

where y is the class, Y_{rp} represents the right prediction for that class and PNI is the number of images per patient.

F-score is defined as follows:

$$F = 2 * (P * R)/(P + R) \tag{3}$$

where P and R are precision and recall respectively. Results shown in Tables 1 and 2 show how the model generalizes with respect to the training set: accuracy and F-score are similar for each one of the cross-validation configurations.

In the task of PJI detection, experts in the field emphasize the significance of focusing on regions in images adjacent to the bones, as these areas often hold discriminatory information. Building upon this expert knowledge, a Gradient-weighted Class Activation Mapping (Grad-Cam) analysis [22] was conducted on each test image using the trained model, aiming to enhance the interpretability of the model's decisions. Through this analysis, deeper insights into the interpretability and clinical relevance of the heatmaps generated were gained. The examination of Grad-Cam heatmaps enabled to identify discriminative features near the bones, which exhibited a strong correlation with accurate predictions. This correlation is exemplified by the green-bordered example shown in Fig. 3. Conversely, when the model placed emphasis on multiple features dispersed throughout the image, the prediction was incorrect.

5   Conclusion and Future Works

In this paper an early detection technique for hip periprosthetic joint infections on computed tomography images was presented. This was a first in the state of the art to the best of author's knowledge. To this aim, hip CT images were collected and labeled by experts. A dedicated pre-processing pipeline was developed and a ResNeSt CNN solution was trained. The proposed pipeline demonstrated to achieve strong performances (in terms of accuracy and F-score) in detecting infected and aseptic images of a specific patient.

In order to strengthen the robustness and generalizability of our proposed model, future research will focus on expanding the dataset through collaboration with other institutions to obtain a more diverse and representative dataset, enabling our model to perform effectively across various clinical settings.

Furthermore, our research will explore alternative forms of input, such as compressed images, and the implementation of additional automated techniques to further enhance the prediction of prosthesis-related joint infections. Additionally, we will investigate the outliers in the collected dataset to better understand their impact on the overall results.

Acknowledgements. Alessia Rondinella is a PhD candidate enrolled in the National PhD in Artificial Intelligence, XXXVII cycle, course on Health and life sciences, organized by Università Campus Bio-Medico di Roma.

Experiments were carried out thanks to the hardware and software granted and managed by iCTLab S.r.l. - Spinoff of University of Catania (https://www.ictlab.srl).

References

1. Workgroup Convened by the Musculoskeletal Infection Society and others: New definition for periprosthetic joint infection. J. Arthroplast. **26**(8), 1136–1138 (2011)
2. Sculco, T.P.: The economic impact of infected total joint arthroplasty. Instr. Course Lect. **42**, 349–351 (1993)
3. Ting, N.T., Della Valle, C.J.: Diagnosis of periprosthetic joint infection-an algorithm-based approach. J. Arthroplast. **32**(7), 2047–2050 (2017)
4. Kuo, F.-C., Hu, W.-H., Hu, Y.-J.: Periprosthetic joint infection prediction via machine learning: comprehensible personalized decision support for diagnosis. J. Arthroplast. **37**(1), 132–141 (2022)
5. Klemt, C., et al.: Machine learning models accurately predict recurrent infection following revision total knee arthroplasty for periprosthetic joint infection. Knee Surgery, Sports Traumatology, Arthroscopy. 1–9 (2021)
6. Cyteval, C., Hamm, V., Sarrabère, M.P., Lopez, F.M., Maury, P., Taourel, P.: Painful infection at the site of hip prosthesis: CT imaging. Radiology **224**(2), 477–483 (2002)

7. Kapadia, B.H., Berg, R.A., Daley, J.A., Fritz, J., Bhave, A., Mont, M.A.: Periprosthetic joint infection. The Lancet **387**(10016), 386–394 (2016)
8. Conoci, S., Traina, F.: Image classification method, in particular medical images, for example radiographic images. I.T. Patent 102020000031289, June 2022
9. Rondinella, A., et al.: Boosting multiple sclerosis lesion segmentation through attention mechanism. Comput. Biol. Med. **161**, 107021 (2023)
10. Zhang, H., et al.: ResNeSt: split-attention networks. In: Proceedings of the IEEE/CVF Conference on Computer Vision and Pattern Recognition (CVPR) Workshops, pp. 2736–2746, June 2022
11. Bülow, E., Hahn, U., Andersen, I.T., Rolfson, O., Pedersen, A.B., Hailer, N.P.: Prediction of early periprosthetic joint infection after total hip arthroplasty. Clin. Epidemiol. 239–253 (2022)
12. Tibshirani, R.: Regression shrinkage and selection via the Lasso. J. Roy. Stat. Soc. Ser. B (Methodol.) **58**(1), 267–288 (1996)
13. Galley, J., Sutter, R., Stern, C., Filli, L., Rahm, S., Pfirrmann, C.W.A.: Diagnosis of periprosthetic hip joint infection using MRI with metal artifact reduction at 15 T. Radiology **296**(1), 98–108 (2020)
14. Isern-Kebschull, J., et al.: Value of multidetector computed tomography for the differentiation of delayed aseptic and septic complications after total hip arthroplasty. Skeletal Radiol. **49**, 893–902 (2020)
15. Andrä, H., et al.: Structural simulation of a bone-prosthesis system of the knee joint. Sensors **8**(9), 5897–5926 (2008)
16. Tao, Y., Hu, H., Li, J., Li, M., Zheng, Q., Zhang, G., Ni, M.: A preliminary study on the application of deep learning methods based on convolutional network to the pathological diagnosis of PJI. Arthroplasty **4**(1), 49 (2022)
17. Morar, L., et al.: Analysis of CBCT bone density using the Hounsfield scale. Prosthesis **4**(3), 414–423 (2022)
18. Suzuki, S., et al.: Topological structural analysis of digitized binary images by border following. Comput. Vision Graph. Image Process. **33**(1), 32–46 (1985)
19. He, K., Zhang, X., Ren, S., Sun, J.: Deep residual learning for image recognition. In: Proceedings of the IEEE Conference on Computer Vision and Pattern Recognition, pp. 770–778 (2016)
20. Xie, S., Girshick, R., Dollár, P., Tu, Z., He, K.: Aggregated residual transformations for deep neural networks. In: Proceedings of the IEEE Conference on Computer Vision and Pattern Recognition, pp. 1492–1500 (2017)
21. Hoffman, J., Roberts, D.A., Yaida, S.: Robust learning with Jacobian regularization. arXiv preprint arXiv:1908.02729 (2019)
22. Selvaraju, R.R., Cogswell, M., Das, A., Vedantam, R., Parikh, D., Batra, D.: Gradcam: Visual explanations from deep networks via gradient-based localization. In: Proceedings of the Ieee International Conference on Computer Vision, pp. 618–626 (2017)

Many-to-Many Metrics: A New Approach to Evaluate the Performance of Structural Damage Detection Networks

Piercarlo Dondi[1]([✉])(iD), Ilaria Senaldi[2](iD), Luca Lombardi[1](iD),
and Marco Piastra[1](iD)

[1] Department of Electrical, Computer and Biomedical Engineering, University of
Pavia, Via Ferrata 5, 27100 Pavia, Italy
{piercarlo.dondi,luca.lombardi,marco.piastra}@unipv.it
[2] EUCENTRE Foundation, European Centre for Training and Research in
Earthquake Engineering, Via Ferrata 1, 27100 Pavia, Italy
ilaria.senaldi@eucentre.it

Abstract. In the last years, Computer Vision and Deep Learning techniques have proved to be useful in supporting structural inspections of buildings and civil infrastructures. Particularly, in the case of post-disaster structural safety assessment, automated damage detection algorithms can accelerate the analysis of survey images and thus contribute to a fast screening of impacted areas. The identification of the various types of damage can be seen as a special case of object detection in which the goal is to identify sub-parts of a large object (e.g., cracks on a building). However, in this scenario, traditional evaluation metrics for object detection tend to underestimate the actual performance of the detector, since they considered a one-to-one match between a ground-truth box and a predicted box. Such approach could be sub-optimal for damage detection: for example, a crack can be labeled as a single entity in the ground truth but detected as two small cracks in inference or vice versa. To compensate this issue and better asses the performance of the detector, we introduce a new set of metrics called Many-to-Many. We tested these metrics using a YOLO network on two datasets containing images of damaged bridges and civil structures, and we collected evidence of an improved evaluation capability.

Keywords: Evaluation Metrics · Structural Damage Detection · Deep Learning · YOLO · Post-Earthquake Intervention

1 Introduction

Buildings and infrastructures are potentially subject to several types of damage during their service life, due to excessive loading conditions, progressive aging, environmental exposure, or disastrous events, like earthquakes. The latter case is particularly critical, since a vast number of structures needs to be inspected

G. L. Foresti et al. (Eds.): ICIAP 2023, LNCS 14234, pp. 144–155, 2023.
https://doi.org/10.1007/978-3-031-43153-1_13

in a relatively short time to rapidly identify situations endangering people and first responders.

In the last years, Computer Vision, and Deep Learning techniques in particular, proved to be very helpful in supporting the traditional structural inspections of building and civil structures [13,22]. Automated damage detection algorithms are especially useful: they can be used to perform a fast preliminary screening of the images acquired on the field, reducing the amount of data that experts have to check.

This damage detection task can be considered as a special case of object detection in which the goal is to identify sub-parts of a large object (e.g., cracks on a building). However, the standard evaluation metrics used for object detection tend to underestimate the actual performance of such detectors. Normally the quality of the detection is evaluated using Intersection over Union (IoU) with a one-to-one match between a ground truth and a predicted box. This approach works well when dealing with clearly and unambiguously defined objects (e.g. a car or a human), but it can fail when multiple valid bounding boxes are acceptable. For example, a crack can be annotated as a single object in the ground truth and detected as two separate cracks by the network, or vice versa. Both the results are correct, but they can be marked as errors. The same problem can occur with other common types of structural damage, like leaching or spalling [2], thus, even if a detector found all the damaged areas in an image its performance can still appear low. This fact could be particularly undesirable when damage detection is performed for screening the health of buildings and structures, where the overall condition is more important than individual detections.

Segmentation can overcome this issue, since it works at pixel level. However, annotating dataset for segmentation is definitely more complex and time consuming than bounding box annotation. Segmentation approaches proposed in literature generally focus only on few types of damage (e.g., only cracks and spalling [2]), use very small datasets of few hundred images (e.g. [18]) or very specialized ones (e.g., only concrete cracks [7]). Even works that deal with multiclass damage segmentation adopt relatively small datasets (e.g., [15]). This is one of the reasons why object detection is still widely used in the field [13,22].

This paper proposes Many-to-Many, an alternative to the traditional object detection evaluation metrics based on IoU. The proposed approach aims to take into account all the nuances of the problem to obtain a more accurate estimation of the Precision and Recall of the detector. We tested our idea using a YOLO network on two datasets: (i) CODEBRIM, an open access dataset developed by Mundt et al. [20], that contains only images of damaged bridges; and (ii) IDEA, a new dataset introduced in this paper, that contains images of damaged bridges and civil buildings acquired during post-earthquake surveys in Italy.

The remaining of the article in structured as follows: Sect. 2 provides a brief overview of previous works in the field; Sect. 3 describes the proposed metrics; Sect. 4 illustrates the two datasets; Sect. 5 presents the conducted experiments and the achieved results; finally, Sect. 6 draws the conclusions and proposes possible future developments.

2 Previous Works

Within the context of structural health monitoring (SHM), several Machine Learning (ML) and Deep Learning (DL) applications have been developed for the assessment of the structural damage and for safety [13,16,22]. Among them, image-based damage detection methods have surely an important role.

Davoudi et al. developed a set of SVM-based models for flexure and shear crack detection of RC beams and slabs [10,11]. Their models were trained on two datasets of cracks pattern images captured from tested specimens. Ye et al. [25] proposed a Convolutional Neural Network (CNN) called Ci-Net for crack detection of concrete beams. Their model was trained on crack pattern images and verified with the experimental test of RC beams. Cha et al. developed a CNN model for surface crack detection of concrete structures [5] and then a Faster region-based CNN algorithm for detecting different types of surface damage such as concrete cracks, delamination, and (medium and high) corrosion in structural steel and bolts [6]. Li et al. used a fully CNN model to detect different types of damage in concrete structures, as cracks, spalling and efflorescence [19]. Chow et al. instead, tried a different approach, employing a convolutional autoenconder to identify the presence of cracks and spalling [8].

Besides concrete structures, CNN have been also employed to detect cracks in masonry walls [9], road pavements [21,23], and steel gusset plates [12]. Yang et al. [24] developed a fully CNN model capable of reducing training time for both crack detection and measurement, through semantic segmentation. Semantic segmentation was also adopted by Hoskere et el, who developed MaDnet, a network able to detect various types of structural damage (cracks, exposed rebar, spalling and corrosion) as well as various types of materials (concrete, steel, asphalt) [15]. Specialized segmentation networks, focused of cracks, have been instead proposed by Kang et al. [18] and Choi and Cha [7].

Recently, Bai et al. [2] developed DL methods for automated classification and segmentation on a variety of structures affected by extreme events, including damage classes like cracks, spalling and collapse type (partial or full). Their classification also considered the structural member (beam, column, wall, or other), the damage severity and the damage type (flexural, shear or combined).

Of course, a critical aspect in the development of a DL solution is the availability of a large dataset of annotated images. In this context, unfortunately, large open-access datasets are still limited. Most of the existing ones have been collected either from experiments carried out in laboratory or for monitoring purposes, while only few have been acquired during post-event reconnaissance surveys [4]. Furthermore, most of the datasets are either focused on a limited number of structural typologies or damage (e.g., ROAD2020 [1] or COCO-Bridge 2021+ [3]), or are constituted by classified images, rather than annotated ones (e.g., PEER Hub ImageNet [14]). The few ones that consider multiple classes of damage and/or buildings contain only few thousands annotated images (e.g., CODEBRIM [20]).

To the best of the authors knowledge, to date, the IDEA dataset is one of the most complex described in literature, as both variety of subjects and acquisition conditions.

3   Our Proposal: Many-to-Many Metrics

As explained in the Introduction, standard metrics might be sub-optimal to properly assess the effectiveness of a damage detection network in the considered scenario. In a standard object detection problem, the targets are clearly defined (Fig. 1), while in our case (Fig. 2), the same damage can be labeled in different ways even by human experts. For example, one expert can label two nearby cracks as different entities, and another one as a single damage.

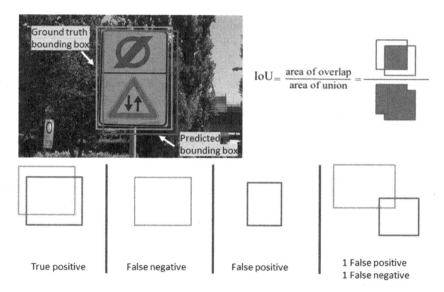

$$IoU = \frac{\text{area of overlap}}{\text{area of union}} =$$

True positive | False negative | False positive | 1 False positive
                                                                 1 False negative

Fig. 1. Example of Intersection over Union (IoU) as metrics for object detection: green box ground truth, red box prediction. (Color figure online)

This ambiguity is also reflected by the trained network that can show the same behavior as can be seen by the predictions in Fig. 2. Since our goal is finding the presence of damage, both the results are fine, yet standard metrics may fail to measure them properly. This is due to the rules used for matching the Predicted Bounding Boxes (PB) to the Ground Truth Bounding Boxes (GTB):

- *One-to-One matching*, namely one PB can only match with one GTB.
- *Intersection over Union (IoU) threshold*, namely a PB is considered a True Positive (TP) only if its match with a GTB is over a certain threshold of IoU (see Fig. 1, bottom for the possible outcomes).

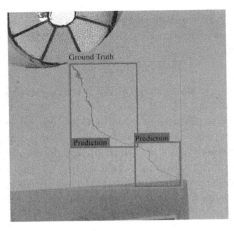

Fig. 2. Examples of different possible valid ground truth (in green) and prediction (in red) boxes for the same damage. (Color figure online)

Thus, in our example, we expect for both cases a correct match, but using the standard metrics we obtain instead: in the first case (Fig. 2, left) 1 TP and 1 False Negative (FN), since the prediction is associated only with the first of the two ground truth boxes and so the second one has no match; while in the second case (Fig. 2, right) 1 TP and 1 False Positive (FP), since only the first prediction is matched with the ground truth while the second one has both no other GTB to match and also a low score in IoU.

Therefore, we have two problems to address. First, we need a method that could match multiple prediction boxes to the same ground truth box (and vice versa). This justify the desire for a new metric more flexible than IoU, since the latter is too sensible to area differences for our scenario.

To find a solution we need to consider various requirements. While, with the standard IoU metrics, the value of TP/FP/TP/TN are used for measuring both Precision and Recall, we think that, in the case in point, they should be evaluated separately. Figure 3 shows four examples from the scenario considered, where all bounding boxes are associated to the same class. In the first case (Fig. 3(a)) the prediction covers only a small part of the ground truth, so we consider this to be a good match for Precision (i.e., 1 TP) but not quite so for Recall, since a large part of the ground truth is missing (i.e., 1 FN). The second case (Fig. 3(b)) shows the opposite situation, in which the prediction box includes the whole ground truth, which is more representative for Recall (i.e., 1 TP), but it also includes a large non-relevant area which is not satisfactory for Precision (i.e., 1 FP). In the third case (Fig. 3(c)) we have a large prediction box that includes multiple ground truth boxes thus we can consider this as 1 TP for Precision and 2 TPs for Recall. Finally, in the last case (Fig. 3(d)) we count 3 TPs for Precision and only 1 TP for Recall since, overall, the three prediction boxes cover a sufficiently large portion of the ground truth.

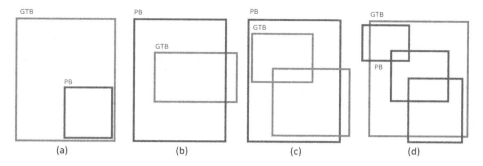

Fig. 3. Examples of possible ground truth (GTB, in green) and predicted (PB, in red) boxes. (Color figure online)

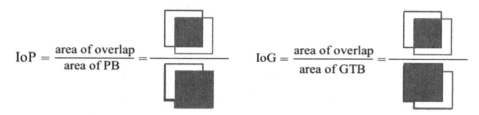

Fig. 4. The proposed new metrics: Intersection over Prediction (on the left) and Intersection over Ground Truth (on the right).

In our proposed solution, we define two new metrics that substitute IoU: *Intersection over Prediction* (IoP) (Fig. 4, left) for Precision, and *Intersection over Ground Truth* (IoG) (Fig. 4, right) for Recall. Note that for IoP the area of overlap may be composed by multiple GTBs referring to the same PB (as in Fig. 3(c)), while for IoG the area of overlap may be composed by multiple PBs referring to the same GTB (as in Fig. 3(d)).

Applying our new metrics to the initial example in Fig. 2, we can now obtain, in the first case, 2 TPs for Recall and 1 TP for Precision, while, in the second, 1 TP for Recall and 2 TPs for Precision.

Summarizing, our Many-to-Many metrics follow these three rules:

- *Many-to-Many matching*, namely the same GTB can be associated to multiple PBs, and vice versa.
- *Intersection over Prediction (IoP) threshold*, used to evaluate Precision only.
- *Intersection over Ground Truth (IoG) threshold*, used to evaluate Recall only.

4   Datasets

We considered two datasets, briefly described in the following: CODEBRIM [20] and IDEA, introduced in this work for the first time.

4.1 CODEBRIM

CODEBRIM (COncrete DEfect BRidge IMage dataset) is an open-access dataset developed by Mundt et al. [20]. It contains almost 1600 images of 30 unique damaged bridges. Five classes of damage were manually labeled: crack, spalling, efflorescence, exposed bars and corrosion stain. This dataset is intended for monitoring purposes; thus, the level of damage is lower than those that can occur after an earthquake.

4.2 IDEA

IDEA (Image Database for Earthquake damage Annotation) is a new dataset by the EUCENTRE Foundation. It consists of images of bridges and civil structures acquired during post-earthquake surveys in occasion of the last three major earthquake sequences in Italy: L'Aquila (2009), Emilia (2012) and Central Italy (2016–2017).

Fig. 5. Some sample images from the IDEA dataset.

Most of the acquisitions have been performed with digital cameras during traditional inspection procedures, while the remaining images have been collected with the use of Unmanned Aircraft Systems (UAS) to reach areas difficult to inspect closely (like portion of bridges).

The resulting dataset is characterized by a high variety (Fig. 5) in terms of image resolution, lighting conditions (depending on meteorological conditions and if the image is shot outdoor or indoor), different points of view and geometrical scales (structural elements, structural portions, or structure as a whole).

Overall, the IDEA dataset contains many thousands of high-resolution photos, whose analysis is still ongoing. Currently, around 2300 images have been manually annotated with bounding boxes. Four common types of damage have been considered: cracks, spalling, corrosion/exposed rebar and leaching. The number of occurrences for each class is as follows:

- Cracks: 2540
- Spalling: 2606
- Leaching: 696
- Corrosion/Exposed rebar: 1292

The EUCENTRE Foundation plans to release this dataset in a later stage of the project when more images and more damage classes will be labeled.[1]

5   Experimental Results

We conducted our experiments using the YOLOv5 network developed by Ultralytics [17]. The network is available in five configurations (nano, small, medium, large, and extra-large) each with an increasing number of parameters. We performed some preliminary tests to find the best configuration for the considered datasets. We obtained the best results using the YOLOv5 large model with an image size of 1080 × 1080. Both datasets where randomly splitted 80–20 between training and testing. Threshold for IoU, IoP and IoG was set to 0.5.

Figure 6 shows the Precision/Recall curves obtained by using both the standard and our Many-to-Many metrics, for CODEBRIM. We can notice a clear difference in the performance estimation: our Many-to-Many metrics returns higher values for both Precision and Recall.

The same behavior appears also for the IDEA dataset (Fig. 7), even if the values are lower than those for CODEBRIM. This is expected since IDEA is a more complex dataset. CODEBRIM includes only images of bridges, mainly framed closed-up, while IDEA contains images of different types of structures, framed at various distance and under different climatic conditions (see Fig. 5).

---

[1] The IDEA dataset will be made available on the EUCENTRE Foundation web site (https://www.eucentre.it/). Access will be granted upon request.

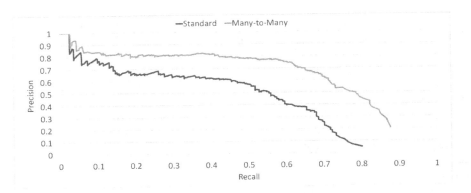

Fig. 6. Precision/Recall curve using the standard (red) and the proposed Many-to-Many (green) metrics for CODEBRIM. (Color figure online)

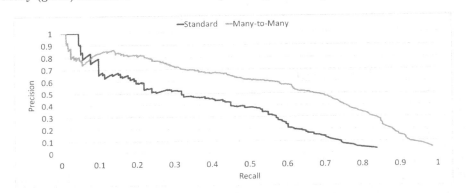

Fig. 7. Precision/Recall curve using the standard (red) and the proposed Many-to-Many (green) metrics for IDEA. (Color figure online)

Figure 8 shows some comparative examples of ground truth and prediction for the IDEA dataset. We can notice how the trained network found all damage although the prediction boxes are different in both number and size from the ground truth boxes. This is a typical case in which traditional metrics underestimate the quality of the detection.

Ground Truth   Prediction

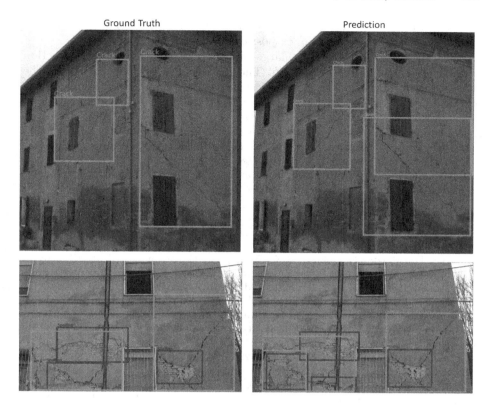

Fig. 8. Comparison between ground truth (on the left) and prediction (on the right) boxes on some sample images of the IDEA dataset.

6   Conclusions

This paper introduces the new Many-to-Many metrics, designed to evaluate the performance of damage detection methods in which the bounding boxing process has unavoidable ambiguities. Experimental evidence has shown how the proposed metrics are more adequate than the standard ones in assessing the performance of a damage detection network. Of course, unlike standard metrics, our Many-to-Many metrics cannot be applied in any scenario, since the latter is very specific and can be used only when dealing with problems similar to the one described in this work. However, with respect to screening images and videos for structural health monitoring and post-event safety assessment, the proposed metrics could yield better results.

Future steps will involve additional experiments with other datasets, and, possibly, with synthetic images. Hopefully, as the availability of annotated data will grow, we plan to further refine and adapt the Many-to-Many metrics to better deal with the complexities that can arise. We are also considering the integration of our new metrics directly in the loss function for the training of the network and not only for an ex-post evaluation.

This work is a first step in a large project dedicated to damage detection during post-earthquake intervention. The final goal is to help experts on the field during the survey operations, providing them a preliminary screening of the conditions of the area, reporting the most likely damaged building and bridges.

Acknowledgments. The current work has been carried out under the financial support of the project Teamaware (European Union's Horizon 2020 research and innovation program, grant agreement No 101019808).

We would like to thank Paul-Maxence Baraton, Michel Bastien and Jamy Lafenetre for their help during the experimental phase.

References

1. Arya, D., Maeda, H., Ghosh, S.K., Toshniwal, D., Sekimoto, Y.: RDD2020: an annotated image dataset for automatic road damage detection using deep learning. Data Brief **36**, 107133 (2021). https://doi.org/10.1016/j.dib.2021.107133
2. Bai, Y., Zha, B., Sezen, H., Yilmaz, A.: Engineering deep learning methods on automatic detection of damage in infrastructure due to extreme events. Struct. Health Monit. **22**(1), 338–352 (2023). https://doi.org/10.1177/14759217221083649
3. Bianchi, E., Hebdon, M.: COCO-Bridge 2021+ Dataset, University Libraries. Virginia Tech (2021). https://doi.org/10.7294/16624495.v1
4. Bianchi, E., Hebdon, M.: Visual structural inspection datasets. Autom. Constr. **139**, 104299 (2022). https://doi.org/10.1016/j.autcon.2022.104299
5. Cha, Y.J., Choi, W., Büyüköztürk, O.: Deep learning-based crack damage detection using convolutional neural networks. Computer-Aided Civil and Infrastructure Engineering **32**(5), 361–378 (2017). https://doi.org/10.1111/mice.12263
6. Cha, Y.J., Choi, W., Suh, G., Mahmoudkhani, S., Büyüköztürk, O.: Autonomous structural visual inspection using region-based deep learning for detecting multiple damage types. Comput. Aid. Civil Infrastruct. Eng. **33**(9), 731–747 (2018). https://doi.org/10.1111/mice.12334
7. Choi, W., Cha, Y.J.: SDDNet: real-time crack segmentation. IEEE Trans. Industr. Electron. **67**(9), 8016–8025 (2020). https://doi.org/10.1109/TIE.2019.2945265
8. Chow, J., Su, Z., Wu, J., Tan, P., Mao, X., Wang, Y.: Anomaly detection of defects on concrete structures with the convolutional autoencoder. Adv. Eng. Inform. **45**, 101105 (2020). https://doi.org/10.1016/j.aei.2020.101105
9. Dais, D., İhsan Engin Bal, Smyrou, E., Sarhosis, V.: Automatic crack classification and segmentation on masonry surfaces using convolutional neural networks and transfer learning. Autom. Constr. **125**, 103606 (2021). https://doi.org/10.1016/j.autcon.2021.103606
10. Davoudi, R., Miller, G.R., Kutz, J.N.: Data-driven vision-based inspection for reinforced concrete beams and slabs: quantitative damage and load estimation. Autom. Constr. **96**, 292–309 (2018). https://doi.org/10.1016/j.autcon.2018.09.024
11. Davoudi, R., Miller, G.R., Kutz, J.N.: Structural load estimation using machine vision and surface crack patterns for shear-critical RC beams and slabs. J. Comput. Civ. Eng. **32**(4), 04018024 (2018). https://doi.org/10.1061/(ASCE)CP.1943-5487.0000766

12. Dung, C.V., Sekiya, H., Hirano, S., Okatani, T., Miki, C.: A vision-based method for crack detection in gusset plate welded joints of steel bridges using deep convolutional neural networks. Autom. Constr. **102**, 217–229 (2019). https://doi.org/10.1016/j.autcon.2019.02.013

13. Flah, M., Nunez, I., Ben Chaabene, W., Nehdi, M.L.: Machine learning algorithms in civil structural health monitoring: a systematic review. Arch. Comput. Methods Eng. **28**, 2621–2643 (2021). https://doi.org/10.1007/s11831-020-09471-9

14. Gao, Y., Mosalam, K.M.: PEER Hub ImageNet: a large-scale multiattribute benchmark data set of structural images. J. Struct. Eng. **146**(10), 04020198 (2020). https://doi.org/10.1061/(ASCE)ST.1943-541X.0002745

15. Hoskere, V., Narazaki, Y., Hoang, T., Spencer, B.: MaDnet: multi-task semantic segmentation of multiple types of structural materials and damage in images of civil infrastructure. J. Civ. Struct. Heal. Monit. **10**(5), 757–773 (2020). https://doi.org/10.1007/s13349-020-00409-0

16. Hsieh, Y.A., Tsai, Y.J.: Machine learning for crack detection: review and model performance comparison. J. Comput. Civ. Eng. **34**(5), 04020038 (2020). https://doi.org/10.1061/(ASCE)CP.1943-5487.0000918

17. Jocher, G.: YOLOv5 by Ultralytics (2020). https://doi.org/10.5281/zenodo.3908559,https://github.com/ultralytics/yolov5

18. Kang, D., Benipal, S.S., Gopal, D.L., Cha, Y.J.: Hybrid pixel-level concrete crack segmentation and quantification across complex backgrounds using deep learning. Autom. Constr. **118**, 103291 (2020). https://doi.org/10.1016/j.autcon.2020.103291

19. Li, S., Zhao, X., Zhou, G.: Automatic pixel-level multiple damage detection of concrete structure using fully convolutional network. Comput. Aided Civ. Infrastruct. Eng. **34**(7), 616–634 (2019). https://doi.org/10.1111/mice.12433

20. Mundt, M., Majumder, S., Murali, S., Panetsos, P., Ramesh, V.: Meta-learning convolutional neural architectures for multi-target concrete defect classification with the concrete defect bridge image dataset. In: 2019 IEEE/CVF Conference on Computer Vision and Pattern Recognition (CVPR), pp. 11188–11197. IEEE Computer Society, Los Alamitos, CA, USA (2019). https://doi.org/10.1109/CVPR.2019.01145

21. Nguyen, N.H.T., Perry, S., Bone, D., Le, H.T., Nguyen, T.T.: Two-stage convolutional neural network for road crack detection and segmentation. Expert Syst. Appl. **186**, 115718 (2021). https://doi.org/10.1016/j.eswa.2021.115718

22. Sony, S., Dunphy, K., Sadhu, A., Capretz, M.: A systematic review of convolutional neural network-based structural condition assessment techniques. Eng. Struct. **226**, 111347 (2021). https://doi.org/10.1016/j.engstruct.2020.111347

23. Yang, Q., Shi, W., Chen, J., Lin, W.: Deep convolution neural network-based transfer learning method for civil infrastructure crack detection. Autom. Constr. **116**, 103199 (2020). https://doi.org/10.1016/j.autcon.2020.103199

24. Yang, X., Li, H., Yu, Y., Luo, X., Huang, T., Yang, X.: Automatic pixel-level crack detection and measurement using fully convolutional network. Comput. Aided Civ. Infrastruct. Eng. **33**(12), 1090–1109 (2018). https://doi.org/10.1111/mice.12412

25. Ye, X.W., Jin, T., Chen, P.Y.: Structural crack detection using deep learning-based fully convolutional networks. Adv. Struct. Eng. **22**(16), 3412–3419 (2019). https://doi.org/10.1177/1369433219836292

Deepfakes Audio Detection Leveraging Audio Spectrogram and Convolutional Neural Networks

Taiba Majid Wani$^{(\boxtimes)}$ ⓘ and Irene Amerini ⓘ

Sapienza University of Rome, Rome, Italy
{majid,amerini}@diag.uniroma1.it

Abstract. The proliferation of algorithms and commercial tools for the creation of synthetic audio has resulted in a significant increase in the amount of inaccurate information, particularly on social media platforms. As a direct result of this, efforts have been concentrated in recent years on identifying the presence of content of this kind. Despite this, there is still a long way to go until this problem is adequately addressed because of the growing naturalness of fake or synthetic audios. In this study, we proposed different networks configurations: a Custom Convolution Neural Network (cCNN) and two pretrained models (VGG16 and MobileNet) as well as end-to-end models to classify real and fake audios. An extensive experimental analysis was carried out on three classes of audio manipulation of the dataset FoR deepfake audio dataset. Also, we combined such sub-datasets to formulate a combined dataset FoR-combined to enhance the performance of the models. The experimental analysis shows that the proposed cCNN outperforms all the baseline models and other reference works with the highest accuracy of 97.23% on FoR-combined and sets new benchmarks for the datasets.

Keywords: Audio Deepfakes · FoR dataset · CNN · VGG16 · MobileNet

1 Introduction

Nowadays, we are said to be living in the "post-truth" era, which refers to a time where malicious actors can sway public opinion through disinformation in society. Disinformation is an active measure that can cause great damage, such as the manipulation of elections, the creation of conditions that could lead to war, the slandering of any individual, and so on. Recently, substantial advancements have been made in the development of deepfakes. This technology has the potential to be utilized in the dissemination of false information and may soon provide a significant risk in the form of fake news. Deepfakes are videos and audio that have been synthesized and created by Artificial Intelligence. Deepfakes are having an increasingly negative impact on people's ability to maintain their privacy and social security, as well as their authenticity [1]. Recent research has been centered on the identification of deepfake video, which has resulted in an adequate detection accuracy [2].

G. L. Foresti et al. (Eds.): ICIAP 2023, LNCS 14234, pp. 156–167, 2023.
https://doi.org/10.1007/978-3-031-43153-1_14

The identification of audio deepfake has received significantly less attention than the detection of video deepfakes. Utilizing deep learning algorithms, audio deepfakes focus on the production, editing, or synthesis of the target speaker's voice. The goal of such manipulations is to depict the speaker as saying something they have not actually spoken. Over the course of the past few years, voice manipulation has also developed into a very advanced art form [3]. Not only does creating synthetic voices present a threat to automated speaker verification systems, but they also present a threat to voice-controlled devices that have been developed for use in Internet of Things (IoT) contexts. Text-to-speech synthesis (also known as TTS) and voice conversion (VC) are two methods that can be used to generate fake voices [4]. A technology known as text-to-speech (TTS) synthesis may recreate the authentic-sounding voice of any speaker by modeling it after a text that is provided. Voice Conversion (VC) is the process of transforming the audio waveform of a source speaker into one that more closely resembles the speech of a target speaker. Voice synthesis using TTS and VC both produce computer-generated voices that are totally synthetic but are almost unrecognizable from real human speech. In [5] is presented a possible threat to biometric voice devices since the most recent speech synthesis algorithms can produce voices that have a high degree of similarity to a particular speaker. There is a significant risk that voice cloning may undermine public trust and provide criminals with the ability to influence corporate interactions or conversations that are private over the phone. It is anticipated that the incorporation of voice cloning into deepfakes will present a fresh difficulty for the identification of deepfakes [6]. Therefore, it is essential that, in contrast to the trend methodologies, which principally concentrate on identifying visual signal alterations, audio forgeries should also be investigated.

The ASVspoof datasets [7] are being utilized extensively in most of the research that is currently going on audio deepfake detection. However, using these datasets has several drawbacks, the most significant of which is that they do not contain any audio that was generated by the most recent text-to-speech algorithms. These algorithms produce audio that sounds more like human speech and may be unrecognizable to the ears of humans. It is possible that a more difficult issue will arise when attempting to differentiate such audio from true human-generated audio, which calls for the creation of reliable solutions. For this study, we made use of the FoR deepfake dataset [8] as it contains examples of audio generated by the most recent text-to-speech algorithms as well as original utterances. This dataset is the largest publicly available dataset and is selected in this study because it provides the true labels indicating whether the audio file is real or fake. These labels are used during model training and help the model to learn the patterns and features of deepfake audio.

The identification of audio deepfakes has made use of several different machine learning (ML) techniques in the literature. ML-based methods to detect deepfakes follow the conventional pipeline such as feature generation, extraction, and then classification. While approaches that are based on deep learning (DL) need less effort from humans to be put into feature engineering and have obtained very accurate results in detecting audio deepfakes, traditional methods still have their uses. In recent years, the DL approach based on Convolutional Neural Networks (CNNs) has been shown to exhibit remarkable performance in image processing benchmarking competitions, and computer vision, because of its powerful learning capabilities [9]. CNNs are strong in their capacity to grasp spatiotemporal correlations and automatically learn data representations by utilizing numerous feature extraction phases.

For this reason, to leverage the CNN powerfulness, we have designed a Custom Convolution Neural Network (cCNN) detection model composed of four convolutional layers and two fully connected layers to prevent overfitting. We demonstrate the effectiveness of the proposed methodology through a comprehensive experimental evaluation over FoR deepfake dataset. We implemented pre-trained transfer learning models, VGG16 and MobileNet and in addition, trained these models end-to-end for extensive evaluation. The presented models take as input the mel spectrogram (a spectrogram where the frequencies are converted to the mel scale) generated from the three sub-datasets, for-norm, for-2-s, and for-rerecording, constituting the FoR dataset. Mel spectrograms are generated to capture the frequency content of an audio signal and the models can concentrate on the most crucial elements of the audio signal for the classification task. We also combined these sub-datasets into one dataset and named it FoR-combined. Data argumentation is performed on FoR-combined and use it for evaluation process.

2 Related Works

Recent developments in TTS and VC techniques have made audio deepfakes an increasingly dangerous threat to voice biometric interfaces and society. There are a few strategies within the realm of audio forensics to recognize them, but the existing studies are not completely efficient. In this section, we have reviewed recent works that have leveraged FoR dataset employing different ML and DL algorithms.

A novel DL architecture namely, DeepSonar based on layer-by-layer neuron behavior was proposed by Wang et. al. [10]. The model used binary classification for detection of fake and real speeches. A total of three datasets were used, FoR dataset in its original form and two datasets created by authors, Sprocket-VC (in English), made by using open-sourced tool sprocket and MC-TTS (in Chinese) using ancient Chinese poetry. The proposed model achieved an accuracy of 98.1% and EER of 2%.

Camacho et. al., presented a two-stage model for the recognition of fake speech [11]. The first stage included the transformation of raw data to scatter plots and modelling of data using CNN was carried out in other stages. CNN was trained on for-original version of FoR dataset and achieved an accuracy of 88% with 11% of EER. Kochare et.al., [12] implemented several machine learning techniques and two deep learning techniques, a temporal convolutional network (TCN) and a spatial transformer network (STN) for the detection of audio deepfakes using for-original dataset only. TCN and STN achieved an accuracy of 92% and 80% respectively.

The architecture of the proposed cCNN model is simpler and easier to interpret as compared to the state-of-art models [11] and [12], with each layer having a specific function. The balance of convolutional layers and the dropout helps to mitigate the risk of overfitting and hence the robustness of model.

Iqbal et. al., [13] proposed an approach based on selecting the best machine learning algorithm and optimal feature engineering. The feature preprocessing involved feature normalization and feature selection. The three sub-datasets of FoR, for-norm for-rerec and for-2-s were used for training six machine learning techniques. eXtreme Gradient Boosting (XGB) obtained the highest average accuracy of 93%. Hamza et. al., [14] carried out detailed experiments on FoR dataset. Different ML algorithms using mel-frequency cepstral coefficients (MFCC) features were trained on 3 sub datasets of FoR dataset.

In this work, we designed a Custom Convolutional Neural Network (cCNN) using mel spectrograms as input features. The novelty of the presented architecture compared to the state-of-art works lies in the specific configuration of the layers and their hyperparameters. Mel spectrograms provide a visual representation of the intensity of different frequencies, allowing cCNNs to recognize various patterns and features to classify real and fake audio. Mel spectrograms also speed up training time and reduce overfitting, improving the models' overall performance. Further, the study has given the possibility of leveraging pre-trained models that were learnt from the well-known dataset ImageNet. These models are then transferred to the specific task of detecting real and fake audios though the use of the FoR dataset.

3 Methodology

The proposed methodology for the detection of real and fake audios is depicted in Fig. 1. Initially, the audio files present in the datasets are pre-processed and undergo a series of transformations. The processed audio files are converted into mel spectrograms facilitating an image-based approach for the required classification process. Finally, the generated mel spectrograms are given as input to the presented models such as the cCNN, VGG16 and MobileNet to detect the real and fake audios. VGG16 and MobileNet are used as pre-trained transfer learning networks as well as models trained end-to-end.

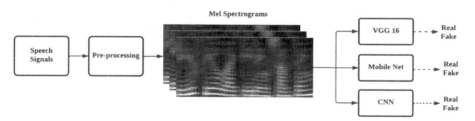

Fig. 1. Proposed methodology for the detection of real and fake audios.

3.1 Pre-processing

Preprocessing is a crucial step in the classification problems, as it converts unstructured data into structured format. FoR dataset consists of duplicate and 0-bit files, and these files affect the model's training and hence performance. We preprocessed the dataset and removed these files. The remaining files are converted into mel spectrograms.

Mel spectrogram applies a frequency-domain filter bank to audio signal that are windowed in time. Mel spectrogram employs the mel scale to simulate the non-linear frequency perception of the human auditory system, which is logarithmic and collects the most perceptually significant information present in audios. Mel spectrogram is generated by dividing the audio signals into overlapping frames using the windowing function [15]. Short-term frequency transform (STFT) is calculated for obtaining the spectrograms, followed by using mel-scale filter banks to convert frequency axis to mel scale, as given in Eq. 1, where m is the mel scale and f is the frequency in Hertz. Lastly, the logarithm of filter bank energies is calculated to acquire the required melspectrogram.

$$m = 2595 * \log 10(1 + f/700) - 1 \tag{1}$$

In this study, we generated mel spectrograms of size $224 \times 224 \times 3$, using the Hanning window with size of 2048 and hop length of 512. The number of mel filter banks used was 224. In addition, logarithmic scaling and normalization were used to make the mel spectrograms more resistant to aberrations and background noise. Figure 2a and 2b represent the mel spectrograms of real and fake audio samples taken from the dataset. Mel spectrograms, with time on the x-axis and frequency on y-axis in Hertz (Hz), are expressed in decibels (dB) as they represent the logarithmic scale of the power of a signal.

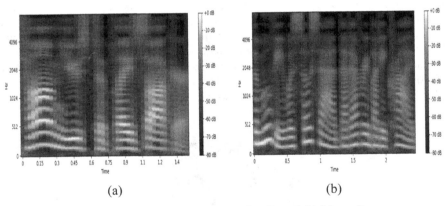

(a)   (b)

Fig. 2. Mel spectrograms of (a) real audio and (b) fake audio.

It can be noticed in Fig. 2a and 2b, that the mel spectrogram of real audio exhibits a consistent and richer spectral content with well-defined format pattern across different frequency bands while the mel spectrogram of fake audio has inconsistencies and unnatural spectral peaks and formats. During training, the network should learn to recognize these variations and efficiently classify audio as real or fake.

3.2 Custom Convolutional Neural Network (cCNN)

In computer vision, Convolutional Neural Networks are the most widely used deep learning technique because of their scalability and stability. In this work, we have proposed a Custom Convolutional Neural Network (cCNN). cCNNs specific layer structure, which includes small filter sizes and max pooling, as well as the addition of a dropout layer to prevent overfitting, is the basis of its efficacy for distinguishing between real and fake audio. The mel spectrograms generated from the three sub-datasets of FoR dataset are taken as input and fed to the first layer of cCNN. The first convolution layer has a kernel size of (3×3) with 64 filter, second and third layer have 128 filters of kernel size (5×5) and last convolution layer has 256 filters with kernel size of (5×5). Each convolution layer is followed by a ReLU activation unit and pooling layer of size (2×2) with same padding and stride of 2. The dimensionality of the feature maps is significantly reduced by using smaller filter sizes of (3×3) and (5×5) and max pooling layers of ($2x2$) as compared to bigger filter sizes and pooling layers. Batch normalization is carried out after every convolution layer to increase the training speed and hence stability. The output of the last pooling layer is fed to the flatten layer in which the 3D volume is converted to a 1D vector. The flatten layer is followed by two fully connected layers with 512 and 1024 neurons respectively. A dropout layer is added after the first fully connected layer with a dropout ratio of 25% to avoid overfitting. This improves the model's ability to generalize to new data and hence performance. The last fully connected layer consists of SoftMax activation function performing the task of classification of real and fake audios (Fig. 3).

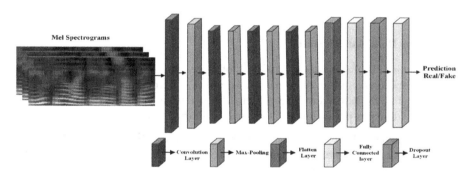

Fig. 3. Proposed Custom-Convolutional Neural Network

3.3 Transfer Learning Approach

Transfer learning is a machine learning technique in which CNNs that have been trained for one task are utilized as the foundation for a model on a different task. We can initialize the weights by employing a pre-trained network that has been trained on large-labeled datasets, such as public image datasets, etc., rather than starting the training process from the very beginning by randomly assigning values to the weights. Pre-trained models trained on ImageNet have demonstrated strong performance and generalization capabilities on various visual recognition tasks. By leveraging a pre-trained model from ImageNet, it can be beneficial to learn rich features representations from a large scale image dataset. In this work, we have used two pre-trained transfer learning models VGG16 [16] and MobileNet [17] for classifying real and fake audios.

Both models' parameters that had been trained using the ImageNet dataset were frozen and loaded in the initial few layers when they were first created [16–18]. The final classification task includes the addition of a flatten layer, which is then followed by a fully connected layer with 256 neurons and ReLU activation, a dropout layer with a dropout ratio of 50%, and a dense layer with 512 neurons and a SoftMax activation function for the final classification task are added. The VGG16 and MobileNet networks were fine-tuned using transfer learning. This enabled efficient learning with a smaller dataset.

4   Experimental Results and Discussion

For the detection of real and fake audios, several experiments are performed using cCNN, VGG16 and MobileNet trained over the sub-datasets of FoR deepfake dataset, for-norm, for-2 s and for-rerecording. We trained VGG16 and MobileNet end-to-end evaluating the performance of such networks against the proposed cCNN on FoR-combined dataset, a combination of the previous three sub-datasets. The experiments do not explicitly include noise or transmission error scenarios. A detailed description of dataset, experimental setup and results are given in the following sub-sections and then compared with the other methods.

4.1   Fake or Real Dataset (foR)

FoR is an audio deepfake dataset consisting of more than 111,000 real utterances collected from speech recording of humans (all genders) and more than 87,000 fake utterances created from 7 different TTS systems. This dataset is divided into 4 different versions based on pre-processing, for-original, for-norm, for-2 s and for-rerecording:

1. for-original: This dataset consists of 195,541 original utterances collected from different sources and is unbalanced in terms of genders and classes i.e., unequal number of real and fake audio sample.
2. for-norm: It is the normalized version of for-original. The audio files are in WAV format and are normalized to 0 dB FS (decibels relative to full scale). It consists of 69,400 utterances and is balanced in terms of genders and classes.
3. for-2 s: This version is similar to for-norm except all the files are trimmed at 2 s. It contains 17,870 utterances.
4. for-rerecording: This dataset consists of files of for-2 s, that have been re-recoded, simulating a real-world attack. 13,268 utterances are present in this version.

In this paper, three versions of FoR dataset, for-norm, for-2 s and for-rerecording have been used. These datasets have already been divided into training, validation and testing by the authors [6] and were used as such in the evaluation process. The total samples present in the dataset and number of samples used for the experiment analysis are given in Table 1.

Additionally, we combined these three sub-datasets and named it as FoR-combined, for extensive experimental analysis. We performed data augmentation on FoR-combined by the introduction of various modifications like height and width shift ranges of 0.2, zoom range of 0.2, horizontal flip, rotation ranges of 30 degrees, and shear range of 0.2. Table 2 shows the number of samples in the FoR dataset. After the data augmentation, the training data was increased to 41,870.

Table 1. Utterances in three sub-datasets of FoR dataset

Dataset		Total samples		Samples considered	
		Fake	Real	Fake	Real
for-norm	**Train**	26,927	26,941	12,015	12,015
	Val	5398	5400	5398	5400
	Test	2370	2264	2370	2264
for-2 s	**Train**	5104	5104	5104	5104
	Val	1143	1101	1143	1101
	Test	408	408	408	408
for-recording	**Train**	6978	6978	6978	6978
	Val	1413	1413	1413	1413
	Test	544	544	544	544

Table 2. Utterances present in FoR-combined dataset.

	Total no of samples considered	
	Fake	Real
Train	24,097	24,097
Val	7,955	7,955
Test	5,703	5,703

4.2 Experimental Setup

The hardware setup is given in Table 3. Mel spectrograms were fed as an input to all models. For all models the batch size was kept to 32. The models were trained with two different sets of epochs, 20 and 50. For the optimization, Adam optimizer is used, cross entropy as loss function and learning rate for cCNN is fixed to 0.0001 and to 0.001 for VGG16 and MobileNet. The models' performance was assessed in terms of their accuracy throughout training, validation, and testing. Accuracy metric measures the proportion of correctly classified samples among all examples in the dataset. To enable model comparison, we present the results for each classification algorithm in tabular form.

Table 3. Hardware Specifications.

CPU	AMD Ryzen 7 5800X 8-Core 16-Thread Processor 3.80 GHz
GPU	Nvidia G-force RTX 2060 (12 GB)
RAM	16 GB
Hard disk	2 TB SSD

4.3 Experiments Using Proposed cCNN

The proposed cCNN was trained and tested on for-norm, for-2 s, for-rerecording, and for-combined. Table 4 shows the testing accuracy of all the considered datasets. With training epochs of 50, cCNN achieved the highest value of accuracy 97.23% on FoR-combined dataset and 96.32% on for-norm dataset. Furthermore, for-rerecording obtains lower accuracy compared to other datasets with 91.86% and 93.4% with 20 and 50 epochs respectively. Since for-rerecording simulates a real-world attack, it is the most difficult case and a decrement in accuracy is expected.

4.4 Experiments Using Transfer Learning Models

The testing accuracy of VGG16 and MobileNet are depicted in Table 4, with 20 and 50 epochs. With training epochs of 50 VGG16 achieved an accuracy of 95.32% and 94.60% on FoR-combined and for-norm respectively, likewise MobileNet achieved the highest accuracy for FoR-combined and for-norm with 96.13% and 95.18% respectively. However, MobileNet performed better than VGG16 on every dataset. VGG16 achieved the accuracy of 89.8% for for-rerecording with 20 training epochs while MobileNet achieved slightly better accuracy of 90.1% than VGG16 on same configuration and dataset.

Table 4. Performance of the presented models.

Dataset	Custom Models					
	cCNN		VGG 16		Mobile Net	
	20 Epochs %	50 Epochs %	20 Epochs %	50 Epochs %	20 Epochs %	50 Epochs %
for Norm	93.86	96.32	92	94.6	92.8	95.1
for 2 s	92.4	94.1	90.1	92.7	90.9	92.9
for rerecording	91.86	93.4	89.8	91.61	90.1	92.4
FoR-combined	95.8	97.32	93.4	95.32	93.8	96.13

From Table 4, it can be clearly seen that among all datasets, for-rerecording achieved lower results, since it mimics a real-world attack, and is composed of less sample respect the others. Deep learning models require large amount data for training and learning the data patterns and thus make accurate predictions. FoR-combined achieved the highest accuracies as it is composed of a large number of samples, providing more diverse and representative data for the model to learn from, reducing overfitting, and improving generalization. Overall, the proposed cCNN performed better than the pre-trained models, VGG16 and MobileNet on every dataset used. This demonstrates that a lower number of levels in the network increases the robustness of the model, and it is better suited for the task.

4.5 Experiments Using VGG16 and MobileNet

VGG16 and MobileNet were trained end-to-end using FoR-combined dataset with data augmentation, so that these models could potentially benefit from learning features relevant to the new dataset from the ground up. VGG16 and MobileNet achieved 96.24% and 97.18% of testing accuracy respectively, as shown in Table 5. Also, the EER (Equal Error Rate) is reported in Table 5.

Table 5. Performance of models trained end-to-end.

Models	Accuracy	EER
VGG16	96.24%	0.0376
MobileNet	97.18%	0.0202

On comparing the results of FoR-combined from Table 4 and Table 5, it can be concluded that such models performed marginally better than the pre-trained models, while the proposed cCNN outperforms all the models when trained with 50 epochs.

4.6 Benchmarking

To the best of our knowledge, the considered three FoR sub-datasets, consisting of an increasing level of difficulty due to various post-processing applied to the audio files, have not been the subject of any previous research using CNN. The authors in [13] employed various machine learning algorithms trained on three subsets of FoR datasets, using hand-crafted features. The highest average accuracy of 93% was achieved by the machine learning model eXtreme Gradient Boosting (XGB). Therefore, we also calculated the average accuracy of the presented models for the comparative analysis of the two works.

The effectiveness of the proposed cCNN can be seen in Table 6, obtaining an average boost of 1.60% in the accuracy respect to the state of the art, establishing a new benchmark for the three subsets, for-norm, for-2-seceond and for-rerecording datasets.

Table 6. Comparison with the state-of art

Study	Models	Accuracy %
[13]	XGB	93
Proposed Models	cCNN	94.60
	Pre-trained VGG16	92.97
	Pre-trained MobileNet	93.49

5 Conclusion

The trustworthiness of audio data is crucial because it serves as an essential tool for strengthening security against spoofing and fraud. In this paper, we proposed a Custom Convolution Neural Network to distinguish a fake audio from an original one demonstrating its validity on three versions of FoR datasets, for-norm, for-2-s and for-rerecording datasets and on a for-combined dataset, a combination of the three sub-datasets. All the audio files have been converted to mel spectrograms and used as input to the proposed cCNN, and to the two pretrained transfer learning models, VGG16 and MobileNet. Such networks have also been employed end-to-end on FoR-combined dataset using data augmentation. All the models achieved high accuracy when trained on FoR-combined and lower accuracy when trained on for-rerecording. VGG16 and MobileNet performed slightly better when trained end-to-end. cCNN achieves an accuracy of 97.23% on the combined dataset, performing better than the other considered methods, demonstrating its validity. In the future, we plan to use features fusion and a continual leaning approach for the detection of audio deepfakes to increase robustness and generalization.

Acknowledgements. This study has been partially supported by SERICS (PE00000014) under the MUR National Recovery and Resilience Plan funded by the European Union – NextGenerationEU and Sapienza University of Rome project 2022–2024 "EV2" (003 009 22).

References

1. Masood, M., Nawaz, M., Malik, K.M., Javed, A., Irtaza, A., Malik, H.: Deepfakes generation and detection: State-of-the-art, open challenges, countermeasures, and way forward. Appl. Intell. **53**(4), 3974–4026 (2023)
2. Akhtar, Z.: Deepfakes generation and detection: a short survey. J. Imaging **9**(1), 18 (2023)
3. Malik, K.M., Malik, H., Baumann, R.: Towards vulnerability analysis of voice-driven interfaces and countermeasures for replay attacks. In: 2019 IEEE Conference on Multimedia Information Processing and Retrieval (MIPR), pp. 523–528. IEEE (2019)
4. Khanjani, Z., Watson, G., Janeja, V.P.: Audio deepfakes: a survey. Front. Big Data **5**, 1001063 (2023). https://doi.org/10.3389/fdata.2022.1001063
5. Aljasem, M., et al.: Secure automatic speaker verification (SASV) system through SM-ALTP features and asymmetric bagging. IEEE Trans. Inf. Forensics Secur. **16**, 3524–3537 (2021)

6. Firc, A., Malinka, K., Hanácek, P.: Deepfakes as a threat to a speaker and facial recognition: an overview of tools and attack vectors. Heliyon **9**(4), e15090 (2023). https://doi.org/10.1016/j.heliyon.2023.e15090

7. Todisco, M., et al.: ASVspoof 2019: future horizons in spoofed and fake audio detection. arXiv preprint arXiv:1904.05441 (2019)

8. Reimao, R., Tzerpos, V.: For: A dataset for synthetic speech detection. In: 2019 International Conference on Speech Technology and Human-Computer Dialogue (SpeD), pp. 1–10. IEEE (2019)

9. Khan, A., Sohail, A., Zahoora, U., Qureshi, A.S.: A survey of the recent architectures of deep convolutional neural networks. Artif. Intell. Rev. **53**, 5455–5516 (2020)

10. Wang, R., et al.: Deepsonar: towards effective and robust detection of ai-synthesized fake voices. In: Proceedings of the 28th ACM International Conference on Multimedia, pp. 1207–1216 (2020)

11. Camacho, S., Ballesteros, D.M., Renza, D.: Fake speech recognition using deep learning. In: Figueroa-García, J.C., Díaz-Gutierrez, Y., Gaona-García, E.E., Orjuela-Cañón, A.D. (eds.) Applied Computer Sciences in Engineering: 8th Workshop on Engineering Applications, WEA 2021, Medellín, Colombia, October 6–8, 2021, Proceedings, pp. 38–48. Springer International Publishing, Cham (2021). https://doi.org/10.1007/978-3-030-86702-7_4

12. Khochare, J., Joshi, C., Yenarkar, B., Suratkar, S., Kazi, F.: A deep learning framework for audio deepfake detection. Arab. J. Sci. Eng. **47**(3), 3447–3458 (2021). https://doi.org/10.1007/s13369-021-06297-w

13. Iqbal, F., Abbasi, A., Javed, A.R., Jalil, Z., Al-Karaki, J.: Deepfake Audio Detection via Feature Engineering and Machine Learning (2022)

14. Hamza, A., et al.: Deepfake audio detection via MFCC features using machine learning. IEEE Access **10**, 134018–134028 (2022)

15. Guha, S., Das, A., Singh, P.K., Ahmadian, A., Senu, N., Sarkar, R.: Hybrid feature selection method based on harmony search and naked mole-rat algorithms for spoken language identification from audio signals. IEEE Access **8**, 182868–182887 (2020)

16. Simonyan, K., Zisserman, A.: Very deep convolutional networks for large-scale image recognition. arXiv preprint arXiv:1409.1556 (2014)

17. Howard, A.G., et al.: Mobilenets: Efficient convolutional neural networks for mobile vision applications. arXiv preprint arXiv:1704.04861 (2017)

18. Alabdulmohsin, I., Maennel, H., Keysers, D.: The impact of reinitialization on generalization in convolutional neural networks. arXiv preprint arXiv:2109.00267 2021

A Deep Natural Language Inference Predictor Without Language-Specific Training Data

Lorenzo Corradi[1(✉)], Alessandro Manenti[1], Francesca Del Bonifro[1], Francesco Setti[2], and Dario Del Sorbo[1]

[1] Data Science Team, Lutech S.p.a., Cinisello Balsamo, Italy
l.corradi@lutech.it
[2] Department of Engineering for Innovation Medicine, University of Verona, Verona, Italy

Abstract. In this paper we present a technique of NLP to tackle the problem of inference relation (NLI) between pairs of sentences in a target language of choice without a language-specific training dataset. We exploit a generic translation dataset, manually translated, along with two instances of the same pre-trained model—the first to generate sentence embeddings for the source language, and the second fine-tuned over the target language to mimic the first. This technique is known as Knowledge Distillation. The model has been evaluated over machine translated Stanford NLI test dataset, machine translated Multi-Genre NLI test dataset, and manually translated RTE3-ITA test dataset We also test the proposed architecture over different tasks to empirically demonstrate the generality of the NLI task. The model has been evaluated over the native Italian ABSITA dataset, on the tasks of Sentiment Analysis, Aspect-Based Sentiment Analysis, and Topic Recognition. We emphasise the generality and exploitability of the Knowledge Distillation technique that outperforms other methodologies based on machine translation, even though the former was not directly trained on the data it was tested over.

Keywords: Natural Language Inference · Knowledge Distillation · Domain adaptation

1 Introduction

Natural Language Processing (NLP) has gained huge improvements and importance in the last years. It has many different applications as it helps in many ways human language productions understanding and analysis in an automated manner. Natural Language Inference (NLI) is one of these applications: it is the task of determining the inference relation between two short texts written in natural language, usually defined as *premise* and *hypothesis* [4,21]. This implies the extraction of the meaning of the two texts and then evaluating if the *Premise* (P) entails the *Hypothesis* (H) (*entailment* situation), if the *premise* and the *hypothesis* are in contradiction between each other (*contradiction* situation), or if none of these two situations happen and there is no inference relation among the two texts (*neutral* situation). This is a challenging task that requires understanding the nuances of language and context, as well as the ability to reason and make logical implications. The relevance of this task can be easily understood by highlighting some of its possible applications. Common tasks based on NLI are Aspect-Based Sentiment Analysis (ABSA), Sentiment Analysis (SA), and Topic Recognition (TR) described in Sect. 4. All these tasks, when approached with NLI strategy, are tackled by comparing an input text (e.g., "We really enjoyed the food, it is tasty and cheap, the staff was very nice and kind. However the restaurant is very hard to reach.") and an hypothesis about the input text (e.g., "The position of the restaurant is difficult to be reached") and predicting if the input text either entails, contradicts, or is not related to the hypothesis ("Entailment" is the correct prediction in the previous example). A common problem for many NLP tasks is the fact that the developed models usually require a big amount of natural language productions data, and usually they are made available in the English language. There are many languages that are underrepresented in these NLP dataset contexts and this made interesting tools hard to develop in these other languages, and there is the need to solve this to make these advance in tech available for low represented languages too. Data scarcity may be tacked with different strategies and this work describe some of them relatively to the NLI task in the Italian language. The goal of this research is to build a model with the following traits:

(a) it can perform the NLI task in a specific language;
(b) based on the sentence embedding operation, such as in [16];
(c) it is able to understand another language, in this case Italian. ;
(d) it is able of being general and not requiring any re-train for each specific industrial task (ABSA, SA, TR);

We can state the Research Question (RQ) which drive our effort is

> **RQ: It is possible to build a NLI model with acceptable performances on NLI related downstream tasks in Italian language compliant with *(a)*, *(b)*, *(c)*, *(d)* constraints, without requiring a language-specific dataset?**

The main differences among the proposed models to achieve these aims is the training approach: one is based on Knowledge Distillation (KD) [17] a technique which aim to transfer knowledge from a *Teacher* model (English-based NLI model) to *Student* model (that will handle the Italian language). The other model includes a step for the dataset translation from English to Italian by means of a Machine Translation model [18]. The approach exploiting KD has been demonstrated to have NLI capability in the target language, namely Italian, without being exposed to a NLI training dataset in Italian. This model has been named **I-SPIn** (Italian-**S**entence **P**air **In**ference) and is available at this link along with all instructions for usage.

The remainder of this paper is structured as follows: in Sect. 2 we report literature and datasets discussion, Sect. 3 describes the two implemented approaches,

Sect. 4 reports the settings and results of the performed experiments, Sect. 5 report discussion and conclusions about this work.

2 Related Work

Common approaches to tackle NLI include Neural Networks, such as Recurrent Neural Networks or Transformer based methods [5, 6]. [6] presents an architecture based both on learning *Hypothesis* and *Premise* in a dependent way using bidirectional LSTM and Attention mechanism [1] to extract the text pair representation needed for final classification. The obtained results on SNLI [4] validation set gives a 89% accuracy. [5] describes language representation model (BERT) building a model based on Transformers [20, 22] and pre-training it in a bidirectional way, this has the aim to serve as a pre-trained model that can be fine-tuned on several different tasks including NLI. BERT is fine-tuned and tested on MNLI dataset [21] and reaches around 86% accuracy. Recent researches [15, 23] demonstrate that Transformers models [20] are more suited for the NLI task, consistently surpassing neural models [22]. All of these high performance approaches mainly hold for English language as it is the language in which there is the higher data availability. Some multi-language NLI approaches are proposed in [24], where cross-lingual training, multilingual training, and meta learning are attempted using a dataset extracted from Open Multilingual WordNet. The best model resulted to be the one exploiting meta learning and reached 76% accuracy on True/False classification task of text pairs for the Italian language. [10] represents another work on multi-language NLI where the Excitement Open Platform is presented as open source software for experimenting in NLI related tasks. It has many linguistics and entailment components based on transformations between *Premise* (P) and *Hypothesis* (H), edit distance algorithms, and a classification using features extracted from P and H. Italian Language is tested on a manually translated RTE-3 dataset [7] and the best model has 63% accuracy. In the context of an Italian Textual Entailment competition the task Recognizing Textual Entailment (RTE) is proposed. It is similar to NLI task but it only contains two Entailment Yes/No classes. The competition's winner model is described in [3] and it is based on a open source software EDITS based on edit distance reaching 71% accuracy on the convention EVALITA2009 dataset which is extracted from Wikipedia. [13] presents a model based on translation. The input texts can be in any language and are translated into English using a standalone machine translation system. The authors show that machine translation can be used to successfully perform the NLI related tasks or when P and H are provided in different languages. For Italian, it uses Bing translation and it is tested on EVALITA2009 dataset reaching 66% accuracy.

Datasets. The datasets used in this work are described in this paragraph and examples can be found in Appendix A.

The Stanford NLI (SNLI) [4] corpus is a collection of 570k human-written English sentence pairs manually labeled for balanced classification with the labels "entailment", "contradiction", and "neutral", supporting the task of NLI. The SNLI dataset presents the canonical dataset split—consisting of train, validation, and test sets.

Multi-Genre NLI (MNLI) [21] corpus is a crowd-sourced collection of 433k sentence pairs annotated with textual entailment information. The corpus is modeled on the SNLI corpus, but differs in that covers a range of *genres* of spoken and written text. The train set is composed of sentences with the same genres: "Telephone", "Fiction", "Government", "Slate" and "Travel". The MNLI dataset supports a distinctive cross-genre generalisation evaluation. There is a matched validation set which is derived from the same source as those in the training set, and a mismatched validation set which do not closely resemble any genres seen at training time.

RTE datasets (RTE3-ITA and RTE2009) are English-native NLI datasets, manually translated by a community of researchers. The Italian version RTE3-ITA refers to the third refinement of this dataset[1]. Instead, RTE2009 was submitted for the EVALITA 2009 Italian campaign [2][2]. These datasets are only used for testing, since they contain too few observations to be suitable for training. The RTE3-ITA dataset contains 1600 observations, whereas RTE-2009 contains 800 observations. Unlike classical NLI, these datasets present only two labels: "Entailment" and "No-Entailment".

TED2020 [14] is a generic translation dataset. The option (English–Italian) has been selected for training among more than a hundred possible languages. The dataset consists of more than 400k parallel sentences. The transcripts have been translated by a community of volunteers. This dataset is used to make a model understand different languages [17], starting from a language known to the model.

3   Method

Three different architectures will be detailed throughout the section. In Sect. 3 the objective is to obtain the a model that is able to perform NLI in English. Starting from this model, we propose two parallel approaches to perform NLI in the target language. One is detailed in Sect. 3, and the other is detailed in Sect. 3. Both approaches attempt a domain adaptation and generalisation in the target language—namely, Italian, while lacking a language-specific dataset. The models' parameters were selected among few different possibilities suggested by online informal documentation and literature. No cross-validation or grid-search analyses have been performed for computational constraints. Therefore, no guarantees on the optimality of the parameters can be made. To reduce computational complexity during the inference phase for the models described in Sects. 3 and 3, we recommend to split the model to obtain independent instances of encoder and classifier. The proposed methodology is the following: transform all the sentence pairs in vectorial forms—with the encoder—first; in a second phase, the classifier will receive the embeddings to return an inference relation.

---

[1] The validation and test datasets can be downloaded at this link.

[2] The validation and test datasets can be found at this link.

NLI Training in the Source Language. The proposed solution makes use of a Transformer [20]. The Transformer lately has become the state-of-the-art architecture for NLP, as detailed in Sect. 2. The first step of our methodology is to retrieve a sentence encoder model, based on Transformers. This sentence encoder model is already fine-tuned for general purposes over different languages. The encoder of choice to transform sentences in vectors was Sentence-BERT [16]. It is a fine-tuning of BERT [5], that is a word embedding Transformer model, tailored for the task of sentence embedding. It has the ability to perform sentence embedding faster than BERT as detailed in [16][3], by means of a Siamese training approach [19]. Referring to this model with the term Sentence-BERT is inappropriate, since it has been fine-tuned on RoBERTa [9], that is a larger counterpart of BERT. Hence, the name Sentence-RoBERTa would be more appropriate. In this paper we will adopt the name Sentence-BERT to refer to any siamese structure accepting a sentence pair as input, including the instance of Sentence-RoBERTa to be fine-tuned. Since Transformers are computationally expensive to train from scratch, we decided to test a multilingual version of Sentence-BERT and fine-tune it on SNLI and MNLI merged together to create a single NLI dataset. After a fine-tuning session over the merged NLI dataset, the result is a model based on Transformers, that can proficiently address the NLI task—only in English though, despite being originally trained on multiple languages. More information about this work available at [11].

The output of the fine-tuned Sentence-BERT is composed of an embeddings pair, containing a vectorial representation of the premise and the hypothesis. Note that the sentence encoder model is invoked two separate times for this operation, for complexity optimisation reasons. The Sentence-BERT output embeddings have been further transformed to maximise and emphasise the relevant information for our task. In detail, the following operations have been applied:

- Element-wise product. Captures similarity of the two embeddings, and highlights components of the embeddings that are more relevant than others.
- Difference. Asymmetric operation; captures the direction of implication. We want the hypothesis to imply the premise, and not vice-versa.

The two transformed embeddings were concatenated and passed as input to a fully-connected Feed Forward architecture of six (6) layers, detailed in Appendix B, with three (3) outputs ("Entailment", "Neutral", "Contradiction"), to predict the probability of the sentence pair to belong to each NLI class. Finally, a softmax function was applied to the three-dimensional vector to obtain the class probabilities (Fig. 1).

Execution-wise, the NLI fine-tuning task on a Tesla P100–PCIE–16GB GPU was completed in approximately six (6) hours on the merged NLI training dataset composed of an ensemble of SNLI and MNLI datasets, accounting for more than 1M observations. The main parameters can be found in Appendix B. In our work, we want to enable a multilingual Transformers-based model, previously

---

[3] Sentence-BERT was downloaded from this link.

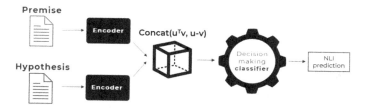

Fig. 1. Model structure. Two sentences are transformed in embeddings. The embeddings are compared with a classifier to get the prediction for the sentence pair.

fine-tuned for a specific task only in one specific language, to proficiently address that specific task in another language.

Knowledge Distillation in the Target Language. The second step of our methodology is to employ a training without language-specific NLI training data and we selected the Knowledge Distillation (KD) [17] approach.

KD was born as a model compression technique [8], where knowledge is transferred from the teacher model to the student by minimizing a loss function, in which the target is the distribution of class probabilities predicted by the teacher model. KD is a powerful technique since it can be used for a variety of multiple tasks. In our experiments, we employed KD to perform NLI in the target language, with the objective of forcing a translated sentence to have the same embedding—i.e. location in the vector space—as the original sentence. The soft targets of the teacher model constitute the labels to be compared with the predictions returned by the student model. The task at hand may fall in the domain adaptation problem sphere.

We require a teacher model (encoder) T, that maps sentences in the source language to a vectorial representation. Further, we need parallel (translated) sentences $D = ((source_1, target_1), ..., (source_n, target_n))$ with $source_j$ being a sentence in the source language and $target_j$ being a sentence in the target language. We train a student encoder model S such that $T(source_j) \approx S(target_j)$. For a given mini-batch B, we minimise the Mean Squared Error loss function:

$$MSE_{(S,T,D=(source_j, target_j))} = \frac{1}{|B|} \sum_{j \in |B|} (T(source_j) - S(target_j))^2 \quad (1)$$

Two instances of the encoder described in Sect. 3 have been taken for the experiment. One acts as teacher encoder model T, the other as a student encoder model S. The application of KD has the objective to share the domain knowledge of the teacher encoder model to the student encoder model, and at the same time learn a new vectorial representation for the target language. A schematic representation is provided in Fig. 2.

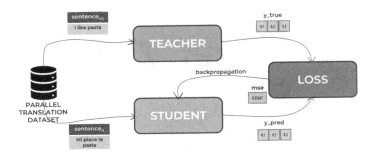

Fig. 2. Knowledge Distillation. Teacher encoder model receives source sentences, student model receives target sentences. Student encoder model is updated with new information from the teacher.

The obtained NLI classifier, able to understand Italian, accepts a sentence pair to output a NLI label. Execution-wise, the KD task on a Tesla P100–PCIE–16GB GPU was completed in approximately five (5) hours on the TED2020 (English–Italian) dataset consisting of more than 400k parallel sentences. The main parameters can be found in Appendix B.

Machine Translation in the Target Language. As an alternative method for our second step, we employ a Large Language Model named No Language Left Behind (NLLB) [18] to address the lack of language-specific NLI training data. To the best of our knowledge, it was not possible to find a comprehensive NLI dataset in Italian. The RTE3-ITA and RTE-2009 datasets, both detailed in Sect. 1, together present about 2500 observations, too few to train a Deep Learning model. Therefore, the dataset used to fine-tune this architecture is the same as in Sect. 3, with an alteration: we perform a translation of the dataset. In fact, the simplest way to perform NLI in a language other than English is to machine translate the ensemble NLI dataset, consisting in SNLI and MNLI merged together. Note that, for memory and performance optimisation, the ensemble NLI training dataset was dynamically translated during execution by invoking the NLLB model for each mini-batch. Execution-wise, this fine-tuning task over the target language on a Tesla P100–PCIE–16GB GPU was completed in approximately ten (10) hours on the translated ensemble NLI dataset, consisting of more than 1M sentence-pairs. The main parameters can be found in Appendix B.

4   Experiments

NLI Results in the Source Language. The architecture discussed as in Sect. 3 has been tested over the standard NLI task in English. For SNLI the accuracy reached 80.69% while for MNLI a 77.00% accuracy is reached.

Table 1. NLI (IT) results.

Dataset	Task	Acc.	Min F1	Macro-Avg F1
SNLI (IT)	NLI	74.21 (−1.83%)	67.19% (−4.34%)	74.08% (−4.94%)
MNLI-Mismatch (IT)	NLI	72.74% (**+1.09%**)	64.53% (**+0.55%**)	72.78% (**+1.37%**)
RTE3-ITA	RTE	67.50% (**+4.75%**)	60.12% (**+5.55%**)	66.35% (**+4.85%**)
RTE-2009	RTE	59.00% (−0.75%)	31.09% (−2.65%)	50.96% (−1.46%)

NLI Results in the Target Language. The architecture discussed as in Sect. 3—that is the main focus of this paper—has been tested over the standard NLI task in Italian, and compared with the alternative architecture based on Machine Translation. The underlying model, an open-source machine translation model, developed by Facebook, named No Language Left Behind [18], was also exploited to obtain a comprehensive Italian NLI dataset, suitable for testing. Results for the SNLI and the MNLI test sets (both translated in Italian) are detailed in Table 1.

SNLI results in Table 1 are not far from the theoretical accuracy cap these models have—presented for NLI in source language. This could be interpreted as a success for the training of both architectures. The Min F1-Score metric captures the most misclassified class. The Neutral class, in general, has been the most challenging to classify, as translation biases may slightly change the connotation of a sentence. Note that this test is biased towards the Machine Translation-based architecture. Remember that this architecture has been fine-tuned over the translated NLI dataset in the target language; The KD-based architecture, instead, had never seen the NLI dataset in the target language. This suggests that the KD-based architecture may have relevant learning capabilities over this task. Differently from the SNLI dataset, we briefly remark that MNLI datasets are divided into genres, and supports a distinctive cross-genre generalisation evaluation by means of the mismatched validation set. A higher accuracy on the mismatched validation set corresponds to a better generalisation of the model. In the same way as before, also for this test the Machine Translation-based architecture had an objective advantage, by being trained on the same dataset it was tested over. Nonetheless, the KD-based architecture performed better in this test. This dataset tests the generalisation capability and the ability to understand a wide range of contexts of a model, as it contains multiple genres. This could be a motivation to consider the KD-based architecture the most powerful architecture of the two.

In addition to the tests above, the architecture has been tested over the RTE datasets. We briefly remind that, unlike classical NLI, these datasets present only two labels: "Entailment" and "No-Entailment". Both our models produce three labels—"Entailment", "Neutral", "Contradiction"—as they were trained on SNLI and MNLI. The two-label mapping for this task maps both *Neutral* and *Contradiction* to *No-Entailment* as this maximises the accuracy on the validation set. Results for the RTE3-ITA and RTE-2009 test sets are reported in Table 1 too.

The performance difference between the two architectures may be explained by the difference in quality of the target language the two architectures have been exposed to during training. In fact, Machine Translation-based architecture has been trained over a machine translated dataset, whereas the KD-based architecture was trained over a manually translated dataset. This supposition can be made because this dataset has been manually translated in Italian, hence presents a better language quality than the NLI datasets translated in the target language.

ABSA Results. Aspect-Based Sentiment Analysis at EVALITA (ABSITA), detailed in [2], is an ABSA dataset. Contains Italian hotel reviews that may touch different topics (such as price, location, cleanliness, etc.) and a sentiment associated to each topic (knowing that sentiments for different topics may be contrasting). By choosing arbitrary NLI hypotheses, this dataset may emulate a total of three (3) different tasks, namely SA, TR, and ABSA. The core idea behind this setting comes from the desire to query a text—in NLI, a set of premises (e.g. a set of reviews), in an unsupervised way, to receive specific answers from a predefined list of answers (e.g. the presence of a topic from a list of topics). In the case of open answers, a question-answer architecture would have been more suitable.

Table 2. ABSITA results, over the Sentiment Analysis and Topic Recognition tasks

Dataset	Balancing	Task	Acc	Min F1	Macro-Avg F1
ABSITA	1:1	SA	88.12% (+**3.08%**)	86.89% (+**3.19%**)	88.02% (+**3.09%**)
ABSITA	1:1	TR	68.09% (−3.1%)	65.75% (−3.01%)	67.97% (−3.16%)
ABSITA	1:7	TR	71.11% (+**5.27%**)	37.94% (−0.77%)	59.56% (+**2.04%**)
ABSITA	1:1	ABSA	94.03% (+**6.24%**)	93.90% (+**6.65%**)	94.02% (+**6.35%**)
ABSITA	1:15	ABSA	78.42% (+**11.39%**)	37.66% (+**8.3%**)	62.30% (+**8.37%**)

Sentiment Analysis (SA) is the task to recognise the overall sentiment of a sentence. As detailed above, we would like to exploit the models to apply SA in an unsupervised manner—to do this, we fix a hypothesis arbitrarily. We assume that the hypothesis we have chosen captures the logical implication that is the core of NLI. Follow results for the ABSITA dataset, detailed in Table 2. Note that the hypothesis has been arbitrarily set to "Sono soddisfatto" ("I feel satisfied"), hence "Entailment" refers to the model predicting positive sentiment. The two-label mapping for this task maps *Neutral* to *Entailment*.

Topic Recognition (TR) is the task to recognise whether or not a sentence is about a topic. As detailed above, we would like to exploit the models to apply TR in an unsupervised manner—to do this, we fix a hypothesis arbitrarily. We assume that the hypothesis we have chosen captures the logical implication that is the core of NLI. Follow results for the ABSITA dataset, detailed in Table 2. The seven (7) in the "Balancing" column stands for the number of different topics in the dataset. The 1:1 balancing has been obtained by randomly sampling sentences from the seven (7) classes that do not compose the target. The two scenarios have been proposed to extensively test the generalisation capability of the models. Note that the hypothesis has been arbitrarily set to "Parlo di pulizia" ("I'm talking about cleanliness"), hence "Entailment" refers to the model predicting the label "cleanliness". The two-label mapping for this task maps *Neutral* to *Entailment*.

Aspect-Based Sentiment Analysis (ABSA) is the task to recognise the sentiment about each sub-topic in a sentence. As detailed above, we would like to exploit the models to apply ABSA in an unsupervised manner—to do this, we fix a hypothesis arbitrarily. We assume that the hypothesis we have chosen captures the logical implication that is the core of NLI. Follow results for the ABSITA dataset, detailed in Table 2. Note that the hypothesis has been arbitrarily set to "La camera é pulita" ("The room is clean"), hence "Entailment" refers to the model predicting positive sentiment and "cleanliness" label. The two-label mapping for this task maps *Neutral* to *Contradiction*.

5   Conclusions and Discussions

To interpret the apparently decent results for NLI in the source language, listed in Sect. 4, we need to consider the fact that, during training, sentence encoders do not look at both inputs simultanously, hence generating good but not top-tier performances. Potentially, we could have obtained slightly better results by making use of a word encoder instead of a sentence encoder, at the cost of a large computational overhead. To address various industrial tasks, we decided to prioritise scalability and responsiveness. The discussed architecture, based on KD, demonstrated to perform better than the other architecture—that was directly trained over machine translated NLI datasets—despite having an objective disadvantage. We stress the fact that the proposed architecture was never directly trained over any kind of Italian NLI data. Compared to the other methodology, the KD presents the following advantages:

1. Easier to extend models: we just require few samples for the new languages.
2. Lower hardware requirements: machine translation—that is an expensive task—is not needed as an intermediate step.

To test our model performances over SA, TR, and ABSA, we employed arbitrary hypotheses. We tried our best to avoid any biases (e.g. hypotheses were chosen by colleagues that had never taken a look at the datasets), but we acknowledge that some bias may have been introduced. This is currently considered an open problem.

Different architectures have been tested showing that it is possible to obtain reasonable accuracies over different NLP tasks by fine-tuning a single architecture based on sentence embeddings over the NLI task. We showed that various NaLP problems may be mapped into a NLI task—in this way, we empirically proved the generality of the NLI task. We would like to stress over the lack of need to re-train any models to obtain the results over each specific task. Moreover, lately NLI models find an important academic usage for boosting the consistency and accuracy of NLP models without fine-tuning or re-training [12]. This is because models should demonstrate internal self-consistency, in the sense that their predictions across inputs should imply logically compatible beliefs about the world—NLI models are trained to achieve that understanding.

A   Dataset Examples

Examples from the benchmark Stanford NLI dataset are shown in Table 3 to show its standard structure.

Table 3. Stanford NLI dataset. A label is produced based on the logical interaction between two short texts.

Premise	Hypothesis	Label
"A soccer game with multiple males playing"	"Some men are playing a sport"	Entailment
"An older and younger man smiling"	"Two men are smiling and laughing at the cats playing on the floor"	Neutral
"A man inspects the uniform of a figure in some East Asian country"	"The man is sleeping"	Contradiction

Examples from the benchmark Multi-Genre NLI dataset are shown in Table 4 to show its standard structure.

Table 4. Multi-Genre NLI dataset. A label is produced based on the logical interaction between two short texts.

Premise	Hypothesis	Label
"It has a staff of about 100 employees, including attorneys and support staff, in 10 branch offices."	"The 10 branches had close to 100 employees."	Entailment
"Theoretically scale economies in delivery are not firm specific."	"Scale economies are flexible."	Neutral
"Mrs. Cavendish is in her mother-in-law's room. "	"Mrs. Cavendish has left the building."	Contradiction

Examples from the RTE3-ITA NLI dataset are shown in Table 5. Note the table will be proposed in the Italian language.

Examples from the TED2020 translation dataset are shown in Table 6.

Table 5. RTE3-ITA dataset. A label is produced based on the logical interaction between two short texts.

Premise	Hypothesis	Label
"All'uscita del gioco Final Fantasy III nella versione per la console Super Nintendo, il nome di Bigg era Vicks."	"Final Fantasy III venne prodotto per la console Super Nintendo."	Entailment
"La signora Minton lasciò l'Australia nel 1961 per proseguire i suoi studi a Londra."	"La signora Mintonè nata in Australia."	Contradiction

Table 6. TED2020 dataset (English–Italian version).

sentence$_{en}$	sentence$_{it}$
"I gave my speech, then went back to the airport to fly back home."	"Io feci il mio discorso, poi andai all'aeroporto per tornare."
"He romanticised the idea they were star-crossed lovers."	"Lui fantasticava sull'idea di loro come amanti sfortunati."
"In Japan, a game of ping-pong is really like an act of love."	"In Giappone, una partita di ping-pong é come un atto d'amore."

B   Models Parameters

Fully-connected Feed Forward architecture used for classification in Sect. 3 is reported here:

```
(layers): ModuleList(
    (0): Linear(in=1536, out=1024, activation=GELU())
    (1): Linear(in=1024, out=512, activation=GELU())
    (2): Linear(in=512, out=256, activation=GELU())
    (3): Linear(in=256, out=128, activation=GELU())
    (4): Linear(in=128, out=64, activation=GELU())
    (5): Linear(in=64, out=3, activation=GELU())
)
```

Hyper-parameters for NLI model training in the source language Sect. 3 are listed here:

```
- batch_size = 8
- max_sentence_length = 256
- max_tokens_length = 128
- epochs = 1
- learning_rate = 2e-5
- epsilon = 1e-8
- weight_decay = 0
- accumulation_step = 8
```

Hyper-parameters for NLI model training in the source language Sect. 3 are listed here:

```
- batch_size = 24
- max_sentence_length = 256
- max_tokens_length = 128
- epochs = 6
- learning_rate = 2e-5
- epsilon = 1e-6
- weight_decay = 1e-2
- accumulation_step = 4
```

Hyper-parameters for NLI model training in the source language Sect. 3 are listed here:

```
- batch_size = 8
- max_sentence_length = 256
- max_tokens_length = 256
- epochs = 5
- learning_rate = 4e-5
- epsilon = 1e-16
- weight_decay = 1e-4
- accumulation_step = 4
```

References

1. Bahdanau, D., Cho, K., Bengio, Y.: Neural machine translation by jointly learning to align and translate. ArXiv 1409 (2014)
2. Basile, P., Croce, D., Basile, V., Polignano, M.: Overview of the Evalita 2018 aspect-based sentiment analysis task (Absita). In: Proceedings of the 6th evaluation campaign of Natural Language Processing and Speech tools for Italian (EVALITA 2018) (2018). https://ceur-ws.org/Vol-2263/paper003.pdf
3. Bos, J., Zanzotto, F.M., Pennacchiotti, M.: Textual entailment at Evalita 2009 (2009)
4. Bowman, S.R., Angeli, G., Potts, C., Manning, C.D.: A large annotated corpus for learning natural language inference. arXiv:1508.05326 (2015). https://doi.org/10.48550/arXiv.1508.05326
5. Devlin, J., Chang, M.W., Lee, K., Toutanova, K.: BERT: pre-training of deep bidirectional transformers for language understanding. arXiv:1810.04805 (2019). https://doi.org/10.48550/arXiv.1810.04805
6. Ghaeini, R., et al.: DR-BILSTM: dependent reading bidirectional LSTM for natural language inference. arXiv:1802.05577 (2018). https://doi.org/10.48550/arXiv.1802.05577
7. Giampiccolo, D., Magnini, B., Dagan, I., Dolan, B.: The third PASCAL recognizing textual entailment challenge. In: Proceedings of the ACL-PASCAL Workshop on Textual Entailment and Paraphrasing, pp. 1–9. Association for Computational Linguistics, Prague, June 2007. https://aclanthology.org/W07-1401
8. Hinton, G., Vinyals, O., Dean, J.: Distilling the knowledge in a neural network. arXiv:1503.02531 (2015). https://doi.org/10.48550/arXiv.1503.02531
9. Liu, Y., et al.: Roberta: a robustly optimized BERT pretraining approach. arXiv:1907.11692 (2019). https://doi.org/10.48550/arXiv.1907.11692

10. Magnini, B., et al.: The excitement open platform for textual inferences. In: Proceedings of 52nd Annual Meeting of the Association for Computational Linguistics: System Demonstrations, pp. 43–48. Association for Computational Linguistics, Baltimore, Maryland, June 2014. https://doi.org/10.3115/v1/P14-5008, https://aclanthology.org/P14-5008

11. Manenti, A., Braunstein, A.: Deep learning techniques for natural language processing: A multilingual encoder model for nli task. Politecnico di Torino, Corso di laurea magistrale in Physics Of Complex Systems (2022). https://webthesis.biblio.polito.it/24750/

12. Mitchell, E., et al.: Enhancing self-consistency and performance of pre-trained language models through natural language inference. arXiv:2211.11875 (2022). https://doi.org/10.48550/arXiv.2211.11875

13. Pakray, P., Neogi, S., Bandyopadhyay, S., Gelbukh, A.: Recognizing textual entailment in non-English text via automatic translation into English. In: Mexican International Conference on Artificial Intelligence (2012)

14. Qi, Y., Sachan, D., Felix, M., Padmanabhan, S., Neubig, G.: When and why are pre-trained word embeddings useful for neural machine translation? In: Proceedings of the 2018 Conference of the North American Chapter of the Association for Computational Linguistics: Human Language Technologies, vol. 2 (Short Papers) (2018). https://doi.org/10.18653/v1/N18-2084

15. Raffel, C., et al.: Exploring the limits of transfer learning with a unified text-to-text transformer. arXiv:1910.10683 (2019). https://doi.org/10.48550/arXiv.1910.10683

16. Reimers, N., Gurevych, I.: Sentence-BERT: sentence embeddings using Siamese BERT-networks. arXiv:1908.10084 (2019). https://doi.org/10.48550/arXiv.1908.10084

17. Reimers, N., Gurevych, I.: Making monolingual sentence embeddings multilingual using knowledge distillation. arXiv:2004.09813 (2020). https://doi.org/10.48550/arXiv.2004.09813

18. Team, N., et al.: No language left behind: scaling human-centered machine translation. arXiv:2207.04672 (2022). https://doi.org/10.48550/arXiv.2207.04672

19. Utkin, L.V., Kovalev, M.S., Kasimov, E.M.: An explanation method for Siamese neural networks. arXiv:1911.07702 (2019). https://doi.org/10.48550/arXiv.1911.07702

20. Vaswani, A., et al.: Attention is all you need. arXiv:1706.03762 (2017). https://doi.org/10.48550/arXiv.1706.03762

21. Williams, A., Nangia, N., Bowman, S.R.: A broad-coverage challenge corpus for sentence understanding through inference. arXiv:1704.05426 (2017). https://doi.org/10.48550/arXiv.1704.05426

22. Wolf, T., et al.: Transformers: state-of-the-art natural language processing. arXiv:1910.03771 (2020). https://doi.org/10.48550/arXiv.1910.03771

23. Yang, Z., Dai, Z., Yang, Y., Carbonell, J., Salakhutdinov, R., Le, Q.V.: XLNet: generalized autoregressive pretraining for language understanding. arXiv:1906.08237 (2019). https://doi.org/10.48550/arXiv.1906.08237

24. Yu, C., Han, J., Zhang, H., Ng, W.: Hypernymy detection for low-resource languages via meta learning. In: Proceedings of the 58th Annual Meeting of the Association for Computational Linguistics, pp. 3651–3656. Association for Computational Linguistics, Online, July 2020. https://doi.org/10.18653/v1/2020.acl-main.336, https://aclanthology.org/2020.acl-main.336

Time-Aware Circulant Matrices for Question-Based Temporal Localization

Pierfrancesco Bruni[1], Alex Falcon[1]($\boxtimes$) (ID), and Petia Radeva[2] (ID)

[1] University of Udine, Udine, Italy
{pierfrancesco.bruni,falcon.alex}@spes.uniud.it
[2] University of Barcelona, Barcelona, Spain
petia.ivanova@ub.edu

Abstract. Episodic memory involves the ability to recall specific events, experiences, and locations from one's past. Humans use this ability to understand the context and significance of past events, while also being able to plan for future endeavors. Unfortunately, episodic memory can decline with age and certain neurological conditions. By using machine learning and computer vision techniques, it could be possible to "observe" the daily routines of elderly individuals from their point of view and provide customized healthcare and support. For example, it could help an elderly person remember whether they have taken their daily medication or not. Therefore, considering the important impact on healthcare and societal assistance, this problem has been recently discussed in the research community, naming it Episodic Memory via Natural Language Queries. Recent approaches to this problem mostly rely on the literature related to similar fields, but contextual information from past and future clips is often unexplored. To address this limitation, in this paper we propose the Time-aware Circulant Matrices technique, which aims at introducing awareness of the surrounding clips into the model. In the experimental results, we present the robustness of our method by ablating its components, and confirm its effectiveness on the Ego4D public dataset, achieving an absolute improvement of more than 1% on R@5.

Keywords: Natural language query for temporal localization · Cross-modal understanding

1 Introduction

The ability to remember events that occur in our lives is a fundamental aspect of human cognition, known as episodic memory [27]. As humans, we can remember past experiences and recall specific details about them, such as when and where they occurred, who was present, and what happened. For instance, to follow a balanced diet, we may want to prepare dinner depending on what we had for lunch, which entails our ability to precisely recall the ingredients in order to compute their macro nutrients. Similarly, elderly people may need to ingest several medications throughout the day, but are they able to recall whether they took everything they need or not? To support the users in these situations by means of an intelligent system, two main components are needed. First, augmented reality glasses and similar vision systems can be used to capture the environment and the interactions with it over time, through a first-person perspective. Second, the system needs to process these visual information and then, once the user asks a question, provide the correct answer by contextualizing them to the contents of the question. Given the need to process both visual and auditory data, this second component requires the use of a variety of artificial intelligence techniques. One recent approach to it is Episodic Memory via Natural Language Queries (NLQ) [15], in which the questions are expressed in textual form to leverage the recent advancements in natural language processing.

To address the new problem, two baselines were utilized, drawing inspiration from previous works on temporal activity localization (TAL), which required to identify and localize simple actions in a video [33,34]. Compared to TAL, NLQ is more difficult as it requires to localize the moment in time from which the answer to an input question can be deduced. Nonetheless, there are several shared problems between TAL and NLQ, including the length of the untrimmed videos and the need to capture multimodal interactions. To deal with these problems, Ge et al. in [12] discovered both textual and visual concepts and used them to ease activity localization, using both the sentence/video embeddings and the concepts embeddings, e.g. verb-object textual pairs and high level concepts extracted from pretrained deep networks. Wu et al. in [28] proposed Multimodal Circulant Fusion, which allowed for multimodal interactions between the visual features and the circulant matrix of textual features, and vice versa, leading to improved localization accuracy. Recently, Zhang et al. in [35] simultaneously explored intra- and inter-modal relations through a multimodal interaction graph. While these methods propose solutions that allow for an improved understanding of the data under analysis, the temporal relations between frames and short clips are often neglected in later parts of the network architecture. To address this limitation, we introduce the Time-aware Circulant Matrices technique, enabling an improved intra-modal reasoning by injecting temporal awareness into later parts of the network. We confirm the effectiveness of our method by testing it on the Ego4D dataset [15], in which we improve the baseline performance in all the metrics under consideration. Moreover, we perform ablation studies and experiments to support our design choices.

The main contribution of this work can be summarized as follows:

- we propose to address the NLQ task by introducing the Time-aware Circulant Matrices technique, which injects temporal awareness by modelling the local context and taking it into account when performing the analysis of the visual features;
- by testing our solution in the Ego4D benchmark, we show that our proposed method achieves considerable improvements in all the metrics under consideration.

After this introduction, the related work is described in Sect. 2. Then, Sect. 3 presents and motivates the proposed Time-aware Circulant Matrices technique.

The experimental results are presented in Sect. 4 and, finally, Sect. 5 concludes the manuscript.

2   Related Work

2.1   Episodic Memory via Natural Language Queries

This challenging problem aims at identifying the moments in a video which contain relevant information to provide the correct answer to a given question. Note that the answer is found within the video (e.g., where did I forget the car keys?), hence why it is called *episodic*; in contrast, *factual/semantic* memory refers to the ability of recalling the correct answer from external knowledge bases (e.g., does Italy share a border with Spain?). Recent advancements on this topic are mostly related to the homonymous Ego4D benchmark track. The initial baselines, VSLNet [33] and 2D-TAN [34], were inspired from previous works on language grounding in video. The former implements at its core a 2D map of adjacent moment candidates, which are then queried by the sentence representation to obtain the best matching one; whereas the latter directly regresses the start and end boundaries from the input visual and textual features, supporting this process by means of a query-guided highlighting module. Building upon these works, several solutions were recently proposed. Lin et al. [18] used VSLNet on top of pretrained EgoVLP features [19]. To tackle the low amount of videos, ReLER [21] proposed data augmentation techniques on top of a multi-scale Transformer-based encoder for VSLNet and several pre-extracted visual and textual features [9,13,24]. Hou et al. in [16] proposed a three-stage approach consisting of feature filtering, using a pre-trained video-language model (EgoVLP [19]), moment proposal with Moment-DETR in [17] extended with inter-windows contrastive learning, and finally a novel intra-windows fine-grained ranking strategy. Mo et al. in [23] used a simple Transformer-based method called ActionFormer [32]. A foundational model, InternVideo in [4], recently obtained state-of-the-art results on dozens of challenges and datasets, and was used as a backbone for VSLNet.

Differently from them, we focus on the modelling aspects and propose a novel technique, which we call Time-aware Circulant Matrices, to analyze the visual features and peek into the surrounding context to discover underlying patterns. Circulant matrices were also used in a previous work dealing with TAL [28] to compute additional relations between multiple sources of information. In this work, we further extend this technique by integrating contextual awareness in its framework.

2.2   Temporal Action Localization and Video Moment Retrieval

Temporal action localization and video moment retrieval are similar tasks to the problem under analysis. *Temporal action localization* requires to identify each action instance in the video, predict its temporal boundaries, and categorize it in a finite set of classes. This problem is typically tackled either in a two-stage or a single-stage fashion. The former starts by first generating coarse video segments as action proposals (e.g., by using anchor windows [3,8] or action boundaries [14,36]), and then by classifying them with action recognition models. In the latter strategy, the proposed approaches try to simultaneously locate and classify the actions without relying on generated action proposals or external classifiers [5,20]. *Video moment retrieval* (VMR) is even closer to Episodic memory via Natural Language Queries, since it requires to localize and retrieve the moments which are described by an input textual query. As in the previous case, one-stage and two-stage approaches have been proposed for VMR. In two-stage approaches, the input video is first split into multiple candidate moments (e.g., by using a sliding window approach) which are then ranked to select the best matching ones [10,34]; in the one-stage scenario, no predefined candidate moments are used and each frame is a possible candidate to represent the initial or final frame of the moment [6,31].

However, both these problems present some fundamental differences with NLQ, leading to models with different capabilities and goals. In fact, while TAL requires to identify and localize simple actions in a video, and VMR requires to retrieve all the moments which can be described by an input textual query, NLQ requires to locate a precise moment depicting certain visual cues from which it is possible to infer the answer to a given question. For instance, this means that if we are trying to recall *the color of the dress worn by the person we spoke with*, in VMR we either need to locate all the moments in which we *interact/speak with someone*, resulting in a coarse selection which would need further processing, or we need to already know the answer, i.e., the color of the dress, and insert it into the query. Therefore, while both these problems are related to the NLQ, they aim at solving different tasks.

3   Proposed Method

An overview of the method is shown in Fig. 1. We start by briefly describing the overall procedure (Sect. 3.1), then we focus on the details of the proposed Time-aware Circulant Matrices technique in Sect. 3.2.

3.1   Overview of the Procedure

Starting from the visual and textual features, $V \in \mathbb{R}^{n \times f_v}$ and $Q \in \mathbb{R}^{m \times f_t}$, a convolutional layer is applied to project the heterogeneous features to the same dimension d, i.e., $V' \in \mathbb{R}^{n \times d}$, $Q' \in \mathbb{R}^{m \times d}$. Then, a Feature Encoder made of a single Transformer Encoder is used to learn for both of them an independent representation in a common space, resulting in $\tilde{V}$ and $\tilde{Q}$.

To model the cross-modal interactions and discover more patterns in the underlying visual data, while also leveraging temporal relations in their progression over time, we use the following equation, based on Context-Query Attention

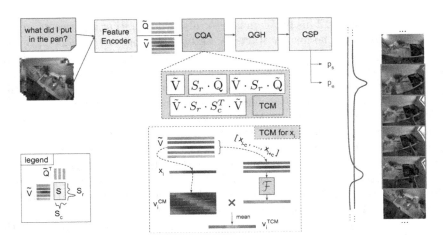

Fig. 1. Overview of the proposed method, Time-aware Circulant Matrices (TCM). The legend shows how S_r and S_c are computed from $\tilde{V}$ and $\tilde{Q}$. More details about the proposed method, TCM, and the other components, CQA, QGH, and CSP can be found in Sect. 3. Best viewed in color.

(CQA) [33] and inspired from previous works, e.g., [29,30]:

$$V^Q = FFN([\tilde{V}, V^{TCM}, A, \tilde{V} \circ A, \tilde{V} \circ B]) \tag{1}$$

where FFN is a linear layer and the other components are obtained as follows. V^{TCM} is derived through the novel three-step process which we name Time-aware Circulant Matrices and explain in Sect. 3.2. A and B are obtained by attending $\tilde{Q}$ and $\tilde{V}$ through S_r and S_c, which are the row-wise and column-wise softmax normalization of S, i.e., their similarity matrix, $S = \tilde{V} \cdot \tilde{Q}^T, S \in \mathbb{R}^{n \times m}$. Specifically, $A = S_r \cdot \tilde{Q}, A \in \mathbb{R}^{n \times d}$ and $B = S_r \cdot S_c^T \cdot \tilde{V}, B \in \mathbb{R}^{n \times d}$. The element-wise multiplication is depicted with $\circ$. The features in V^Q are then combined with h_Q, i.e., a sentence-level representation of the word features $\tilde{Q}$ obtained through content-based attention [1], to form $\hat{V}^Q = [[v_1^Q; h_Q], [v_2^Q; h_Q], \ldots, [v_n^Q; h_Q]]$.

Then, the Query-Guided Highlighting (QGH) module introduced in [33] is responsible for discriminating "foreground" moments, i.e., those which are relevant to the target, from the "background" ones, while also allowing for some flexibility in the boundaries: by labeling each moment with a binary label (0 for background, 1 for foreground), QGH reduces to a binary classification problem which helps highlighting important features obtained as $\tilde{V}^Q = S_h \cdot \hat{V}^Q$, where S_h is the highlighting score, computed as $\sigma(Conv1D(\hat{V}^Q))$.

Finally, the prediction of the boundaries is done by the Conditioned Span Predictor (CSP), which uses a shared Transformer Encoder with 4 layers, followed by two convolutional-based networks, to predict the start and end probability distributions, $p_s, p_e \in \mathbb{R}^n$, starting from the $\tilde{V}^Q$ features. The model is trained by means of the following joint loss function:

$$\mathcal{L} = \mathcal{L}_{span} + \mathcal{L}_{QGH}$$

$$\mathcal{L}_{span} = \frac{1}{2}\big(f_{CE}(p_s, y_s) + f_{CE}(p_e, y_e)\big) \tag{2}$$

$$\mathcal{L}_{QGH} = f_{BCE}(S_h, Y_h)$$

where f_{CE} is the Cross Entropy loss computed between the groundtruth and the predicted probability distributions for start, y_s and p_s, and end boundaries, y_e and p_e; and f_{BCE} is the Binary Cross Entropy loss used by the QGH module to compute the loss between the predicted and groundtruth highlighting scores, S_h and Y_h.

3.2 Time-Aware Circulant Matrices Technique

The proposed Time-aware Circulant Matrices technique consists of a three-steps procedure used to discover additional temporal intra-modal relations in the visual features.

First of all, given the video features $V = [x_1, x_2, \ldots, x_N]$, each clip vector $x_i \in \mathbb{R}^{1 \times d}$ is transformed into its circulant matrix, $v_i^{CM} \in \mathbb{R}^{d \times d}$. This is obtained by the following equation:

$$v_i^{CM} = (\overrightarrow{x_i}^0 \overrightarrow{x_i}^1 \ldots \overrightarrow{x_i}^{d-1})^T \tag{3}$$

where $\overrightarrow{x_i}^j = [x_{(d-1)-j+1}, x_{(d-1)-j+2}, \ldots, x_0, \ldots, x_{(d-1)-j}]$, i.e., $\overrightarrow{x_i}^j$ represents a shift in x_i by j.

Secondly, the resulting v_i^{CM} is multiplied by $\mathcal{F}([x_{i-c}, \ldots, x_i, \ldots, x_{i+c}])$ to establish further relations between each visual feature and the surrounding clips. By doing so, temporal awareness is injected into the model and, using a context of size c, the contextual information is obtained through the aggregator function $\mathcal{F}$. Formally:

$$Z_i = v_i^{CM} * \mathcal{F}([x_{i-c}, \ldots, x_{i+c}]) \tag{4}$$

where $\mathcal{F}([x_{i-c}, \ldots, x_{i+c}]) \in \mathbb{R}^{1 \times d}$ is broadcast to all the rows in v_i^{CM}. To implement $\mathcal{F}$, we consider a linear transformation of the concatenated context as in the following equation:

$$\mathcal{F}([x_{i-c}, \ldots, x_{i+c}]) = W_{lin}[x_{i-c}, \ldots, x_{i+c}] + b_{lin} \tag{5}$$

where W_{lin} and b_{lin} are learned at training time.

Lastly, the column-wise average of Z_i is utilized to collate the newly identified information, leading to v_i^{TCM}:

$$v_i^{TCM} = \frac{1}{d} \sum_{j=1}^{d} z_i^{(j)} \tag{6}$$

which represents the i-th vector of $V^{TCM} \in \mathbb{R}^{N \times d}$, the output of the proposed Time-aware Circulant Matrices technique.

By performing these three steps, we obtain different combinations of the visual features, possibly leading to the discovery of additional relations which were not previously considered. Note that the use of circulant matrices was also considered in [28] for better video-language understanding capabilities. However, up to our knowledge they have not been used for video temporal modelling.

4   Experimental Results

In this section, we first discuss the dataset and evaluation metrics, and the implementation details. Then, we present the experimental results related to the temporal context modeling and width, a study on the use of an asymmetric context, an ablation study on the proposed technique, and a comparison with the state of the art.

4.1   Dataset and Evaluation Metrics

The Ego4D dataset [15] is a large-scale collection of egocentric perspective videos that comprises over 3000 hours of footage. The videos are divided into clips which are 8 min long and annotated by a short narration (6–8 words), yielding roughly 3.85 million annotations. About 17000 queries for training and validation are created from these annotations for the Episodic Memory via Natural Language Queries task, using different templates such as "where is X after Y?", where X is an object and Y an event, and "who did I talk to in Z?", where Z is a location. The main evaluation metric used is Recall@k, IoU=m, which measures the proportion of instances where the intersection-over-union (IoU) between the ground truth interval and at least one of the top k predictions is greater than or equal to m. The values of k and m used are $k = 1, 5$ and $m = 0.3, 0.5$.

4.2   Implementation Details

To implement the proposed method, we start from the official codebase provided for VSLNet[1]. The PyTorch version is 1.11.0. The training procedure lasts for 200 epochs and the best model on the validation set is selected. The optimizer used is AdamW with a learning rate of 0.0001. The batch size is 32 and we used 512 as the maximum for the positional embedding in the Feature Encoder and the CSP. We use BERT [7] for the textual features and InternVideo for the visual ones [4].

4.3   Temporal Context Modeling and Width

We consider three possibilities for the function $\mathcal{F}$ which is used to aggregate the temporal context $[x_{i-c}, \ldots, x_{i+c}]$ as detailed in Eq. 4. These include: the linear transformation of the concatenated context ($cat+lin$), which is part of

---

[1] https://github.com/EGO4D/episodic-memory/tree/main/NLQ/VSLNet.

the proposed method; a simple *mean* pooling, in which $\mathcal{F}([x_{i-c}, \ldots, x_{i+c}]) = \frac{1}{2*c+k} \sum_{j=i-c}^{i+c} x_j$, where c is the size of the context (see Sect. 3) and $k = 1$ if x_i is used, otherwise $k = 0$; and finally a *GRU* model, in which $\mathcal{F}([x_{i-c}, \ldots, x_{i+c}]) = h_{2*c+k}$, where h_{2*c+k} is the last hidden state computed by the GRU.

Figure 2 reports the results obtained by the different models as the size of the considered temporal context increases. Specifically, on the top line R@1 is shown (respectively, with IoU = 0.3 and IoU = 0.5), whereas the bottom line displays R@5 values. In each of the four plots, the result achieved by VSLNet is reported for reference. Overall, it can be seen that introducing the contextual information obtained from the surrounding clips can be helpful. Specifically, the *cat+lin* strategy achieves good results when the context is small, e.g., it achieves better R@1 (IoU=0.5) and R@5 performance than the mean pooling when $c = 2$ (19.7% and 14.6% respectively for IoU=0.3 and IoU=0.5); its performance decreases as c increases, most likely because each time an element is added to the context, the weight parameter becomes bigger, possibly leading to higher memorization and lower generalization. The mean pooling leads to generally good results, with two best solutions: when x_i is not included in the context, $c = 20$ leads to the highest metrics, e.g., it achieves 12.1% R@1 IoU=0.3 (+1.3% than the baseline) and 7.4% R@1 IoU=0.5 (+0.4%); whereas $c = 10$ is preferred when x_i is included, e.g., it achieves similar R@1 in both IoU thresholds, but better R@5 (19.4% R@5 IoU=0.3, +0.9% than the baseline, and 14.5% R@5 IoU=0.5, +1.3%). Finally, the GRU solution does not lead to improvements when compared to the baseline.

The solution obtained by the concatenation and the linear leads to the best performance across most of the considered metrics, therefore it was chosen for the proposed method.

4.4   Asymmetric Context

In Sect. 3, the context which is aggregated by the function $\mathcal{F}$ has a size determined by the hyperparameter c and consists of the surrounding clips, both from past and future ones. In this experiment, we aim to investigate the effect of an asymmetric context, that is by using two values of c, c_p and c_f, and vary the amount of past and future information used. Figure 3 reports the mR@1 (Fig. 3 left) and the mR@5 (Fig. 3 right), i.e., the average value computed at IoU=0.3 and IoU=0.5 for R@1 and R@5, obtained on the validation set.

It can be seen that completely removing either of them (past or future clips) leads to generally worse solutions, e.g., up to 9.4% mR@1 and 16.6% mR@5 is obtained when either $c_p = 0$ or $c_f = 0$, whereas 9.7% and 17.1% are obtained when $c_p = c_f = 2$. In contrast, when reducing either of them, but keeping at least both one past and one future clip leads to less conclusive statements: for instance, it leads to up to 9.3% mR@1 and 16.5% mR@5 when, respectively, $c_p = 1, c_f = 1$ whereas with $c_p = c_f = 2$ it achieves 9.7% and 17.1%. However, compared to $c_p = c_f = 3$, a c_p of 2 leads to slightly better performance (e.g., 16.9% mR@5 compared to 16.7%). These results confirm that the model learns how to effectively make use of the contextual information from both preceding

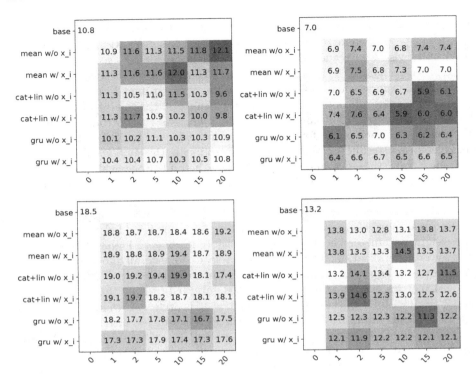

Fig. 2. Recall@1 (**top**) and Recall@5 (**bottom**) for two levels of IoU (**left**: IoU=0.3, **right**: IoU=0.5) computed on the validation set as the temporal model and the size of the context vary. VSLNet (*base*) is reported for reference. The other models are the Mean pooling, concatenation followed by a linear (*cat+lin*), and the GRU. Each model is tested both with (*w/ x_i*) and without x_i in the context (*w/o x_i*). Details in Sect. 4.3. Color scale from red (worse than *base*) to blue (better). (Color figure online)

and subsequent clips, although increasing the context too much might lead to worse generalization, due to an increased number of trainable parameters.

4.5   Ablation Study

In this ablation study, we show that the addition of the temporal awareness and the use of the circulant matrices are both important for the model. The results are reported in Table 1. The first line reports the performance achieved by the proposed method. In the second line, we remove the contextual awareness provided in Eq. 4, that is $\mathcal{F}([x_{i-c}, \ldots, x_{i+c}]) = x_i$ is used. The experimental results confirm that providing the model the information from the surrounding clips is useful, as in fact all the metrics decrease. Then, if the circulant matrices

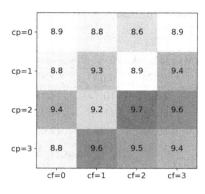 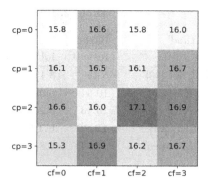

Fig. 3. Mean of Recall@1 (**left**) and Recall@5 (**right**) computed on the validation set as the amount of past, c_p, and future, c_f, clips in the context varies. VSLNet is reported for reference at $c_p = 0, c_f = 0$. Details in Sect. 4.4. Color scale from red (worse than the baseline) to blue (better).

Table 1. Ablation study reporting best results on validation set.

Method	R@1			R@5	
	IoU=0.3	IoU=0.5	Mean r@1	IoU=0.3	IoU=0.5
TCM	**11.72**	**7.64**	**9.68**	**19.70**	**14.56**
w/o T	11.33	7.12	9.23	19.02	13.99
w/o CM	10.76	7.02	8.89	18.48	13.16

are also removed, the VSLNet baseline is again obtained: as in the previous case, the metrics indicate the usefulness of the features obtained through the addition of the circulant matrix in Eq. 1.

4.6 Comparison with State-of-the-Art

As a final experimental result, we compare the performance achieved by the proposed method to that of several published works. VSLNet [33] is the baseline method and the results on the test set are taken from [15]. MSRA-AIM3 used a different set of pre-extracted visual features, made of both Swin Transformer and CLIP [22,24], which are first encoded by a Transformer-based Feature Encoder and then by a cross-modal encoder using multiple Transformer layers working on both visual and textual inputs [37]. EgoVLP is a pretraining strategy based on Frozen-in-Time [2] which used two separate Transformers to encode visual and textual inputs directly from the raw data, and pretrained them with a customized task based on multiple choice question answering and a loss function designed to specialize the selection of the samples used for the contrastive loss [18]. Finally, ReLER used a multi-scale cross-modal Transformer to model the complex interactions between video and text, two data augmentation techniques to reduce overfitting issues, and additional loss functions [21].

Table 2. Results obtained on the test set.

Method	R@1			R@5		
	IoU=0.3	IoU=0.5	Mean r@1	IoU=0.3	IoU=0.5	Mean r@5
VSLNet [33]	5.45	3.12	4.28	10.74	6.63	8.68
MSRA-AIM3 [37]	10.34	6.09	8.22	**18.01**	10.71	14.36
EgoVLP [18]	10.46	6.24	8.35	16.76	11.29	14.02
ReLER [21]	**12.89**	**8.14**	**10.51**	15.41	9.94	12.67
TCM (**ours**)	11.64	6.84	9.24	17.43	**11.39**	**14.41**

Table 2 presents the comparison. It can be seen that by using the proposed technique, we achieve better performance than VSLNet, MSRA-AIM3, and EgoVLP. Compared to ReLER we achieve better Recall@5, meaning that the top 5 candidates predicted by our model are generally more precise than those predicted by ReLER; on the other hand, ReLER achieves better Recall@1, meaning that their first candidate is generally more precise than ours. This may be due to the additional samples "generated" by the data augmentation techniques.

There are also some very recent works which tackle this challenging task [25], but the results are difficult to compare since an updated version of the annotations for the dataset, almost doubling the total amount of annotations (around 27k queries in place of the 17k that we used), has been released. In future work, the new version of the dataset will be considered.

5   Conclusions

Machines often lack the ability to remember past events involving other people, the interactions both with objects and other people, and the locations, i.e., episodic memory, which, on the other hand, is a fundamental aspect of human cognition. Considering the important impact on healthcare, societal assistance, and education, a novel problem has been recently proposed, called Episodic Memory via Natural Language Queries [15]. Previous works from the literature address this problem by leveraging methods inspired from other research domains (temporal activity localization and video moment retrieval), although in most of them the usage of contextual information from adjacent clips is limited to early layers of the network and often unexplored. In this paper, we address this limitation by proposing the Time-aware Circulant Matrices technique, which injects contextual awareness and intra-modal reasoning in later parts of the model. The experimental results motivate the design choices of the proposed method, present its robustness by means of an ablation study, and also confirm its effectiveness by comparing it to other state of the art methods from the literature. As a future work, we aim at further exploring its effectiveness on other datasets from the video-language grounding literature [11,26], and extend its intra-modal reasoning to enable the understanding of cross-modal interactions.

Acknowledgements. This work was supported by the Department Strategic Plan (PSD) of the University of Udine-Interdepartmental Project on Artificial Intelligence (2020-25). This work was partially funded by the Horizon EU project MUSAE (No. 01070421), 2021-SGR-01094 (AGAUR), Icrea Academia'2022 (Generalitat de Catalunya), Robo STEAM (2022-1-BG01-KA220-VET-000089434, Erasmus+ EU), DeepSense (ACE053 /22/000029, ACCIÓ), DeepFoodVol (AEI-MICINN, PDC2022-133642-I00), and CERCA Programme/Generalitat de Catalunya.

References

1. Bahdanau, D., Cho, K., Bengio, Y.: Neural machine translation by jointly learning to align and translate. In: International Conference on Learning Representations (2015)
2. Bain, M., Nagrani, A., Varol, G., Zisserman, A.: Frozen in time: a joint video and image encoder for end-to-end retrieval. In: Proceedings of the IEEE/CVF International Conference on Computer Vision, pp. 1728–1738 (2021)
3. Buch, S., Escorcia, V., Shen, C., Ghanem, B., Carlos Niebles, J.: SST: single-stream temporal action proposals. In: Proceedings of the IEEE Conference on Computer Vision and Pattern Recognition, pp. 2911–2920 (2017)
4. Chen, G., et al.: InternVideo-Ego4D: a pack of champion solutions to Ego4D challenges. arXiv preprint: arXiv:2211.09529 (2022)
5. Cheng, F., Bertasius, G.: TallFormer: temporal action localization with a long-memory transformer. In: Avidan, S., Brostow, G., Cisse, M., Farinella, G.M., Hassner, T. (eds.) ECCV 2022. Lecture Notes in Computer Science, vol. 13694, pp. 503–521. Springer, Cham (2022). https://doi.org/10.1007/978-3-031-19830-4_29
6. Cui, R., et al.: Video moment retrieval from text queries via single frame annotation. In: Proceedings of the 45th International ACM SIGIR Conference on Research and Development in Information Retrieval, pp. 1033–1043 (2022)
7. Devlin, J., Chang, M.W., Lee, K., Toutanova, K.: BERT: pre-training of deep bidirectional transformers for language understanding. In: Proceedings of the 2019 Conference of the North American Chapter of the Association for Computational Linguistics: Human Language Technologies, Volume 1 (Long and Short Papers), pp. 4171–4186 (2019)
8. Escorcia, V., Caba Heilbron, F., Niebles, J.C., Ghanem, B.: DAPs: deep action proposals for action understanding. In: Leibe, B., Matas, J., Sebe, N., Welling, M. (eds.) ECCV 2016. LNCS, vol. 9907, pp. 768–784. Springer, Cham (2016). https://doi.org/10.1007/978-3-319-46487-9_47
9. Feichtenhofer, C., Fan, H., Malik, J., He, K.: SlowFast networks for video recognition. In: Proceedings of the IEEE/CVF International Conference on Computer Vision, pp. 6202–6211 (2019)
10. Gao, J., Sun, X., Xu, M., Zhou, X., Ghanem, B.: Relation-aware video reading comprehension for temporal language grounding. In: Proceedings of the 2021 Conference on Empirical Methods in Natural Language Processing, pp. 3978–3988 (2021)
11. Gao, J., Sun, C., Yang, Z., Nevatia, R.: TALL: temporal activity localization via language query. In: Proceedings of the IEEE International Conference on Computer Vision, pp. 5267–5275 (2017)
12. Ge, R., Gao, J., Chen, K., Nevatia, R.: MAC: mining activity concepts for language-based temporal localization. In: 2019 IEEE Winter Conference on Applications of Computer Vision (WACV), pp. 245–253. IEEE (2019)

13. Girdhar, R., Singh, M., Ravi, N., van der Maaten, L., Joulin, A., Misra, I.: Omnivore: a single model for many visual modalities. In: Proceedings of the IEEE/CVF Conference on Computer Vision and Pattern Recognition, pp. 16102–16112 (2022)
14. Gong, G., Zheng, L., Mu, Y.: Scale matters: temporal scale aggregation network for precise action localization in untrimmed videos. In: 2020 IEEE International Conference on Multimedia and Expo (ICME), pp. 1–6. IEEE (2020)
15. Grauman, K., et al.: Ego4D: around the world in 3,000 hours of egocentric video. In: Proceedings of the IEEE/CVF Conference on Computer Vision and Pattern Recognition, pp. 18995–19012 (2022)
16. Hou, Z., et al.: An efficient coarse-to-fine alignment framework@ Ego4D natural language queries challenge 2022. arXiv preprint: arXiv:2211.08776 (2022)
17. Lei, J., Berg, T.L., Bansal, M.: Detecting moments and highlights in videos via natural language queries. In: Advances in Neural Information Processing Systems, vol. 34, pp. 11846–11858 (2021)
18. Lin, K.Q., et al.: Egocentric video-language pretraining@ Ego4D challenge 2022. arXiv preprint: arXiv:2207.01622 (2022)
19. Lin, K.Q., et al.: Egocentric video-language pretraining. In: Advances in Neural Information Processing Systems, vol. 35, pp. 7575-7586 (2022)
20. Lin, T., Zhao, X., Shou, Z.: Single shot temporal action detection. In: Proceedings of the 25th ACM International Conference on Multimedia, pp. 988–996 (2017)
21. Liu, N., Wang, X., Li, X., Yang, Y., Zhuang, Y.: Reler@ zju-alibaba submission to the Ego4D natural language queries challenge 2022. arXiv preprint: arXiv:2207.00383 (2022)
22. Liu, Z., et al.: Video Swin transformer. arXiv preprint: arXiv:2106.13230 (2021)
23. Mo, S., Mu, F., Li, Y.: A simple transformer-based model for Ego4D natural language queries challenge. arXiv preprint: arXiv:2211.08704 (2022)
24. Radford, A., et al.: Learning transferable visual models from natural language supervision. In: International Conference on Machine Learning, pp. 8748–8763. PMLR (2021)
25. Ramakrishnan, S.K., Al-Halah, Z., Grauman, K.: NaQ: Leveraging narrations as queries to supervise episodic memory. arXiv preprint: arXiv:2301.00746 (2023)
26. Soldan, M., et al.: MAD: a scalable dataset for language grounding in videos from movie audio descriptions. In: Proceedings of the IEEE/CVF Conference on Computer Vision and Pattern Recognition, pp. 5026–5035 (2022)
27. Tulving, E.: Episodic and semantic memory: where should we go from here? Behav. Brain Sci. **9**(3), 573–577 (1986)
28. Wu, A., Han, Y.: Multi-modal circulant fusion for video-to-language and backward. In: IJCAI, vol. 3, p. 8 (2018)
29. Xiong, C., Zhong, V., Socher, R.: Dynamic Coattention networks for question answering. In: International Conference on Learning Representations (2016)
30. Yu, A.W., et al.: QaNet: combining local convolution with global self-attention for reading comprehension. In: International Conference on Learning Representations (2018)
31. Zeng, R., Xu, H., Huang, W., Chen, P., Tan, M., Gan, C.: Dense regression network for video grounding. In: Proceedings of the IEEE/CVF Conference on Computer Vision and Pattern Recognition, pp. 10287–10296 (2020)
32. Zhang, C.L., Wu, J., Li, Y.: ActionFormer: localizing moments of actions with transformers. In: Avidan, S., Brostow, G., Cisse, M., Farinella, G.M., Hassner, T. (eds.) ECCV 2022. Lecture Notes in Computer Science, vol. 13664, pp. 492–510. Springer, Cham (2022)

33. Zhang, H., Sun, A., Jing, W., Zhou, J.T.: Span-based localizing network for natural language video localization. In: Proceedings of the 58th Annual Meeting of the Association for Computational Linguistics, pp. 6543–6554 (2020)
34. Zhang, S., Peng, H., Fu, J., Luo, J.: Learning 2D temporal adjacent networks for moment localization with natural language. In: Proceedings of the AAAI Conference on Artificial Intelligence, vol. 34, pp. 12870–12877 (2020)
35. Zhang, Z., Han, X., Song, X., Yan, Y., Nie, L.: Multi-modal interaction graph convolutional network for temporal language localization in videos. IEEE Trans. Image Process. **30**, 8265–8277 (2021)
36. Zhao, P., Xie, L., Ju, C., Zhang, Y., Wang, Y., Tian, Q.: Bottom-up temporal action localization with mutual regularization. In: Vedaldi, A., Bischof, H., Brox, T., Frahm, J.-M. (eds.) ECCV 2020. LNCS, vol. 12353, pp. 539–555. Springer, Cham (2020). https://doi.org/10.1007/978-3-030-58598-3_32
37. Zheng, S., Zhang, Q., Liu, B., Jin, Q., Fu, J.: Exploring anchor-based detection for ego4d natural language query. arXiv preprint: arXiv:2208.05375 (2022)

Enhancing Open-Vocabulary Semantic Segmentation with Prototype Retrieval

Luca Barsellotti[(✉)][ID], Roberto Amoroso[ID], Lorenzo Baraldi[ID],
and Rita Cucchiara[ID]

University of Modena and Reggio Emilia, Modena, Italy
{luca.barsellotti,roberto.amoroso,lorenzo.baraldi,
rita.cucchiara}@unimore.it

Abstract. Large-scale pre-trained vision-language models like CLIP exhibit impressive zero-shot capabilities in classification and retrieval tasks. However, their application to open-vocabulary semantic segmentation remains challenging due to the gap between the global features extracted by CLIP for whole-image recognition and the requirement for semantically detailed pixel-level features. Recent two-stage methods have attempted to overcome these challenges by generating mask proposals that are agnostic to specific classes, thereby facilitating the identification of regions within images, which are subsequently classified using CLIP. However, this introduces a significant domain shift between the masked and cropped proposals and the images on which CLIP was trained. Fine-tuning CLIP on a limited annotated dataset can alleviate this bias but may compromise its generalization to unseen classes. In this paper, we present a method to address the domain shift without relying on fine-tuning. Our proposed approach utilizes weakly supervised region prototypes acquired from image-caption pairs. We construct a visual vocabulary by associating the words in the captions with region proposals using CLIP embeddings. Then, we cluster these embeddings to obtain prototypes that embed the same domain shift observed in conventional two-step methods. During inference, these prototypes can be retrieved alongside textual prompts. Our region classification incorporates both textual similarity with the class noun and similarity with prototypes from our vocabulary. Our experiments show the effectiveness of using retrieval to enhance vision-language architectures for open-vocabulary semantic segmentation.

Keywords: Open-Vocabulary · Semantic Segmentation · Retrieval

1   Introduction

Semantic segmentation is a widely studied Computer Vision task, involving the partitioning of an image into regions that correspond to specific object classes with semantic meaning. However, obtaining precise annotations for this task can be costly, hindering scalability to large datasets. In addition, conventional

G. L. Foresti et al. (Eds.): ICIAP 2023, LNCS 14234, pp. 196–208, 2023.
https://doi.org/10.1007/978-3-031-43153-1_17

semantic segmentation models [5,7] are typically trained on a finite set of classes, making them unable to recognize novel or unexpected objects. To overcome these challenges, recent studies have focused on developing open-vocabulary semantic segmentation models [9,12,22,27] that can recognize a variable number of classes, including previously unseen or out-of-domain samples. These models offer greater flexibility and applicability to various real-world scenarios, such as robotics, autonomous driving, and medical image analysis [4,8].

The growing interest in open-vocabulary semantic segmentation can be attributed to the emergence of large-scale pre-trained vision-language models, such as CLIP [24] and ALIGN [16]. These models have been trained on billions of image-text training examples, enabling them to learn rich multi-modal features. Notably, they exhibit the ability to embed a vast vocabulary and, consequently, excellent zero-shot capabilities when applied to downstream tasks such as classification [3,10] and image retrieval. However, transferring this knowledge to dense prediction tasks presents challenges, as the model must not only identify the object classes within an image but also precisely localize them.

Two-stage approaches have emerged as effective methods for addressing the open-vocabulary segmentation task and tackling the localization problem. These methods involve two stages: first, a mask proposer generates class-agnostic mask proposals. Then, the image regions corresponding to the generated masks are extracted, and a CLIP model is used to perform open-vocabulary classification on each region. Although the class-agnostic proposer demonstrates strong generalization to arbitrary categories [23], the bottleneck in the performance is represented by the inability of CLIP in recognizing the masked and cropped image regions [18]. This limitation stems from the domain shift between the images provided to CLIP during training and those used in this setup. Resizing, masking and cropping the object image adversely affect its positioning in the feature space with respect to text embeddings. However, fine-tuning CLIP on a closed-vocabulary annotated dataset to compensate for this domain shift may interfere with its generalization capabilities on unseen classes.

To address the challenge introduced by the domain shift *without fine-tuning*, we propose a pre-processing step that involves creating a visual vocabulary that associates a given word with a series of reference CLIP visual feature embeddings. These embeddings are generated by collecting region proposals extracted from an image-caption dataset and by applying a clustering algorithm on top of them. Thus, the resulting cluster centroids incorporate the same domain shift while providing a rich variety of visual characteristics of the corresponding word. Alongside CLIP's open-vocabulary classification on each region, the vocabulary visual reference embeddings can be retrieved to augment the segmentation process, thereby improving its robustness and accuracy.

Our experiments demonstrate the effectiveness of integrating retrieval methods to enhance the two-stage architecture without the need for further fine-tuning. The combination of the visual vocabulary reference embeddings and the two-step segmentation approach yields enhanced performance, highlighting the potential of utilizing pre-existing knowledge and domain adaptation techniques to address the domain shift challenge in open-vocabulary segmentation.

2   Related Works

Semantic Segmentation is a fundamental dense prediction task in Computer Vision that aims to assign a label to each pixel of an image. The field is primarily driven by two main lines of research: one that treats it as a pixel-level classification problem [1,2,5], and another that decouples it into a two-subtask problem [7], involving the proposal of regions of interest and the subsequent classification of these proposed regions. Both approaches have shown excellent performance in closed-vocabulary scenarios under the supervised learning paradigm.

Zero-Shot Semantic Segmentation has gained significant attention in recent years, driven by the high costs associated with annotating masks for a wide range of categories. In this setting, models are trained on a set of seen classes and are then expected to generalize their knowledge to unseen classes. While early works predominantly relied on discriminative [26] and generative methods [14], recent advancements have shifted towards the decoupling paradigm [9,28]. These methods aim to enhance the generalization capabilities of the class-agnostic mask proposer, enabling it to accurately identify novel objects. Additionally, they leverage the power of large-scale pre-trained vision-language models to assign appropriate labels to each region proposal, further improving the overall performance of zero-shot semantic segmentation. Our proposed method is closely related to zero-shot semantic segmentation as it harnesses pre-existing knowledge encoded in the visual vocabulary and employs reference embeddings to enhance the segmentation performance for previously unseen categories.

Open-Vocabulary Segmentation is a generalized zero-shot learning task that aims to establish a method for arbitrary recognition of an unlimited number of object classes, even with the use of additional training data. LSeg [17] aligns dense per-pixel and textual embeddings in the same semantic space, whereas OpenSeg [12] and GroupViT [27] propose to group pixels before learning visual-semantic alignments. Some methods, such as MaskCLIP [30] and PACL [22], investigate the capability of CLIP itself in producing dense predictions already aligned with text embeddings. Two-stage approaches have proven remarkable performance in open-vocabulary segmentation, compensating for the poor localization ability of CLIP. Their main bottleneck is given by the domain shift between the masked regions and the images on which CLIP has been trained. To bridge this gap, ZSSeg [28] proposes a textual prompt-learning approach, whereas OVSeg [18] exploits the usage of learnable tokens to replace blank areas of the masked regions. In our proposed method, we tackle the domain shift issue by constructing a visual vocabulary that aligns with the preprocessing steps applied to the input images. This alignment effectively incorporates the domain shift and improves the robustness of the model.

3   Method

Open vocabulary semantic segmentation involves the task of assigning a label from a set of arbitrary categories to each pixel in an image. In two-stage meth-

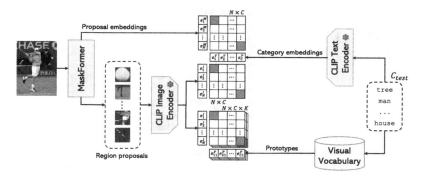

Fig. 1. Overview of our proposed method, VOCSeg, for two-stage open-vocabulary semantic segmentation enhanced by visual prototype retrieval.

ods [9,18,28], this task is reformulated into dividing the image into coherent regions and assigning each region a category.

In our proposed open-vocabulary semantic segmentation architecture depicted in Fig. 1, we introduce a novel approach to tackle these challenges. The architecture comprises three main components: a mask proposer, an enhanced CLIP model with retrieval capabilities, and a visual vocabulary. The mask proposer generates region proposals within the image, while the CLIP model extracts embeddings for these proposed regions. These embeddings serve as representations for independent open vocabulary classification of each region. However, it is essential to consider the domain shift introduced by cropping and masking regions, as it deviates from the training images of CLIP. To mitigate this domain shift, we introduce the concept of *visual prototypes*. Firstly, we employ a two-stage segmentation method on a dataset consisting of image-text pairs to obtain region proposals for a diverse range of words. These proposals collectively form the visual vocabulary, which encapsulates the domain shift resulting from the cropping and masking process. Subsequently, we generate visual prototypes for each word by clustering the corresponding set of collected regions. These prototypes serve as representative embeddings within the feature space.

At inference time, we leverage textual category embeddings and retrieved prototypes for each category. These prototypes reside in the same feature space as the embeddings and allow us to incorporate both textual and visual similarities using only the CLIP model, avoiding an increase in computational effort.

3.1 Prototype Extraction from Image-Caption Pairs

Collecting a Visual Vocabulary. In our approach to open-vocabulary segmentation, it is crucial to utilize prototypes that capture both the distinctive features of each category and the domain shift resulting from masking the regions. These prototypes play a pivotal role in classifying the proposed regions by identifying visually similar correspondences. However, collecting regions for a large vocabulary represents a challenge, making the use of pre-annotated segmentation datasets

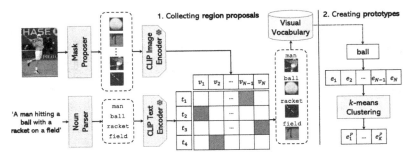

Fig. 2. Overview of the approach for collecting region proposals starting from image-caption pairs and of the clustering process used to generate prototypes.

infeasible due to their limited category coverage. To tackle this challenge, we adopt a *self-labeling* strategy for constructing an open-vocabulary collection of regions. This strategy involves extracting regions from a dataset of image-caption pairs, associating them with a vocabulary based on their corresponding captions, and subsequently generating prototypes through the clustering of similar embeddings, as shown in Fig. 2.

Specifically, we extract nouns from each caption, incorporate them into a text prompt, and provide them as input to the Text Encoder of a CLIP model. Subsequently, we obtain mask proposal embeddings using the Image Encoder of the same CLIP model and match the mask proposals with each noun using their respective computed embeddings. Although this matching process may introduce some noise, the presence of the noun in the caption ensures that one of the masks must be related to the corresponding object. Finally, we singularize the extracted nouns and store the CLIP embeddings of each match in a visual vocabulary.

Generating Prototypes. Finally, we perform a k-means clustering on the set of collected region embeddings for each noun in the vocabulary to generate a set of prototypes, represented by the cluster centroids. The k-means algorithm groups similar features, forming representative prototypes for each noun category. In this way, we ensure that our prototypes capture a wide range of visual characteristics.

Handling Rare Nouns. There are cases where the number of collected embeddings may not be sufficient to perform k-means clustering effectively, either due to a limited correspondence in the captions or arbitrary test categories that do not match entries in the visual vocabulary. For these rare nouns, we employ a k-nearest neighbors algorithm. This algorithm matches the textual embeddings extracted using CLIP with the most similar words present in the vocabulary. Subsequently, we perform k-means clustering on the embeddings of the N neighbors to generate prototypes. We increment the value of N until we have an adequate number of embeddings to perform the k-means clustering effectively.

3.2   Two-Stage Open-Vocabulary with Prototype Retrieval

The objective of two-stage open-vocabulary semantic segmentation is to identify a pair of mappings $(\mathcal{S}, \mathcal{L})$ for an input image $I \in \mathbb{R}^{H \times W \times 3}$ across C_{test} arbitrary categories. In this task, $\mathcal{S}$ partitions I into a set P of T regions, defined as follows:

$$P = \{P_i,\}_{i=1}^{T} \quad \text{with} \quad P_i \subseteq I, \cup_{i=1}^{T} P_i = I, \forall i,j : i \neq j, P_i \cap P_j = \emptyset , \quad (1)$$

whereas $\mathcal{L}$ assigns a category $c \in C_{\text{test}}$ to each region $P_i \subseteq I$, where $i = 1, \ldots, T$.

Extracting Mask Proposals Embeddings. To obtain class-agnostic mask proposals, we utilize MaskFormer [7]. This model is trained on a set of classes C_{train}, nevertheless, as reported by Xu et al. [28], it can generate T high-quality mask proposals $\{M_i\}_{i=1}^{T}$ and their corresponding mask embeddings, even for unseen classes. Each mask proposal $M_i \in \mathbb{R}^{H \times W}$ is converted into a binary mask $M_i^B \in {0,1}^{H \times W}$ by applying a sigmoid function followed by thresholding. The binary mask indicates the location of the object in the input image.

In the original MaskFormer [7] architecture, the mask embedding is a C_{train}-dimensional distribution that represents the probability of each training class. To extend the model to an open-vocabulary setting, inspired by [18,28], we modify MaskFormer in such a way that each mask generates an F-dimensional embedding, where F is the embedding dimension of a CLIP model. This adaptation ensures compatibility between the mask embeddings and the CLIP textual embeddings, which are extracted from the nouns of various semantic classes, thus enabling open-vocabulary capabilities. We include an additional F-dimensional learnable embedding for `no-object`.

Further, we also employ the CLIP image encoder to extract an additional set of embeddings from the proposed regions, which complements the ones generated for each region by MaskFormer. In particular, for each binary mask M_i^B, we erase the unused background, crop around a bounding box, that incorporates entirely the foreground area, and resize to the input resolution of CLIP. Then, the region is fed to CLIP to produce an embedding that can be used to compute similarity against the textual category embeddings.

Assigning Proposals to Classes. For each category in C_{test}, we retrieve a set of K reference prototype embeddings from a visual vocabulary. To compute the final similarities between region proposals and categories, we combine two terms: one which exploits textual category labels and one that exploits the reference prototype embeddings. In particular, for each category $c_j \in C_{\text{test}}$ we extract an embedding e_j^T with CLIP using the Textual Encoder, we retrieve a set of prototypes $\{e_{jk}^P\}_{k=1\ldots K}$, and for each region P_i we extract an embedding e_i^I with the Image Encoder of CLIP and an embedding e_i^M with MaskFormer. First, we aggregate the prototype similarities by considering the average of the maximum similarity with the K prototypes assigned to c_j and the mean similarity with all of them. This is a trade-off between considering the nearest reference embedding which is the most significant for the current region and the robustness offered by a single average embedding representative for the whole concept:

$$s_{i,j}^P = \frac{1}{2} \max_k \text{sim}(e_i^I, e_{jk}^P) + \frac{1}{2K} \sum_{k=1}^{K} \text{sim}(e_i^I, e_{jk}^P), \quad (2)$$

where $i = 1 \ldots T, j = 1 \ldots |C_{\text{test}}|, k = 1 \ldots K$ and $\text{sim}(\cdot, \cdot)$ is the cosine similarity.

Then, since both the prototype similarities and the textual similarities are computed in the same feature space, we fuse them using a linear combination with weights α and $(1 - \alpha)$. This ensembling strategy rewards the situations in which the textual and prototype similarities agree, whereas penalizes cases of disagreement. Formally, the resulting aggregated similarity is defined as

$$\tilde{s}_{i,j} = \alpha s_{i,j}^P + (1 - \alpha) \cdot \text{sim}(e_i^I, e_j^T). \tag{3}$$

The probability vector over classes $\tilde{p}$ is computed through the softmax function with a temperature τ.

Fusing with MaskFormer Predictions. Since MaskFormer is trained on C_{train}, its performance is biased towards categories belonging to this set. When the object contained in the region P_i is not recognized as a category of C_{train}, MaskFormer produces an embedding similar to the no-object embedding. Hence, when the softmax is applied to its similarities, all the resulting probabilities corresponding to the categories of C_{test} are small, and the one corresponding to no-object is large, which is removed after the softmax. So, the final prediction of P_i and c_j is obtained through the weighted geometric mean, with weights β and $(1 - \beta)$, between the probability $\tilde{p}$ of the visual-text branch and the probability $\hat{p}$ resulting from MaskFormer, in such a way that the prediction of MaskFormer is enhanced only when it is confident about it (*i.e.*, when c_j belongs to C_{train} too):

$$p_{i,j} = \tilde{p}_{i,j}^{\beta} \cdot \hat{p}_{i,j}^{(1-\beta)}. \tag{4}$$

Computing Semantic Segmentation. Finally, mask predictions and probabilities are aggregated to compute the semantic segmentation. Specifically, the score $z_j(q)$ of a category $c_j \in C_{\text{test}}$ in a pixel q is computed as the sum of each mask activation M_i multiplied for the corresponding probability $p_{i,j}$:

$$z_j(q) = \sum_{i=1}^{T} M_i(q) p_{i,j}. \tag{5}$$

4   Experimental Evaluation

4.1   Datasets

Following Liang *et al.* [18], we train our MaskFormer backbone on COCO-Stuff [6] using the all available 171 categories. We conduct experiments on five sets of test categories, obtained upon three datasets: PASCAL-VOC 2012 [11], ADE20k [29] (150 and 847 categories), and PASCAL-Context [21] (59 and 459 categories).

COCO-Stuff is an extension of the MS COCO [19] dataset for semantic segmentation. It contains annotations for 171 classes on 118,287 training images and 5,000 validation images. Due to its high-quality annotations, we use it as the training dataset for the mask proposer. As reported in [18,28], MaskFormer trained on a set of seen classes can produce high-quality masks on unseen classes.

PASCAL-VOC 2012 contains annotations for 20 classes on 11,185 training images and 1,449 validation images. Its classes exhibit significant overlapping with COCO-Stuff categories (95% overlap). This overlap makes it interesting to evaluate performance on known objects sampled from a distribution that differs from the distribution of the training dataset.

ADE20k is a challenging segmentation dataset containing several indoor and outdoor scenes. It is partitioned into 20,000 training images, 3,000 test images, and 2,000 validation images. In the original setting, it contains 150 classes ($\sim$ 45% overlap with COCO-Stuff), but its full version comprises more than 3,000 classes. Following [7], we evaluate the performance on the set containing 847 classes.

PASCAL-Context is an extension of the PASCAL-VOC 2010 dataset. It contains 4,998 training images and 5,005 validation images in two settings, one with the most frequently used 59 classes ($\sim$ 83% overlap with COCO-Stuff) and one with the whole 459 classes.

4.2 Experimental Setup

We train the modified MaskFormer model on the COCO-Stuff dataset, according to [18], with the Swin-B [20] backbone. We follow the original training settings of MaskFormer [7]. We use the OpenCLIP [15] implementation of CLIP with ViT-L/14 backbone trained on LAION2B [25]. To embed the category names with CLIP, we surround them with the text prompts proposed in the original CLIP [24] and in ViLD [13]. To obtain a diverse set of prototypes, we utilize COCO Captions [6]. We collect 15,000 unique nouns from the dataset. To extract binary masks we apply a threshold of 0.4 after the sigmoid.

4.3 Ablation Studies

Masking Strategy. We investigate the impact of three different masking strategies for extracting the regions detected by the mask proposer. In particular, MaskFormer generates N mask proposals denoted as $M_i \in \mathbb{R}^{H \times W}$. These proposals indicate the activation level of each position in the image with respect to the detected region. In our main pipeline, referred to as *binary* strategy, we consider the binarized masks $\{M_i^B\}_{i=1}^N$. In order to isolate the foreground object and eliminate the potential interference of surrounding context noise on the open-vocabulary classification of the region through CLIP, we erase the background information, keeping solely the foreground object. However, we also acknowledge that in certain cases, the background can provide crucial information for accurately recognizing the object. To address this, we explore two alternative strategies: one in which we crop the region without erasing the background (which we name *none*), and one, instead, in which we attenuate the background by multiplying the image pixels with a normalized heatmap derived from the originally proposed mask (termed *heatmap*). This allows us to retain some contextual information while still emphasizing the foreground object of interest.

Table 1. Ablation on three different masking strategies, in terms of mIoU score.

	Masking Strategy		
Dataset	None	Heatmap	Binary
ADE-150	17.7	17.7	**22.5**
PAS-20	82.51	85.0	**93.4**

Table 2. Ablation on similarity ensembling, in terms of mIoU score.

	Similarity		
Dataset	Text	Visual	Ensembling
ADE-150	21.0	20.1	**22.5**
PAS-20	92.6	93.2	**93.4**

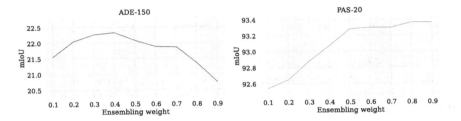

Fig. 3. Ablation on different values of the ensembling weight α.

Our experimental results, as reported in Table 1, demonstrate that the *binary strategy* provides the best mIoU scores. We argue that the noise introduced by the background overwhelms any potential advantage gained from the contextual information when it comes to clarifying the foreground object.

Ensembling. In our method, we introduce the usage of CLIP for both image-to-text and image-to-image similarities to leverage their benefits concurrently. In Table 2, we present a comparison between the individual usage of these similarities, as well as their ensembling. The results show a significant improvement of +1.5 mIoU on the ADE-150 dataset and +0.2 on the PAS-20 dataset compared to the baseline that considers only visual similarity. We argue that the reason behind this observed improvement is the complementary nature of the two types of similarities provided by CLIP. Image-to-text similarity captures the semantic understanding of the textual information associated with the images, while image-to-image similarity focuses on the shared visual content between images.

In Fig. 3, we present the trend of the mIoU as a function of the ensemble weight, for both ADE-150 and PAS-20 datasets. Notably, we observe that the performance trends differ between the two datasets, with ADE-150 performing better when assigning a larger weight to the text similarity, while PAS-20 performs better with a larger weight assigned to the visual similarity. We hypothesize that this discrepancy is influenced by the number of arbitrary categories in (C_{test}) and the quality of the vocabulary employed. Factors such as the number of samples collected for a specific word, the accuracy of matching region with words, the distribution of the embeddings in the feature space, and their representativeness of the semantic concept all play significant roles. These observations emphasize the need for an adaptation phase specific to the set of arbitrary classes, by tuning the value of the ensemble weight to obtain the best performance.

Table 3. Comparison with other state-of-the-art two-stage models.

Method	Training Dataset	Frozen CLIP	Similarity Text	Similarity Visual	PAS 20	ADE 150	ADE 847	PC 59	PC 459
GroupViT [27]	GCC+YFCC	✓	✓	✗	52.3	–	–	22.4	–
ZegFormer [9]	COCO-Stuff-156	✓	✓	✗	80.7	16.4	–	–	–
OpenSeg [16] (R-101) [a]	COCO Panoptic	✗	✓	✗	60.0	15.3	4.0	36.9	6.5
ZSSeg [28] (R-101)	COCO-Stuff-171	✗	✓	✗	88.4	20.5	7.0	47.7	–
OVSeg [18] (R-101)	COCO-Stuff-171	✗	✓	✗	89.2	24.8	7.1	53.3	11.0
OVSeg [18] (Swin-B)	COCO-Stuff-171	✗	✓	✗	94.5	29.6	9.0	55.7	12.4
VOCSeg	COCO-Stuff-171	✓	✓	✓	93.4	22.5	8.1	47.3	10.8

[a]OpenSeg uses ALIGN as the pre-trained vision-language model instead of CLIP.

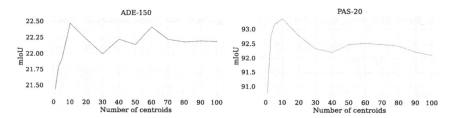

Fig. 4. Ablation on the number of clusters used in the k-means algorithm.

Number of Reference Prototypes. Figure 4 illustrates the trend of the mIoU as the number of clusters k in the k-means algorithm increases. We observe that the mIoU reaches its peak at $k = 10$ for both datasets and shows a tendency to stabilize as k further increases. The variation in mIoU can be attributed to the frequency of word occurrences in the captions. We theorize that as k increases, the noise incorporated in the reference embeddings also increases. On the other hand, when using a small value of k, the variety of representations offered by the vocabulary becomes limited. This limitation hampers the ability to embed different visual concepts under the same word, leading to decreased performance in capturing the multitude of nuances in the objects.

4.4   Comparison with State-of-the-Art Methods

We conduct a comparison with other open-vocabulary architectures based on a two-stage approach: GroupViT [27], ZegFormer [9], OpenSeg [16], ZSSeg [28] and OVSeg [18]. The results can be observed in Table 3. The "Similarity" column highlights the uniqueness of our approach in leveraging the similarities between image embeddings to bridge the gap between the images used to train CLIP and the regions extracted in two-stage approaches. Despite introducing a pre-processing step without additional parameters or fine-tuning CLIP, our method outperforms ZSSeg, which utilizes learnable tokens in the textual prompts, on both the ADE-150 and ADE-847 settings by $+2$ and $+1.1$ mIoU respectively and on PAS-20 by 5 mIoU. It also surpasses OpenSeg on all benchmark datasets, obtaining a +7.2 on ADE-150, +4.1 on ADE-847, +23.4 on PAS-20, +10.4 on P-59 and +4.3 on P-459. Furthermore, it outperforms OVSeg with a ResNet-101 backbone on ADE-150 by +4.2 and ADE-847 by +1.0. These architectures achieve high performance through fine-tuning or learnable tokens on a limited set of annotated segmentation data, which limits their generalization ability. In contrast, our method provides comparable results while allowing the extension of the visual vocabulary without compromising the quality of previously collected prototypes. Moreover, our VOCSeg largely outperforms ZegFormer and GroupViT, which operate in the same setting (*i.e.*, without fine-tuning CLIP). Our best performance is achieved using $k = 10$ in the k-means algorithm, $N = 10$ in the k-nearest neighbors algorithm, α equal to 0.8, 0.35, 0.2, 0.9, and 0.1 on, respectively, PAS-20, ADE-150, ADE-847, PAS-59 and PAS-459, and β equal to 0.7 on ADE-150 and ADE-847, and 0.6 on PAS-20, PAS-59, and PAS-459.

5   Conclusions

Our solution introduces the concepts of visual vocabulary and visual prototypes. These prototypes, extracted through clustering techniques, are a collection of reference embeddings in the vision-language space containing visual features common to the object they refer to. Through extensive experiments, we have shown that it is possible to retrieve these prototypes at inference time to enhance the recognition of the proposed regions without additional learnable parameters and without fine-tuning the large-scale vision-language model.

Acknowledgments. Research partly funded by PNRR - M4C2 - Investimento 1.3, Partenariato Esteso PE00000013 -"FAIR - Future Artificial Intelligence Research" - Spoke 8 "Pervasive AI", funded by the European Commission under the NextGeneration EU programme.

References

1. Amoroso, R., Baraldi, L., Cucchiara, R.: Assessing the role of boundary-level objectives in indoor semantic segmentation. In: Tsapatsoulis, N., Panayides, A., Theocharides, T., Lanitis, A., Pattichis, C., Vento, M. (eds.) CAIP 2021. LNCS, vol. 13052, pp. 455–465. Springer, Cham (2021). https://doi.org/10.1007/978-3-030-89128-2_44
2. Amoroso, R., Baraldi, L., Cucchiara, R.: Improving indoor semantic segmentation with boundary-level objectives. In: Rojas, I., Joya, G., Català, A. (eds.) IWANN 2021. LNCS, vol. 12862, pp. 318–329. Springer, Cham (2021). https://doi.org/10.1007/978-3-030-85099-9_26
3. Bruno, P., Amoroso, R., Cornia, M., Cascianelli, S., Baraldi, L., Cucchiara, R.: Investigating bidimensional downsampling in vision transformer models. In: Sclaroff, S., Distante, C., Leo, M., Farinella, G.M., Tombari, F. (eds.) ICIAP 2022. Lecture Notes in Computer Science, vol. 13232, pp. 287–299. Springer, Cham (2022)

4. Cancilla, M., et al.: The DeepHealth toolkit: a unified framework to boost biomedical applications. In: ICPR (2021)
5. Chen, L.C., Papandreou, G., Kokkinos, I., Murphy, K., Yuille, A.L.: DeepLab: Semantic image segmentation with deep convolutional nets, Atrous convolution, and fully connected CRFs. TPAMI **27**, 834–848 (2017)
6. Chen, X., et al.: Microsoft coco captions: data collection and evaluation server. arXiv preprint: arXiv:1504.00325 (2015)
7. Cheng, B., Schwing, A., Kirillov, A.: Per-pixel classification is not all you need for semantic segmentation. In: NeurIPS (2021)
8. Cipriano, M., et al.: Deep segmentation of the mandibular canal: a new 3D annotated dataset of CBCT volumes. IEEE Access **10**, 11500–11510 (2022)
9. Ding, J., Xue, N., Xia, G.S., Dai, D.: Decoupling zero-shot semantic segmentation. In: CVPR (2022)
10. Dosovitskiy, A., et al.: An image is worth 16x16 words: transformers for image recognition at scale (2021)
11. Everingham, M., Van Gool, L., Williams, C.K.I., Winn, J., Zisserman, A.: The PASCAL visual object classes challenge 2012 (VOC2012) Results
12. Ghiasi, G., Gu, X., Cui, Y., Lin, T.Y.: Scaling open-vocabulary image segmentation with image-level labels. In: Avidan, S., Brostow, G., Cisse, M., Farinella, G.M., Hassner, T. (eds.) ECCV 2022. Lecture Notes in Computer Science, vol. 13696, pp. 540–557. Springer, Cham (2022). https://doi.org/10.1007/978-3-031-20059-5_31
13. Gu, X., Lin, T.Y., Kuo, W., Cui, Y.: Open-vocabulary object detection via vision and language knowledge distillation. arXiv preprint: arXiv:2104.13921 (2021)
14. Gu, Z., Zhou, S., Niu, L., Zhao, Z., Zhang, L.: Context-aware feature generation for zero-shot semantic segmentation. In: ACM Multimedia (2020)
15. Ilharco, G., et al.: OpenCLIP. Zenodo **4**, 5 (2021)
16. Jia, C., et al.: Scaling up visual and vision-language representation learning with noisy text supervision. In: ICML (2021)
17. Li, B., Weinberger, K.Q., Belongie, S., Koltun, V., Ranftl, R.: Language-driven semantic segmentation. In: ICLR (2022)
18. Liang, F., et al.: Open-vocabulary semantic segmentation with mask-adapted CLIP. In: CVPR (2023)
19. Lin, T.Y., et al.: Microsoft COCO: common objects in context. In: Fleet, D., Pajdla, T., Schiele, B., Tuytelaars, T. (eds.) ECCV 2014. LNCS, vol. 8693, pp. 740–755. Springer, Cham (2014). https://doi.org/10.1007/978-3-319-10602-1_48
20. Liu, Z., et al.: Swin transformer: hierarchical vision transformer using shifted windows. In: CVPR (2021)
21. Mottaghi, R., et al.: The role of context for object detection and semantic segmentation in the wild. In: CVPR (2014)
22. Mukhoti, J., et al.: Open vocabulary semantic segmentation with patch aligned contrastive learning. In: CVPR 2023 (2022)
23. Qi, L., et al.: Open world entity segmentation. TPAMI (2022)
24. Radford, A., et al.: Learning transferable visual models from natural language supervision. In: ICML (2021)
25. Schuhmann, C., et al.: LAION-5B: an open large-scale dataset for training next generation image-text models. In: NeurIPS Datasets and Benchmarks Track (2022)
26. Xian, Y., Choudhury, S., He, Y., Schiele, B., Akata, Z.: Semantic projection network for zero-and few-label semantic segmentation. In: CVPR (2019)
27. Xu, J., et al.: GroupViT: semantic segmentation emerges from text supervision. In: CVPR (2022)

28. Xu, M., et al.: A simple baseline for open-vocabulary semantic segmentation with pre-trained vision-language model. In: Avidan, S., Brostow, G., Cisse, M., Farinella, G.M., Hassner, T. (eds.) ECCV 2022. Lecture Notes in Computer Science, vol. 13689, pp. 736–753. Springer, Cham (2022). https://doi.org/10.1007/978-3-031-19818-2_42

29. Zhou, B., Zhao, H., Puig, X., Fidler, S., Barriuso, A., Torralba, A.: Scene parsing through ADE20K dataset. In: CVPR (2017)

30. Zhou, C., Loy, C.C., Dai, B.: Extract Free Dense Labels from CLIP. In: Avidan, S., Brostow, G., Cisse, M., Farinella, G.M., Hassner, T. (eds.) ECCV 2022. Lecture Notes in Computer Science, vol. 13688, pp. 696–712. Springer, Cham (2022)

LBKENet:Lightweight Blur Kernel Estimation Network for Blind Image Super-Resolution

Asif Hussain Khan[1]([✉])[ID], Rao Muhammad Umer[2][ID], Matteo Dunnhofer[1], Christian Micheloni[1], and Niki Martinel[1]

[1] Department of Mathematics, Computer Science and Physics, University of Udine, 33100 Udine, Italy
khan.asifhussain@spes.uniud.it,
{matteo.dunnhofer,christian.micheloni,niki.martinel}@uniud.it
[2] Institute of AI for Health, Helmholtz Zentrum München - German Research Center for Environmental Health, Neuherberg 85764, Germany

Abstract. Blind image super-resolution (Blind-SR) is the process of leveraging a low-resolution (LR) image, with unknown degradation, to generate its high-resolution (HR) version. Most of the existing blind SR techniques use a degradation estimator network to explicitly estimate the blur kernel to guide the SR network with the supervision of ground truth (GT) kernels. To solve this issue, it is necessary to design an implicit estimator network that can extract discriminative blur kernel representation without relying on the supervision of ground-truth blur kernels. We design a lightweight (LBKENet) approach for blind super-resolution (Blind-SR) that estimates the blur kernel and restores the HR image based on a deep convolutional neural network (CNN) and a deep super-resolution residual convolutional generative adversarial network. Since the blur kernel for blind image SR is unknown, following the image formation model of the blind super-resolution problem, we first introduce a neural network-based model to estimate the blur kernel. This is achieved by (i) a Super Resolver that, from a low-resolution input, generates the corresponding SR image; and (ii) an Estimator Network generating the blur kernel from the input datum. The output of both models is used in a novel loss formulation. The proposed network is end-to-end trainable. The methodology proposed is substantiated by both quantitative and qualitative experiments. Results on benchmarks demonstrate that our computationally efficient approach ($12\times$ fewer parameters than the state-of-the-art models) performs favorably with respect to approaches that have less number of parameters and can be used on devices with limited computational capabilities.

Keywords: blind image super-resolution (Blind-SR); · isotropic blur kernel · anisotropic blur kernels

© The Author(s), under exclusive license to Springer Nature Switzerland AG 2023
G. L. Foresti et al. (Eds.): ICIAP 2023, LNCS 14234, pp. 209–222, 2023.
https://doi.org/10.1007/978-3-031-43153-1_18

1   Introduction

Single image super-resolution (SISR) is used to generate high-resolution (HR) images from low-resolution (LR) images and has applications in a variety of disciplines [3, 8, 9, 15, 20, 26–29]. Existing literature typically employs a bicubic downsampling kernel to generate LR images [16], but due to the mismatch with actual degradation settings, this method may produce distinct results. In order to surmount this limitation, blind super-resolution (Blind-SR) methods following the degradation process have been developed.

$$\mathbf{Y} = (\mathbf{k} \circledast \hat{x}) \downarrow_S + \eta \tag{1}$$

where $\hat{x}$ and $\mathbf{Y}$ are the HR and LR images, $\circledast$ is the convolution operation, $\mathbf{k}$ is the blur kernel, and η is the additive white Gaussian noise. $\downarrow_S$ is a downsampling operator with scale factor S. In the real world, η also includes factors that can alter the image acquisition process, including inherent sensor noise, stochastic noise, compression artifacts, and possible mismatches between the forward observation model and the camera device.

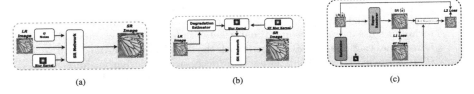

(a)   (b)   (c)

Fig. 1. The illustration of different blur kernel estimators. (**a**) Non-blind SR methods directly use predefined degradation information to guide SR networks. (**b**) Many Blind-SR methods estimate the blur kernel explicitly with the supervision of ground-truth blur kernels. (**c**) Our proposed approach (LBKENet) can estimate the blur kernel implicitly to guide SR without a ground-truth blur kernel.

The approaches that assume a known blur kernel $\mathbf{k}$ are named *non-blind image SR* (Non-Blind-SR see Fig. 1a) and have been extensively studied in literature [13, 32, 39]. Super Resolution is an ill-posed problem, and the success of image priors and deep neural networks (DNNs) in solving these challenges has proven to be successful approaches [16, 18]. While image prior-based algorithms have yielded promising results, they require solutions to difficult optimization problems. DNN-based algorithms that learn the mapping from LR to HR images beat image prior-based approaches significantly [16], but they assume known blur kernels, which is not true in real-world circumstances. As a result, we present a Blind-SR method that differs from earlier methods that use known/predetermined blur kernels.

For Non-Blind-SR, the blur kernel is the bicubic interpolation kernel [2, 37]. This is used to synthesize a large-scale dataset for model training. Real LR images might have a large disparity with the bicubic-generated ones since blur kernels are often more complex.In recent years, deep neural network-based methods have achieved remarkable results in SISR [6, 7, 23, 44, 46]. This pushed the literature to focus on SR in the presence of unknown blur kernels. Methods that assume the blur kernel **k** as unknown are named *blind image SR* (Blind-SR). Different Blind-SR techniques have been developed to recover HR images from blur kernels in LR input images that are unknown [12,24,38,40,41]. Traditional methods require estimations of the blur kernel and the latent HR restoration, which are constrained by the priors of statistical images. Deep CNN-based methods have been developed to solve the Blind-SR problem without using statistical image priors. However, some of them need difficult optimization techniques or result in artifacts from blur kernel error. Several techniques have been proposed to address the Blind-SR problem by first estimating blur kernels using statistical priors (e.g., patch self-similarity [31]) or deep neural networks (e.g., [4,12]) and then applying traditional SR techniques assuming a known kernel (e.g., [18,34]). Our approach has the same spirit as [12,24] with some relevant differences. In [12,24] the basic properties of the blind image SR problem are not adequately modeled by separately estimating the blur kernels and latent HR images, thus affecting the final latent HR image restoration. In addition, both such methods are extremely time-consuming and computationally very expensive. Differently, our approach employs a trainable end-to-end network with limited parameters, thus opening to memory and computationally constrained devices.

Existing Blind-SR methods have two main problems: they have to estimate blur kernels explicitly with the help of a GT kernel (Blind-SR see Fig. 1b), and they are hard to use on devices with limited computing power [21,42]. We introduce a novel Blind-SR method that uses a lightweight convolutional neural network (CNN) to predict the unknown blur kernel and restore a super-resolved image simultaneously. We proposed two different modules: the Estimator and the Super Resolver. The Estimator predicts the blur kernel **k** based on the LR($\tilde{x}$) input image and the Super Resolver restores the SR image ($\hat{x}$). The two outputs (i.e., **k** and $\hat{x}$) are then combined in a joint loss function to encourage an image as close as possible to high-quality ground truth HR datum.

The main contributions are:

- We introduce a Blind-SR approach that estimates the blur kernel **k** and the SR image ($\hat{x}$) simultaneously. The blur kernel is implicitly estimated, hence not requiring the supervision of a ground truth kernel;
- Our proposed network (LBKENet) compares favorably with respect to state-of-the-art approaches that have a similar number of learnable parameters;
- We provide an end-to-end architecture for the proposed algorithm and extensively analyze its performance via quantitative and qualitative evaluation on benchmark datasets.

2   Method

2.1   Problem Formulation

According to (1), estimating $\hat{x}$ from **Y** is primarily based on the variational strategy for combining observation and prior knowledge. This requires solving

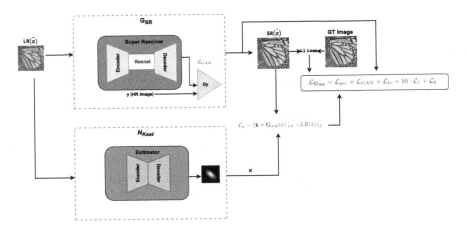

Fig. 2. An overview of the proposed method (LBKENet) .

the following minimization problem:

$$\hat{E}(\hat{x}) = \arg\min_{\hat{x}} \frac{1}{2}\|\mathbf{Y} - \mathbf{k}\hat{x}\|_2^2 + \lambda R_W(\hat{x}) \tag{2}$$

where $\frac{1}{2}\|\mathbf{Y} - \mathbf{k}\hat{x}\|_2^2$ is the data fidelity term related to the model likelihood. It measures how closely the solution matches the observations. $R_W(\hat{x})$ is a regularization term related to image priors, and λ is a trade-off parameter that controls how closely the solution resembles the observations. It is interesting to note that the variational technique directly relates to the Bayesian approach. The generated solutions can be categorized as either maximum a posteriori (MAP) estimates [10] or penalized maximum likelihood estimates. Due to strong prior capabilities, we adopt the generator network [38] for super-resolution learning and a simple CNN-based novel kernel estimator network, which predicts the kernel as closest to the original one. We trained both networks end-to-end by exploiting a GAN framework to minimize the energy-based objective function (2) with the discriminative and residual learning approaches, considering the estimated kernel.

2.2   Kernel Estimation

For training kernel estimation network (N_{kest}), we provided low-resolution input $\tilde{x}$, which is exploited by the network (N_{kest}) and predicts the kernel $\mathbf{k}$, which is further utilized for super resolver learning, as shown in Fig. 2.

2.3   Super Resolver

The estimated blur kernel through the kernel estimator network is shown in Fig. 2. In the training process, we use the same low-resolution input $\tilde{x}$ for super-resolver G_{SR} and our proposed estimator N_{kest}. The super-resolved image $\hat{x}$

obtained from super-resolver G_{SR} is convolved with the predicted blur kernel **k** obtained from the N_{kest} (estimator). The blur kernel **k** predicted by N_{kest} (estimator)is used to calculate the novel loss $\mathcal{L}_k$ after the convolution with the super-resolved image. We calculate the $\mathcal{L}_1$ from the super-resolved image $\hat{x}$ and ground truth. The loss $\mathcal{L}_k$ is added to the $\mathcal{L}_1$ and other network losses (see Sect. 2.5) to calculate the final loss $\mathcal{L}_{\mathbf{G_{SR}}}$ as given in (9).

2.4   Network Architectures

The network architectures of the Generator ($\mathbf{G}_{SR}$), Discriminator ($\mathbf{D}_y$), and Kernel Estimator ($\mathbf{N}_{kest}$) are depicted in Fig. 3. The letters s, c, and k denoted stride size, number of filters, and kernel size, respectively.

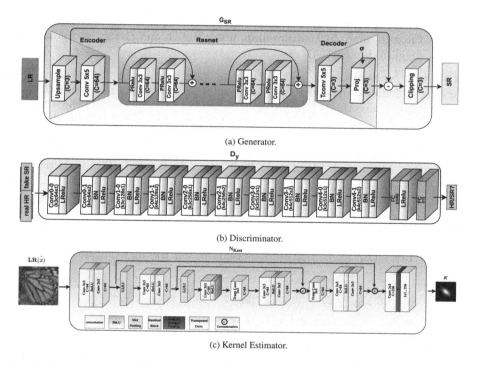

(a) Generator.

(b) Discriminator.

(c) Kernel Estimator.

Fig. 3. The architectures of Generator, Discriminator, and Kernel Estimator networks.

Generator Network ($\mathbf{G}_{SR}$): As illustrated in Fig. 3a, the Encoder and Decoder include 64 feature maps, $C \times H \times W$ tensors, a 5×5 kernel, and C input channels. The LR input $\tilde{x}$ is upsampled ($\times 2$, $\times 4$) using Bilinear kernel $\mathbf{H}^T$. There are 5 residual blocks, 2 pre-activation convolutional layers with 64 feature maps each, and 3×3 kernel. Pre-activations are parametrized rectified linear units (PReLU) that support 64 feature maps. To achieve super-resolution, the image was first subjected to transposed convolution in the decoder block, which resulted in an upsampling of the image to the desired size and project layer (Proj) [19] in the decoder determines the proximal map using standard deviation (σ), which accounts for the prior terms and data fidelity term. The α parameter in Proj is fine-tuned with the back-propagation in training. In addition, the Resnet block located in-between the encoder and the decoder is where the noise is estimated. The input LR image is then subtracted from the estimated residual image provided by the Decoder. Finally, the clipping layer is responsible for complying with the valid intensity of the image ranging from 0 to 255. In training, reflection padding is utilized to lag the convolutional layers, which makes the consistent change in the input image.

Discriminator Network ($\mathbf{D}_y$): Figure 3b shows the discriminator network architecture. This aims to classify if the input is a generated SR image $\hat{x}$ (i.e., *fake*) or a *real* HR image y. The discriminator comprises 3×3 kernel, 4×4 kernel, 64 feature map, 512 feature map, leaky ReLU, and Batch Normalization (BN) as suggested in SRGAN [18].

Estimator Network ($\mathbf{N}_{kest}$): Estimating the blur kernels from given LR images is difficult because blur and downsampling operations result in information loss. Existing methods typically require sophisticated priors and the solution of complex optimization problems [31]. Bell-Kligler et al. estimate the blur kernel from a single LR image using a generative adversarial network in Ref. [4]. Unlike [4,31], we develop a simple deep CNN model ($\mathbf{N}_{kest}$) that ingests an LR image to estimate its blur kernel. This network, shown in Fig. 3c, is trained using (8).

2.5 Loss Calculation

Texture Loss ($\mathcal{L}_{GAN}$): Texture loss is based on the high output frequency, which is defined as follows.

$$
\begin{aligned}
\mathcal{L}_{GAN} = \mathcal{L}_{RaGAN} &- \mathbb{E}_y[\log(1 - \mathbf{D_y})(\mathbf{y}, \mathbf{G_{SR}}((\hat{x})))] \\
&- \mathbb{E}_{\hat{y}}[\log(\mathbf{D_y}(\mathbf{G}_{SR}(\hat{x}), y))]
\end{aligned} \tag{3}
$$

where $\mathbb{E}_y$ and $\mathbb{E}_{\hat{y}}$ are used for the average of real (y) and fake ($\hat{y}$) data, respectively. We use a discriminator that gives the GAN score of a real image (HR) and a fake image (SR) as used in Ref. [43]. It is defined as follows.

$$
\mathbf{D}_y(y, \hat{y})(C) = \sigma(C(y) - \mathbb{E}[C(\hat{y})]) \tag{4}
$$

The sigmoid function and the output of the raw discriminator have been represented by σ and C, respectively, as shown in Fig. 3b.

Perceptual Loss ($\mathcal{L}_{per}$): Perceptual loss is used to measure the perceptual quality of the output, which is defined as follows.

$$\mathcal{L}_{per} = \frac{1}{N} \sum_i^N \mathcal{L}_{VGG} \tag{5}$$

$$= \frac{1}{N} \sum_i^N \|\phi(\mathbf{G}_{SR}(\hat{x}_i) - \phi(y_i)\|_1$$

where ϕ denotes extracted feature from VGG-19 pretrained network as specified in Ref. [43].

TV (Total-Variation) Loss ($\mathcal{L}_{tv}$): TV or total variation loss is defined as follows.

$$\mathcal{L}_{tv} = \frac{1}{N} \sum_i^N (\|\nabla_h \mathbf{G}_{SR}(\hat{x}_i) - \nabla_h(y_i)\|_1$$

$$+ \|\nabla_v \mathbf{G}_{SR}(\hat{x}_i) - \nabla_v(y_i)\|_1) \tag{6}$$

where ∇_v and ∇_h represent the vertical and horizontal gradients of the images.

Content Loss ($\mathcal{L}_1$): Content loss is defined as follows.

$$\mathcal{L}_1 = \frac{1}{N} \sum_i^N \|\mathbf{G}_{SR}(\hat{x}) - y_i\|_1 \tag{7}$$

where N represents the batch size.

Estimator Loss($\mathcal{L}_k$): We computed the estimator loss in such a manner that we combined two outputs (i.e., $\mathbf{k}$ and $\hat{x}$) such that $\hat{x}$ should be as close as possible to the ground-truth HR image, and it is convolved (with $\hat{x}$) and the down-scaled version matches the LR input.

$$\mathcal{L}_k = \|\mathbf{k} \circledast \mathbf{G}_{SR}(\hat{x}) \downarrow_S -LR(\tilde{x})\|_2 \tag{8}$$

where S is a downscaling factor, $\mathbf{k}$ is the estimated kernel, and $\mathbf{G}_{SR}(\hat{x})$ is the super-resolved imaged.

Total loss function ($\mathcal{L}_{\mathbf{G_{SR}}}$) formulation is defined as:

$$\mathcal{L}_{\mathbf{G_{SR}}} = \mathcal{L}_{GAN} + \mathcal{L}_{per} + \mathcal{L}_{tv} + 10 \cdot \mathcal{L}_1 + \mathcal{L}_k \tag{9}$$

3   Experiments

3.1   Datasets

We followed a common protocol [12,24,33,38] and used 3450 high-resolution (HR) images from DIV2K [1] and Flickr2K [36] for model training. For a fair comparison with existing approaches, we followed [12,24] and trained/evaluated our approach with the following degradation settings. We follow the protocol in Ref. [12] and set the kernel size to 21. For scale factors 4 and 2, the kernel width is uniformly sampled during training in the ranges of [0.2, 4.0] and [0.2, 2.0]. Evaluation is conducted on popular benchmark HR datasets, such as Set5 [5], Set14 [45], Urban100 [14], BSD100 [25], and Manga109 [30]. For a fair comparison with existing methods, during testing, we adopted the same approach of Ref. [12] and uniformly selected 8 kernels from the ranges [1.8, 3.2] and [0.80, 1.60] for scale factors 4 and 2, respectively. The HR images are first blurred using the selected blur kernels, then downsampled to generate synthetic test images.

3.2   Model Optimization

We trained our model with 32×32 LR patches for 51,000 iterations. To minimize (9), we used 16 samples per batch with the Adam optimizer [17] having $\beta_1 = 0.9$, $\beta_2 = 0.999$, and $\epsilon = 10^{-8}$ without weight decay for both generator and discriminator. We initially set the learning rate to 10^{-4}, then reduce it by a factor of 2 after 5 K, 10 K, 20 K, and 30 K iterations. The projection layer parameter σ (standard deviation) is estimated according to Ref. [22] from the input LR image. We initialize the projection layer parameter α on log-scale values from $\alpha_{max} = 2$ to $\alpha_{min} = 1$ and then further fine-tune during the training via a back-propagation. Using a GAN framework in Ref. [11] and the following loss functions, we fine-tune the SRResCGAN network to learn the super-resolution. (i.e., pre-trained $\mathbf{G}_{SR}$) [38].

Random vertical and horizontal flipping and 90 deg rotations are used as data augmentation strategies.

3.3   Evaluation Metrics

For a fair comparison, we evaluated the trained model under the Peak Signal-to-Noise Ratio (PSNR) and Structural Similarity (SSIM) [47] metrics. The PSNR and SSIM are distortion-based measures. The RGB color space is used to evaluate the quantitative SR results.

Table 1. Quantitative SR (×2, ×4) results comparison of state-of-the-art non-blind (*) and blind image SR methods on the benchmark datasets with different unknown Gaussian blur kernels and model parameters.

Methods	Scale	Set5 PSNR/SIM	Set14 PSNR/SIM	B100 PSNR/SIM	Urban100 PSNR/SIM	Manga109 PSNR/SIM	#Params (M)
Bicubic	×2	28.65/0.84	26.70/0.77	26.26/0.73	23.61/0.74	25.73/0.84	/
RCAN*		29.73/0.86	27.65/0.79	27.07/0.77	24.74/0.78	27.64/0.87	15.59
ZSSR*		29.74/0.86	27.57/0.79	26.96/0.76	24.34/0.77	27.10/0.87	0.22
MZSR*		29.88/0.86	27.32/0.79	26.96/0.77	24.12/0.77	27.24/0.87	0.22
KernelGAN+ZSSR		26.02/0.77	20.19/0.58	21.42/0.60	19.55/0.61	24.22/0.78	0.52
KernelGAN+MZSR		29.39/0.88	23.94/0.72	24.42/0.73	23.39/0.77	28.38/0.89	0.52
IKC		33.62/0.91	29.14/0.85	28.46/0.82	26.59/0.84	30.51/0.91	9.05
DAN		34.55/0.92	29.92/0.86	29.66/0.85	27.96/0.87	33.82/0.95	4.33
LBKENet (Ours)		**31.02/0.89**	**27.87/0.80**	**27.67/0.79**	**25.11/0.80**	**28.44/0.89**	**0.38**
Bicubic	×4	24.49/0.69	23.01/0.59	23.64/0.59	20.58/0.57	21.97/0.70	/
RCAN*		24.95/0.71	23.33/0.61	23.65/0.62	20.73/0.61	23.30/0.76	15.59
ZSSR*		24.77/0.70	23.32/0.60	23.72/0.61	20.74/0.59	22.75/0.74	0.22
MZSR*		24.99/0.70	23.45/0.61	23.83/0.61	20.92/0.61	23.25/0.76	0.22
KernelGAN+ZSSR		17.59/0.42	19.20/0.49	17.14/0.40	16.95/0.47	19.40/0.61	0.52
KernelGAN+MZSR		23.08/0.66	22.24/0.61	21.51/0.56	19.37/0.58	22.05/0.70	0.52
IKC		27.84/0.80	25.02/0.67	24.76/0.65	22.41/0.67	25.37/0.81	9.05
DAN		27.64/0.80	25.46/0.69	25.35/0.67	23.21/0.71	27.04/0.85	4.33
LBKENet (Ours)		**25.94/0.73**	**23.83/0.62**	**24.12/0.60**	**21.15/0.59**	**22.15/0.71**	**0.38**

3.4 Comparison with the State-of-Art Methods

We compared our proposed Blind-SR method with state-of-the-art approaches such as IKC [12] DAN [24], and KernelGAN [4], as well as non-Blind-SR approaches like RCAN [48], ZSSR [34], and MZSR [35]. Table 1 presents the quantitative evaluation of benchmark datasets on scale factors 2 and 4, showing that the proposed method outperformed most of the considered methods. While IKC and DAN have higher PSNRs and SSIMs, they require more parameters than the proposed method. Non-blind SR approaches like RCAN and MZSR produced results with considerable blur impact Fig. 4c–e), whereas KernelGAN produced artifacts due to inaccurate blur kernels. IKC [12] and DAN [24] produced visually fine results but with many parameters. On the other hand, our proposed method implicitly estimates blur kernels and HR images simultaneously using a lightweight (less number of parameters) end-to-end trainable network and the super resolver network to restore the HR images. Figures 4 and 5 show that our generated SR images are much better than state-of-the-art methods (e.g., Fig. 4d–f), however Fig. 5g, h outperforms ours.

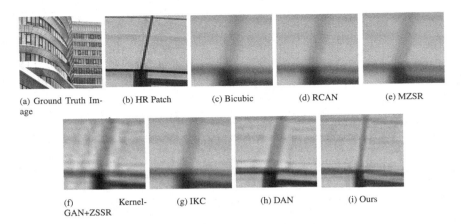

Fig. 4. Visual comparison of our method with other state-of-art methods.

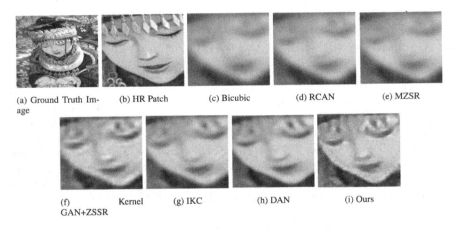

Fig. 5. Visual comparisons (×4) on the Set14 dataset.

Table 2. This table displays that the DIV2K validation set (100 images with unknown blur kernels) was used for our ablation investigation.

Methods	w/o ($\mathbf{N}_{kest}$)	LBKENet
PSNR/SSIM	25.63/0.69	25.80/0.71

Table 3. Quantitative performance of proposed approach on Set5(x4) with different blur kernels width.

Kernel Width	1.0	2.5	3.0	3.5	4.0
Bicubic	25.20/0.72	24.38/0.69	23.70/0.66	23.18/0.63	22.80/0.62
ZSSR [34]	26.30/0.76	25.06/0.72	24.11/0.68	23.44/0.65	22.95/0.62
IKC [12]	28.12/0.82	28.32/0.82	28.29/0.81	27.90/0.80	24.26/0.69
Ours	**26.31/0.76**	**26.05/0.73**	**25.09/0.70**	**24.17/0.66**	**23.38/0.63**

3.5 Ablation Study

In Table 2, we analyze the impact of the kernel estimation module. Results are computed on the DIV2K validation set with unknown blur kernels. It shows that without using ($\mathbf{N}_{kest}$) it does not generate good SR images in terms of PSNR/SSIM, and with our proposed LBKENet, PSNR/SSIM increased by +0.17/+0.02, respectively. To study the effects of different blur kernels, we compute the results in Table 3. These indicate that our method can generalize better than a Non-Blind-SR method (i.e., Ref. [34]) while maintaining competitive performance with a Blind-SR method (i.e., Ref. [12]) requiring 7.5× more learnable parameters.

4   Conclusions

In this paper, we have introduced an effective, lightweight, and implicit blur kernel estimation end-to-end approach for blind image super-resolution (blind-SR). Our proposed approach is based on a deep convolutional neural network (CNN) named estimator and a deep super-resolution residual convolutional generative adversarial network named super resolver. The Estimator module implicitly estimates the blur kernels from the LR input without the supervision of the ground truth kernel. The Super Resolver module restores the SR image by exploiting a GAN framework to minimize the energy-based objective function with the discriminative and residual learning approaches, considering the estimated kernel. The whole architecture is trained in an end-to-end fashion. Results on different benchmark datasets show that our approach achieves better performance than state-of-the-art methods having a similar number of learnable parameters, enabling it to work on devices with limited computational capacity.

References

1. Agustsson, E., Timofte, R.: Ntire 2017 challenge on single image super-resolution: dataset and study. In: IEEE CVPRW, pp. 126–135 (2017)
2. Ahn, N., Kang, B., Sohn, K.A.: Fast, accurate, and lightweight super-resolution with cascading residual network. In: ECCV, pp. 252–268 (2018)

3. Bansal, V., Foresti, G.L., Martinel, N.: Cloth-changing person re-identification with self-attention. In: WACVW, pp. 602–610 (2022)
4. Bell-Kligler, S., Shocher, A., Irani, M.: Blind super-resolution kernel estimation using an internal-gan. In: NeurIPS, vol. 32 (2019)
5. Bevilacqua, M., Roumy, A., Guillemot, C., Alberi-Morel, M.L.: Low-complexity single-image super-resolution based on nonnegative neighbor embedding (2012)
6. Bhat, G., et al.: Ntire 2021. In: IEEE Computer Society Conference on Computer Vision and Pattern Recognition Workshops, pp. 613–626 (2021)
7. Bhat, G., Danelljan, M., Timofte, R., et al.: NTIRE 2021 challenge on burst super-resolution: methods and results. In: CVPRW (2021)
8. Dunnhofer, M., Martinel, N., Micheloni, C.: Improving MRI-based knee disorder diagnosis with pyramidal feature details. In: International Conference on Medical Imaging with Deep Learning, vol. 143, pp. 131–147. PMLR (2021)
9. Dunnhofer, M., Martinel, N., Micheloni, C.: Weakly-supervised domain adaptation of deep regression trackers via reinforced knowledge distillation. IEEE Rob. Autom. Let. 6(3), 5016–5023 (2021)
10. Figueiredo, M., Bioucas-Dias, J.M., Nowak, R.D.: Majorization-minimization algorithms for wavelet-based image restoration. IEEE Trans. Image Process. 16(12), 2980–2991 (2007)
11. Goodfellow, I., et al.: Generative adversarial networks. Commun. ACM 63(11), 139–144 (2020)
12. Gu, J., Lu, H., Zuo, W., Dong, C.: Blind super-resolution with iterative kernel correction. In: IEEE/CVF CVPR, pp. 1604–1613 (2019)
13. Haris, M., Shakhnarovich, G., Ukita, N.: Deep back-projection networks for super-resolution. In: IEEE, pp. 1664–1673 (2018)
14. Huang, J.B., Singh, A., Ahuja, N.: Single image super-resolution from transformed self-exemplars. In: CVPR, pp. 5197–5206 (2015)
15. Huang, S., Teo, R., Leong, W., Martinel, N., Foresti, G.L., Micheloni, C.: Coverage control of multiple unmanned aerial vehicles: a short review. Unmanned Syst. 6(2), 131–144 (2018)
16. Kim, J., Lee, J.K., Lee, K.M.: Accurate image super-resolution using very deep convolutional networks. In: IEEE CVPR, pp. 1646–1654 (2016)
17. Kingma, D.P., Ba, J.: Adam: a method for stochastic optimization. arXiv preprint arXiv:1412.6980 (2014)
18. Ledig, C., et al.: Photo-realistic single image super-resolution using a generative adversarial network. In: IEEE CVPR, pp. 4681–4690 (2017)
19. Lefkimmiatis, S.: Universal denoising networks: a novel cnn architecture for image denoising. In: IEEE CVPR, pp. 3204–3213 (2018)
20. Leong, W.L., Martinel, N., Huang, S., Micheloni, C., Foresti, G.L., Teo, R.S.H.: An intelligent auto-organizing aerial robotic sensor network system for urban surveillance. J. Intell. Rob. Syst. 102(2), 33 (2021)
21. Liang, J., Zhang, K., Gu, S., Van Gool, L., Timofte, R.: Flow-based kernel prior with application to blind super-resolution. In: IEEE/CVF CVPR, pp. 10601–10610 (2021)
22. Liu, X., Tanaka, M., Okutomi, M.: Single-image noise level estimation for blind denoising. IEEE TIP 22(12), 5226–5237 (2013)
23. Lugmayr, A., Danelljan, M., Timofte, R.: NTIRE 2020 challenge on real-world image super-resolution: methods and results. In: CVPRW (2020)
24. Luo, Z., Huang, Y., Li, S., Wang, L., Tan, T.: Unfolding the alternating optimization for blind super resolution. In: NeurIPS, vol. 33 (2020)

25. Martin, D., Fowlkes, C., Tal, D., Malik, J.: A database of human segmented natural images and its application to evaluating segmentation algorithms and measuring ecological statistics. In: ICCV, vol. 2, pp. 416–423. IEEE (2001)
26. Martinel, N., Dunnhofer, M., Foresti, G.L., Micheloni, C.: Person re-identification via unsupervised transfer of learned visual representations. In: Proceedings of the 11th International Conference on Distributed Smart Cameras, pp. 151–156. ACM (2017)
27. Martinel, N., Dunnhofer, M., Pucci, R., Foresti, G.L., Micheloni, C.: Lord of the rings: hanoi pooling and self-knowledge distillation for fast and accurate vehicle reidentification. IEEE Trans. Ind. Inf. **18**(1), 87–96 (2022)
28. Martinel, N., Foresti, G.L., Micheloni, C.: Deep pyramidal pooling with attention for person re-identification. IEEE Trans. Image Process. **29**, 7306–7316 (2020)
29. Martinel, N., Micheloni, C., Foresti, G.L.: A pool of multiple person re-identification experts. Pattern Recogn. Lett. **71**, 23–30 (2016)
30. Matsui, Y., et al.: Sketch-based manga retrieval using manga109 dataset. Multimedia Tools Appl. **76**(20), 21811–21838 (2017)
31. Michaeli, T., Irani, M.: Nonparametric blind super-resolution. In: IEEE ICCV, pp. 945–952 (2013)
32. Muhammad Umer, R., Luca Foresti, G., Micheloni, C.: Deep iterative residual convolutional network for single image super-resolution. In: ICPR (2021)
33. Umer, R.M., Micheloni, C.: Deep cyclic generative adversarial residual convolutional networks for real image super-resolution. In: Bartoli, A., Fusiello, A. (eds.) ECCV 2020. LNCS, vol. 12537, pp. 484–498. Springer, Cham (2020). https://doi.org/10.1007/978-3-030-67070-2_29
34. Shocher, A., Cohen, N., Irani, M: "Zero-shot" super-resolution using deep internal learning. In: IEEE CVPR, pp. 3118–3126 (2018)
35. Soh, J.W., Cho, S., Cho, N.I.: Meta-transfer learning for zero-shot super-resolution. In: IEEE/CVF CVPR, pp. 3516–3525 (2020)
36. Timofte, R., Agustsson, E., Van Gool, L., Yang, M.H., Zhang, L.: Ntire 2017 challenge on single image super-resolution: methods and results. In: CVPRW, pp. 114–125 (2017)
37. Umer, R.M., Foresti, G.L., Micheloni, C.: Deep super-resolution network for single image super-resolution with realistic degradations. In: ICDSC, pp. 21:1–21:7 (2019)
38. Umer, R.M., Foresti, G.L., Micheloni, C.: Deep generative adversarial residual convolutional networks for real-world super-resolution. In: IEEE/CVF CVPRW, pp. 438–439 (2020)
39. Umer, R.M., Micheloni, C.: Rbsricnn: raw burst super-resolution through iterative convolutional neural network. arXiv preprint arXiv:2110.13217 (2021)
40. Umer, R.M., Micheloni, C.: Real image super-resolution using gan through modeling of lr and hr process. In: AVSS, pp. 1–8. IEEE (2022)
41. Umer, R.M., Munir, A., Micheloni, C.: A deep residual star generative adversarial network for multi-domain image super-resolution. In: SpliTech, pp. 01–05. IEEE (2021)
42. Wang, L., Wang, Y., Dong, X., Xu, Q., Yang, J., An, W., Guo, Y.: Unsupervised degradation representation learning for blind super-resolution. In: IEEE/CVF CVPR, pp. 10581–10590 (2021)
43. Wang, X., et al.: Esrgan: enhanced super-resolution generative adversarial networks. In: ECCVW (2018)
44. Wei, P., Lu, H., Timofte, R., Lin, L., Zuo, W., et al.: AIM 2020 challenge on real image super-resolution: methods and results. In: ECCVW (2020)

45. Zeyde, R., Elad, M., Protter, M.: On single image scale-up using sparse-representations. In: Boissonnat, J.-D., Chenin, P., Cohen, A., Gout, C., Lyche, T., Mazure, M.-L., Schumaker, L. (eds.) Curves and Surfaces 2010. LNCS, vol. 6920, pp. 711–730. Springer, Heidelberg (2012). https://doi.org/10.1007/978-3-642-27413-8_47
46. Zhang, K., et al.: AIM 2020 challenge on efficient super-resolution: methods and results. In: ECCVW, pp. 5–40 (2020)
47. Zhang, R., Isola, P., Efros, A.A., Shechtman, E., Wang, O.: The unreasonable effectiveness of deep features as a perceptual metric. In: IEEE CVPR, pp. 586–595 (2018)
48. Zhang, Y., Li, K., Li, K., Wang, L., Zhong, B., Fu, Y.: Image super-resolution using very deep residual channel attention networks. In: ECCV, pp. 286–301 (2018)

Hierarchical Pretrained Backbone Vision Transformer for Image Classification in Histopathology

Luca Zedda, Andrea Loddo[⊠], and Cecilia Di Ruberto

Department of Mathematics and Computer Science, University of Cagliari, via Ospedale 72, 09124 Cagliari, Italy
{luca.zedda,andrea.loddo,dirubert}@unica.it

Abstract. Histopathology plays a crucial role in clinical diagnosis, treatment planning, and research by enabling the examination of diseases in tissues and organs. However, the manual analysis of histopathological images is time-consuming and labor-intensive, requiring expert pathologists. To address this issue, this work proposes a novel architecture called Hierarchical Pretrained Backbone Vision Transformer for automated histopathological image classification, a critical tool in clinical diagnosis, treatment planning, and research. Current deep learning-based methods for image classification require a large amount of labeled data and significant computational resources to be trained effectively. By leveraging pretrained Visual Transformer backbones, our approach can classify histopathology images, achieve state-of-the-art performance, and take advantage of the pretrained backbones' weights. We evaluated it on the Chaoyang histopathology dataset, comparing it with other state-of-the-art Visual Transformers. The experimental results demonstrate that the proposed architecture outperforms the others, indicating its potential to be an effective tool for histopathology image classification.

Keywords: Histopathology · Deep Learning · Hierarchical ViT

1 Introduction

Histopathology is a critical tool in clinical diagnosis, treatment planning, and research, enabling the study of disease in tissues and organs. Histopathological images provide a wealth of information about the morphology and cellular structure of tissues and organs and can reveal important insights into the underlying pathophysiology of diseases. However, analyzing histopathological images is time-consuming and labor-intensive, requiring expert pathologists to examine and classify the images based on their visual features.

Image classification is a computer-aided technique that can automate the analysis of histopathological images by automatically identifying regions of interest and classifying them into different categories. It helps pathologists by reducing their workload, improving accuracy, and enabling large-scale disease studies.

ⓒ The Author(s), under exclusive license to Springer Nature Switzerland AG 2023
G. L. Foresti et al. (Eds.): ICIAP 2023, LNCS 14234, pp. 223–234, 2023.
https://doi.org/10.1007/978-3-031-43153-1_19

In this context, deep learning (DL)-based methods have achieved remarkable performance also because they can automatically learn and extract features from raw data, even identifying features that are not easily discernible to humans. Although their advantages, the black-box nature of DL models causes pathologists to hesitate when adopting them in high-stakes environments. In order to comply with regulations and facilitate a feedback loop that integrates model diagnosis and refinement in the development process, there is an increasing need for explainable deep learning [18].

This work proposes a novel DL architecture called Hierarchical Pretrained Backbone Vision Transformer (HPB-ViT) for image classification. The HPB-ViT architecture is based on the Vision Transformer (ViT) model [3], which has shown state-of-the-art performance in computer vision (CV) tasks [3,12]. Nevertheless, they require large amounts of labeled data to train effectively. This aspect is even more pronounced in medical imaging because obtaining labeled data can be challenging due to the need for expert annotation and the potential for variability in annotations between different experts.

Pretraining has become a popular technique in DL to address this challenge. Models are first trained on large-scale datasets, i.e., ImageNet [2] or COCO [7] before being fine-tuned on the target dataset. Pretraining improves the performance of DL models by enabling them to learn generalizable features that can be then transferred to different tasks and datasets [6].

Although pretraining ViT on large-scale datasets is effective, it can be time-consuming and computationally expensive, which hinders their practical use in real-world applications. To tackle this challenge, the HPB-ViT architecture integrates various pretrained off-the-shelf ViT backbones, resulting in faster training with better performance. By leveraging the capability of pretrained ViT backbones, HPB-ViT can effectively learn to classify histopathology images with smaller amounts of labeled data while still achieving state-of-the-art performance.

We evaluated the performance of HPB-ViT on the Chaoyang histopathology dataset [21] and compared it with other state-of-the-art models. The experimental results showed that HPB-ViT achieved superior performance, demonstrating the effectiveness of our proposed architecture for the task at hand.

The rest of this work is organized as follows. Firstly, we review related works in Sect. 2, specifically addressing histopathology and its current issues. Next, we present our proposed approach in Sect. 3. We then report and discuss the experimental results in Sect. 4. Finally, the conclusions are drawn in Sect. 5.

2   Background Concepts and Related Work

When analyzing whole-slide digital pathology images, several challenges need to be addressed. These images are extremely large and are measured in terms of gigapixels, which makes it necessary to break them down into smaller tiles to be processed effectively. Additionally, different magnifications are required for specific tasks and to combine information from multiple scales. Predicting survival can be challenging since there may only be weak slide-level labels available, and the most crucial areas of the image may not always be obvious. Annotations can also be complex due to the variability in disease subtypes, which requires the expertise of highly trained pathologists. Cell-based methods involve detecting and characterizing thousands of objects, which can pose a challenge. DL architectures have become increasingly adapted to this task to tackle these issues, and new approaches specifically designed for digital pathology are emerging [11]. These approaches have replaced traditional handcrafted methods [10].

Deep Learning in Histopathology. CNNs have been the reference approach since the release of AlexNet. However, several promising architectures, such as T2T-ViT [17], Swin [8,9], DeepViT [19], CvT [15] have emerged after the introduction of Transformer architecture [14] and the advent of ViT [3].

One of the earliest studies that employed ViT for histopathology was conducted by Zhou et al. [20]. They proposed a hybrid model that combined convolutional residual networks [5] and ViT mechanisms to create a network with inductive solid bias capabilities and scale and rotation invariance, obtaining state-of-the-art performance in a brain biopsies dataset.

Building upon this, Chen et al. [1] also proposed a hybrid convolutional and vision transformer network similar to inception networks [13]. They achieved state-of-the-art results on the HE-GHI-DS dataset, further highlighting the effectiveness of the hybrid approach in histopathology classification.

Issues in Deep Learning for Histopathology. Although the promising results indicate that ViTs have the potential to revolutionize the field of histopathology, some limitations and challenges must be considered. First, the typical extensive size of these images can lead to high computational requirements. Some studies addressed this problem by dividing the original high-resolution images into smaller patches [11]. However, selecting the appropriate magnification level requires a profound understanding of the analyzed task [4].

Also, there is a shortage of adequately labeled images for training. To tackle this problem, Xu et al. [16] developed a generative approach that employs ViTs to generate synthetic data for training. The results were promising, particularly for the underrepresented minority classes, with $\approx 5\%$ performance gain.

3   The Proposed Architecture

In this section, we present the proposed HPB-ViT architecture and its modules. Section 3.1 describes the concept of attention in Transformers [14]. In Sect. 3.2, we describe the approach to standardize the backbone's output. Section 3.3 discusses the module which learns patch pooling. Finally, in Sect. 3.4, we provide a detailed description of the overall architecture.

3.1   Attention Mechanism

Attention mechanisms have become increasingly popular due to the advancements in transformer architecture and multi-head self-attention (MHSA) [14].

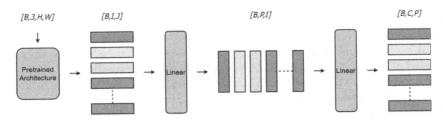

Fig. 1. From left to right: example feature map derived from the backbone pretrained Architecture, Liner transformation, second linear transformation after feature transpose.

The main goal of attention mechanisms is to allow models to focus on important parts of the input while disregarding irrelevant information.

In the MHSA formulation, the attention mechanism is defined as:

$$MultiHead(Q, K, V) = Concat(head_1, \ldots, head_h)W^O \tag{1}$$

where Q, K, and V represent the query, key, and value matrices. The W_i^Q, W_i^K, and W_i^V matrices represent the learned projection matrices for the ith head, and W^O represents the learned output projection matrix. h is the number of heads, and d_k is the dimensionality of the key vectors.

The attention mechanism calculates a weighted sum of their values to prioritize important input elements and ignore irrelevant ones. The weights are determined by comparing the queries and keys, which creates a soft alignment between the query and key vectors. This then allows the corresponding value vectors to be weighted accordingly. Our proposed architecture employs the attention mechanism in the *Transformer layer* shown in Fig. 3.

3.2 Backbone Encapsulator

The Backbone Encapsulator (BE) module receives an input batch of images of size $[B, 3, H, W]$, where B is the batch size, 3 is the number of channels that forms the images, while H and W are height and width, respectively. Then, BE passes the images through a selected pre-trained backbone, obtaining features of size $[B, I, J]$, where I is the number of channels of the output feature map provided by the backbone, and J is the backbone's feature map size. Then, BE applies two different linear transformations to the backbone output features. The first linear transformation expands the last dimension of the features to the predefined size. Then, the features are transposed and passed through a second linear layer to expand the number of patches to the predefined size. This transformation serves the purpose of reorganizing the backbone output to match the expected input size of the HPB model, i.e., $[B, C, P]$, where C is the final number of channels and P is the final feature map size.

In summary, the Backbone Encapsulator module is used to make the output of the pre-trained backbone consistent with the HPB model. This condition is

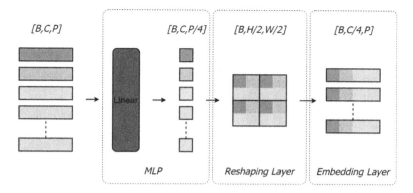

Fig. 2. Representation of the PPHR module. From left to right: example feature map derived from the image, patch representation of the image (MLP block), linear transformation (MLP block), patch representation of the image after linear transformation (Reshaping layer), patch reorganization (Embedding layer).

achieved by modifying the backbone output to match the expected input size of the HPB model. A schematic illustration is shown in Fig. 1.

3.3 Patch Pooling and Hierarchical Reconstruction Module

The Swin Transformer [9] is a state-of-the-art transformer-based architecture for image classification. It features a hierarchical structure that efficiently handles large images by gradually reducing the feature map's resolution while increasing the transformer layers' receptive field.

The Patch Pooling and Hierarchical Reconstruction (PPHR) module works by merging nearby patches into larger ones, reducing the number of patches, and increasing the receptive field of the following transformer layers.

Some distinctions exist between the Swin Transformer's patch merging layer and our PPHR module. The PPHR module comprises three primary components: a *multi-layer perceptron* (MLP), a *reshaping layer*, and an *embedding layer*. A schematic illustration is shown in Fig. 2.

The MLP takes a series of patches as input, with a $[B, C, P]$ size. Here, B represents the batch size, C is the number of patches, and P denotes the size of each patch squared. These patches are then projected into a space of size $[B, C, E]$, where E represents the expansion embedding dimension. Finally, the patches are further projected into a lower-dimensional space of size $[B, C, P/4]$. These patches are normalized using a *Layer Normalization* layer before passing through the linear layer.

The output of the MLP is then passed through the Reshaping layer, which reshapes it into a 2D grid of patches employing the *rearrange function*. Finally, the grid is rearranged into shape $[B, C/4, P]$.

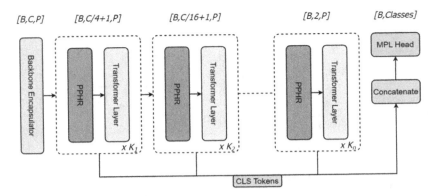

Fig. 3. The architecture of the proposed Hierarchical Pretrained Backbone Vision Transformer over different stages.

After the reshaping layer, the embedding layer applies a linear and another Layer Normalization layer. The final result is an output tensor of size $[B, C/4, K]$, where K is the same as P in our approach.

3.4 Overall Architecture

Our proposal's overall architecture involves several stages that lead to a single vector of image features. It is illustrated in Fig. 3.

Each stage is composed of a PPHR module (see Sect. 3.3) and a set of Transformer Layers [14], which are repeated K_n times. The value of K_n is chosen, and n is the current stage's index.

Our approach involves using learnable class tokens, similar to traditional ViTs. However, we use them separately for each stage instead of simultaneously like conventional ViTs.

At the beginning of each stage, a learnable positional encoding is added, and class tokens with size $[B, 1, P]$ are concatenated. After the final stage, the class tokens from each stage are concatenated and used as input to an MLP head, which produces the scores for each possible class. The concatenation of the class tokens for each stage produces a hierarchical dense representation of the image, resulting in higher accuracy for the current task.

Ultimately, the proposed architecture combines the strengths of the PPHR module and Transformer Layers to learn a hierarchical representation of the input image, resulting in improved performance on image classification tasks.

4 Experiments

The goal of the experiments is to determine if the proposed method can enhance the accuracy of histopathological image classification concerning off-the-shelf Transformer-based architectures. We will begin by introducing the dataset

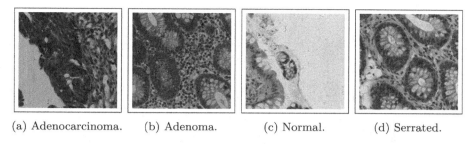

(a) Adenocarcinoma.   (b) Adenoma.   (c) Normal.   (d) Serrated.

Fig. 4. Visual representation of the four classes included in the Chaoyang dataset.

(Sect. 4.1), outlining the experimental setup (Sect. 4.2), and then analyzing the results in Sect. 4.3.

4.1   Chaoyang Dataset

The Chaoyang dataset is a histopathology image collection containing 512×512 sample patches of four different types of colon-rectum tissue conditions: normal, serrated, adenocarcinoma, and adenoma (see Fig. 4). The dataset was constructed in a realistic and practical context, meaning that it may contain noise and other imperfections that can make the classification task more challenging. The authors provide the dataset in two sets: a training set with 4,022 samples and a test set with 2,039 samples.

The class distribution in the dataset is shown in Fig. 5. As can be observed, the dataset is imbalanced, with a more significant number of adenocarcinoma samples compared to the other classes. Overall, the Chaoyang dataset provides a challenging and realistic benchmark for evaluating the performance of image classification models on histopathology images of the colon rectum. The dataset's imbalance and potential imperfections make it a more realistic representation of real-world applications.

4.2   Experimental Setting

We validated our approach using the Chaoyang dataset [21]. To provide a fair comparison between our proposed architecture and several off-the-shelf Vision Transformers, i.e., ViT [3] and SwinV2 [8], we reported accuracy, F1-score, precision, and recall as performance metrics, calculated using the macro average. Additionally, we provided information on the backbone used and the number of parameters for each employed configuration.

Training Details. To help avoid overfitting, we reserved 10% of the training set with the actual class distribution to build a validation set.

Then, we implemented an oversampling strategy on the training set, comprising two steps to address the class imbalance. Firstly, we repeated the underrepresented samples until we had equal samples for each class. Secondly, we augmented

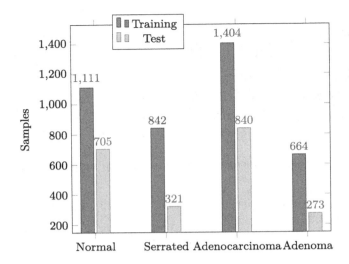

Fig. 5. Class distribution in the Chaoyang dataset. It is imbalanced, with a meager presence of serrated and adenoma classes.

Table 1. Image augmentation parameters adopted for models training.

Augmentation	Parameters	Probability
HorizontalFlip	–	0.5
VerticalFlip	–	0.5
RandomRotate90	[90, 180, 270] degrees	0.5
RandomResizedCrop	[0.5, 1.0] of original size	0.5

the repeated samples online to enhance the diversity of the images' representations with various geometric transformations while preserving important visual features. The augmentations are indicated in Table 1. No color augmentations were considered to preserve the staining procedure and obtain a first baseline.

Our augmentation pipeline effectively increased the diversity of our training set and improved our model's ability to handle variations in input data.

Implementation Details. The experiments were conducted on a workstation with an Intel(R) Core(TM) i5-9400 @ 4.10 GHz CPU, 32 GB RAM, and an NVIDIA RTX 3060 GPU with 12 GB memory. Every method was trained with the following hyperparameters: AdamW was set as the optimizer with a weight decay of 1×10^{-2} and momentum of 0.9. The initial learning rate was 1×10^{-4} across a total of 100 epochs. Dropout was set to 0.2.

HPB Configuration. As for HPB, SwinV2 Tiny [8] and ViT base [3] was chosen as the backbones to experiment with the architecture with two different kinds of image representations: hierarchical provided by SwinV2 and standard by ViT base. Each stage had a different number of repeats, with 2, 2, 4, and 2, respectively. We incorporated 8 attention heads with a size of 64 for each transformer layer. For the embedding dimension, we opted for a small value of 384. Lastly, we set the expansion factor for the feed-forward layer inner dimension to 3.

4.3   Experimental Results

Table 2 presents the results obtained by our proposed methods in its two versions: HPB-ViT using ViT Base [3] as the backbone, from now on referred to as **HPB-Base**, and **HPB-Tiny**, HPB-ViT with the SwinV2 Tiny [8] backbone, and some state-of-the-art methods on the Chaoyang dataset. For fairness, the table is divided into three sections: the first is, to the best of our knowledge, the current baseline on this dataset [21]; the second represents ViT-based architectures, and the last includes SwinV2-based architectures. We point out that we applied the same oversampling and augmentation strategy for every experimented approach, except for the baseline, which was reported from the work of Zhu et al. [21].

In particular, the method of Zhu et al. [21], referred to in our table as *NSHE+CNN*, uses ResNet-34 CNN [5] as the classification model. In addition, the authors proposed the *noise suppressing and hard enhancing* (NSHE) technique to make the classification model resistant to possible noise interference in the images, as the Chaoyang dataset was collected from real-world settings and the main goal was to create a method robust to noise. According to the authors, their approach achieved an accuracy of 0.83 and an F1-score of 0.77. In addition, we also used ViT [3] and SwinV2 [8] for comparison purposes. ViT comes in two versions: *Base* and *Large*, while SwinV2 has *Tiny* and *Small* versions.

Both proposed versions result in significant performance improvements across all reported metrics, with some differences. The HPB-Base version outperforms the baseline work and ViT architectures in every metric. Despite having slightly more parameters than ViT Base, HPB-Base is a suitable compromise between ViT Base and ViT Large in terms of performance gains.

On the other hand, HPB-Tiny performs better than both SwinV2 models and is the top performer among all the tested models. It is based on the SwinV2 Tiny backbone and has fewer parameters than HPB-Base but slightly more than SwinV2 Small. However, it still has double the parameters of SwinV2 Tiny. Overall, HPB-Tiny strikes a good balance between performance and complexity.

Overall, the results demonstrate that the proposed HPB models outperformed the baseline on the Chaoyang dataset [21], highlighting the effectiveness of the proposed approach, even compared to off-the-shelf Transformers architectures. Additionally, we observed that our proposed HPB architecture outperformed the work of Zhu et al. [21], specifically designed to take into account the noise interference commonly found in real-world acquired images. This aspect highlights the robustness and effectiveness of our approach and underscores its potential for handling noise interference images with improved accuracy and reliability.

Table 2. Experimental results obtained on the Chaoyang dataset [21]. The reported performance metrics, calculated using the macro average, include accuracy, F1-score, precision, and recall. Additionally, we provided information on the backbone used and the number of parameters for each model tested. The best results are highlighted in bold, while the second best are underlined.

Model	Backbone	Image size	Acc↑	F1↑	Pre↑	Rec↑	Params↓
NSHE+CNN [21]	ResNet-34	224	0.83	0.77	0.78	0.75	–
ViT Base [3]	–	224	0.84	0.79	0.78	0.79	85.8M
ViT Large [3]	–	224	0.84	0.79	0.79	0.78	303.3M
HPB-Base (Our)	ViT Base [3]	224	0.85	0.80	0.81	0.80	112.3M
SwinV2 Tiny [8]	–	256	0.84	0.79	0.80	0.79	27.6M
SwinV2 Small [8]	–	256	0.84	0.79	0.80	0.78	51.3M
HPB-Tiny (Our)	SwinV2 Tiny [8]	256	0.86	0.81	0.82	0.80	54M

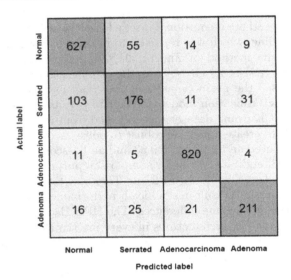

Fig. 6. Confusion matrix obtained with HBP-Tiny. It shows the misclassification issues between normal and serrated classes.

Limitations. Despite its positive outcomes, the proposed HPB architecture has some limitations. One is the accurate differentiation between normal and serrated classes. Even seasoned pathologists struggle with this classification, which could result in incorrect labeling. Therefore, this issue must be addressed appropriately to ensure the architecture's practical application in certain circumstances.

We use the confusion matrix in Fig. 6 to illustrate this aspect. It shows misclassifications for the serrated and normal classes in the best-proposed architecture (HPB-Tiny). Further refinements are necessary to accurately classify these classes.

5   Conclusions

The HPB-ViT architecture has shown great potential in automating the classification of histopathological images. This result is achieved by utilizing the pretrained SwinV2 Tiny backbone, which allows the HPB architecture to learn how to classify these images with less labeled data, yet still perform at a state-of-the-art level. The effectiveness of this proposed architecture is demonstrated through its evaluation of the Chaoyang histopathology dataset for image classification.

There are several ways to improve the performance, practicality, and classification accuracy of the proposed HPB-ViT architecture. One possible direction is to include new pretrained ViT backbones like T2T-ViT, DeepViT, and CvT. Additionally, further integrating CNNs as backbones in the HPB architecture could improve its performance. Testing the approach on different datasets may help generalize the proposed architecture's effectiveness in various fields and show its potential against noise interference. Finally, incorporating multiple backbones into the HPB architecture could improve classification accuracy.

Acknowledgments. We acknowledge financial support under the National Recovery and Resilience Plan (NRRP), Mission 4 Component 2 Investment 1.5 - Call for tender No.3277 published on December 30, 2021 by the Italian Ministry of University and Research (MUR) funded by the European Union – NextGenerationEU. Project Code ECS0000038 – Project Title eINS Ecosystem of Innovation for Next Generation Sardinia – CUP F53C22000430001- Grant Assignment Decree No. 1056 adopted on June 23, 2022 by the Italian Ministry of University and Research (MUR)"

References

1. Chen, H., et al.: Gashis-transformer: a multi-scale visual transformer approach for gastric histopathological image detection. Pattern Recogn. **130**, 108827 (2022)
2. Deng, J., Dong, W., Socher, R., Li, L.-J., Li, K., Li, F.-F.: A large-scale hierarchical image database. In: Imagenet (2009)
3. Dosovitskiy, A., et al.: An image is worth $16{\times}16$ words: transformers for image recognition at scale. In: 9th International Conference on Learning Representations, ICLR 2021, Virtual Event, Austria, 3–7 May 2021 (2021)
4. Glotsos, D., et al.: Improving accuracy in astrocytomas grading by integrating a robust least squares mapping driven support vector machine classifier into a two level grade classification scheme. Comput. Methods Progr. Biomed. **90**(3), 251–261 (2008)
5. He, K., Zhang, X., Ren, S., Sun, J.: Deep residual learning for image recognition. In: Proceedings of the IEEE Conference on Computer Vision and Pattern Recognition, pp. 770–778 (2016)
6. Hendrycks, D., Lee, K., Mazeika, M.: Using pre-training can improve model robustness and uncertainty. In: Chaudhuri, K., Salakhutdinov, R. (eds.) Proceedings of the 36th International Conference on Machine Learning, ICML 2019, Long Beach, California, USA, 9–15 June 2019, vol. 97 of Proceedings of Machine Learning Research, pp. 2712–2721. PMLR (2019)

7. Lin, T.-Y., et al.: Microsoft COCO: common objects in context. In: Fleet, D., Pajdla, T., Schiele, B., Tuytelaars, T. (eds.) ECCV 2014. LNCS, vol. 8693, pp. 740–755. Springer, Cham (2014). https://doi.org/10.1007/978-3-319-10602-1_48

8. Liu, Z., et al.: Swin transformer V2: scaling up capacity and resolution. In: IEEE/CVF Conference on Computer Vision and Pattern Recognition, CVPR 2022, New Orleans, LA, USA, 18–24 June 2022, pp. 11999–12009. IEEE (2022)

9. Liu, Z., et al.: Swin transformer: hierarchical vision transformer using shifted windows. In: 2021 IEEE/CVF International Conference on Computer Vision, ICCV 2021, Montreal, QC, Canada, 10–17 October 2021, pp. 9992–10002. IEEE (2021)

10. Putzu, L., Fumera, G.: An empirical evaluation of nuclei segmentation from h&e images in a real application scenario. Appl. Sci. **10**(22), 7982 (2020)

11. Srinidhi, C.L., Ciga, O., Martel, A.L.: Deep neural network models for computational histopathology: a survey. Medical Image Anal. **67**, 101813 (2021)

12. Steiner, A.P., Kolesnikov, A., Zhai, X., Wightman, R., Uszkoreit, J., Beyer, L.: How to train your vit? data, augmentation, and regularization in vision transformers. In: Transactions on Machine Learning Research (2022)

13. Szegedy, C., et al.: Going deeper with convolutions. In: IEEE Conference on Computer Vision and Pattern Recognition, CVPR 2015, Boston, MA, USA, 7–12 June 2015, pp. 1–9. IEEE Computer Society (2015)

14. Vaswani, A., et al. Attention is all you need. Adv. Neural Inf. Process. Syst. **30** (2017)

15. Wu, H., et al.: Introducing convolutions to vision transformers. In: Cvt (2021)

16. Xu, X., Kapse, S., Gupta, R., Prasanna, P.: Vit-dae: transformer-driven diffusion autoencoder for histopathology image analysis. CoRR, abs/2304.01053 (2023)

17. Li, Y., et al.: Training vision transformers from scratch on imagenet. In: Tokens-to-Token Vit (2021)

18. Zhang, X., Chan, F.T.S., Mahadevan, S.: Explainable machine learning in image classification models: an uncertainty quantification perspective. Knowl. Based Syst **243**, 108418 (2022)

19. Zhou, D., et al.: Towards deeper vision transformer. In: Deepvit (2021)

20. Zhou, X., Tang, C., Huang, P., Tian, S., Mercaldo, F., Santone, A.: Asi-dbnet: an adaptive sparse interactive resnet-vision transformer dual-branch network for the grading of brain cancer histopathological images. Interdisc. Sci. Comput. Life Sci. **15**(1), 15–31 (2023)

21. Zhu, C., Chen, W., Peng, T., Wang, Y., Jin, M.: Hard sample aware noise robust learning for histopathology image classification. IEEE Trans. Med. Imaging **41**, 881–894 (2021)

On Using rPPG Signals for DeepFake Detection: A Cautionary Note

Alessandro D'Amelio[1]([✉])[iD], Raffaella Lanzarotti[1][iD], Sabrina Patania[1][iD], Giuliano Grossi[1][iD], Vittorio Cuculo[2][iD], Andrea Valota[1], and Giuseppe Boccignone[1][iD]

[1] PHuSe Lab, Universitá degli Studi di Milano, Milan, Italy
{alessandro.damelio,raffaella.lanzarotti,sabrina.patania,giuliano.grossi,
andrea.valota,giuseppe.boccignone}@unimi.it
[2] AImageLab, Universitá degli Studi di Modena e Reggio Emilia, Modena, Italy
vittorio.cuculo@unimore.it

Abstract. An experimental analysis is proposed concerning the use of physiological signals, specifically remote Photoplethysmography (rPPG), as a potential means for detecting Deepfakes (DF). The study investigates the effects of different variables, such as video compression and face swap quality, on rPPG information extracted from both original and forged videos. The experiments aim to understand the impact of face forgery procedures on remotely-estimated cardiac information, how this effect interacts with other variables, and how rPPG-based DF detection accuracy is affected by these quantities. Preliminary results suggest that cardiac information in some cases (e.g. uncompressed videos) may have a limited role in discriminating real videos from forged ones, but the effects of other physiological signals cannot be discounted. Surprisingly, heart rate related frequencies appear to deliver a significant contribution to the DF detection task in compressed videos.

Keywords: Deepfake detection · rPPG · Video forensics · Physiological signals

1 Introduction

Fake videos generated through deep learning techniques (Deepfakes [1], DF) blur the line between truth and deception (surmising, for simplicity, that such a line can be drawn). This very fact might lead to a responsible use in the service of many realms (entertainment, education [2], advertising, or privacy protection via de-identification [3]). However, the major concern lies in that DFs might pave the way to the murky realm of fake identity creation for unethical and malicious applications, posing a variety of threats to individuals (e.g. fake porn), organizations (e.g. blackmail to managers to stop sharing their compromising DFs), and politicians (e.g. fake news to sabotage government leaders) [4].

It is no surprise that, since the paradigmatic generation of a synthesized version of Obama [5] in 2017, increasing efforts have been devoted to develop

© The Author(s), under exclusive license to Springer Nature Switzerland AG 2023
G. L. Foresti et al. (Eds.): ICIAP 2023, LNCS 14234, pp. 235–246, 2023.
https://doi.org/10.1007/978-3-031-43153-1_20

DF detection (DFD) methods that can differentiate between real and forged videos [6,7]. Yet, DFD raises subtle issues that, in spite of the flourishing of published methods achieving good metrics performance on public datasets, have been hitherto most often neglected, an attitude that can negatively impinge on DFD generalization from the lab to real-world contexts. A chief concern of this note is to make a step forward in unveiling some of such issues.

DFD methods might be coarsely grouped according to the kind of artifacts that a DF technique can introduce into the forged video (but see [4]), from spatial artifacts and disarranged temporal coherence to anomalies in human behaviours and semantic inconsistencies (such as those between visemes and phonemes).

A class of methods that have recently gained currency is that relying on virtual measurements of physiological data. The idea is simple and straightforward: DF techniques are likely to disrupt physiological signals (such as heart rate, blood flow, and breathing) that can be detected in a contactless way from the RGB video stream. One such case is represented by measurements (e.g., heart rate, HR, and respiration rhythm, RR) derived from the Blood Volume Pulse (BVP) signal estimated via remote photoplethysmography (rPPG, [8]).

In this note we investigate rPPG-based DFD in order to eventually gauge the effects of different variables (e.g. video compression, rPPG method, etc.) on the rPPG information extracted from both original and forged videos in a controlled experimental setting. Specifically, we address the following research questions:

1. (*RQ1*) How a face forgery procedure impinges the remotely-estimated cardiac information?
2. (*RQ2*) How such effect interacts with other factors, such as video compression and DF quality?
3. (*RQ3*) How rPPG-based DF detection accuracy is overall affected by the above factors?

Preliminary results so far achieved suggest that, in discriminating real videos from forged ones, cardiac information displays a nuanced role but depending on a number of factors, while effects of physiological signals other than the cardiac one embedded within the BVP cannot be completely ruled out. Under such circumstances, the application of rPPG-based DFD in real-world contexts should be carefully designed and weighed up.

2 Background and Related Works

A variety of techniques have been developed to efficiently and effectively estimate vital signs relying solely on standard cameras [8–10]. DF detection techniques exploiting physiological information rely on the assumption that face manipulation would produce a (partial) corruption of such information, thus introducing physiological artifacts. Several methods relying on remotely estimated cardiac information have proven effective in DFD, yet, to the best of our knowledge, an analysis of the determinants of such effectiveness is still missing. Notoriously, rPPG approaches are fraught with hurdles, requiring specific conditions to be met, one among many, the availability of uncompressed (or lightly compressed) videos [11]. Interestingly enough, remarkable DFD results have been achieved even on heavily compressed datasets using rPPG information.

DeepRythm [12], employs motion-magnified spatial-temporal maps to highlight chrominance spatio-temporal signals and a dual-spatial-temporal attention network to reduce interference. Another approach, described in [13], uses a two-stage network to identify rhythmic patterns in PPG signals that persist in deepfakes. DeepFakesonPhys [14] relies on a convolutional attention network composed of two parallel CNNs to extract and combine spatio-temporal information from videos. FakeCatcher method [15] exploits an SVM to train on 126-dimensional feature vectors computed from rPPG-derived signals extracted from three facial regions of interest by two rPPG methods. To improve performance, a CNN classifier is trained on PPG maps. Similarly, in [16] the authors use PPG maps and show that various manipulation techniques produce distinct patterns of heartbeats. In [17] a simple and explainable approach is proposed, which relies on gauging both intra-patch complexity measures and inter-patch coherence of rPPG signals. In [18], Spatial-Temporal Filtering Network (STFNet) is considered for rPPG filtering together with a Spatial-Temporal Interaction Network (STINet) accounting for the interaction of PPG signals. More recently, a Multi-scale Spatial-Temporal PPG map has been used to exploit cardiac signals extracted from multiple facial regions [19]. In order to capture both spatial and temporal inconsistencies, the authors laid down a two-stage network consisting of a Mask-Guided Local Attention module together with a Temporal Transformer.

3 Material and Methods

Over the last few years a number of benchmark datasets have been proposed with the aim of scoring different DF detection approaches (e.g. [20–23]). Typically, these benchmarks are built by crawling YouTube videos and subsequently adopting DF techniques to swap identities, reenact faces or manipulate attributes. Clearly, these datasets have been conceived for direct methods comparison and not to understand the effects of face forgery techniques on rPPG signals, which is the problem we are addressing here. To this end, a physiological ground truth is mandatory to perform appropriate comparisons. Moreover, albeit some benchmarks provide a distinction between high quality and low quality videos, uncompressed video is not in general available for the task of DF detection. Under such circumstances, to lay down a controlled setup, we opted to build our own artificially manipulated dataset (UBFC1-forged Dataset, UBFC1-F) as described in the following.

3.1 The UBFC1-Forged Dataset

UBFC1-F is constructed from the publicly available UBFC1 dataset [24]. The latter is typically employed to assess the quality of rPPG methods. It is composed of 8 videos (about 16500 frames) recorded using a low-cost webcam (Logitech

C920 HD Pro) at a resolution of 640×480 pxl in uncompressed 8-bit RGB format, with a frame rate of 30 fps. The ground truth PPG data, including the PPG waveform and heart rates, were obtained using a CMS50E transmissive pulse oximeter. The recordings were made while the subjects sat in front of the camera, positioned approximately 1 m away, with their face visible. The participants were instructed to remain motionless; even so, some of the recorded videos exhibit noticeable movement.

UBFC1-F dataset was generated from UBFC1 as follows. The freely available *FaceSwap* tool[1] has been employed in order to perform face forgery. This method replaces a face in a target sequence with one observed in a source sequence using a simple method based on neural image style transfer [25].

Each subject identity was swapped with every other individual and for each video a variant with different compression rates: 1) *No Compression*, 2) *High Compression*. Compressed videos were obtained via H.264 encoding with a Constant Rate Factor (CRF) set to 30 (High Compression).

Moreover, in order to simulate 1) *Low Quality* and 2) *High Quality* forged videos, training iterations of the *FaceSwap* model were stopped to 25000 in the former case and brought to convergence in the latter.

3.2   Physiological Estimation and Analysis

For each video in the UBFC1-F, either real of forged, the *Blood Volume Pulse* (BVP) signal is estimated from displayed faces via rPPG by exploiting the pyVHR framework[2] [9,10]. pyVHR implements well established pipelines allowing to derive cardiac information using either classic signal processing or learning-based approaches. Here, the classic pipeline is adopted due to its simplicity and higher level of explainability. Though Deep Learning based approaches exhibit superior performance on many benchmark datasets, their generalization abilities are still subject to some controversy [9,10,26].

The RGB videos are used to estimate hidden physiological information, starting with the detection of the face of a possibly manipulated subject, and the automatic tracking of a set of patches on the face around the cheeks area (which typically contains pixels belonging to the swapping area). For each of the P patches, the color intensities of the pixels within it at time t, denoted by $\{p_i^j(t)\}_{j=1}^{N_i}$ ($i = 1, \ldots, P$), are averaged to create P RGB traces. Let $q_i(t)$ be the RGB trace obtained from the i-th patch of length N_i:

$$q_i(t) = \frac{1}{N_i} \sum_{j=1}^{N_i} p_i^j(t), \qquad i = 1, \ldots, P. \tag{1}$$

These traces are then split into K overlapping time-windows, $q_i^k(t) = q_i(t)w(t - k\tau F_s)$, $k = 0, \ldots, K-1$, F_s being the video frame rate, $\tau < 1$ the fraction of overlap, and w a rectangular window. For each patch i and at each

---

[1] https://github.com/deepfakes/faceswap.
[2] https://github.com/phuselab/pyVHR.

time frame k, the BVP signal is estimated using either the GREEN [27] or the POS [28] method. Denote $\mathcal{M} := \{\text{GREEN}, \text{POS}\}$; the BVP estimate for the i-th patch and the k-th time frame is obtained as:

$$x_i^k(t) = \mathcal{M}\left[q_i^k(t)\right]. \tag{2}$$

SNR Analysis. In order to address ($RQ1$) and ($RQ2$), a *Signal-to-Noise Ratio* (SNR) analysis is performed. The goal is to assess to what extent physiological information in forged videos is either disrupted or maintained. At the same time, we gauge the effects of both video compression and forgery method's quality.

The SNR can be operationalised according to [29]; namely, the ratio of the power around the reference HR frequency (i.e., the ground truth HR frequency) plus the first harmonic of the estimated pulse-signal and the remaining power contained in the spectrum of the estimated BVP:

$$\text{SNR} = \frac{1}{K}\sum_k 10\log_{10}\left(\frac{\sum_v \left(U^k(v)S^k(v)\right)^2}{\sum_v \left(1 - U^k(v)\right)S^k(v))^2}\right), \tag{3}$$

where $S^k(v)$ is the power spectral density of the estimated BVP in the k-th time window and $U^k(v)$ is a binary mask that selects the power contained within ± 12 BPM around the reference HR and its first harmonic. The SNR is computed for each BVP signal estimated in each video of the UBFC1-F dataset in order to highlight differences in *real vs. forged* identities, *uncompressed vs. compressed* videos and *Low Quality vs. High Quality* scenarios. Results are visualised in Fig. 1.

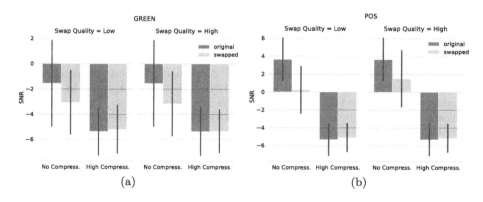

Fig. 1. SNR values obtained on the rPPG estimates on original and swapped videos using the GREEN **(a)** and POS **(b)** methods with different iterations of the *FaceSwap* DF approach. (Color figure online)

Quantitatively, statistical significance is gauged via the BEST statistical analysis (a Bayesian version of the t-Test) [30]; results are presented in Fig. 2, where the posterior probability distributions of the difference of means between Signal-to-Noise Ratio (SNR) values obtained from original and swapped videos of the UBFC1-F dataset are shown. The results are demonstrated for both the GREEN and POS rPPG methods and for scenarios where video compression is applied or not.

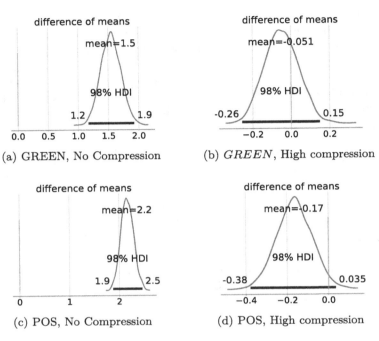

Fig. 2. Result of the BEST test for statistical comparison of two groups. Posterior probability distributions of the difference of means between SNR values obtained from original and swapped videos of the UBFC1-F dataset in case video compression is applied or not (for both GREEN and POS rPPG methods). If the 0 difference value falls outside the highest density interval (98% HDI) of the posterior, it can be deemed an implausible value (the two distributions can be considered distinct). The test reveals significant differences in uncompressed videos ((a) and (c)) and non-significant differences in compressed ones ((a) and (d)). Figure shows results for the *High Quality* swap case; similar results are obtained for the *Low Quality* case. (Color figure online)

3.3   DeepFake Detection Analysis

As to (*RQ3*), the BVP signals estimated on the UBFC1-F dataset are employed to gauge the effect of the same variables considered above on the DF classification accuracy. To this end, the DF detection method proposed in [17] and depicted at a glance in Fig. 3 is exploited. Two sets of features are used as predictors of the presence of faking interventions:

- *Intra-Patch BVP Complexity Measures*: A set of features quantifying the entropy rate of BVP signals
- *Inter-Patch BVP Coherence Measures*: A set of features measuring the degree of consistency between the BVP estimates across the patches.

Feature extraction is followed by an SVM binary classifier for the final DFD step.

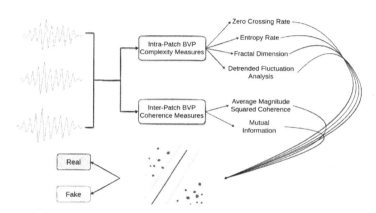

Fig. 3. Overview of the DF detection approach presented in [17] adapted for the analyses presented here

As with the SNR analysis we quantify the effect of video compression and *FaceSwap* level of convergence (method quality) on the DF detection task. In addition, to gauge the role played by heart-rate (HR) related information, we exploit two different filters (*ftype*) applied on the Power Spectral Density (PSD) of the BVP signal:

1. *Band-stop*: A band-stop filter removing the HR-related frequencies (0.75– 4.0 Hz)
2. *Bandpass*: A band-stop filter keeping only the HR related frequencies (0.75– 4.0 Hz)

The DFD accuracy levels achieved on the UBFC1-F dataset by varying the above factors are reported in Fig. 4.

Further statistical hypothesis testing to quantify significant differences between accuracy levels is performed according to [31]. We evaluate the potential improvement of one scenario over another by utilizing the Bayesian Sign-Rank Test [32], a Bayesian non-parametric method that extends the Wilcoxon signed-rank test.

Figure 5 depicts the results of the analysis by reporting the posterior samples for the Bayesian Sign-Rank Test on the simplex. Each vertex of the triangle represents the case where a scenario is either more probable to yield higher DF detection accuracy w.r.t the other or equivalent (probability over the Region of Practical Equivalence, $P(\text{ROPE})$).

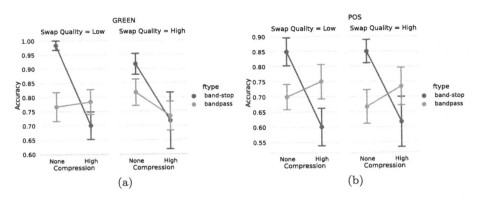

Fig. 4. Accuracy levels for DF detection using the GREEN **(a)** and POS **(b)** rPPG methods with varying compression rates, filter types and *FaceSwap*'s swap quality. (Color figure online)

4   Discussion

SNR Analysis. This analysis clearly shows that on uncompressed videos displaying real identities, physiological information is baldly present; this is particularly evident when using the POS rPPG method, which notoriously exhibits more robust performances if compared to the baseline GREEN method (cfr. Fig. 1(b)). Interestingly enough, when *FaceSwap* is applied, the SNR drastically drops, albeit betraying some residual original cardiac information (especially when the swapping method is trained to convergence). Notably, on compressed videos the SNR reveals weak presence of cardiac information with non significant differences between real and forged videos either using the GREEN or POS rPPG method.

Indeed, from a statistical standpoint, BEST analysis (Fig. 2) shows that in the case of uncompressed videos, significant differences are observed between the SNR values of original and forged videos (Fig. 3, panels **(a)** and **(c)**), whereas non-significant differences are observed in compressed videos (panels **(b)** and **(d)**). Presented results refer to the "*High Quality* swap" case, but the same conclusion can be drawn for the "*Low Quality*" scenario.

DFD Analysis. Results reported in Fig. 4(a) show that the rPPG signals estimated by the GREEN method in general yield the highest accuracy levels on uncompressed videos. This result is rather surprising, considering that, on average, GREEN's signals exhibited negative SNR (cfr. Fig. 1(a)), which in principle should result in poor capabilities of capturing BVP information. Video compression, significantly cuts down the DFD performance of the method. However, at high compression, cardiac information (selected via bandpass filtering) yields higher DFD accuracy than that obtain from other PSD bands (bandstop filtering) when *FaceSwap* simulates a *Low Quality* swapper, and compa-

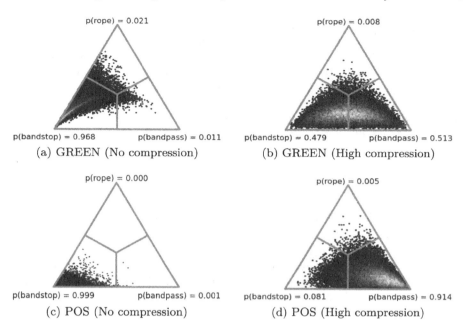

Fig. 5. Posterior samples for the Bayesian Sign-Rank Test on the simplex comparing the distributions of accuracy levels when choosing a *band-stop* vs. *bandpass* filtering of a BVP signal estimated using either GREEN or POS rPPG methods. Each plot shows the probability for a particular scenario to yield higher accuracy if compared with another.

rable accuracy for *High Quality* swapping. Such results are statistically confirmed via the Bayesian Sign-Rank Test (Fig. 5, panels (a) and (b)). Further, when the POS method is considered, such trend becomes crystal clear, and the no compression/high compression factor swaps the roles played by information obtained after bandpass/band-stop filtering to achieve DFD accuracy (Fig. 3(b) and Fig. 5(c) and (d))

Overall, results can be summarized as follows: at zero compression, if we consider the classification obtained from *band-stop* filtered vs. *bandpass* filtered rPPG signals, the former is significantly better for all methods. However, with compressed videos results are reversed: when considering the classification accuracy obtained from *band-stop* filtered vs. *bandpass* filtered rPPG signals, the latter is generally better for both rPPG methods (GREEN and POS). Therefore, *prima facie*, it can be surmised that starting from uncompressed videos, the DF method introduces artifacts that are better captured by the PSD bands of the BVP that do not strictly involve cardiac information. However, other physiological signals may be still present in the filtered BVPs and hence eventually be recruited for the classification. As a matter of fact, BVP signals carry a variety of vital signs (e.g. respiratory signals, blood oxygen levels, blood pressure etc.) that can be extracted (see, for instance, [8, 26]). Video compression, though, seems to smooth out these artifacts that no longer provide enough information, while the HR-related frequency band "withstand" compression as opposed to other frequency bands, in spite of the low SNR. This is an interesting result because, if we look at the SNR of the estimated BVPs (Fig. 1), which is significantly different between the original (high SNR) and fake (low SNR) cases at zero compression, it becomes negative with no significant differences between original and fake videos in compressed videos (cfr. Fig. 2). In other words, under conditions of high compression, the noise is higher than the signal, but the signal, markedly the HR-related component, is still present and is likely to be used.

5   Conclusions

The effective role played by HR information in DFD is the result of complex interactions between the original quality of the video stream and its compression, the power of the DF method, the quality of the estimated signal and the method adopted for BVP estimation. Such interactions can lead to counterintuitive results. Although the current study is a simple simulation-based approach and has potential limitations as it solely relies on the *FaceSwap* technique to generate deepfakes, it provides insights into how different variables can impact remotely estimated BVP signals. It's worth remarking that, the reported classification results depend on the specific DF detection method [17] employed here; the dependence of the analysed quantities on other DF methods and classification approaches will be addressed in a future research. Preliminary results in such direction (not reported here) show that, in the outlined scenario, different BVP estimation methods can also lead to different outcomes over different datasets, thus impinging on DFD generalization when a cross-dataset evaluation is performed. To sum up, the obtained results suggest that the role of cardiac information in distinguishing genuine videos from manipulated ones may be limited in certain situations, such as with uncompressed videos. However, the impact of other physiological signals should not be disregarded. Interestingly, though, frequencies associated with heart rate seem to play a significant role in detecting manipulated videos, particularly in compressed formats.

All in all, in spite of the promising results reported "in-the-lab" condition, HR information for DFD should be handled with care when applications "in-the-wild" are targeted.

References

1. Lee, S.-H., Yun, G.-E., Lim, M.Y., Lee, Y.K.: A study on effective use of bpm information in deepfake detection. In: 2021 International Conference on Information and Communication Technology Convergence (ICTC), pp. 425–427. IEEE (2021)
2. Lee, D.: Deepfake Salvador Dalí takes selfies with museum visitors. The Verge (2019)

3. Bursic, S., D'Amelio, A., Granato, M., Grossi, G., Lanzarotti, R.: A quantitative evaluation framework of video de-identification methods. In: 2020 25th International Conference on Pattern Recognition (ICPR), pp. 6089–6095. IEEE (2021)
4. Mirsky, Y., Lee, W.: The creation and detection of deepfakes: a survey. ACM Comput. Surv. (CSUR) **54**(1), 1–41 (2021)
5. Suwajanakorn, S., Seitz, S.M., Kemelmacher-Shlizerman, I.: Synthesizing obama: learning lip sync from audio. ACM Trans. Graph. (ToG) **36**(4), 1–13 (2017)
6. Tolosana, R., Vera-Rodriguez, R., Fierrez, J., Morales, A., Ortega-Garcia, J.: Deepfakes and beyond: a survey of face manipulation and fake detection. Inf. Fusion **64**, 131–148 (2020)
7. Nguyen, T.T., Nguyen, C.M., Nguyen, D.T., Nguyen, D.T., Nahavandi, S.: Deep learning for deepfakes creation and detection: a survey. arXiv preprint arXiv:1909.11573 (2019)
8. McDuff, D.: Camera measurement of physiological vital signs. ACM Comput. Surv. **55**(9), 1–40 (2023)
9. Boccignone, G., Conte, D., Cuculo, V., D'Amelio, A., Grossi, G., Lanzarotti, R.: An open framework for remote-PPG methods and their assessment. IEEE Access **8**, 216083–216103 (2020)
10. Boccignone, G., et al.: pyvhr: a python framework for remote photoplethysmography. PeerJ Comput. Sci. **8**, e929 (2022)
11. McDuff, D.J., Blackford, E.B., Estepp, J.R.: The impact of video compression on remote cardiac pulse measurement using imaging photoplethysmography. In: 2017 12th IEEE International Conference on Automatic Face & Gesture Recognition (FG 2017), pp. 63–70. IEEE (2017)
12. Qi, H., et al.: DeepRhythm: exposing deepfakes with attentional visual heartbeat rhythms. In: Proceedings of the 28th ACM International Conference on Multimedia, pp. 4318–4327 (2020)
13. Liang, J., Deng, W.: Identifying rhythmic patterns for face forgery detection and categorization. In: 2021 IEEE International Joint Conference on Biometrics (IJCB), pp. 1–8 (2021)
14. Hernandez-Ortega, J., Tolosana, R., Fierrez, J., Morales, A.: DeepFakesON-Phys: deepfakes detection based on heart rate estimation. arXiv preprint arXiv:2010.00400 (2020)
15. Ciftci, U.A., Demir, I., Yin, L.: FakeCatcher: detection of synthetic portrait videos using biological signals. IEEE Trans. Pattern Anal. Mach. Intell. (2020)
16. Ciftci, U.A., Demir, I., Yin, L.: How do the hearts of deep fakes beat? deep fake source detection via interpreting residuals with biological signals. In: 2020 IEEE International Joint Conference on Biometrics (IJCB), pp. 1–10. IEEE (2020)
17. Boccignone, G., et al.: Deepfakes have no heart: a simple rppg-based method to reveal fake videos. In: Image Analysis and Processing-ICIAP 2022: 21st International Conference, Lecce, Italy, 23–27 May 2022, Proceedings, Part II, pp. 186–195. Springer, Heidelberg (2022). https://doi.org/10.1007/978-3-031-06430-2_16
18. Liang, J., Deng, W.: Identifying rhythmic patterns for face forgery detection and categorization. In: 2021 IEEE International Joint Conference on Biometrics (IJCB), pp. 1–8. IEEE (2021)
19. Wu, J., Zhu, Y., Jiang, X., Liu, Y., Lin, J.: Local attention and long-distance interaction of rppg for deepfake detection. Visual Comput., 1–12 (2023)
20. Rössler, A., Cozzolino, D., Verdoliva, L., Riess, C., Thies, J., Nießner, M.: Face-Forensics++: learning to detect manipulated facial images. In: International Conference on Computer Vision (ICCV) (2019)

21. Dolhansky, B., et al.: The deepfake detection challenge (DFDC) dataset. arXiv preprint arXiv:2006.07397 (2020)
22. Li, Y., Yang, X., Sun, P., Qi, H., Lyu, S.: Celeb-df: a large-scale challenging dataset for deepfake forensics. In: IEEE Conference on Computer Vision and Patten Recognition (CVPR) (2020)
23. Jiang, L., Li, R., Wu, W., Qian, C., Loy, C.C.: DeeperForensics-1.0: a large-scale dataset for real-world face forgery detection. In: CVPR (2020)
24. Bobbia, S., Macwan, R., Benezeth, Y., Mansouri, A., Dubois, J.: Unsupervised skin tissue segmentation for remote photoplethysmography. Pattern Recogn. Lett. **124**, 82–90 (2019)
25. Liu, M.-Y., Breuel, T., Kautz, J.: Unsupervised image-to-image translation networks. Adv. Neural Inf. Process. Syst. **30**, 1–9 (2017)
26. Boccignone, G., D'Amelio, A., Ghezzi, O., Grossi, G., Lanzarotti, R.: An evaluation of non-contact photoplethysmography-based methods for remote respiratory rate estimation. Sensors **23**(7), 3387 (2023)
27. Verkruysse, W., Svaasand, L.O., Nelson, J.S.: Remote plethysmographic imaging using ambient light. Opt. Express **16**(26), 21434–21445 (2008)
28. Wang, W., den Brinker, A.C., Stuijk, S., De Haan, G.: Algorithmic principles of remote PPG. IEEE Trans. Biomed. Eng. **64**(7), 1479–1491 (2016)
29. De Haan, G., Jeanne, V.: Robust pulse rate from chrominance-based rPPG. IEEE Trans. Biomed. Eng. **60**(10), 2878–2886 (2013)
30. Kruschke, J.K.: Bayesian estimation supersedes the t test. J. Exp. Psychol. Gener. **142**(2), 573 (2013)
31. Benavoli, A., Corani, G., Demšar, J., Zaffalon, M.: Time for a change: a tutorial for comparing multiple classifiers through bayesian analysis. J. Mach. Learn. Res. **18**(1), 2653–2688 (2017)
32. Benavoli, A., Corani, G., Mangili, F., Zaffalon, M., Ruggeri, F.: A Bayesian Wilcoxon signed-rank test based on the Dirichlet process. In: International Conference on Machine Learning, pp. 1026–1034. PMLR (2014)

An Unsupervised Learning Approach to Resolve Phenotype to Genotype Mapping in Budding Yeasts Vacuoles

Vito Paolo Pastore[1,3,4(✉)], Paolo Didier Alfano[1,2], Ashwini Oke[4,5],
Sara Capponi[3,4], Daniel Eltanan[5], Xavier Woodruff-Madeira[5], Anita Nguyen[5],
Jennifer Carol Fung[4,5], and Simone Bianco[3,4,6]

[1] MaLGa - DIBRIS, University of Genoa, Genoa, Italy
Vito.Paolo.Pastore@unige.it, Paolodidier.alfano@edu.unige.it
[2] Istituto Italiano di Tecnologia, Genoa, Italy
[3] IBM Almaden Research Center, San Jose, CA, USA
sara.capponi@ibm.com
[4] Center for Cellular Construction, San Francisco, CA, USA
[5] Department of Obstetrics, Gynecology and Reproductive Sciences, University of
California, San Francisco, CA, USA
{Ashwini.Oke,Jennifer.Fung}@ucsf.edu,xwoodruf@ucsc.edu, atn017@ucsd.edu
[6] Altos Labs, Redwood City, CA, USA
sbianco@altoslabs.com

Abstract. The relationship between the genotype, the set of instructions encoded into a genome, and the phenotype, the macroscopic realization of those instructions, has not been fully explored. This is mostly due to the general absence of tools capable of uncovering this relationship. In this work, we develop an unsupervised learning framework relating changes in cellular morphology to genetic modifications. We focus on yeast organelles called vacuoles, which are cellular compartments that vary in size and shape as a response to various stimuli. Our approach can be applied extensively for live fluorescence image analysis, potentially unveiling the basic principles relating genotypic variation to vacuole morphology in yeast cells. This can, in turn, be a first step for the inference of cell design principles of cellular organelles with a desired morphology.

Keywords: Unsupervised learning · Yeast vacuoles image analysis · transfer learning

1 Introduction

Cellular organelles like the nucleus, mitochondria, and vacuoles are specialized compartments that evolved to contain complex biochemical reactions. Their shape and structure are related to the biological functions they mediate. Hence, one may assume that the design of a specific organelle would influence the biochemistry of the processes it contains [5]. While much work has been devoted to

© The Author(s), under exclusive license to Springer Nature Switzerland AG 2023
G. L. Foresti et al. (Eds.): ICIAP 2023, LNCS 14234, pp. 247–258, 2023.
https://doi.org/10.1007/978-3-031-43153-1_21

using mathematical models that investigate the design principles of living organisms, the same is not true for a purely data-driven approach. Specifically, the data-driven design of living organisms must be achieved through a precise characterization of the relationship between genetic modifications and the resulting changes in cellular morphology. This is a complicated problem of multiplicity: many distinct morphologies can be associated with one or more driving mutations. Thus, a map of the relationships linking desirable structures to a set of mutations capable of generating those structures is a first and important stepping stone towards more complex problems, like the effect of combinations of mutations or environmental perturbations.

Thus, in this paper, we focus on the problem of mapping cellular morphology to genetic perturbations in model organisms, using budding yeasts vacuoles as test organelles. This is not a new problem, and a few methods have been recently developed, facing the phenotype-to-genotype mapping problem using supervised deep learning and machine learning techniques [17,18]. However, applying deep learning in a supervised fashion requires a significant amount of training data [1]. Obtaining an extensive and comprehensive unbiased annotated training set is a potential bottleneck towards applying such algorithms to the task at hand. Annotations are in fact complicated because of the sheer biological variability, and expensive in terms of time and resources. To overcome this issue, we present an unsupervised framework capable of providing automatic morphotype detection starting from unlabeled budding yeast vacuoles' images. We propose to use the obtained clusters as a starting point for an end-to-end mapping between inferred organelle morphologies and genetic mutations in budding yeast (see Fig. 1). Specifically, our results show that different genetic perturbations correspond to different prevalences in vacuole morphotypes. Although we use yeast vacuoles as test organelles, our pipeline (and the extracted features) is not specific for vacuole morphology and can be extended to other organelles.

To summarize, the main contributions of this work are (i) the development of an unsupervised framework to reveal the relationship between phenotype and genotype in an accurate and unbiased way; (ii) an extensive comparison between hand-crafted features and deep features extracted by means of ImageNet pretrained convolutional neural networks and transformers, with the aim of obtaining the best performances on the test set; (iii) the introduction of a publicly available dataset of budding yeasts vacuoles images, post-processed by exploiting a custom depth-dependent color encoding to incorporate 3D shape information into 2D images.

The remainder of the paper is organized as follows. First, we describe the data. Then we provide a comprehensive overview of the proposed unsupervised pipeline. Finally, we present and discuss our results.

2   Related Works

Mapping cellular morphology to genetics in model organisms like yeast is a fundamental problem for efficient synthetic engineering. This is not itself a new problem, and various methods have started to be proposed for the task at hand. Recently, methods that allow to *paint the cell* with many morphological fluorescence labels have been introduced and widely adopted to define cell morphology elements [22]. These techniques are called *cell painting* [9]. In [18], a supervised deep-learning approach is used to predict the phenotype corresponding to chemical compounds for drug discovery applications. In this work, features are extracted from chemical structures using deep neural networks, and cell painting is used to identify morphological elements in yeasts, bacteria, and human cells. A similar approach is exploited in [27], where the authors make use of deep neural networks trained in a supervised framework to predict the mechanisms of action of chemical compounds exploiting cell painting and genetic information. Another relevant work is described in [17], where the authors use yeast endocytic pathways as models to relate mutations to phenotype at a single-cell level. They identified hundreds of genes that affect one endocytic compartment, exploiting live-cell fluorescence and artificial neural networks single-cell analysis. A final set of 21 phenotypes focusing on vacuoles and other yeast organelles is obtained and mapped back to a sequence of genetic insults. In this work, the authors use a supervised deep learning-based approach to classify the identified phenotypes prior to the mapping. The described works tackle the phenotype-to-genotype mapping problem in a supervised fashion. However, getting high-quality and unbiased annotations is not trivial when dealing with biological data, including budding yeast vacuoles. Moreover, morphology in the cell is inherently three-dimensional, but only two-dimensional images are used in the previously cited works. For this reason, in this paper, we first exploit a depth-dependent color encoding to include 3D information in our 2D vacuoles images. Furthermore, we introduce an unsupervised learning pipeline to support phenotype-to-genotype mapping in budding yeast vacuoles. Our procedure provides output clusters based on the inferred morphology, that can later be used to reveal convenient genetic perturbations, as well as the co-occurrence of specific morphotypes.

3   Proposed Method

In this section, we provide a detailed description of the designed approach. Figure 1 shows a schematic representation of the proposed unsupervised framework. The workflow includes a features engineering step, with the computation of a set of features to be used as input for the unsupervised partitioning module, consisting of a clustering algorithm. The output clusters correspond to vacuole morphotypes automatically identified with the designed algorithm. At this stage, the distribution of output clusters per well is computed. Each well in our experiments has a different genetic configuration. Thus, the genetic insults applied to the wells with the most desirable prevalence of morphotypes can be selected and repeated for future experiments. This step is our preliminary idea for phenotype-to-genotype mapping in budding yeast vacuoles. We explore two variants for the features engineering step (red square in Fig. 1). The first approach includes a deep-learning-based vacuoles segmentation algorithm and the computation of a set of hand-crafted features. The second variant exploits a deep neural network pre-trained on ImageNet as a feature extractor, to provide a set of pre-trained deep features. Our aim is to determine the best descriptors in terms of clustering purity (see Sect. 3.5). An accurate clustering is, in fact, a crucial requirement for our pipeline. The more precise the clustering, the more reliable the process of using clusters' occurrence percentages in the wells to select the most convenient genetic perturbations. In the next paragraphs, we provide details of each component of the pipeline.

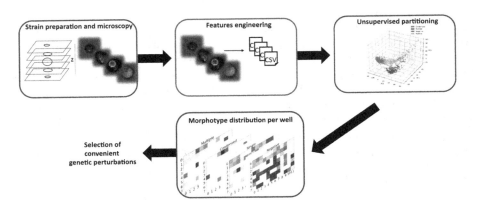

Fig. 1. Schematic overview of the proposed pipeline. Two variants were explored for feature engineering (red box). (Color figure online)

3.1   Strain Preparation and Microscopy

We use the background yeast strain BY4741 containing a vacuolar membrane protein coupled to a green fluorescent protein. VPH1-GFP-HIS3 [6] is used as the base strain for the yeast deletion library. We then incorporate the VPH1-GFP into the library which represents different deleted genes by yeast transformation. Each well in a 96-well plate is given a strain with VPH1-GFP with a different genotype, characterized by a complete deletion of one gene. Intuitively, the aforementioned procedure is used to ensure that each well contains yeasts with distinct genetic perturbations. The well plates are arranged into a 8×12 grid, as represented in Fig. 3, and imaged using a GE InCell 6000 imaging platform (Molecular Devices, San Jose, CA). A 2D brightfield image is taken to use for yeast segmentation. Forty-eight $0.2\,\mu m$ sections of VPH1-GFP fluorescence are used to acquire a 3D vacuole structure ($60\times$ magnification). A random selection from cells in the 96 well plate is imaged for a total of 998 vacuole single cell images, which are used as input for the features engineering module of the proposed pipeline. Each image is 80×80 pixels in size and contains a single cell. In order to retain the 3D information within a 2D image, depth-dependent color encoding is performed. This is done by normalizing the image intensity from 0 to

1 and assigning a unique color to each z slice. Therefore, the vacuoles result to be colored to encode for depth, with blue being at the bottom and red being at the top (see Fig. 2 for an example). Our vacuoles dataset is open-source and available at https://github.com/CCCofficial/Vacuoles-dataset-unsupervised-learning.git.

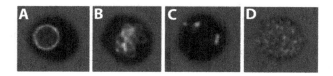

Fig. 2. Representative vacuole images of the 4 classes associated by our experts to the clusters identified by the Partition Coefficient in our pipeline (PC): A) single, B) multiple, C) condensed vacuole, and D) negative or dead cell

3.2   Features Engineering: Hand-Crafted Descriptors Computation

Image Segmentation. The computation of the engineered set of hand-crafted features requires the segmentation of vacuoles in our images. We exploit a deep-learning-based segmentation algorithm. First, we use ImageJ [25] to annotate a subset of 250 images randomly extracted from the complete set of vacuole images. We later train an ImageNet pre-trained customized version of the U-Net architecture [23] on the annotated data. Finally, we use the trained model for the segmentation of the entire dataset composed of 998 yeast vacuole images, which we exploit for the computation of the hand-crafted features.

Hand-Crafted Features. The output of the segmentation network is an image where only vacuoles are present. Starting from such masked images, we engineer a set of 131 morphological features to describe the shape and texture of the yeast vacuoles. The designed set is similar to the one introduced in [21] for plankton species discovery and includes 14 geometric features (e.g. area, perimeter, circularity, and eccentricity), 7 moments-based features (using Hu-moments [14]) and 25 Zernike moments (up to order 5 [28]). In addition, we extract 10 Fourier Descriptors (FD) from the vacuoles contour [20], 54 Local Binary Pattern (LBP), and 13 Haralick descriptors [10]. A set of 5 features is extracted from the gray-values histogram (mean value, standard deviation, kurtosis, skewness, and entropy) [19]. Finally, we use the ratios between the mean pixel intensity values of the three color channels as features to complete the set of 131 engineered descriptors. It is worth underlining that the designed set of features is not specific to the vacuoles morphology description problem, but is generic and can be extended and generalized to other cell organelles.

3.3   Features Engineering: Deep Pre-trained Features

We adopt a transfer-learning approach, exploiting a CNN model pre-trained on ImageNet as a features extractor. Each vacuole image is fed to the CNN and the output of the last convolutional layer is used as a corresponding set of features. As an ablation study, we consider five CNNs and one transformer as feature extractors, comparing the resulting descriptors in terms of test purity (see Sect. 3.5). We choose five CNNs with different depths and complexity: (i) a DenseNet201 [13]; (ii) a ResNet50 [11]; (iii) an InceptionResNetv2 [26] (iv) an Xception [7] and (v) a MobileNet [12]. The CNNs are pre-trained on ImageNet1K, a subset of ImageNet introduced in [24] containing approximately 1.3 million of images belonging to 1000 different classes. Recent works have suggested that the quality of deep pre-trained features increases with the number of classes and images of the pre-training dataset [15,16]. Thus, we further consider a ViT-B/16 transformer model pre-trained on ImageNet22K [8], the complete version of ImageNet containing around 14 million of images belonging to 21841 classes.

3.4   Unsupervised Partitioning

We employ a fuzzy K-means [4] clustering algorithm in the unsupervised partitioning module of our pipeline. Moreover, we exploit the Partition Coefficient (PC) algorithm [3] to infer automatically the number of different clusters (i.e., the morphological classes) from our training data.

$$PC = \frac{1}{N} \sum_{i=1}^{N} \sum_{j=1}^{K} u_{ij} \tag{1}$$

The PC is defined in Eq. 1, where u_{ij} is the degree of membership of the sample i to the cluster j, N is the total number of samples, and K the maximum number of considered clusters. The PC varies between 0 and 1, with higher uncertainty in clustering for lower PC values. Thus, a good estimation of the number of classes corresponds to the PC peak with respect to j.

3.5   Evaluation Metrics

We evaluate the accuracy of the clustering algorithm by means of a purity metric [2], defined as:

$$purity = \frac{1}{N} \sum_{i} \max_{j} |\lambda_i \cap c_j| \tag{2}$$

where $\lambda_1, ..., \lambda_K$ are the computed clusters and $c_1, ..., c_n$ are the ground-truth classes. Every cluster i is assigned the most overlapping ground truth class j. A purity equal to 1 corresponds to a perfect overlap between clusters and ground truth classes.

4   Experiments

4.1   Experiments Details

In this section, we present technical details about the conducted experiments.

We split our dataset of 998 vacuole images using an 80:20 ratio for training and testing, with a hold-out approach. The test set is stored for the evaluation of our pipeline. We perform the PC computation on the training set for ten iterations, recomputing the fuzzy K-means each time. We then use the mode of the peak on the ten runs as an estimation of the number of clusters.

To assess the robustness of our results, we perform a 10-fold analysis on the training set for all the experiments included in this work.

Images are resized to 224×224 and standardized using ImageNet mean and standard deviation for deep feature extraction. To extract the deep features we test six different pre-trained models: DenseNet (DN), InceptionResNet (IRN), MobileNet (MN), ResNet (RN), Xception (XN), Vision Transformer Base/16 (ViT-B/16), in their tensorflow keras implementations. The upper-bound supervised accuracies presented in Table 3, are computed by fine-tuning the six aforementioned pre-trained models, modified by adding 3 fully connected layers on top of the convolutional layers. Every pre-trained model is entirely fine-tuned for 100 epochs with a 10^{-3} learning rate. The batch size is set to 32.

4.2   Clustering Accuracy with the Engineered Descriptors

First, we adopt the Partition Coefficient (PC) based algorithm on our set of hand-crafted features, as explained in Sect. 3.4. The PC reveals a peak corresponding to four clusters, meaning that four classes best divide the training data set. The 4 detected clusters have been later assigned to 4 specific morphological classes by our experts in the field (see Fig. 2): single vacuole, multiple (2 or more) vacuoles, condensed vacuoles, and dead cells, which we call negative. We evaluate the quality of the unsupervised partitioning module of our pipeline, comparing the representation provided by our two different types of features, in terms of clustering purity: (i) the set of 131 hand-crafted features; (ii) the sets of deep features extracted by means of each one of the five ImageNet pre-trained CNNs and the transformer considered in our study. At this stage, we apply the fuzzy K-means algorithm to cluster each investigated representation.

Table 1 shows the obtained results in terms of clustering purity. The purity metric is computed using our team's expert annotations, based on the semantics of the four automatically extracted clusters, as ground truth. The deep pre-trained features have a dimensionality of up to two orders of magnitude higher than the hand-crafted features. Such a high dimensionality could negatively affect the clustering accuracy, resulting in an unfair comparison with respect to the hand-crafted features. Thus, we also apply a Principal Components Analysis (PCA) algorithm to the deep pre-trained features, selecting 131 principal components, that is the number of hand-crafted descriptors. As we can see in Table 1, the dimensionality reduction by means of PCA brings better performances with respect to the original features for all the pre-trained deep neural networks considered in this study. The ImageNet22K pre-trained ViT-B/16 features provide the best performances, with a test purity of 0.942. Furthermore, the ImageNet pre-trained CNNs generally outperform the hand-crafted features, with a maximum purity of 0.911 corresponding to the usage of a ResNet152 model.

Table 1. Clustering performances using different sets of features as input. Hand-crafted features, and deep features extracted by means of pre-trained CNNs and a transformer.

Input	PCA	#features	Purity
DenseNet201	✗	94080	0.797 ± 0.007
ResNet152	✗	100352	0.885 ± 0.008
MobileNet	✗	50176	0.817 ± 0.009
InceptionResNetV2	✗	38400	0.884 ± 0.003
Xception	✗	100352	0.895 ± 0.005
ViT-B/16	✗	768	0.941 ± 0.002
DenseNet201	✓	131	0.810 ± 0.003
ResNet152	✓	131	0.911 ± 0.010
MobileNet	✓	131	0.866 ± 0.004
InceptionResNetV2	✓	131	0.893 ± 0.002
Xception	✓	131	0.906 ± 0.003
ViT-B/16	✓	131	$\mathbf{0.942 \pm 0.002}$
Hand-crafted features	✗	131	0.849 ± 0.002

In this work, differently from the state-of-the-art on phenotype-to-genotype mapping, we exploit a custom depth-encoding procedure to retain 3D morphological information into 2D images. We perform an ablation study to prove the effectiveness of the color-based depth encoding on the clustering test accuracy. We exploit two commonly adopted 2D variants of the acquired images, obtained by considering: (i) the max projection among each of the vacuoles' 3D volume (MaxProj); (ii) the center slice of the 3D volume (CenterSlice). Table 2 summarizes the obtained results, using our best-performing configuration, that is 131 PCA components computed on top of the pre-trained ViT-B/16 features. Our custom depth-encoding procedure provides the best test purity, with an average improvement of 4% and 6%, with respect to the MaxProj and CenterSlice, respectively.

Table 2. Clustering performances using different variants of the vacuoles dataset where depth encoding is performed (Color-based) or not (MaxProj and CenterSlice.

Dataset version	Depth Encoding	Purity
MaxProj	✗	0.903 ± 0.002
CenterSlice	✗	0.880 ± 0.007
Color-based (ours)	✓	$\mathbf{0.942 \pm 0.002}$

At this stage, to better frame the performances of our unsupervised pipeline, we evaluate an upper bound in accuracy by fine-tuning the same pre-trained deep neural networks used in this study in a supervised fashion, using the ground truth provided by our experts in the field. Table 3 summarizes the obtained results. The best performances are obtained with the adoption of an Xception model, with an average test accuracy of 0.982. Our unsupervised procedure with ViT-B/16 features corresponds to a marginal drop of 4% with respect to the best-performing fully supervised algorithm, proving the high accuracy of the proposed unsupervised framework. Since the ViT-B/16 features provide the best embedding for our yeast vacuoles, we use the clusters computed on top of these descriptors in the final step of our pipeline.

Table 3. Upper bound supervised test accuracy for the models included in this work: DenseNet (DN); InceptionResNet (IRN); MobileNet (MN); ResNet (RN); Xception (XN); Vision Transformer Base 16(ViT).

DN	IRN	MN	RN	XN	ViT
0.971 ± 0.009	0.976 ± 0.012	0.974 ± 0.012	0.932 ± 0.072	$\mathbf{0.982 \pm 0.004}$	0.967 ± 0.006

4.3 Relating Vacuole Morphology to Different Genetic Conditions

Morphotypes Distribution per Well. Our pipeline assigns each of the cells' images to one of the 4 automatically detected clusters, later identified by our experts in the field to 4 different vacuole morphotypes (see Sect. 4.2).

Figure 3 reports the resulting inferred number of cells belonging to each morphological class as a percentage with respect to the total amount of cells for each of the 96 wells considered in our experimental protocol. Each panel in Fig. 3 represents a class. The elements of the represented matrix correspond to the 96 wells in the acquisition plate (see Sect. 3.1) and, thus, to distinct genetic mutations. The color intensity represents the relative fraction of a particular morphological type. It is worth noticing that the number of yeast vacuole images acquired is not the same for each well and that some of the wells did not produce valid vacuole cells (they are missing and appear as white pixels in Fig. 3). Our results show that different genetic perturbations result in similar and specific prevalent morphological types, as defined by our methodology. For instance, many of the wells in rows E, G, and H, show a prevalence of negative morphotypes, that is dead cells. As budding yeast vacuoles can be used to store products in synthetic engineering applications, we suggest that the corresponding genetic perturbations are not convenient and should be disregarded. In the same way, wells in rows A and B show the prevalence of single and multiple vacuoles, which are convenient morphotypes. Thus, the corresponding genetic perturbations could be considered interesting and further explored. Another relevant example can be seen in wells at row A and columns 10 and 11, and in well at row D and column 2, where the single vacuole is prevalent. It is also possible to appreciate the emergence of associated morphotypes. Looking at the wells with high percentages of condensed cells, we can notice that they are also associated with high percentages of multiple vacuole cells, and a negligible appearance of the other morphotypes. Interestingly, this is not true in the other direction, as wells with a high percentage of multiple vacuoles do not necessarily show condensed cells (e.g., row D and column 3, wells in row F). This preliminary analysis, including the possibility to reveal the co-occurrence of two or more phenotypes in specific wells, and the presence or the absence of a strong phenotype in some well, suggests the potential for our pipeline to reveal a link between mutations (i.e., the genotype) and specific phenotypes.

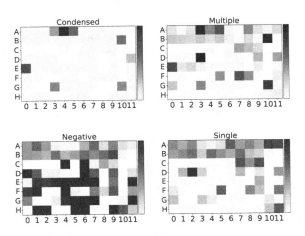

Fig. 3. Relative fraction of the inferred morphotypes as a function of 96 well plates for each class.

5   Conclusions

In this paper, we focus on the problem of relating phenotype to genotype in budding yeast vacuoles. Obtaining high-quality and unbiased annotations is cumbersome for the task at hand. Thus, we introduce an unsupervised pipeline to provide clusters later exploited to reveal the relationship between phenotype and genotype in an accurate and unbiased way. We evaluate two different sets of descriptors, namely a collection of 131 hand-crafted features and a set of deep pre-trained features, obtained by exploiting ImageNet pre-trained deep neural networks as feature extractors. We use a partition coefficient-based algorithm to automatically infer the number of clusters, obtaining four clusters from our set of 998 vacuole images. Our experts assigned the obtained clusters to 4 specific morphological classes (single vacuole, multiple vacuoles, condensed vacuoles, and dead cells, which we call negative). Our results show that the ImageNet pre-trained ViT-B/16 features provide the best representation, resulting in an average clustering purity of 0.942. We finally exploit our method to identify the percentage of vacuole morphotypes in the 96 acquired wells, each of which is associated with distinct genetic mutations. We find a systematic co-occurrence of some vacuole morphotypes (e.g., condensed and multiples) and different wells when the prevalence of one class emerges. While the mechanistic design of cells and cellular structures is still a long-term scientific goal, introducing methods capable of inferring biological design principles is an important first step toward achieving a high level of predictability in bioengineering. The yeast vacuole is an example of an organelle that can be used as a biological reactor for chemical production. Even if further work is necessary to support our findings, we propose that our method may represent a first step toward the fundamental task of accurately establishing the link between cellular phenotype and genotype in a bioengineering experiment setting, in an accurate and unbiased way.

Acknowledgments. This material is partially based upon work supported by NSF grant No. DBI-1548297. VPP was supported by FSE REACT-EU-PON 2014–2020, DM 1062/2021.

References

1. Adadi, A.: A survey on data-efficient algorithms in big data era. J. Big Data **8**(1), 24 (2021). https://doi.org/10.1186/s40537-021-00419-9
2. Alfano, P.D., Rando, M., Letizia, M., Odone, F., Rosasco, L., Pastore, V.P.: Efficient unsupervised learning for plankton images (2022). https://doi.org/10.48550/ARXIV.2209.06726
3. Bezdek, J.C.: Numerical taxonomy with fuzzy sets. J. Math. Biol. **1**(1), 57–71 (1974). https://doi.org/10.1007/BF02339490
4. Bezdek, J.C.: Pattern Recognition with Fuzzy Objective Function Algorithms. Springer, Heidelberg (2013). https://doi.org/10.1007/978-1-4757-0450-1
5. Bianco, S., Chan, Y.H.M., Marshall, W.F.: Towards computer-aided design of cellular structure. Phys. Biol. **17**(2), 023001 (2020). https://doi.org/10.1088/1478-3975/ab6d43
6. Chan, Y.H.M., Marshall, W.F.: Organelle size scaling of the budding yeast vacuole is tuned by membrane trafficking rates. Biophys. J. **106**(9), 1986–1996 (2014). https://doi.org/10.1016/j.bpj.2014.03.014
7. Chollet, F.: Xception: deep learning with depthwise separable convolutions. In: Proceedings of the IEEE Conference on Computer Vision and Pattern Recognition, pp. 1251–1258 (2017)
8. Deng, J., Dong, W., Socher, R., Li, L.J., Li, K., Fei-Fei, L.: Imagenet: a large-scale hierarchical image database. In: 2009 IEEE Conference on Computer Vision and Pattern Recognition, pp. 248–255 (2009). https://doi.org/10.1109/CVPR.2009.5206848
9. Gustafsdottir, S.M., et al.: Multiplex cytological profiling assay to measure diverse cellular states. PloS One **8**(12), e80999 (2013)
10. Haralick, R.: Statistical and structural approaches to texture. Proc. IEEE **67**, 786–804 (1979). https://doi.org/10.1109/PROC.1979.11328
11. He, K., Zhang, X., Ren, S., Sun, J.: Deep residual learning for image recognition. In: Proceedings of the IEEE Conference on Computer Vision and Pattern Recognition, pp. 770–778 (2016)

12. Howard, A.G., et al.: Mobilenets: efficient convolutional neural networks for mobile vision applications. arXiv preprint arXiv:1704.04861 (2017)

13. Huang, G., Liu, Z., Van Der Maaten, L., Weinberger, K.Q.: Densely connected convolutional networks. In: Proceedings of the IEEE Conference on Computer Vision and Pattern Recognition, pp. 4700–4708 (2017)

14. Huang, Z., Leng, J.: Analysis of hu's moment invariants on image scaling and rotation, vol. 7, pp. 7–476 (2010). https://doi.org/10.1109/ICCET.2010.5485542

15. Huh, M., Agrawal, P., Efros, A.A.: What makes imagenet good for transfer learning? arXiv preprint arXiv:1608.08614 (2016)

16. Maracani, A., Pastore, V.P., Natale, L., Rosasco, L., Odone, F.: In-domain versus out-of-domain transfer learning in plankton image classification. Sci. Rep. **13**(1), 10443 (2023)

17. Mattiazzi Usaj, M., et al.: Systematic genetics and single-cell imaging reveal widespread morphological pleiotropy and cell-to-cell variability. Molec. Syst. Biol. **16**(2), e9243 (2020). https://doi.org/10.15252/msb.20199243

18. Moshkov, N., et al.: Predicting compound activity from phenotypic profiles and chemical structures. Nat. Commun. **14**(1), 1967 (2023)

19. Pastore, V.P., Zimmerman, T., Biswas, S.K., Bianco, S.: Establishing the baseline for using plankton as biosensor. In: Imaging, Manipulation, and Analysis of Biomolecules, Cells, and Tissues XVII, vol. 10881, p. 108810H. International Society for Optics and Photonics (2019)

20. Pastore, V.P., Zimmerman, T.G., Biswas, S.K., Bianco, S.: Annotation-free learning of plankton for classification and anomaly detection. Sci. Rep. **10**(1), 12142 (2020). https://doi.org/10.1038/s41598-020-68662-3

21. Pastore, V.P., Megiddo, N., Bianco, S.: An anomaly detection approach for plankton species discovery. In: Sclaroff, S., Distante, C., Leo, M., Farinella, G.M., Tombari, F. (eds.) ICIAP 2022. LNCS, pp. 599–609. Springer, Cham (2022). https://doi.org/10.1007/978-3-031-06430-2_50

22. Rohban, M.H., et al.: Systematic morphological profiling of human gene and allele function via cell painting. Elife **6**, e24060 (2017)

23. Ronneberger, O., Fischer, P., Brox, T.: U-Net: convolutional networks for biomedical image segmentation. In: Navab, N., Hornegger, J., Wells, W.M., Frangi, A.F. (eds.) MICCAI 2015. LNCS, vol. 9351, pp. 234–241. Springer, Cham (2015). https://doi.org/10.1007/978-3-319-24574-4_28

24. Russakovsky, O., et al.: ImageNet large scale visual recognition challenge. Int. J. Comput. Vision (IJCV) **115**(3), 211–252 (2015). https://doi.org/10.1007/s11263-015-0816-y

25. Schindelin, J., et al.: Fiji: an open-source platform for biological-image analysis. Nat. Methods **9**(7), 676–682 (2012). https://doi.org/10.1038/nmeth.2019

26. Szegedy, C., Ioffe, S., Vanhoucke, V., Alemi, A.A.: Inception-v4, inception-resnet and the impact of residual connections on learning. In: Thirty-First AAAI Conference on Artificial Intelligence (2017)

27. Way, G.P., et al.: Morphology and gene expression profiling provide complementary information for mapping cell state. Cell Syst. **13**(11), 911–923 (2022)

28. Yang, Z., Fang, T.: On the accuracy of image normalization by Zernike moments. Image Vision Comput. **28**(3), 403–413 (2010). https://doi.org/10.1016/j.imavis.2009.06.010

Food Image Classification: The Benefit of In-Domain Transfer Learning

Larbi Touijer, Vito Paolo Pastore$^{(\boxtimes)}$, and Francesca Odone

MaLGa - DIBRIS, University of Genoa, Genoa, Italy
larbi.touijer@edu.unige.it, {vito.paolo.pastore,francesca.odone}@unige.it

Abstract. Monitoring food intake and calories may be fundamental for a healthy lifestyle and preventing nutrition-related illnesses. Recently, deep-learning approaches have been extensively exploited to provide an automatic analysis of food images. However, food image datasets have peculiar challenges, including fine granularity with a high intra-class and low inter-class variability. In this work, we focus on training strategies considering the typical scenario where data availability and computational resources are limited. Exploiting convolutional neural networks, we show that in-domain source datasets provide a better representation with respect to only using ImageNet, bringing a significant increase in test accuracy. We finally show that ensembling different CNN models further improves the learned representation.

Keywords: Food image classification · transfer learning · ensemble of convolutional neural networks

1 Introduction

Monitoring food intake and calories challenge everyone who is willing to eat healthy and balanced. It is motivated by several reasons, starting from an overall healthy lifestyle, all the way to severe cases of illnesses. A number of studies highlight the relevance of counting the calories and macros as life-saving for cancer patients [16,18]. The challenge comes from the tedious calculations that one has to keep track of every time there is a meal to prepare. Consequently, there is a growing interest in the design and development of automatic methods for food intake estimation, covering different specific challenges. The U.S. Department of Agriculture proposes the *Food and Nutrient Database for Dietary Studies*(FNDDS) [1], that allows the correspondence between food and a list of nutritional values. This database has also allowed for the development of Food Ontologies like FoodKG [7], a food knowledge graph that gives the consumer information about how healthy the food is, and its provenance. To effectively address this challenge, there is also a line of research involving automatic food image classification. Several works have shown interest in this domain, starting from the one introducing the ETH Food 101 food image dataset [4]. Likewise, the recent dataset Recipe1M+ enables the retrieval of food images with ingredients

G. L. Foresti et al. (Eds.): ICIAP 2023, LNCS 14234, pp. 259–269, 2023.
https://doi.org/10.1007/978-3-031-43153-1_22

and cooking instructions. Beyond research, the market of mobile applications contains a fair amount of Apps, such as *Foodvisor-Nutrition & Diet* with more than one million downloads. It relies on taking pictures with the mobile camera and predicting the recipe's name and nutritional values. In spite of the amount of research in this direction, there is still a lack of analysis of appropriate machine- and deep-learning solutions for this very specific task.

The aim of this work is to carry out an empirical study, reasoning on the specific challenges of the training process for the application domain of food image recognition, with the aim to maximize models' accuracy. Food image datasets present their own peculiar challenge regarding image classification. It is a fine-grained problem, manifesting distinct shapes and textures, as seen in pasta dishes or oven-baked food. In addition, food datasets typically show high intra-class variability among the same food recipe and a small inter-class distance. Confirming this specificity, over the past ten years, food datasets have seen a notable evolution, with several proposed datasets different in terms of the number of images and classes [4–6, 9, 10, 14, 15, 17, 19]. There has also been an evolution in granularity, with datasets providing information on the ingredients and the nutritional values.

In the described context, we study and discuss the potential of different training strategies in a transfer-learning framework [12]. Our case study focuses on situations where training a model from scratch is not possible, either for data scarcity or limited computational resources. We focus on the efficacy of out-of-domain pre-trained features, the benefit of in-domain fine-tuning, and the generalization of in-domain features as we apply them to different datasets of the same nature. The experimental results demonstrate that using in-domain datasets improves the representation of target food datasets, leading to higher test accuracy. To summarize, our contributions include: (i) a complete analysis of training strategies with a comparison between ImageNet and in-domain transfer-learning on food datasets; (ii) the introduction of two CNNs' ensemble approaches, to increase test accuracy for automatic food image classification.

The remainder of the paper is organized as follows: Sect. 2 presents the works related to this research, Sect. 3 summarises the adopted computational pipelines, Sect. 4 presents and discusses in detail the experimental analysis and the results.

2 Related Works

The early works on food image recognition can be traced back to the work of UECFOOD100 [15], where they performed the image classification using engineered features. Starting from an image, they compute candidate regions, which then are aggregated, to apply feature extractors like SIFT or HoG, to finally classify with MKL-SVM. The authors of ETH Food 101 [4] have used a multi-stage process to classify the food images. It starts with Random Forest to cluster superpixels of the image, then select the best components to be fed to an SVM classifier. With the advancement in CNNs, all the works on food images have been relying on deep learning. Food101 has become one of the benchmark datasets for testing new CNN architectures, like in the case of EfficientNet [21]. Opposite to UECFOOD100's original paper, the authors of [3] have applied an Ensemble of ResNext-101 and DenseNet-161, reporting an increase in accuracy using the ensemble setting. The author of Food524 fine-tuned their dataset on a ResNet-50, pre-trained on ImageNet. As a test, they performed an image retrieval on the UNICT-FD1200 dataset and noted an increase in accuracy using the model fine-tuned on their dataset. In FoodX251, they conducted two experiments on ResNet-101: first, fine-tuning only the last layer; second, fine-tuning all the layers. Fine-tuning all the layers out-performed fine-tuning only one layer. In Recipe1M+ [14], the authors adopted two different approaches to represent the text data and the images. They relied on multiple LSTM for each of the recipes, ingredients, and cooking instructions, and on ResNet-50 to embed the images. The author of Food2K [17] also proposed their own architecture, called PRENet, on which they train their own dataset. Differently from the cited works, in this paper, we investigate training strategies, evaluating whether in-domain source datasets can provide a better representation of food datasets, leading to an increase in accuracy. We use three different CNN architectures, comparing ImageNet pre-training with an in-domain transfer learning strategy, where the ImageNet pre-trained models are first fine-tuned on an in-domain source dataset, and then used as a feature extractor to classify a target food dataset.

3   Methods

Food datasets are typically fine-grained with a large intra-class and a small inter-class variation. We adopt a transfer-learning approach to address such peculiarities in the case of limited resource availability. We start with an ImageNet pre-trained CNN to classify our food datasets and explore two different transfer-learning scenarios: (i) in one case we exploit the pre-trained CNN as a feature extractor, training a classifier on top of the deep pre-trained features; (ii) in the other, we fine-tune the entire CNN on an in-domain source dataset and then apply the fine-tuned model to a target dataset of the same nature (Fig. 1 shows a schematic overview of the latter). Indeed, even if ImageNet classes include food images, it has already been suggested in the literature that fine-grained datasets' benefit from ImageNet pre-trained networks may be limited [2,11]. Moreover, food datasets have peculiar properties. For instance, the background is generally little or no informative, and only a small portion of the images, usually (but not always) the center, may indicate the corresponding class.

Thus, we investigate whether in-domain features could allow us to capture the relevant image properties better than only using ImageNet pre-training. This should be beneficial in all the circumstances in which a classical fine-tuning on the target dataset is not applicable, either for limited data availability or for stringent computational requirements. In fact, the number of images per class is limited for most of the publicly available food datasets, bringing poor performances when a CNN model is trained from scratch on a target food image dataset. To address this goal, we consider ImageNet pre-trained models and fine-tune them on food datasets. The fine-tuned model should be able to capture general image structures, as well as specific food patterns. We use each obtained model as an in-domain feature extractor, training a classifier on top of the deep features extracted.

Our experimental analysis includes several CNN architectures. In the end, we discuss the benefits of building an ensemble of fine-tuned CNNs, to further improve the test accuracy. In order to exploit the full potential of the ensemble, we propose two strategies: (i) a straightforward averaging of the final logits; (ii) a learnable ensemble approach, where the averaging weights are learned by a classifier, to exploit each of the ensemble models better. We design and compare two different ensemble learning strategies, described in detail in Sect. 4.2.

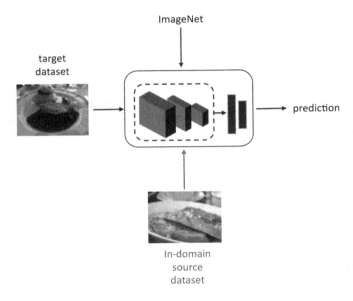

Fig. 1. Schematic overflow of the proposed pipeline.

4   Experiments

4.1   Datasets

Our experiments include the datasets ETH Food 101, FoodX251, and UEC-FOOD100 (see Sect. 2 for further details). Table 1 reports the number of classes, i.e. recipes, together with the total number of available images, and the type of images contained in the dataset, with Miscellaneous referring to a mixture of cuisines. Figure 2 shows representative examples of the images included in the datasets. Generally, the subject dish is at the center of the image, with the background occupied by the serving plate. Depending on the recipe, there might be one or more instances of the dish in the same image. All the images are taken in the wild, though since the data are gathered from the internet, it happens that some images either contain people or other non-food objects. Food101 has a perfect distribution for the number of images per class, while FoodX251 is strongly imbalanced. UECFOOD100 is slightly imbalanced and it belongs to a different cuisine with respect to the other two, as it contains recipes from Japanese cuisine. Being the smallest, we chose UECFOOD100 to investigate the impact of the dataset size in learning the food representation; as well as the effect of domain shift between different types of cuisines.

Table 1. A description of the Food Datasets adopted in the experiments, reporting the number of classes and images, and the type of cuisine they represent

Dataset	#Classes	#Images	Cuisine
Food101 [4]	101	101,000	Miscellaneous
FoodX251 [9]	251	158,846	Miscellaneous
UECFOOD100 [15]	100	14,361	Japanese

(a) Food101   (b) FoodX251   (c) UECFOOD100

Fig. 2. Sample images from the three datasets

4.2 Experiment Details

The experiments have been conducted using three CNN models with an increasing depth and number of parameters: EfficientNet-b0 [21], EfficientNet-V2-S [22] and ResNet-152 [8], in their Torchvision Library implementation [13]. The Feature Extraction has been performed by running the model's instance in an evaluation mode, removing the final Fully Connected (FC) layers, and using the output of the last convolutional layer as features. We perform pre-processing with data transformation to resize the images to 224×224, shifting their values to $[0 - 1]$, and normalizing them using ImageNet mean $[0.485, 0.456, 0.406]$, and standard deviation $[0.229, 0.224, 0.225]$. We build a classifier on top of the extracted features composed of two fully connected layers, with a batch-normalization layer at the beginning, and a drop-out after the first FC layer. The input size depends on the feature dimension of the architecture. The first hidden layer maps from the input size to 1024, followed by a ReLU activation function and a Dropout of

0.15. The final layer maps from 1024 to the number of classes of each dataset. For all the models, the classifier has been trained for 30 epochs, with a batch size equal to 512, and an Adam optimizer with an Exponential Scheduler and a learning rate starting at $1e - 3$. The training data is split into 75% training and 25% validation. The Feature Extraction and Classification have been performed on a laptop with an NVIDIA RTX3060 of 6Go of memory.

Fine-tuning begins with the same pre-processing strategy. We use data augmentation, with random rotations, and horizontal and vertical flips. The training data is split with a 75/25 ratio for training and validation. The last layer of every model has been replaced with the classifier previously described. The implemented model has been trained for 40 epochs, with an Adam optimizer and an Exponential Scheduler, with a Learning Rate starting at $1e - 3$. To efficiently manage the GPU memory, an Accumulative Gradient strategy has been adopted. Fine-tuning has been performed on a server with an NVIDIA A30 of 24Go of memory.

After separately fine-tuning all models on every single dataset, we implement two different ensemble techniques, which we apply to each dataset:

- An *Average Ensemble*, consisting in averaging the logits of all three CNN models. The weights have a dimension of $[3, 1]$, containing the values $[\frac{1}{3}, \frac{1}{3}, \frac{1}{3}]$. The predicted class corresponds to the maximum logit after the averaging.
- A *Learnable Ensemble*, where the averaging weights are learned through a 1D convolution layer, trained on the logits extracted from the training set. We experiment with different configurations:
  - Weights of dimension $[3, 1]$, obtained with a 1D-Conv layer with a single learnable filter.
  - We Increase the number of filters in the 1D-Conv layer to 10, learning weights with shape $[3, 10]$. This approach requires the usage of a Pooling Layer for obtaining the final predictions. In our experiments, we test both an average pooling and a max pooling strategy.

In both cases, the predicted class is taken as the maximum logit from the last layer.

4.3   Results

Out-of-Domain ImageNet Features. We start by defining a baseline in terms of test accuracy, corresponding to the use of an ImageNet pre-trained CNN as a feature extractor, coupled with a shallow neural network as a classifier (see sect. 4.2). The first rows of every sub-table in Table 2 summarize the obtained results for the three CNNs included in this work. Coherently in the three cases: (i) the results of Food101 as a target dataset are acceptable in terms of test accuracy, outperforming the original paper [15]; (ii) the deep features extracted with ImageNet work reasonably well for UECFOOD100, considering the peculiarity of the target domain; (iii) instead, we have poor performances for FoodX251, a larger, challenging, and imbalanced dataset.

Table 2. Transfer learning performances for the three models and the three food datasets included in our work. The source indicates the pre-training source dataset: we consider one out-of-domain baseline (the classical ImageNet) and each one of the three in-domain food datasets. The target represents the destination food dataset in our transfer learning procedure.

ResNet-152

Source \ Target	Food101	FoodX251	UECFOOD100
ImageNet	68.935	48.315	57.950
Food101	-	61.781	66.444
FoodX251	81.129	-	66.583
UECFOOD100	73.354	55.503	-

EfficientNet-b0

Source \ Target	Food101	FoodX251	UECFOOD100
ImageNet	69.129	53.860	62.100
Food101	-	60.777	65.024
FoodX251	78.931	-	65.776
UECFOOD100	71.295	60.867	-

EfficientNet-V2-S

Source \ Target	Food101	FoodX251	UECFOOD100
ImageNet	71.145	52.385	57.533
Food101	-	59.646	64.300
FoodX251	79.343	-	64.133
UECFOOD100	72.515	55.661	-

Fine-tuning on the Target Dataset. We compare the obtained results with a complete fine-tuning of each model on the target datasets. As we can see in Table 3, fine-tuning leads to a noticeable increase in accuracy with respect to the baseline. The obtained results are in line with state-of-the-art works as [2,11], reporting that ImageNet pre-trained networks may benefit significantly from fine-tuning, in the case of fine-grained datasets (as the food ones are).

Table 3. Fine-tuning performance of the three models on the three source datasets.

Dataset	RestNet152	EffNet-b0	EffNet-V2-S
Food101	**85.804**	84.176	84.410
FoodX251	**86.724**	78.569	77.635
UECFOOD100	67.271	66.294	**67.708**

In-Domain Features. We test whether in-domain fine-tuning can provide a better representation of food datasets. After fine-tuning our ImageNet pre-trained

CNNs on each one of the three food datasets we consider, we use each obtained model to extract features from the remaining two datasets. The in-domain deep features are used to train the similarly designed classifiers as in the previous experiments (see Sect. 4.2). Table 2 shows the obtained results. As we can see, there is always a benefit in adopting in-domain features, with an average improvement of 6.5% among the datasets and the architectures. The models fine-tuned on FoodX251 led to a higher increase in accuracy with respect to Food101 and UECFOOD100. These results suggest that the number of classes and images are fundamental properties for an in-domain source dataset to be used in a transfer-learning scenario for food datasets. Although UECFOOD100 is from a different cuisine, the experiments demonstrate that the features representation from in-domain food datasets transfer well between the cuisines, further consolidating the benefits of in-domain transfer learning in the case of food images.

Double In-Domain Fine-Tuning. With the in-domain fine-tuning reaching higher results, we test whether fine-tuning the model on a target dataset starting from an in-domain fine-tuned model would result in an accuracy increase. In particular, we consider UECFOOD100 as a target dataset, as it is the smallest in terms of the number of images and thus the one that should benefit more from such a procedure. We indeed find a small improvement in test accuracy, as reported in Table 4 confirming that a double fine-tuning procedure brings a better representation.

Table 4. Test accuracy for the double in-domain fine-tuning experiment on UEC-FOOD100. The procedure consists in fine-tuning on UECFOOD100 an ImageNet pre-trained model previously fine-tuned on each one of the other two domain source datasets. The baseline refers to the ImageNet pre-trained models directly fine-tuned on the UECFOOD100 dataset.

Target Dataset: UECFOOD100			
Source Dataset \ Model	ResNet-152	EffNet-b0	EffNet-V2-S
FoodX251	68.060	67.196	**68.979**
Food101	67.502	66.889	68.283
Baseline	67.271	66.294	67.708

Adopting Ensemble Classifiers. Considering that each CNN extract features in a peculiar way, we design and test two ensemble strategies of the three CNNs models used in this work, as explained in Sect. 4.2. Table 5 summarizes the obtained results. The learnable ensemble with max pooling corresponds to the best performances, bringing an average improvement of 2.6% with respect to the best single model. Interestingly, we notice that the averaging weights inferred with the learnable ensemble strategy are proportional to the individual model accuracy.

Table 5. Test accuracy for the designed ensemble strategies applied to the three CNN models used in this work. See Sect. 4.2 for more details.

Dataset	Avg Ens	Ens. 1D-Conv	1D-Conv Max 10	1D-Conv Avg 10
Food101	89.283	89.251	**89.342**	89.275
FoodX251	87.627	88.577	**88.719**	88.602
UECFOOD100	70.398	70.256	**70.565**	70.147

GradCAM Visualization. In our experiments, we noticed that in-domain datasets provide a general improvement of the learned representations for the target datasets. To further investigate the effect of in-domain fine-tuning, in Fig. 3 we report an example of GradCAM-based [20] visualizations for the last feature map of the EfficientNet-b0 model. This experiment is meant to provide a qualitative understanding of the effect of in-domain fine-tuning. We consider the ImageNet pre-trained model and the same model fine-tuned on each one of the available food datasets. The 4 sample images in Fig. 3 belong to the Food101 dataset. For this reason, the output of the Food101 dataset can be regarded as a reference to indicate which part of the images the models should consider more for providing the prediction. The in-domain fine-tuned models focus more on the food in the image, while the ImageNet pre-trained model does not show a specific focus or is more activated by other objects in the image.

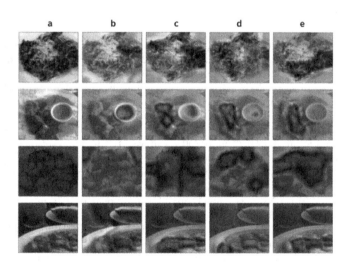

Fig. 3. GradCAM overlay, visualizing the last feature map of the EfficientNet-b0 applied to the image on the left, (from Food101) comparing the pre-training on ImageNet with the fine-tuning on each of the three food datasets. (a) input image; (b)-(e) GradCAM corresponding to different models. (b) ImageNet pre-trained model; (c) ImageNet pre-trained model fine-tuned on Food101; (d) ImageNet pre-trained model fine-tuned on FoodX251; (e) ImageNet pre-trained model fine-tuned on UECFOOD100.

5 Conclusion

In this work, we explored the problem of automatic food image classification for nutrient estimation. We investigated a transfer-learning framework with three CNN models, focusing on a typical case study where training from scratch is not possible. This situation can be related to data scarcity or stringent computational requirements. First, we show that fine-tuning is beneficial when dealing with food datasets, in agreement with state-of-the-art results showing that ImageNet transferability on fine-grained datasets is limited. Later, we investigated whether in-domain fine-tuning can provide better representations for food images. We noticed there is always a benefit in accuracy when using in-domain source datasets. Furthermore, we design a learnable ensemble approach, obtaining an improvement in test accuracy on the target datasets.

Our experiments suggest that the higher the number of classes and images in a source in-domain dataset, the better the benefit in accuracy for the target one. Thus, acquiring and making available large-scale food datasets with mixed recipe domains and enough images per class may be fundamental for increasing the accuracy of automatic food image classification.

Acknowledgments. VPP was supported by FSE REACT-EU-PON 2014–2020, DM 1062/2021.

References

1. U.s. department of agriculture, agricultural research service (2022). usda food and nutrient database for dietary studies 2019–2020, food Surveys Research Group Home Page. http://www.ars.usda.gov/nea/bhnrc/fsrg
2. Alfano, P.D., Pastore, V.P., Rosasco, L., Odone, F.: Fine-tuning or top-tuning? transfer learning with pretrained features and fast kernel methods (2022). arXiv:2209.07932
3. Arslan, B., Memis, S., Battinisonmez, E., Batur, O.Z.: Fine-grained food classification methods on the UEC food-100 database. IEEE Transactions on Artificial Intelligence (2021)
4. Bossard, L., Guillaumin, M., Van Gool, L.: Food-101 – mining discriminative components with random forests. In: Fleet, D., Pajdla, T., Schiele, B., Tuytelaars, T. (eds.) ECCV 2014. LNCS, vol. 8694, pp. 446–461. Springer, Cham (2014). https://doi.org/10.1007/978-3-319-10599-4_29
5. Jing-jing Chen, C.w.N.: Deep-based ingredient recognition for cooking recipe retrival. ACM Multimedia (2016)
6. Ciocca, G., Napoletano, P., Schettini, R.: Learning CNN-based features for retrieval of food images. In: Battiato, S., Farinella, G.M., Leo, M., Gallo, G. (eds.) New Trends in Image Analysis and Processing - ICIAP 2017: ICIAP International Workshops, WBICV, SSPandBE, 3AS, RGBD, NIVAR, IWBAAS, and MADiMa 2017, Catania, Italy, September 11–15, 2017, Revised Selected Papers, pp. 426–434. Springer International Publishing (2017). https://doi.org/10.1007/978-3-319-70742-6_41

7. Haussmann, S., et al.: Foodkg: a semantics-driven knowledge graph for food recommendation. In: The Semantic Web-ISWC 2019: 18th International Semantic Web Conference, Auckland, New Zealand, October 26–30, 2019, Proceedings, Part II 18, pp. 146–162. Springer (2019)

8. He, K., Zhang, X., Ren, S., Sun, J.: Deep residual learning for image recognition. In: Proceedings of the IEEE Conference on Computer Vision and Pattern Recognition, pp. 770–778 (2016)

9. Kaur, P., Sikka, K., Wang, W., Belongie, S., Divakaran, A.: Foodx-251: a dataset for fine-grained food classification. arXiv preprint arXiv:1907.06167 (2019)

10. Kawano, Y., Yanai, K.: Automatic expansion of a food image dataset leveraging existing categories with domain adaptation. In: Proceedings of ECCV Workshop on Transferring and Adapting Source Knowledge in Computer Vision (TASK-CV) (2014)

11. Kornblith, S., Shlens, J., Le, Q.V.: Do better imagenet models transfer better? In: Proceedings of the IEEE/CVF Conference on Computer Vision and Pattern Recognition, pp. 2661–2671 (2019)

12. Maracani, A., Pastore, V.P., Natale, L., Rosasco, L., Odone, F.: In-domain versus out-of-domain transfer learning in plankton image classification. Sci. Rep. **13**(1), 10443 (2023)

13. Marcel, S., Rodriguez, Y.: Torchvision the machine-vision package of torch. In: Proceedings of the 18th ACM International Conference on Multimedia, pp. 1485–1488 (2010)

14. Marin, J., et al.: Recipe1m+: A Dataset for Learning Cross-modal Embeddings for Cooking Recipes and Food Images. IEEE Trans. Pattern Anal. Mach, Intell (2019)

15. Matsuda, Y., Hoashi, H., Yanai, K.: Recognition of multiple-food images by detecting candidate regions. In: Proceedings of IEEE International Conference on Multimedia and Expo (ICME) (2012)

16. Mayne, S.T., Playdon, M.C., Rock, C.L.: Diet, nutrition, and cancer: past, present and future. Nat. Rev. Clin. Oncol. **13**(8), 504–515 (2016)

17. Min, W., et al.: Large scale visual food recognition. IEEE Trans. Pattern Anal. Mach. Intell. **45**(8), 9932–9949 (2023)

18. Ravasco, P.: Nutrition in cancer patients. J. Clin. Med. **8**(8), 1211 (2019)

19. Salvador, A., et al.: Learning cross-modal embeddings for cooking recipes and food images. In: Proceedings of the IEEE Conference on Computer Vision and Pattern Recognition (2017)

20. Selvaraju, R.R., Cogswell, M., Das, A., Vedantam, R., Parikh, D., Batra, D.: Gradcam: Visual explanations from deep networks via gradient-based localization. In: Proceedings of the IEEE International Conference on Computer Vision, pp. 618–626 (2017)

21. Tan, M., Le, Q.: Efficientnet: Rethinking model scaling for convolutional neural networks. In: International Conference on Machine Learning, pp. 6105–6114. PMLR (2019)

22. Tan, M., Le, Q.: Efficientnetv2: Smaller models and faster training. In: International Conference on Machine Learning, pp. 10096–10106. PMLR (2021)

LCMV: Lightweight Classification Module for Video Domain Adaptation

Julian Neubert[1][(✉)] , Mirco Planamente[1,2,3] , Chiara Plizzari[1] ,
and Barbara Caputo[1,3]

[1] Politecnico di Torino, Torino, Italy
julian.neubert@gmx.de,
{mirco.planamente,chiara.plizzari,barbara.caputo}@polito.it
[2] Istituto Italiano di Tecnologia, Genoa, Italy
[3] CINI Consortium, Torino, Italy

Abstract. Video action recognition models exhibit high performance on in-distribution data but struggle with distribution shifts in test data. To mitigate this issue, Unsupervised Domain Adaptation (UDA) methods have been proposed, consisting in training on labeled data from a *source* domain and incorporating unlabeled test data from a *target* domain to reduce the domain gap. This requires simultaneous access to data from both domains, which may not be practical in real-world scenarios due to privacy issues. A more practical approach called Source-Free Domain Adaptation (SFDA) has been recently proposed, which consists in adapting a well-trained source model using only unlabeled target data. However, existing SFDA methods are computationally intensive and designed for specific architectures. In this paper, we propose an approach called Lightweight Classification Module for Video Domain Adaptation (LCMV). LCMV is based on a backpropagation-free prototypical algorithm, which efficiently adapts a source model using unlabeled target data only. Results on two popular datasets, HMDB-UCF$_{full}$ and EPIC-Kitchens-55, show significant improvements of LCMV compared to the previous state-of-the-art SFDA methods, and competitive results when compared to state-of-the-art UDA methods.

Keywords: Source-Free Domain Adaptation · Action Recognition

1 Introduction

Video action recognition is attracting an ever-growing amount of interest in the research community [2,18]. However, despite the advances in the field, models still suffer from the so-called "environmental bias" [24]. In action recognition this problem is amplified by the high-dimensional data and complex environments of videos, causing a significant drop in classification accuracy when evaluating models in new domains (see Fig. 1).

So far, most researchers address this issue in the Unsupervised Domain Adaptation (UDA) setting, using labeled samples from the *source* domain and unlabeled samples from the *target* domain to reduce the models' bias [3,18] and

G. L. Foresti et al. (Eds.): ICIAP 2023, LNCS 14234, pp. 270–282, 2023.
https://doi.org/10.1007/978-3-031-43153-1_23

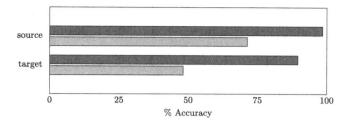

Fig. 1. Evaluating a model in a new environment often leads to significantly reduced performance, highlighting the need for domain adaptation methods. This Figure shows the prediction accuracy on **UCF-HMDB**$_{full}$ and EPIC-Kitchens-55 when evaluating our I3D model on a seen source and unseen target domain without any adaptation.

achieve optimal performance on the target domain. These methods have proven to be successful in improving the robustness of models and enabling them to perform well across different environments. However, a key drawback of UDA is that it requires access to both source and target data during training, which may not be possible in many real-world applications where data sharing is restricted due to privacy concerns. In the Source-Free Domain Adaptation (SFDA) setting, which presents a more challenging yet realistic scenario, the training and adaptation processes are separated, allowing to adapt pretrained off-the-shelf source models using exclusively unlabeled target data. Current SFDA methods for action recognition achieve this by training with an ensemble of complex loss functions, for example, to ensure internal consistency [28] or align feature representations [11] and adapt the model using the weighted sum of many individual loss functions [4, 11, 28].

However, these SFDA approaches present several new challenges. Firstly, determining the individual weight for each component of the loss function typically entails manual selection of hyperparameters, necessitating adjustments on a per-dataset basis. Moreover, conducting backpropagation-based training during the adaptation phase proves computationally demanding, demanding large amounts of high-quality target data, and potentially compromising the performance of the underlying model. Lastly, numerous SFDA methods impose tight architectural constraints, tailored to specific architectures [28], thereby limiting modifications to the source model [4] or necessitating the design of unique architectures [11].

In this work, we propose a Lightweight Classification Module for Video Domain Adaptation (LCMV). Specifically, LCMV replaces the final classification layer with a prototypical-based classification module. Each class is represented by a "class prototype", and classification is done by assigning samples to the nearest prototype in feature space, without requiring any optimization process.

To summarize, our contributions are threefold:

– We develop a backpropagation-free architecture-independent classification solution for Source-Free Domain Adaptation for action recognition;

- Our method preserves the integrity of the original source model, allowing for seamless adaptation to multiple domains and adjustments to continuous domain shift;
- We validate our method on the UCF-HMDB$_{full}$ and EPIC-Kitchens-55 datasets, achieving state-of-the-art results.

2   Related Work

Action Recognition. The goal of Action Recognition (AR) is to classify the action that the subject is performing in a video. Architectures generally consist of convolutional networks utilizing either 2D [16,25,32] or 3D [1,18,20,27] convolutions. Sometimes additional modules are added to improve temporal modeling capabilities [9,16,19,32]. Finally, incorporating information from multiple modalities has proven to improve performance, especially in the cross-domain scenario [4,11,13,18,29].

Unsupervised Domain Adaptation for AR. Domain adversarial networks [10] have emerged as a well-established and widely used approach, integrated into numerous recent UDA methods for action recognition [3,6,18,26,29]. TransVAE [26] attempts to disentangle the domain information from task-specific information during the adaptation. TA^{3}N [3] integrates a temporal relation module for temporal alignment. Other methods use contrastive learning techniques [13,19]. Another research direction focuses on incorporating self-supervised learning as an auxiliary task to improve feature learning [6,18]. Finally, some works use pseudo-labels to guide the adaptation on unlabeled target data [13,19].

Source-Free Domain Adaptation. The application of SFDA methods to action recognition is a relatively new area of investigation. SFTADA [4] proposed to address SFDA in videos by learning temporal consistency to transfer robust centroid-based temporal attention weights from the source domain to the target domain. ATCoN [28] learns a temporal consistency which is composed of both feature and source prediction consistency. Recently, MTRAN [11] presented a loss derived from Mixup [31] to align the internal representation of source and target features, which is applied to their visual transformer-inspired architecture. CleanAdapt [8] proposed a self-training approach that selects the clean samples from the noisy pseudo-labeled target domain ones. To the best of our knowledge, no previous SFDA publication specific to action recognition exist, that is completely backpropagation-free or formulated independently of the underlying architecture.

Prototypical Classification. Prototypical networks have proven successful for zero-shot domain adaptation [5,21]. Subsequently, the use of prototypes has found its way into domain generalization, UDA, and SFDA, often to generate

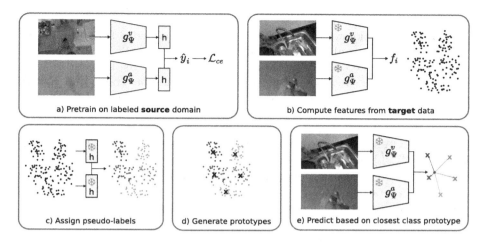

Fig. 2. Overview of LCMV. (a) Before adaptation, the source model (feature extractor and classification layer) is trained on the labeled source domain. (b) The source model is used to generate representations for target data from unlabeled target domain. (c) The source classifier is used to assign pseudo-labels for target. (d) Class prototypes are computed as mean feature vectors. (e) LCMV predicts samples based on the closest class prototype.

pseudo-labels [11, 12, 19, 28] or to robustly estimate attention weights [4]. Inspired by [12], we propose to use prototypes directly as the model's prediction in the SFDA setting, where the larger amount of available data allows for significantly more robust prototypes. To the best of our knowledge, our work represents the first attempt to apply this concept in SFDA for action recognition, and it introduces a backpropagation-free approach to the field.

3 Proposed Method

We propose to replace the standard classification layer of action recognition models with a backpropagation-free prototypical classifier which we call Lightweight Classification Module for Video Action Recognition (LCMV). Given a model pre-trained on source data, we feed it with samples from target and assign a class label to them. We then compute class prototypes as the normalized average feature vectors of target samples from each class. The final classification is performed by assigning each target sample to the closest class prototype. In this section, we formally describe the proposed LCMV and discuss it in detail.

3.1 Multi-Modal Source-Free Domain Adaptation

In the Unsupervised Domain Adaptation (UDA) setting, we are given a source domain $\mathcal{S}$, where $\mathcal{S} = \{(x_{s,i}, y_{s,i})\}_{i=1}^{N_s}$ is composed of N_s source samples with known labels $y_{s,i} \in Y_s$, and a target domain $\mathcal{T} = \{x_{t,i}\}_{i=1}^{N_t}$ of N_t target samples whose labels are unknown. The goal is to transfer knowledge from the labeled source domain to the unlabeled target domain by reducing the distribution discrepancy between the two during training. In the Source-Free Domain Adaptation (SFDA) setting we are only given a source model pre-trained on $\mathcal{S}$ with supervision, but we do not have access to $\mathcal{S}$ for adaptation on target. The key idea is thus to learn discriminative target features while aligning them with the source data distribution embedded within the source classifier.

In our SFDA setting, video clips x_i from both domains consist of two modalities $x_i = \{x_i^v, x_i^f\}$, which are visual appearance (RGB) x_i^v and motion (optical flow) x_i^f respectively. Predictions are made by a two-stream action recognition model which averages the final softmaxed outputs over both modalities to return a fused probability score. For simplicity, we generically refer to both modalities in our equations, unless explicitly stated otherwise.

3.2   Prototype Generation

We are given a model trained on source data, where each modality stream $\Phi = h_{\{w,b\}} \circ g_\Psi$ can be divided into a feature extractor g_Ψ with parameters Ψ and a linear classification layer $h_{\{w,b\}}$ with weights w and bias b (see Fig. 2-a). Prediction $\hat{y}_i$ for the input sample x_i to belong to class k is then given by

$$p(\hat{y}_i = k | x_i, \Psi, w, b) = \frac{\exp(d(w^k, g_\Psi(x_i)) + b^k)}{\sum_j \exp(d(w^j, g_\Psi(x_i)) + b^k)} \tag{1}$$

where w^k and b^k are the k-th row of the weight matrix and bias parameter of $h_{\{w,b\}}$ respectively, and $d(w^k, g_\Psi(x_i)) = w^k \cdot g_\Psi(x_i)$ is the dot product between w^k and $g_\Psi(x_i)$.

Let $\{x_1, \ldots, x_N\}$ be the set of all videos available in the training set of the target domain. Given an action video x_i we sample N_c clips $\{x_{i,1}, \ldots, x_{i,N_c}\}$ it. Using the given source model, we obtain the predictions over all clips $\{\hat{y}_{i,1}, \ldots, \hat{y}_{i,N_c}\}$ and average them to get a more robust prediction $\tilde{y}_i$ ("pseudo-label") for the video (Fig. 2-c). Similarly, we use the normalized mean feature vector f_i over all clip features $\{g_\Psi(x_{i,1}), \ldots, g_\Psi(x_{i,N_c})\}$ to obtain a feature-based representation of video x_i (Fig. 2-b).

Each prototype c^k for each class k is extracted by simply averaging the feature representation f_i of all videos x_i belonging to class $\tilde{y}_i = k$ (Fig. 2-d):

$$c^k = \frac{1}{N_k} \sum_{\substack{i \\ \tilde{y}_i = k}} f_i \tag{2}$$

Following [12], we improve the prototypes quality by selecting for each class k the t videos with the lowest prediction entropy and compute the prototypes in Eq. 2 only on this subset.

3.3   Final Prediction

To classify target samples using the computed prototypes, we simply assign them to the closest class prototype in feature space (Fig. 2-e). Given a sample x_i we compute its class probabilities as

$$p(\hat{y}_i = k | x_i, \Psi) = \frac{\exp(d(c^k, g_\Psi(x_i)))}{\sum_j \exp(d(c^j, g_\Psi(x_i)))} \tag{3}$$

where $d(c^k, g_\Psi(x_i)) = (c^k - g_\Psi(x_i))$ is the Euclidean distance between c^k and $g_\Psi(x_i)$. Note that this formulation is the same as Eq. 1, by substituting w_k with c^k and removing the bias term. Indeed, the rows w_k of the weight matrix of a standard classification layer can be seen as a class prototype [12].

Finally, we iteratively refine the prototypes similar to how the k-means algorithm works. This is accomplished by using the formulation described in Eq. 3 to generate more accurate pseudo-labels $\tilde{y}_i$ for the target videos x_i. The prototypes are then updated accordingly using Eq. 2.

The proposed LCMV offers a distinct advantage: our prototypes are not learned parameters that are susceptible to domain shift. Instead, they are directly computed from the target data, ensuring robustness and reducing the impact of the distribution gap.

4   Experiments

We evaluate the ability of LCMV to perform well on target data by comparing it against baseline and state-of-the-art Unsupervised Domain Adaptation (UDA) and Source-Free Domain Adaptation (SFDA) methods adapted to our setting. We then perform an analysis of the impact of class imbalance and multi-modal learning on our method. Finally, we show ablations on its different components.

4.1   Experimental Setting

Datasets. We evaluate our proposed method on the two most commonly used datasets for domain adaptation for action recognition: **EPIC-Kitchens-55** [7] and **UCF-HMDB**$_{full}$ [14,23]. **EPIC-Kitchens-55** contains egocentric action videos recorded in different kitchens across the world. We follow the setting proposed in [18], including the three largest kitchens by number of samples, each representing one domain (D1, D2, and D3). For **UCF-HMDB**$_{full}$ we follow the setting proposed in [3], which consists of 12 overlapping classes between the UCF101 and HMDB51 datasets. We train the model on one dataset and evaluate it on the other. In the following we use the abbreviation H→U to denote training on HMDB51 and evaluating on UCF101, and vice versa.

Implementation Details. Our model uses a two-stream I3D backbone pre-trained on Kinetics [1] with averaged late fusion to combine multi-modal inputs. Flow frames are extracted using the TV-L1 algorithm [30]. Clips consist of 16 frames following the dense sampling strategy with random cropping, scale jitters, and horizontal flipping as augmentations. We evaluate our model using the average prediction on 5 clips per video. On UCF-HMDB$_{full}$ prototypes are computed using the 20 lowest entropy samples and refined through a single iteration of the k-means algorithm. On EPIC-Kitchens-55 we do not perform any prototype adjustment. Experiments have been run on a workstation with two NVIDIA GeForce GTX 1070 GPUs.

4.2   Comparison with the State-of-the-Art

We compare LCMV results with state-of-the-art works in both UDA and SFDA settings. While UDA methods are included as a reference and are expected to outperform our approach due to the additional availability of labeled target data for adaptation, we primarily focus on evaluating our performance against relevant SFDA methods, including ATCoN [28], SFTADA [4], MTRAN [11], and the concurrent work CleanAdapt [8]. On both datasets, we are able to achieve good results. For **UCF-HMDB**$_{full}$ we improve upon the previous state-of-the-art in SFDA by 1.1% (Table 1) even outperforming recent UDA works. The egocentric nature of its videos makes **EPIC-Kitchens-55** a much more difficult dataset, especially in the cross-domain scenario, yet LCMV achieves results on-par with the concurrent SFDA state-of-the-art method [8]. This emphasizes the effectiveness of using a backpropagation-free prototypical classifier, which allows

Table 1. Comparison with DA and SFDA methods on UCF-HMDB$_{full}$. Best results are marked in **bold** and MM indicates multi-modal results.

Method	Backbone	SFDA	MM	U→H	H→U	Avg
ATCoN [28]	ResNet-101	✓	-	79.7	85.3	82.5
SFTADA [4]	ResNet-101	✓	✓	87.2	91.2	89.2
STCDA [22]	I3D	-	-	83.1	92.1	87.6
CoMix [19]	I3D	-	-	86.7	93.9	90.3
CIA + TA3N [29]	I3D	-	-	91.9	94.6	93.2
TransVAE [26]	I3D	-	-	87.8	99.0	93.4
TA3N [3,6]	I3D	-	-	81.4	90.5	86.0
SAVA [6]	I3D	-	-	82.2	91.2	86.7
MM-SADA [13,18]	I3D	-	✓	84.2	91.1	87.7
CIA [29]	I3D	-	✓	90.6	94.2	92.4
Kim et al. [13]	I3D	-	✓	84.7	92.8	88.8
SHOT [11,15]	I3D	✓	✓	89.7	91.8	90.8
MTRAN [11]	I3D	✓	✓	**92.2**	95.3	93.8
CleanAdapt [8]	I3D	✓	✓	89.8	**99.2**	94.5
Ours	I3D	✓	✓	**92.2**	99.0	**95.6**

to perform equally well (better on 3/6 shifts) than a network trained end-to-end on target. The proposed LCMV demonstrates good performance compared to UDA methods, being only 2.6% behind the UDA state-of-the-art (Table 2).

4.3   Class Imbalance Analysis

We provide a brief analysis of the impact of class imbalance in the **EPIC-Kitchens-55** dataset on domain shift. In fact, unlike many benchmark datasets that are artificially balanced with an equal number of samples per class, this dataset reflects the real-world scenario where class distributions are often uneven.

To demonstrate the influence of class imbalance on performance, we compare the average accuracy with the average per-class accuracy in Table 3. The average accuracy is 3.4% higher than the average per-class accuracy in the supervised setting (*Target-only*), indicating that the model is slightly better at classifying samples from common classes. In the source-only case, this discrepancy is radically increased to 11.4%, with the model achieving 0% class accuracy in some extreme cases. This shows that the model is much less robust with respect to underrepresented classes, especially in cross-domain scenarios. Our proposed LCMV does not explicitly encode any class bias, leading to an average per-class accuracy 4.1% higher than the average accuracy, and showing a significant improvement over the source-only baseline.

Table 2. Comparison with DA and SFDA methods on **EPIC-Kitchens-55**. Best UDA results are marked in **bold**, best SFDA results in **bold blue**, and the MM columns indicates which methods are multi-modal.

Method	SFDA	MM	D2→D1	D3→D1	D1→D2	D3→D2	D1→D3	D2→D3	Avg
DANN [10,19]	-	-	38.3	38.8	37.7	42.1	36.6	41.9	39.2
TA3N [3,26]	-	-	40.9	39.9	34.2	44.2	37.4	42.8	39.9
CoMix [19]	-	-	38.6	42.3	42.9	49.2	40.9	45.2	43.2
TransVAE [26]	-	-	**50.3**	48.0	50.5	**58.0**	**50.3**	**58.6**	**52.6**
MMD [17,29]	-	✓	46.6	39.2	43.1	48.5	48.3	55.2	46.8
MM-SADA [18]	-	✓	48.2	50.9	49.5	56.1	44.1	52.7	50.3
Kim et al. [13]	-	✓	49.5	51.5	50.3	56.3	46.3	52.0	51.0
STCDA [22]	-	✓	49.0	**52.6**	52.0	55.6	45.5	52.5	51.2
CIA [29]	-	✓	49.8	52.2	**52.5**	57.6	47.8	53.2	52.2
SHOT [11,15]	✓	✓	44.1	54.0	40.8	36.5	49.0	45.3	45.0
MTRAN [11]	✓	✓	46.3	**58.2**	42.2	38.1	**52.3**	46.1	47.2
CleanAdapt [8]	✓	✓	46.2	47.8	**52.7**	**54.4**	47.0	52.7	**50.1**
Ours	✓	✓	**52.1**	50.5	45.6	52.3	41.1	**55.1**	49.5

Table 3. Average **accuracy** and **per-class accuracy** on **EPIC-Kitchens-55**.

Method	D2→D1	D3→D1	D1→D2	D3→D2	D1→D3	D2→D3	Avg	
Source-only	46.3	46.8	47.6	55.3	43.0	50.3	48.2	average
Ours	52.1	50.5	45.6	52.3	41.1	55.1	49.5	average
Target-only	64.6	64.6	76.4	76.4	72.9	72.9	71.3	average
Source-only	35.9	33.4	37.2	53.8	23.6	36.8	36.8	per-class
Ours	59.8	53.2	57.0	63.7	40.6	47.5	53.6	per-class
Target-only	68.0	68.0	79.8	79.8	55.8	55.8	67.9	per-class

4.4 Multi-Modal Learning

Figure 3 shows a comparison in terms of accuracy on **UCF-HMDB**$_{full}$ using our method, source-only, target-only, and the multi-modal RGB and Flow approaches, as well as the individual modalities.

Results show that training and evaluating the models on both modalities yield the best performance across all experiments. This indicates the models' ability to leverage the complementary information provided by the two modalities. While the improvement in the target-only case is relatively minor, it becomes more significant in the cross-domain scenario (source-only). Notably, our method shows a substantial improvement when using multi-modal data, achieving an increase of over 5% compared to the RGB-only baseline. This highlights the additional benefits that LCMV gains from incorporating multiple modalities.

4.5 Ablation Study

In this section, we perform an ablation study to analyze the impact of key parameters and design choices in our method. Specifically, we examine the accuracy based on the number of samples used for computing each prototype and the iterative refinement of prototypes. Finally, we add some considerations on memory and time complexity.

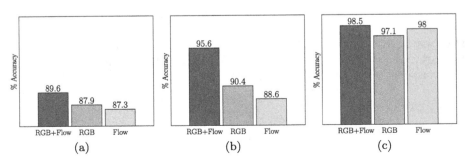

Fig. 3. **Uni-modal** and **multi-modal** accuracy (%) on **UCF-HMDB**$_{full}$ for **source-only** (a), **ours** (b) and **target-only** (c).

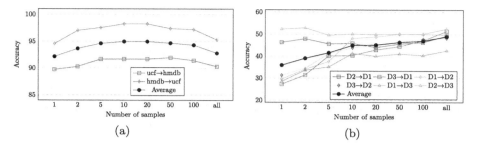

Fig. 4. Accuracy (%) on **UCF-HMDB**$_{full}$ (a) and **EPIC-Kitchens-55** (b) depending of the number t of lowest entropy samples used to compute each prototype.

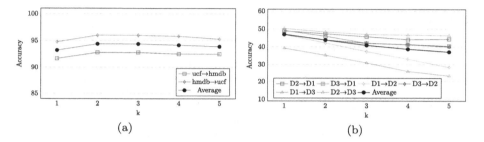

Fig. 5. Accuracy (%) on **UCF-HMDB**$_{full}$ (a) and **EPIC-Kitchens-55** (b) based on the number of iterations for prototype refinement.

Sample Selection. Figure 4 shows an analysis on the number of samples used for computing prototypes. Selecting the 20 lowest entropy features from each class noticeably improves our method's accuracy on **UCF-HMDB**$_{full}$. On the more difficult **EPIC-Kitchens-55**, however, we do not observe this pattern, with accuracy increasing proportionally to the number of features used to compute the prototypes.

Prototypes Refinement. Figure 5 shows that iteratively refining the prototypes works well on **UCF-HMDB**$_{full}$ for the first 2–3 iterations, improving performance by over 1.1%, but deteriorating thereafter. On the other hand, on **EPIC-Kitchens-55** this technique leads to gradually decreasing performance.

Memory and Time Complexity. Our backpropagation-free method is significantly more efficient than previous SFDA methods [11], requiring only 1.5GB of GPU memory (4.5 times less) and exhibiting an average execution time of 2 min and 33 s for **UCF-HMDB**$_{full}$ (3.6x faster) and 12 min and 20 s for **EPIC-Kitchens-55** (2.1x faster).

5   Conclusion

This work proposes a Source-Free Domain Adaptation method which consists in replacing the final classification layer with a prototype-based classifier. We show that the proposed LCMV improves cross-domain performance without modifying the network parameters themselves, making it more efficient than traditional backpropagation-based adaptation methods. Experiments show that LCMV is able to outperform all previous SFDA methods on two of the most important datasets for action recognition, achieving competitive results w.r.t. UDA methods. Moreover, we show that LCMV does not suffer from class imbalance, allowing for higher performance.

References

1. Carreira, J., Zisserman, A.: Quo vadis, action recognition? a new model and the kinetics dataset. In: Proceedings of the IEEE Conference on Computer Vision and Pattern Recognition, pp. 6299–6308 (2017)
2. Chen, C.F.R., et al.: Deep analysis of cnn-based spatio-temporal representations for action recognition. In: Proceedings of the IEEE/CVF Conference on Computer Vision and Pattern Recognition, pp. 6165–6175 (2021)
3. Chen, M.H., Kira, Z., AlRegib, G., Yoo, J., Chen, R., Zheng, J.: Temporal attentive alignment for large-scale video domain adaptation. In: Proceedings of the IEEE/CVF International Conference on Computer Vision, pp. 6321–6330 (2019)
4. Chen, P., Ma, A.J.: Source-free temporal attentive domain adaptation for video action recognition. In: Proceedings of the 2022 International Conference on Multimedia Retrieval, pp. 489–497 (2022)
5. Chen, W.Y., Liu, Y.C., Kira, Z., Wang, Y.C.F., Huang, J.B.: A closer look at few-shot classification. In: International Conference on Learning Representations (2019)
6. Choi, J., Sharma, G., Schulter, S., Huang, J.-B.: Shuffle and attend: video domain adaptation. In: Vedaldi, A., Bischof, H., Brox, T., Frahm, J.-M. (eds.) ECCV 2020. LNCS, vol. 12357, pp. 678–695. Springer, Cham (2020). https://doi.org/10.1007/978-3-030-58610-2_40
7. Damen, D., et al.: Scaling egocentric vision: The epic-kitchens dataset. In: Proceedings of the European Conference on Computer Vision (ECCV), pp. 720–736 (2018)
8. Dasgupta, A., Jawahar, C., Alahari, K.: Overcoming label noise for source-free unsupervised video domain adaptation. In: ICVGIP 2022-Indian Conference on Computer Vision, Graphics and Image Processing, pp. 1–9. ACM (2022)
9. Fan, Q., Chen, C.F.R., Kuehne, H., Pistoia, M., Cox, D.: More is less: Learning efficient video representations by big-little network and depthwise temporal aggregation. Adv. Neural. Inf. Process. Syst. **32**, 2261–2270 (2019)
10. Ganin, Y., Ustinova, E., Ajakan, H., Germain, P., Larochelle, H., Laviolette, F., Marchand, M., Lempitsky, V.: Domain-adversarial training of neural networks. The J. Mach. Learn. Res. **17**(1), 2030–2096 (2016)
11. Huang, Y., Yang, X., Zhang, J., Xu, C.: Relative alignment network for source-free multimodal video domain adaptation. In: Proceedings of the 30th ACM International Conference on Multimedia, pp. 1652–1660 (2022)

12. Iwasawa, Y., Matsuo, Y.: Test-time classifier adjustment module for model-agnostic domain generalization. Adv. Neural. Inf. Process. Syst. **34**, 2427–2440 (2021)

13. Kim, D., et al.: Learning cross-modal contrastive features for video domain adaptation. In: Proceedings of the IEEE/CVF International Conference on Computer Vision, pp. 13618–13627 (2021)

14. Kuehne, H., Jhuang, H., Garrote, E., Poggio, T., Serre, T.: Hmdb: a large video database for human motion recognition. In: 2011 International Conference on Computer Vision, pp. 2556–2563. IEEE (2011)

15. Liang, J., Hu, D., Feng, J.: Do we really need to access the source data? source hypothesis transfer for unsupervised domain adaptation. In: International Conference on Machine Learning, pp. 6028–6039. PMLR (2020)

16. Lin, J., Gan, C., Han, S.: TSM: Temporal shift module for efficient video understanding. In: Proceedings of the IEEE/CVF International Conference on Computer Vision, pp. 7083–7093 (2019)

17. Long, M., Cao, Y., Wang, J., Jordan, M.: Learning transferable features with deep adaptation networks. In: International Conference on Machine Learning, pp. 97–105. PMLR (2015)

18. Munro, J., Damen, D.: Multi-modal domain adaptation for fine-grained action recognition. In: Proceedings of the IEEE/CVF Conference on Computer Vision and Pattern Recognition, pp. 122–132 (2020)

19. Sahoo, A., Shah, R., Panda, R., Saenko, K., Das, A.: Contrast and mix: temporal contrastive video domain adaptation with background mixing. Adv. Neural. Inf. Process. Syst. **34**, 23386–23400 (2021)

20. Singh, S., Arora, C., Jawahar, C.: First person action recognition using deep learned descriptors. In: Proceedings of the IEEE Conference On Computer Vision and Pattern Recognition, pp. 2620–2628 (2016)

21. Snell, J., Swersky, K., Zemel, R.: Prototypical networks for few-shot learning. In: Advances in Neural Information Processing Systems, vol. 30 (2017)

22. Song, X., et al.: Spatio-temporal contrastive domain adaptation for action recognition. In: Proceedings of the IEEE/CVF Conference on Computer Vision and Pattern Recognition, pp. 9787–9795 (2021)

23. Soomro, K., Zamir, A.R., Shah, M.: Ucf101: A dataset of 101 human actions classes from videos in the wild. arXiv preprint arXiv:1212.0402 (2012)

24. Torralba, A., Efros, A.A.: Unbiased look at dataset bias. In: CVPR 2011, pp. 1521–1528. IEEE (2011)

25. Wang, L.: Temporal segment networks: towards good practices for deep action recognition. In: Leibe, B., Matas, J., Sebe, N., Welling, M. (eds.) ECCV 2016. LNCS, vol. 9912, pp. 20–36. Springer, Cham (2016). https://doi.org/10.1007/978-3-319-46484-8_2

26. Wei, P., et al.: Unsupervised video domain adaptation: A disentanglement perspective. arXiv preprint arXiv:2208.07365 (2022)

27. Wu, C.Y., Feichtenhofer, C., Fan, H., He, K., Krahenbuhl, P., Girshick, R.: Long-term feature banks for detailed video understanding. In: Proceedings of the IEEE/CVF Conference on Computer Vision and Pattern Recognition, pp. 284–293 (2019)

28. Xu, Y., Yang, J., Cao, H., Wu, K., Wu, M., Chen, Z.: Source-free video domain adaptation by learning temporal consistency for action recognition. In: Computer Vision-ECCV 2022: 17th European Conference. pp. 147–164. Springer (2022). https://doi.org/10.1007/978-3-031-19830-4_9

29. Yang, L., Huang, Y., Sugano, Y., Sato, Y.: Interact before align: Leveraging cross-modal knowledge for domain adaptive action recognition. In: Proceedings of the IEEE/CVF Conference on Computer Vision and Pattern Recognition, pp. 14722–14732 (2022)

30. Zach, C., Pock, T., Bischof, H.: A duality based approach for realtime TV-L^1 optical flow. In: Hamprecht, F.A., Schnörr, C., Jähne, B. (eds.) DAGM 2007. LNCS, vol. 4713, pp. 214–223. Springer, Heidelberg (2007). https://doi.org/10.1007/978-3-540-74936-3_22

31. Zhang, H., Cisse, M., Dauphin, Y.N., Lopez-Paz, D.: mixup: Beyond empirical risk minimization. arXiv preprint arXiv:1710.09412 (2017)

32. Zhou, B., Andonian, A., Oliva, A., Torralba, A.: Temporal relational reasoning in videos. In: Proceedings of the European Conference on Computer Vision (ECCV), pp. 803–818 (2018)

FEAD-D: Facial Expression Analysis in Deepfake Detection

Michela Gravina[1]([✉]) [ID], Antonio Galli[1] [ID], Geremia De Micco[1],
Stefano Marrone[1] [ID], Giuseppe Fiameni[2] [ID], and Carlo Sansone[1] [ID]

[1] DIETI, University of Naples Federico II, Naples, Italy
{michela.gravina,antonio.galli,geremia.demicco-apple,
stefano.marrone,carlo.sansone}@unina.it
[2] NVIDIA AI Technology Center Italy, Torino, Italy
gfiameni@nvidia.com

Abstract. As the development of deep learning (DL) techniques has progressed, the creation of convincing synthetic media, known as deepfakes, has become increasingly easy, raising significant concern about the use of these videos to spread false information and potentially manipulate public opinion. In recent years, deep neural networks, such as Convolutional Neural Networks (CNNs) and Recurrent Neural Networks (RNNs), have been used for deepfake detection systems, exploiting the inconsistencies and the artifacts introduced by generation algorithms. Taking into account the main limitation of fake videos to realistically reproduce the natural human emotion patterns, in this paper, we present FEAD-D, a publicly available tool for deepfake detection performing facial expression analysis. Our system exploits data from the DeepFake Detection Challenge (DFDC) and consists of a model based on bidirectional Long Short-Term Memory (BiLSTM) capable of detecting a fake video in about two minutes with an overall accuracy of 84.29% on the test set (i.e. comparable with the current state-of-the-art, while consisting of fewer parameters), showing that emotional analysis can be used as a robust and reliable method for deepfake detection.

Keywords: Deepfake · Bidirectional LSTM · Convolutional Neural Network

1 Introduction

Deepfakes refer to synthetic media, including images and videos, that are generated using Artificial Intelligence (AI) techniques to alter the appearance or speech of real individuals [18]. As Deep Learning (DL) approaches have advanced, deepfakes have become increasingly realistic, raising concerns about their potential to spread misinformation and manipulate public opinion. Consequently, the ability to accurately detect deepfakes has become a critical issue. One of the main challenges in this field is to identify effective and robust features that can distinguish between real and fake videos.

G. L. Foresti et al. (Eds.): ICIAP 2023, LNCS 14234, pp. 283–294, 2023.
https://doi.org/10.1007/978-3-031-43153-1_24

Deepfake detection systems usually leverage various computer vision and Machine Learning (ML) approaches to analyze video content. Recent advances in deep learning led to investigations into the use of deep neural networks (DNNs), such as Convolutional Neural Networks (CNNs) and Recurrent Neural Networks (RNNs), for this purpose. CNNs are employed to detect patterns in images that may indicate fake content, such as changes in facial features, skin texture, and lighting conditions, while RNNs capture temporal characteristics and identify unnatural or incoherent variations.

Emotions, in particular, have been emerging as a valuable feature for deepfake detection due to the difficulty of synthesizing realistic emotional expressions, which remains a major limitation of deepfake creation algorithms. Emotions are indeed essential for human communication and can be easily conveyed through facial expressions, voice tone, and body language, providing cues to assess the authenticity of a message. Consequently, the detection of emotional characteristics in videos has received increased attention as a potential approach for identifying manipulated content. Indeed, deepfakes often struggle to accurately reproduce emotive aspects, including micro-expressions, which are brief and involuntary facial movements that reveal emotions. The use of emotions for deepfake detection is predicated on the assumption that synthetic emotional expressions generated by deepfake algorithms can be distinguished from genuine ones due to their lack of realism and variability [9, 11, 13].

In this work, we aim to present FEAD-D (an acronym for Facial Expression Analysis in Deepfake Detection), a system based on facial expressions for deepfake detection. The system has been trained and tested on data coming from the Deepfake Detection Challenge (DFDC) [5], a competition released by Facebook, Microsoft, and the Partnership on AI's Media Integrity Steering Committee, in 2019. Although in recent years increasingly complex systems have been proposed [3, 16] with high performance, our contribution consists of the implementation of a publicly available easy-to-use tool capable of detecting a fake video in about two minutes, showing an overall accuracy equal to 84.29% computed on the released DFDC test set. This result is very interesting, as it shows that the proposed methodology achieves performance comparable with the current state-of-the-art approaches while consisting of fewer parameters (about 59 million, compared with 462 [1], 101 [6], and 89 [19] million), making it potentially usable in low-power devices (such as smartphones).

The rest of the paper is organized as follows: Sect. 2 gives an overview of the state-of-art approaches; Sect. 3 introduces the implemented methodology; Sect. 4 describes the experimental set-up; finally Sect. 5 provides some conclusions.

2   Related Works

In recent years, researchers have developed a variety of techniques to detect deepfake videos using AI [12, 18]. One of the most popular approaches consists in the identification of the human face to detect inconsistencies leading to non-natural patterns [10, 14]. Indeed, the majority of manipulation techniques act on the facial region, producing face swapping or pixel-wise modification.

CNNs provide surprising performance in deepfake identification, focusing on subtle distortions, especially around the eyes and mouth. For example, the solution proposed by Selim. [1], the winner of the DeepFake Detection Challenge (DFDC) [5], exploited EfficientNet B7 to analyze the images resulting from the face detector. Moreover, Bonettini et al. [4] proposed an ensemble of Efficient-Net B4 architectures, improving the classification performance. Other detectors proposed in the literature examine the movement patterns of objects, such as the position and orientation of faces, to detect inconsistencies that could indicate manipulation [2,15]. More recently, different methodologies implemented in [6,19] combine CNNs with Vision Transformers to enhance the characteristics of the facial region denoting deepfakes.

Some works [9,11,13] also analyzed the difference between the emotional patterns between real and fake video, highlighting that expressions are not adequately reproduced in the manipulated content. One common approach involves using neural networks, such as CNNs or Long Short-Term Memory (LSTM), to extract facial features and analyze the dynamics of emotional expressions in deepfake videos. However, as reported in [14], the contextual information surrounding the human face should be considered during the analysis of the inconsistencies resulting from frame manipulation. Taking these aspects into account, our proposal (FEAD-D) differs from the other detector systems based on emotional patterns, since it exploits also the textural characteristics of the extracted facial region.

3   Methods

FEAD-D aims at exploiting the unnatural variation in the facial expressions introduced by the artefacts generated during the video creation. Indeed, a deep-fake may not accurately capture the full range of emotions and facial movements of the original subject, especially when it comes to micro-expressions. Through a comprehensive analysis of a significant sample of fake videos generated by various tools, and by conducting a comparative examination with their corresponding original versions, a temporal inconsistency arising from non-natural sequences of facial expressions was identified. In particular, Fig. 1 shows how the emotional patterns vary across consecutive frames belonging to real and fake videos respectively, highlighting the need to consider temporal evolution.

The implemented framework consists of the following modules, as described in Fig. 2:

- **Face detection**, in which the target video is analyzed to detect one or more faces.
- **Features Extraction**, that analyzes the texture of target images and extracts the emotion from the detected face frame by frame.
- **Features temporal analysis**, in which all the features extracted in the previous stages are analyzed together in a cross-frame fashion to spot incoherent and unnatural patterns in the emotional evolution of the target subject.

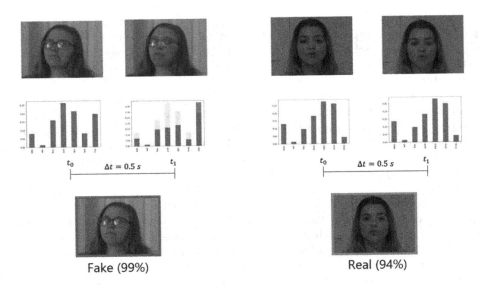

Fig. 1. Illustrative example of how the emotion patterns vary across frames for a fake and for a real video: t_0 and t_1 represent two temporal instants where $t_1 = t_0 + \Delta t$. The histograms report the emotional changes across the frames corresponding to t_0 and t_1 in the video under analysis. It is possible to note that in a fake video, the emotion pattern is characterized by an abrupt change (pink shadow in the histogram computed at t_1).

3.1 Face Detection

Face detection is a crucial stage, as in recent deep fakes the subject's head pose (and thus face) can change a lot during the video and/or the subject moves in the scene (e.g., walks). In particular, the implemented process acts on each frame f_i where i goes from 1 to N, representing the number of frames in a video, exploiting a CNN as a detector to efficiently detect human faces, as summarized in Fig. 3. To facilitate the recognition, the Contrast Limited Adaptive Histogram Equalization (CLAHE) [20] is applied, improving the contrast in images and enhancing the definitions of edges in each region. Then, the resulting frame $\hat{f}_i$ feeds the involved CNN, which provides the box including the face as output. The developed algorithm also mitigates the failure of recognition by implementing different pre-processing operations on the input image. Indeed, if the detector fails in the identification, a second stage is performed, considering the grayscale version of the frame f_i, denoted as f_i^g as input for the network. In the case no bounding box is found, CLAHE is applied to f_i^g, obtaining $\hat{f}_i^g$, before leveraging again the CNN for face detection.

After having analyzed all f_i with this process consisting of three consecutive steps, the information about the face position in each frame is used to crop the image considering a square box that is 1.4 times the size of the bounding box

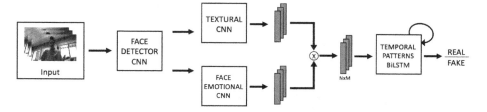

Fig. 2. Implemented methodology for FEAD-D system, consisting of a face detector, two CNNs extracting textural and emotional features respectively, and a bidirectional LSTM based network for the classification.

provided by the detector. However, we also take into account the possibility of failure in the face recognition on $\hat{f}_i^g$, which would result in a lack of information on face position in the corresponding f_i. To address this issue, we implement an interpolation stage that utilizes the spatial coordinates of the nearest detected bounding boxes, as illustrated in Fig. 4. As a result, each video is transformed into a sequence of cropped frames.

3.2 Features Extraction

This step operates considering the face detected in each frame f_i, extracting both textural and emotional features. The aim is to use the textural characteristics for recognizing the artifacts related to the contextual information, while the emotional features to analyze the patterns of the micro-expressions. We use a CNN pre-trained on ImageNet dataset [7] as a textural features extractor, exploiting the knowledge learned from this vast amount of data. We consider transfer learning to use the network capability to extract highly discriminative features that are useful for a wide range of computer vision tasks, considering the vector coming from the layer immediately preceding the output level.

As aforementioned, deepfakes exhibit a non-continuous temporal trend of facial expressions, resulting from the visual artifacts introduced during the generation process. To take into account the emotion patterns, we extract the expression characteristics for each cropped f_i, using a CNN specifically trained for this purpose, whose architecture is reported in Fig. 5. The proposed model consists of four blocks, each comprising a convolutional layer, batch normalization, average pooling, and dropout, followed by a flatten operation and three fully connected layers, which alternate with dropout. The first block presents a features map with 64 output channels, while the others double the number of channels. The network is trained considering data coming from the Facial Expression Recognition (Fer2013) [8] challenge, published at the International Conference on Machine Learning (ICML) in 2013, consisting of images belonging to seven emotional categories (anger, disgust, fear, happiness, sadness, surprise, and neutral). In our specific case, the objective of the proposed model is to associate different emotions with distinct facial expressions. Therefore, it is not strictly necessary

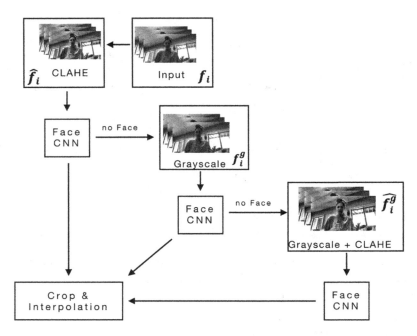

Fig. 3. Schema describing the implemented Face Extraction step, consisting of a CNN and successive pre-processing operations.

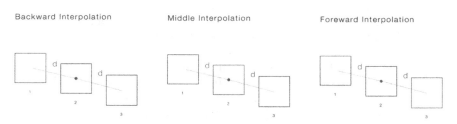

Fig. 4. Process illustrating the interpolation operation consists of backward, middle, and forward cases. In each of the three parts, the red box represents the interpolated bounding box. (Color figure online)

for the predicted emotion to match the actual one, as the primary interest lies in identifying how the visual artifacts introduced in the generated videos disrupt the temporal coherence of the emotional transitions. After a neural network has been trained, the process of extracting expression features from it is performed by considering the output vector obtained after the flatten operation that contains high-level representations of the input data.

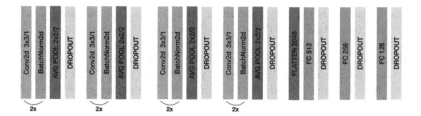

Fig. 5. CNN implemented for facial emotions features extraction, consisting of four blocks, each comprising a convolutional layer, batch normalization, average pooling, and dropout, followed by a flatten operation and three fully connected layers, which alternate with dropout. The details about the filter size and the stride are reported in each convolutional and average pooling layer.

3.3 Features Temporal Analysis

The present stage focuses on analyzing the temporal patterns of the features extracted from the procedure detailed in Sect. 3.2, with the aim of discerning between authentic and fake videos. Given the sequential nature of the frames f_i within a given video, we investigate the variations in textural and emotional characteristics, which deepfakes generation algorithms often fail to replicate due to their difficulty in emulating genuine human behaviour. Specifically, we generate a feature vector by concatenating the representations produced by the associated CNNs, resulting in an $N \times M$ matrix for each video that represents a multivariate time series, where M denotes the size of the feature space. For classification purposes, we employ a Bidirectional LSTM (BiLSTM) network. This model outperforms traditional LSTM approaches by capturing the dependencies in both directions, resulting in higher prediction accuracy and overall improved performance. By considering past and future contexts, BiLSTM is better equipped to represent the sequence data, thereby reducing overfitting and enhancing the model's generalization ability.

The network architecture used in this study comprises three layers of bidirectional LSTM units with 1024, 512, and 512 units, respectively, followed by two fully connected layers containing 512 and 256 elements. The final fully connected layer consists of one output neuron, which represents the probability of a given video being classified as fake.

4 Experiments and Results

In this paper, we present FEAD-D, a publicly available tool for deepfake detection exploiting facial expression temporal analysis. We consider data coming from the DeepFake Detection Challenge (DFDC) [5] which is a public benchmark dataset developed for the purpose of advancing the research in detecting and preventing the use of deepfakes. The dataset consists of over 100000 videos that were generated by several generation methods and edited with various post-production techniques to make them look more convincing. The DFDC dataset comprises three main parts representing the training, the validation, and the test sets and consisting of 119154, 4000, and 10000 videos respectively. The clips in the DFDC dataset are of diverse nature, depicting people of different ages, ethnicities, and genders, in various contexts and scenarios, and containing both famous and anonymous individuals, with a wide range of facial expressions and poses, lighting conditions, and backgrounds. The videos are mostly short in length, typically around 10 s in duration. Each clip is labelled with a binary value indicating whether the video is fake or real. The deepfakes in the dataset are generated using several different methods, such as Generative Adversarial Networks (GANs), autoencoders, and face-swapping techniques, that aim to create fake videos of individuals performing various activities, such as speaking, singing, and showing different facial expressions.

In the Face detection step described in Sect. 3.1, N is set to 300, and in the case of a video with a frames number less than the selected value, a new sequence is added, starting from the flipped version of the clip. We use a CNN provided by the dlib[1] library to detect human faces in each video. We then exploit the Fer2013 dataset [8] to train a CNN able to perform facial expression recognition in the Features Extraction stage. In particular, Fer2013 is an open-source dataset consisting of 35887 grayscale face images of size 48×48 pixels, labeled with 7 distinct emotions, including angry, disgust, fear, happy, sad, surprise, and neutral. In particular, the training set consists of 28709 samples (3995 angry, 436 disgust, 4097 fear, 7215 happy, 4830 sad, 3171 surprise, and 4965 neutral), while the test set contains 7178 images (958 angry, 111 disgust, 1024 fear, 1774 happy, 1247 sad, 831 surprise, and 1233 neutral). We first train our implemented CNN on the Fer2013 [8] training set, and we use the provided test set to assess the performance for emotion recognition, resulting in an overall accuracy equal to 66 % while the accuracy for each class is equal to 58%, 63%, 52%, 85%, 59%, 73%, 62% for angry, disgust, fear, happy, sad, surprise and neutral respectively. Although this may not be considered a good result, it is worth noting that we obtain performance comparable to that shown by the top 5 participants in the Kaggle competition, considering the average between the public and the private leaderboard. Moreover, as aforementioned, our primary interest lies in analyzing the variation of the predicted emotions rather than the correct classification. Then, we leverage the trained CNN to extract the high-level emotional features using the DFDC dataset as input, and considering the vector obtained after the flatten operation with 2048 elements.

We consider the InceptionV3 [17] as a network pretrained on ImageNet [7] to obtain the textural characteristics since it provides a good trade-off between network parameters and classification accuracy. In particular, we extract the 2048 features provided by the flatten operation before the last fully connected layer. As a consequence, the resulting features vector, obtained by concatenating the textural and the emotional characteristics, consists of 4096 elements. Each video is then transformed in a multivariate time series, represented by a matrix of 300×4096, where 4096 is the size of the features space, denoted with M in

---

[1] http://dlib.net/python/index.html.

Sect. 3.3. As reported in the DFDC[2] the log loss L is the performance metric used to evaluate the classification model we have developed. It is defined as follows:

$$L = -\frac{1}{T} \sum_{i=1}^{T} [y_i \log(\hat{y}) + (1 - y_i) \log(1 - \hat{y})] \qquad (1)$$

where T is the total number of elements in the test set, y_i is the real label of the i-th sample, while $\hat{y}_i$ is the model output probability. Therefore the binary cross-entropy (BCE) loss, which exploits the logarithm function to strongly penalize the confident but incorrect predictions, is used to train the implemented BiLSTM-based model. The total number of epochs is 100, the batch size is set to 32, while the learning rate is 10^{-4}. The entire pipeline, including video processing, face extraction, feature merging, and the final training stage, is developed with Python 3.8 and Keras framework using TensorFlow as backend, using the accelerated cluster based on the IBM Power9 architecture and Volta NVIDIA GPUs. Specifically, we utilized a node configuration consisting of 8 V100 SXM2 GPUs.

To assess the performance of FEAD-D, we used half of the test set of the DeepFake Detection Challenge (DFDC) [5]. Indeed, the authors released the correct labels for only half of the samples, resulting in a set of data consisting of 5000 videos, where 50% of them belong to the fake class. As a consequence, it is possible to provide a fair comparison with recent state-of-art approaches considering the same experimental set-up. In particular, we selected the methodologies with a publicly available implementation, following the instruction provided by the authors for the computation of the performance. Table 1 reports the results obtained by the detector presented in this paper, showing the value of the loss L computed according to Eq. 1, together with the accuracy (ACC), F1-score (F1), and Area under the ROC Curve (AUC). We also include the values of performance obtained by [6,19], and the winner of the DFDC competition [1], denoting with - when that metric is not reported by the authors.

Table 1. Performance of the methodology presented in this paper and recent state-of-art approaches.

Approach	ACC	F1	AUC	L
FEAD-D (our)	84.29%	83.57%	91.69%	**0.365**
Coccomini et al. [6]	-	88.02%	95.09%	-
Selim. [1] -	-	**90.56%**	**97.18%**	0.428
Wodajo et al. [19]	-	77.23%	84.39%	-

Our model achieves 84.29%, 83.57%, 91.69% in terms of ACC, F1, AUC respectively, and a value of L equal to 0.365. It is worth noting that, even

---

[2] https://www.kaggle.com/competitions/deepfake-detection-challenge.

if FEAD-D does not outperform all the state-of-art methodologies, it obtains comparable and good results with a reduced number of parameters to train, that is about 59 million, compared with 462, 101, and 89 million of the solution proposed in [1,6], and [19], respectively. Moreover, the obtained values of performance are in line with the results reported by the authors in [13], and [9] exploiting emotional patterns, that are not included in Table 1 since they did not use the public test set for the evaluation.

To make FEAD-D a publicly available tool, we release the notebook and the weights of the models, implementing all the steps of the described methodology at *this link*[3]. We tested the efficiency in Google Colaboratory[4], with an NVIDIA Tesla T4 GPU with 16 GB RAM, measuring an execution time of about 2 min and 40 s for the classification of a single video.

5 Conclusions

The growing sophistication of deepfakes and the potential risks they pose to public opinion have resulted in a critical need for accurate detection systems. A variety of methods have been proposed, utilizing computer vision and machine learning techniques to analyze video content, including inconsistencies resulting from frame manipulation, movement patterns of objects, audio analysis, and most recently, emotional analysis.

This paper presents FEAD-D, a publicly available tool based on facial expressions for deepfake detection, that achieves a 84.29% accuracy on the test set and is able to process a video in just two minutes. Our contribution to the field lies in the development of a fast and reliable system that can be easily adopted by users with minimal technical knowledge. Our approach shows that emotional analysis can be a robust and reliable method of deepfake detection. More in detail, the presented results demonstrate the effectiveness of using facial expressions for deepfake detection and serve as a basis for further research and development in this field. Indeed, while emotional analysis offers a promising approach for detecting deepfakes, it also presents several challenges that must be addressed to ensure its reliability. Emotional expressions can vary across individuals, cultures, and contexts, making it difficult to establish a consistent baseline for what constitutes a genuine emotional expression. Additionally, creation algorithms can be specifically designed to mimic emotional expressions, rendering detection based on emotional inconsistencies unreliable. Further research is necessary to overcome these challenges and establish emotional analysis as a robust and reliable method of deepfake detection.

Future work will include the analysis of the generalization ability of the proposed approach, considering different datasets and the effect of video manipulation, such as resize or blurring operations, on performance. Moreover, we will focus on developing more robust deepfake detection methods that take

---

[3] The code is available here: https://github.com/priamus-lab/FEAD-D_Facial-Expression-Analysis-in-Deepfake-Detection.

[4] https://research.google.com/colaboratory/.

into account the multiple sources of inconsistencies present in fake videos, using advanced computer vision techniques to identify visual artifacts that result from creation algorithms, such as mismatches in lighting and shadows. To improve the detection of fake audio, machine learning models can be trained to analyze both lip movements and speech signals. Also, it may be useful to investigate the effectiveness of combining multiple detection methods, such as those based on visual, audio, and human-based features, to improve the overall performance of deepfake detectors. Finally, we will try to further reduce the number of parameters of the detector model, making its integration on low-power devices easier.

Code base and model weights

The source code is available at *this link*[5], where we also provide the weights of the trained models for video classification.

Acknowledgements. We acknowledge the CINECA award under the ISCRA initiatives, for the availability of high-performance computing resources and support within the projects IsC80_FEAD-D and IsC93_FEAD-DII. We also acknowledge the NVIDIA AI Technology Center, EMEA, for its support and access to computing resources. This work has been supported by BullyBuster - PRIN 2017 Project, funded by MIUR (CUP: E24I19000590001)

References

1. Selim seferbekov: Winner of deepfake detection challenge. https://github.com/selimsef/dfdc_deepfake_challenge
2. Agarwal, S., Hu, L., Ng, E., Darrell, T., Li, H., Rohrbach, A.: Watch those words: Video falsification detection using word-conditioned facial motion. In: Proceedings of the IEEE/CVF Winter Conference on Applications of Computer Vision, pp. 4710–4719 (2023)
3. Almars, A.M.: Deepfakes detection techniques using deep learning: a survey. J. Comput. Commun. **9**(5), 20–35 (2021)
4. Bonettini, N., Cannas, E.D., Mandelli, S., Bondi, L., Bestagini, P., Tubaro, S.: Video face manipulation detection through ensemble of CNNs. In: 2020 25th International Conference on Pattern Recognition (icpr), pp. 5012–5019. IEEE (2021)
5. Dolhansky, B.: Joanna Bitton. The deepfake detection challenge dataset, B.P.J.L.R.H.M.W.C.C.F. (2020)
6. Coccomini, D.A., Messina, N., Gennaro, C., Falchi, F.: Combining efficientnet and vision transformers for video deepfake detection. In: Image Analysis and Processing-ICIAP 2022: 21st International Conference, Lecce, Italy, May 23–27, 2022, Proceedings, Part III. pp. 219–229. Springer (2022). https://doi.org/10.1007/978-3-031-06433-3_19
7. Deng, J., Dong, W., Socher, R., Li, L.J., Li, K., Fei-Fei, L.: Imagenet: A large-scale hierarchical image database. In: CVPR, 2009, pp. 248–255. IEEE (2009)

---

[5] The code is available here: https://github.com/priamus-lab/FEAD-D_Facial-Expression-Analysis-in-Deepfake-Detection.

8. Dumitru, Ian Goodfellow, W.C.Y.B.: Challenges in representation learning: Facial expression recognition challenge (2013)
9. Hosler, B., et al.: Do deepfakes feel emotions? a semantic approach to detecting deepfakes via emotional inconsistencies. In: Proceedings of the IEEE/CVF Conference on Computer Vision and Pattern Recognition (CVPR) Workshops, pp. 1013–1022 (June 2021)
10. Kaur, S., Kumar, P., Kumaraguru, P.: Deepfakes: temporal sequential analysis to detect face-swapped video clips using convolutional long short-term memory. J. Electron. Imaging **29**(3), 033013 (2020)
11. López-Gil, J.M., Gil, R., García, R., et al.: Do deepfakes adequately display emotions? a study on deepfake facial emotion expression. Comput. Intell. Neurosci. **2022** (2022)
12. Maras, M.H., Alexandrou, A.: Determining authenticity of video evidence in the age of artificial intelligence and in the wake of deepfake videos. Int. J. Evidence Proof **23**(3), 255–262 (2019)
13. Mittal, T., Bhattacharya, U., Chandra, R., Bera, A., Manocha, D.: Emotions don't lie: An audio-visual deepfake detection method using affective cues. In: Proceedings of the 28th ACM International Conference on Multimedia, pp. 2823–2832 (2020)
14. Nirkin, Y., Wolf, L., Keller, Y., Hassner, T.: Deepfake detection based on discrepancies between faces and their context. IEEE Trans. Pattern Anal. Mach. Intell. **44**(10), 6111–6121 (2021)
15. Prashnani, E., Goebel, M., Manjunath, B.: Generalizable deepfake detection with phase-based motion analysis. arXiv preprint arXiv:2211.09363 (2022)
16. Rana, M.S., Nobi, M.N., Murali, B., Sung, A.H.: Deepfake detection: A systematic literature review. IEEE Access (2022)
17. Szegedy, C., Vanhoucke, V., Ioffe, S., Shlens, J., Wojna, Z.: Rethinking the inception architecture for computer vision. In: Proceedings of the IEEE Conference on Computer Vision and Pattern Recognition, pp. 2818–2826 (2016)
18. Westerlund, M.: The emergence of deepfake technology: a review. Technol. Innov. Manage. Rev. **9**(11) (2019)
19. Wodajo, D., Atnafu, S.: Deepfake video detection using convolutional vision transformer. arXiv preprint arXiv:2102.11126 (2021)
20. Yadav, G., Maheshwari, S., Agarwal, A.: Contrast limited adaptive histogram equalization based enhancement for real time video system. In: 2014 International Conference on Advances in Computing, Communications and Informatics (ICACCI), pp. 2392–2397 (2014). https://doi.org/10.1109/ICACCI.2014.6968381

Large Class Separation is Not What You Need for Relational Reasoning-Based OOD Detection

Lorenzo Li Lu[1], Giulia D'Ascenzi[1], Francesco Cappio Borlino[1,2(✉)],
and Tatiana Tommasi[1,2]

[1] Politecnico di Torino, Corso Duca Degli Abruzzi 24, 10129 Torino, Italy
{lorenzo.lu,giulia.dascenzi}@studenti.polito.it,
{francesco.cappio,tatiana.tommasi}@polito.it
[2] Italian Institute of Technology, Genova, Italy

Abstract. Standard recognition approaches are unable to deal with novel categories at test time. Their overconfidence on the known classes makes the predictions unreliable for safety-critical applications such as healthcare or autonomous driving. Out-Of-Distribution (OOD) detection methods provide a solution by identifying semantic novelty. Most of these methods leverage a learning stage on the known data, which means training (or fine-tuning) a model to capture the concept of *normality*. This process is clearly sensitive to the amount of available samples and might be computationally expensive for on-board systems. A viable alternative is that of evaluating similarities in the embedding space produced by large pre-trained models without any further learning effort. We focus exactly on such a fine-tuning-free OOD detection setting.

This works presents an in-depth analysis of the recently introduced relational reasoning pre-training and investigates the properties of the learned embedding, highlighting the existence of a correlation between the inter-class feature distance and the OOD detection accuracy. As the class separation depends on the chosen pre-training objective, we propose an alternative loss function to control the inter-class margin, and we show its advantage with thorough experiments.

Keywords: Out-Of-Distribution Detection · Cross-Domain Learning · Relational Reasoning

1 Introduction

In recent years, Deep Neural Networks have seen widespread adoption in multiple computer vision tasks. Still, standard recognition algorithms are typically evaluated under the *closed-set* assumption [27], limiting their prediction ability to the same categories experienced at training time. As most real-world scenarios are very different from the well-defined and controlled laboratory environments, an agent operating in the wild will inevitably face data coming from unknown distributions, thus it should be able to handle novelty which is a task of utmost

G. L. Foresti et al. (Eds.): ICIAP 2023, LNCS 14234, pp. 295–306, 2023.
https://doi.org/10.1007/978-3-031-43153-1_25

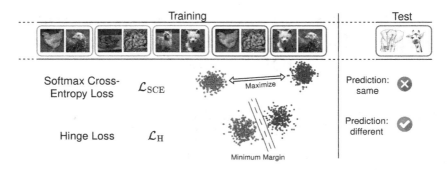

Fig. 1. Schematic overview of a relational reasoning-based OOD method that exploits different training losses. Our work empirically demonstrates that controlling, and in particular reducing, the distance between *same* and *different* classes improves semantic novelty detection.

importance for safety-critical applications. In this regard, Out-Of-Distribution (OOD) detection techniques have gained considerable attention as they enable models to recognize when test samples are *In-Distribution* (ID) with respect to the training ones, or conversely *Out-Of-Distribution* (OOD). Specifically, *Semantic Novelty Detection* [27] refers to the *open-set* case in which the distribution shift originates from the presence of unknown categories in the test set, together with the known *normal* ones already seen during training. Many techniques have been proposed for this task [8,16,18,23]. However, they typically need a significant number of reference known samples for the model to learn the concept of *normality* through either training from scratch or at least a fine-tuning phase. While such approaches generally lead to good performance, they can also be problematic for low-power edge devices with limited computational resources, and anyway become unfeasible if the amount of known data is scarce or their training access is restricted for privacy reasons.

Recently, two studies proposed techniques that enable OOD detection without fine-tuning [1,19]. Both rely on pre-trained models whose data representation can be easily exploited to perform comparisons and identify unknown categories, avoiding further learning effort. In terms of training objectives, they share the choice of moving away from standard classification and highlight the importance of analogy-based learning to better manage open-set conditions. Specifically, ReSeND [1] proposed a new relational reasoning paradigm to learn whether pairs of images belong or not to the same class. Instead, MCM [19] inherits the CLIP model trained on vision-language data pairs to promote multi-modal feature alignment via contrastive learning.

In this work, we are interested in the single modality case to further evaluate its potential and limits. More precisely, we examine the connection between inter-class distance in the features embedding produced via relational reasoning and the ability to perform semantic novelty detection in that space. As highlighted in [11], training objectives that enforce a stronger inter-class separation may cause the learned representations to be less transferable. Thus, we present an extensive analysis of relational reasoning performed with various loss functions that have different control on class separation. Our findings indicate that avoiding to maximize inter-class separations provides more generalizable features, improving the performance of the pre-trained model on the downstream semantic novelty detection task (see Fig. 1). Building on this conclusion, we design a tailored hinge loss function that provides direct control of class separation and increases the OOD results of the relational reasoning-based model.

Finally, we observe that certain OOD detection methods based on classification pre-training and originally intended to be used via fine-tuning may skip the latter learning phase [14,24]. Hence, these approaches can serve as a fair benchmark reference for relational reasoning-based methods.

To summarize, our key contributions are:

- We discuss and evaluate how the feature distributions originating from the use of different pre-training objectives affect the capability of a relational reasoning model for OOD detection;
- We introduce an alternative loss function that provides better control of class-specific feature distributions;
- We run a thorough experimental analysis that demonstrates the advantages of the proposed loss, considering as a reference also the powerful but costly k-NN-based OOD detector [24], re-casted for the first time to work in the fine-tuning-free setting.

2   Related Work

Out-Of-Distribution Detection is the task of determining whether test data belong to the same distribution as the training data or not. A distributional shift may occur due to a change in domain (*covariate shift*) or categories (*semantic shift*). We expect a trustworthy model to detect whether a sample belongs to a new category regardless of the visual domain, thus our main focus is on semantic novelty detection. The first baseline for this task was proposed by Hendrycks et al. [8], who suggested that the Maximum Softmax Probability (MSP) score produced by a classification model trained on ID data should be higher for ID test samples than for OOD ones. Several other approaches followed the same post-hoc strategy enhancing ID-OOD separation, via temperature scaling [16], or by focusing on energy scores [18] and network unit activations [23]. A different family of techniques uses distance metrics to identify OOD samples in the feature space learned on the ID data [14,24].

OOD Detection Without Fine-Tuning. All the OOD detection solutions described in the previous paragraph require training (or at least fine-tuning) on nominal samples in order to learn the concept of normality. However, this learning phase requires a sizable amount of ID data and computational resources, making it expensive and impractical for many real-world applications. Additionally, fine-tuning can hurt the generalization of learned representations [13] as it is susceptible to *catastrophic forgetting* [10]: the model may overfit the fine-tuning dataset, losing the knowledge previously learned on a much larger one. Only some recent work has started addressing this problem, proposing solutions that do not require a fine-tuning stage to perform OOD detection [1,19]. In particular, [1] suggests substituting the standard classification-based pre-training task with relational reasoning, which directs the network's focus on the semantic similarity between two input images to predict a normality score. This pretext objective is less domain- and task-dependent than classification and leads to an embedding space with great transfer capabilities. On the other hand, [19] leverages CLIP [22] to perform zero-shot OOD detection, thus exploiting two modalities (vision and language) rather than one. Finally, we point out how distance-based strategies such as [14,24] could also be used without performing the fine-tuning stage, although this aspect was not addressed in their original works.

Pre-training Loss Functions and Transfer Learning. The possibility to easily inherit and reuse pre-trained models for novel tasks is certainly one of the more appreciated characteristics of deep learning. As discussed above, such a procedure may be relevant even for OOD detection applications in which the representation learned on a large-scale dataset is leveraged to evaluate sample similarity. Still, only a few works have analyzed how the exact choice of the pre-training objective influences the transferability of the extracted knowledge. An implicit hypothesis is that models that perform well on the pre-training task also perform well on the downstream one. However, this is not always the case [12]: for instance, some regularization techniques that provide an improvement on the pre-training task produce penultimate layers features that are worse in generalization. This phenomenon has been described as *supervision collapse* [5]. The R^2 metric introduced in [11] to evaluate intra-class compactness and inter-class separation provides a way to shed light on this behavior: the most advanced strategies to increase accuracy on the pre-training task lead to a greater class separation which however is associated with reduced knowledge transferability. As the use of pre-trained models without fine-tuning for OOD detection is still scarcely explored, we find it relevant to perform an analysis of the role of different pre-training objectives for this downstream task. Specifically, we focus on relational reasoning-based OOD detection performed via different loss functions.

3   Relational Reasoning for OOD Detection

We consider a set of labeled samples $\mathcal{S} = \{x^s, y^s\}$ that we call *support set*, and a set of unlabeled ones $\mathcal{T} = \{x^t\}$ called *test set*. They are drawn from two different distributions and present a category shift, besides also a potential domain shift. The support label set $\mathcal{Y}_\mathcal{S}$ identifies *known* categories. The target label set $\mathcal{Y}_\mathcal{T}$ includes both known and unknown semantic categories: $\mathcal{Y}_\mathcal{S} \subset \mathcal{Y}_\mathcal{T}$. The goal of an OOD detector is to identify all the test samples whose categories do not appear in the support set (i.e., which are *unknown*). Traditional methods require a training or fine-tuning stage on the support set $\mathcal{S}$, while in the fine-tuning-free scenario the support set is only accessed at evaluation time.

In ReSeND [1], the authors presented a relational reasoning-based learning approach specifically designed for OOD detection. The model is trained on sample pairs (x_i, x_j) drawn from a large-scale object recognition dataset and learns to distinguish whether the two images belong to the same category ($l_{ij} = 1$, if $y_i = y_j$) or not ($l_{ij} = 0$, if $y_i \neq y_j$). This task can be cast as binary classification or regression. In both cases, the model learns how to encode in an embedding space the samples' semantic relationship $\boldsymbol{p}_m = r(\boldsymbol{z}_i, \boldsymbol{z}_j)$, where the index m ranges over all the possible sample pairs, and $\boldsymbol{z} = \phi(\boldsymbol{x})$ represents the features extracted via an encoder ϕ from the image $\boldsymbol{x}$. Then, the last network layer converts this information into a scalar similarity value that is compared to the ground truth l_m with a chosen loss function. At inference time, the support set samples are grouped according to their category and their representation is averaged to get per-class prototypes $\bar{\boldsymbol{z}}_y^s$ for $y = \{1, \ldots, |\mathcal{Y}_S|\}$. Each test sample $\boldsymbol{z}^t = \phi(\boldsymbol{x}^t)$ is then compared with every prototype to get the corresponding similarity score. Finally, the vector collecting all the $|\mathcal{Y}_S|$ elements is filtered by a softmax function on which MSP is applied to get the final normality score.

In this framework, by observing the embedding space produced by the penultimate layer of the network, we expect to see pairs of samples of the training dataset organized into two clusters representing the broad *same* and *different* concept classes. Once trained on the large-scale ImageNet-1K dataset, this embedding can be used for OOD detection on a variety of domains without fine-tuning, so its generalization ability is crucial.

4 Relational Reasoning and Class Separation

4.1 Class Compactness and Separation

In order to analyze the learned feature space we focus on the separation between the *same* and *different* classes described above. In particular, we leverage the R^2 index introduced in [11]. This metric is based on the ratio between the average within-class and average global cosine distance for the considered feature vectors, providing a relative measure of the sparsity of the representation of each class in the embedding space. Specifically, the index value is given by:

$$R^2 = 1 - \bar{d}_{within}/\bar{d}_{total} \tag{1}$$

$\bar{d}_{within} = \sum_{k=1}^{K} \sum_{i=1}^{M_k} \sum_{j=1}^{M_k} \frac{1-\text{sim}(\boldsymbol{p}_i^k, \boldsymbol{p}_j^k)}{K M_k^2}$, $\bar{d}_{total} = \sum_{h=1}^{K} \sum_{k=1}^{K} \sum_{i=1}^{M_h} \sum_{j=1}^{M_k} \frac{1-\text{sim}(\boldsymbol{p}_i^h, \boldsymbol{p}_j^k)}{K^2 M_h M_k}$

where the indices $i, j \in \{1, \ldots, M_k\}$ now range on the pairs of samples $\boldsymbol{p}^k$ which belong respectively to the $K = 2$ classes. The relative distance is measured via the cosine similarity: $\text{sim}(\boldsymbol{a}, \boldsymbol{b}) = \boldsymbol{a}^T \boldsymbol{b}/(\|\boldsymbol{a}\|\|\boldsymbol{b}\|)$. The right part of Fig. 3 gives an idea of what high and low R^2 values mean in terms of class separation.

4.2 Relational Reasoning Loss Functions

In the following we review some of the most common loss functions used for binary problems. In all the loss equations we use σ to refer to the score produced as output by the network for a sample pair $\boldsymbol{p}$, while the ground truth label is l.

Binary Cross-Entropy. The Cross-Entropy loss is defined as:

$$\mathcal{L}_{CE} = - \sum_{m=1}^{M} \sum_{k=1}^{K} t_{m,k} \log(\hat{t}_{m,k}) \tag{2}$$

where K is the number of categories, while $t_{m,k}$ and $\hat{t}_{m,k}$ are respectively the target value and the predicted probability of the class k for the sample m. In particular $t_{m,k}$ will assume the value 1 for the ground truth class of the sample ($k = l_m$) and 0 for all the other categories (one-hot encoding). In the binary case (i.e., when $K = 2$), such loss function can be expressed as:

$$\mathcal{L}_{BCE} = - \sum_{m=1}^{M} \left(t_{m,1} \log(1 - \hat{t}_{m,2}) + t_{m,2} \log(\hat{t}_{m,2}) \right) \tag{3}$$

where $\hat{t}_{m,2}$ is generally obtained by applying the logistic sigmoid function to the model output score ($f(\sigma) = 1/(1 + e^{-\sigma})$).
Impact on the class separation: this loss is non-zero even for correctly classified samples. As a result, the intra-class compactness and inter-class separation keep increasing for the whole training procedure.

Softmax Cross-Entropy. The categorical Cross-Entropy loss that is generally adopted for multi-class problems, is obtained by using the Cross-Entropy in Eq. (2) after having applied the softmax function to the model output scores ($f(\sigma)_k = e^{\sigma_k} / \sum_{c=1}^{C} e^{\sigma_c}$). Considering that the labels are one-hot, the overall summation will contain for each sample only the term corresponding to its ground truth label, so we can write the Softmax Cross-Entropy as:

$$\mathcal{L}_{SCE} = - \sum_{m=1}^{M} \log \frac{e^{\sigma_{m,l_m}}}{\sum_{k=1}^{K} e^{\sigma_{m,k}}} \tag{4}$$

where $\sigma_{m,k}$ is the score corresponding to the class k for the sample m and l_m represents its ground truth label. In the binary case we suppose $k, l \in \{1, 2\}$.
Impact on the Class Separation: As in the previous BCE case, this loss is non-zero even for correctly classified samples. It has been shown that the consequent trend of growing intra-class compactness and inter-class separation leads to miscalibrated classifiers providing overconfident predictions [20,26].

Focal Loss. A possible solution for the miscalibration issue mentioned above is to adjust the penalty assigned to a sample based on the network's confidence in predicting its true class [20]. This can be accomplished with the Focal Loss [17]. Starting from the Cross-Entropy formulation (see Eq. 2), such loss function can be expressed as:

$$\mathcal{L}_{focal} = - \sum_{m=1}^{M} \sum_{k=1}^{K} t_{m,k} (1 - \hat{t}_{m,k})^{\gamma} \log(\hat{t}_{m,k}) \tag{5}$$

where γ is a hyperparameter controlling the rescaling strength.

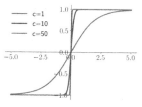

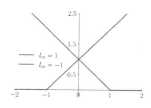

(a) Compressed sigmoid	(b) Hinge loss trend

Fig. 2. (a) Increasing c in the MSE compressed sigmoid transforms it into a Heaviside step function: the loss is zero when the output score has the correct sign. (b) H loss trend for positive ($l_m = 1$) and negative ($l_m = -1$) pairs ($\delta = 1$).

Impact on the Class Separation: by varying γ, it's possible to tune the magnitude of the rescaling, effectively bringing the loss value for correctly classified samples near zero and therefore mitigating the class separation tendency.

MSE with a Compressed Sigmoid. In ReSeND [1] the problem of separating the same and different classes was formalized as a regression task by using the MSE loss computed between the ground truth $l_m \in \{-1, 1\}$ and the output provided by a sigmoid rescaled on the $[-1, 1]$ range and with a modified slope, controlled by a factor c (see Fig. 2 (a)):

$$\mathcal{L}_{MSE} = \sum_{m=1}^{M} (\hat{s}_c(\sigma_m) - l_m)^2 \quad \text{with} \quad \hat{s}_c(\sigma_m) = \frac{2}{1 + e^{-c\sigma_m}} - 1 \qquad (6)$$

Impact on the class separation: by varying the value of c, it is possible to tune the penalty associated with different scores σ. Specifically, for higher c values, the sigmoid function will be more horizontally compressed: as a consequence samples already correctly classified receive a loss value that is almost zero decreasing the need for further class separation.

4.3 Controlling Class Separation: Hinge Loss for Relational Reasoning

As it is clear that class separation is crucial for the problem, we introduce a loss function that allows us to precisely tune it in a simple and straightforward way.

Let's start from the output of the last layer which is a scalar score σ_m and can be positive or negative, indicating the corresponding two classes. We can simply set a threshold at zero and fix a margin δ around it, within which even correct predictions pay a penalty. The loss will cancel out for $\sigma_m > \delta$ on positive samples and $\sigma_m < -\delta$ for negative ones, but would grow linearly if a negative score is assigned to a positive sample and vice-versa (see Fig. 2 (b)).

In this way we keep the two classes separated (which is crucial to retain the model's discriminative power), but the margin is limited and fixed to δ. This formulation corresponds to a hinge loss applied on the scalar score σ_m:

$$\mathcal{L}_H = \sum_{m=1}^{M} \max(0, \delta - l_m \sigma_m) \quad \text{with} \quad l_m \in \{-1, 1\} \qquad (7)$$

5 Experiments

5.1 Experimental Protocol

Our experimental analysis presents an extensive benchmark of fine-tuning-free OOD methods. All of them consist of a pre-training phase on ImageNet-1K [4] with a different objective, followed by a distance-based OOD prediction protocol. For ReSeND [1] the pre-training task is relational reasoning (same vs different) executed with all the loss functions described in the previous Section. The other competitors exploit either supervised classification or self-supervised objectives, with both cross-entropy-based approaches (ResNet [7], ViT [6], CutMix [28]), and contrastive strategies (SimCLR [2], SupCLR [9], CSI [25], SupCSI [25]). We also evaluate Mahalanobis [14] and k-NN [24]. We emphasize that the k-NN approach has never been previously evaluated in a fine-tuning-free setting. We include it in our comparison despite its potentially higher computational cost, as it involves comparing the test sample with each support set instance (which must be stored in memory), rather than with a single prototype per class.

Unless otherwise specified, we always adopt a ResNet-101 backbone, as it includes a comparable number of parameters to ReSeND (44M and 40M, respectively). We publish the code, together with implementation details and additional results in our project page[1].

We adopt two different experimental set-ups, by following [1]. The *intra-domain* setting is designed to evaluate the OOD detection ability of a model when there is a purely semantic distribution shift between the support and the test sets. It is built upon the DomainNet [21] and DTD [3] datasets. In the *cross-domain* setting the support and test set are sampled from different domains so we can evaluate the ability of the OOD methods to focus on semantics and disregard other visual appearance discrepancies. Rather than using the limited PACS dataset [15] as done in [1], we propose a novel benchmark built on top of DomainNet [21]. This choice allows for more statistically significant results.

Following common practice, we report results in terms of Area Under the Receiver Operating Characteristic curve (AUC) and FPR@TPR95 (FPR), which indicates the false positive rate value when the ID true positive rate is 95%.

5.2 Impact of the Training Objective

We evaluate the impact on ReSeND of various loss functions. As the learning objective shapes the structure of the feature space, we can investigate how the distribution of the data in the learned embedding relates to the final OOD performance. For this analysis we focus on the intra-domain setting. The average AUC on the four datasets, along with the corresponding R^2 value, are reported in the scatter plot in the left part of Fig. 3. Detailed per-dataset results can be found on our project page. The results clearly highlight a general trend in which a higher inter-class separation is associated with a lower OOD detection

---

[1] https://github.com/lulor/ood-class-separation.

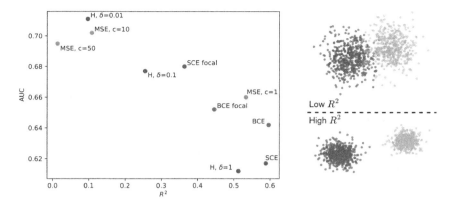

Fig. 3. Analysis of the OOD performance of a relational reasoning-based model trained with different loss functions (average *intra-domain* benchmarks results). The scatter-plot on the left shows that lower OOD results are generally associated with higher R^2 values, which means a stronger class separation as in the point distribution shown in the right-bottom part. On the other hand, more generalizable features and higher OOD performance are obtained with lower R^2 values corresponding to minimal class separation as in the right-top part.

performance. This behavior is even more evident when focusing on a specific loss and looking at how the results change by varying its hyperparameter value (e.g. when changing δ for our H loss or c for the MSE). Of course, there is a limit in the performance gain that can be reached by reducing the inter-class separation: after a certain point, the features start losing their discriminative power. For example with $c \geq 50$ for the MSE case, the training becomes less effective and the OOD detection performance starts slowly decreasing. We can conclude that for relational reasoning it is important to choose a learning objective that allows for a precise margin control. Only the proposed $\mathcal{L}_H$ loss satisfies this condition. Its hyperparameter δ represents a geometrical margin, and when a training sample meets the margin condition, the sample loss no longer affects the learning process. This behavior avoids the overconfidence typical of the standard softmax cross-entropy as highlighted by the normality score distributions represented in Fig. 4: the normality score values provided by the SCE loss are generally higher and have a larger range than those provided by the $\mathcal{L}_H$ loss (see the horizontal axis), but at the same time they provide a weaker ID-OOD separation.

5.3   Intra-Domain and Cross-Domain OOD Results

Intra-Domain analysis. In this setting support and test sets only differ in terms of semantics. Still, with respect to the pre-training dataset (ImageNet-1K), there may be a domain shift of varying magnitude (smaller for the Real case, larger for the others). In Table 1 we collect the results of the original ReSeND formulation ($\mathcal{L}_{MSE}$, with $c = 10$), its version based on the hinge loss that we name

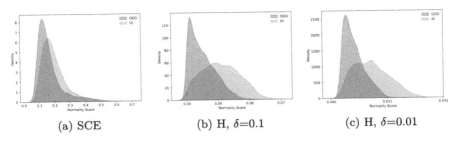

(a) SCE   (b) H, δ=0.1   (c) H, δ=0.01

Fig. 4. Normality Score distributions on the intra-domain Real setting for ReSeND pre-trained with different loss functions. We can see how the hinge loss with low margin pushes the model to provide more conservative scores, which are very close to each other (check the horizontal axis' scale) but more discernible.

Table 1. Intra-domain setting. Best result in bold and second best underlined

Model	Texture		Real		Sketch		Painting		Avg		Avg
	AUC ↑	FPR ↓	AUC ↑	FPR ↓	AUC ↑	FPR ↓	AUC ↑	FPR ↓	AUC ↑	FPR ↓	n.comp ↓
ResNet	0.672	0.897	0.710	0.863	0.554	0.939	0.649	0.919	0.646	0.904	25
ViT	0.537	0.937	0.701	0.829	0.553	0.955	0.673	0.853	0.616	0.894	25
CutMix	0.605	0.925	0.722	0.876	0.544	0.944	0.627	0.929	0.625	0.919	25
SimCLR	0.526	0.942	0.475	0.943	0.489	0.955	0.508	0.959	0.500	0.950	25
SupCLR	0.588	0.921	0.496	0.956	0.481	0.953	0.514	0.959	0.520	0.947	25
CSI	0.627	0.898	0.695	0.850	0.513	0.960	0.613	0.912	0.612	0.905	25
SupCSI	0.662	0.896	0.716	0.864	0.521	0.957	0.640	0.902	0.635	0.904	25
Mahalanobis	0.656	0.911	0.744	0.850	0.590	0.928	0.710	0.857	0.675	0.886	25
ReSeND	0.684	<u>0.847</u>	0.782	0.777	0.610	0.934	<u>0.721</u>	<u>0.826</u>	0.699	0.846	25
ReSeND-H	<u>0.688</u>	0.885	<u>0.798</u>	<u>0.755</u>	<u>0.638</u>	**0.898**	0.719	<u>0.826</u>	<u>0.711</u>	<u>0.841</u>	25
k-NN (k=1)	**0.774**	**0.840**	**0.843**	**0.596**	**0.640**	<u>0.914</u>	**0.800**	**0.757**	**0.764**	**0.777**	4100

ReSeND-H ($\mathcal{L}_H$, with $\delta = 0.01$) as well as the ResNet baseline and several reference approaches. We observe that ReSeND-H obtains a small but meaningful improvement across most of the considered settings, particularly in the Real and Sketch ones. The k-NN method from [24], applied without fine-tuning, achieves the best performance among all the considered techniques. We highlight how this result comes with a significant cost in terms of memory usage. Specifically, we calculated the average number of comparisons (n.comp) per test sample needed at evaluation time and reported the corresponding value in the last table column. Indeed, this introduces an important scalability limitation.

Cross-Domain Analysis. In this setting train and test data differ in semantic content and in visual style. From the results in Table 2 we can see how, despite using the whole support set rather than just the class prototypes, the k-NN method does not achieve the same performance advantage exhibited in the intra-domain case. Indeed, relying on all the support samples appears misleading. As a consequence, this method is less robust to domain shifts compared to ReSeND and ReSeND-H, which instead shows similar performance to the corresponding one in the intra-domain setting.

Table 2. Cross-domain setting. Best result in bold and second best underlined.

Model	Real-Paint.		Real-Sketch		Paint.-Real		Paint.-Sketch		Sketch-Real		Sketch-Paint.		Avg		Avg
	AUC ↑	FPR ↓	AUC ↑	FPR ↓	AUC ↑	FPR ↓	AUC ↑	FPR ↓	AUC ↑	FPR ↓	AUC ↑	FPR ↓	AUC ↑	FPR ↓	n.comp ↓
ResNet	0.596	0.949	0.539	0.938	0.627	0.922	0.546	0.941	0.533	0.929	0.524	0.940	0.561	0.937	25
ViT	0.627	0.921	0.526	<u>0.931</u>	0.618	0.901	0.524	0.946	0.568	0.945	0.591	0.924	0.576	0.928	25
CutMix	0.585	0.940	0.533	0.944	0.630	0.915	0.534	0.949	0.550	0.939	0.530	0.950	0.560	0.940	25
SimCLR	0.499	0.965	0.486	0.949	0.465	0.961	0.489	0.956	0.496	0.961	0.419	0.966	0.476	0.960	25
SupCLR	0.507	0.966	0.471	0.959	0.468	0.962	0.469	0.957	0.524	0.968	0.463	0.965	0.484	0.963	25
CSI	0.585	0.942	0.531	0.943	0.689	<u>0.863</u>	0.503	0.953	0.552	0.867	0.448	0.942	0.551	0.918	25
SupCSI	0.586	0.943	0.492	0.963	0.658	0.898	0.473	0.957	0.490	0.963	0.434	0.973	0.522	0.949	25
Mahalanobis	0.612	0.945	0.564	0.938	0.646	0.943	0.577	0.928	0.577	0.912	0.564	0.919	0.590	0.931	25
ReSeND	**0.666**	<u>0.912</u>	<u>0.572</u>	0.934	<u>0.727</u>	0.878	0.566	0.942	**0.705**	<u>0.860</u>	0.659	<u>0.911</u>	0.649	0.906	25
ReSeND-H	0.639	0.938	**0.583**	0.919	0.720	0.864	**0.590**	**0.899**	<u>0.679</u>	0.895	<u>0.637</u>	0.914	0.641	<u>0.905</u>	25
k-NN (k=1)	<u>0.662</u>	**0.902**	0.560	0.934	**0.754**	**0.781**	<u>0.584</u>	<u>0.908</u>	0.666	**0.836**	0.627	**0.900**	<u>0.642</u>	**0.877**	4800

6   Conclusions

In this work, we focused on the OOD detection task considering methods that do not need a fine-tuning stage on the ID data in order to detect semantic novelties. We analyzed how different learning objectives influence the performance of a relational reasoning-based solution to this problem, showing that a lower inter-class separation leads to better generalization. Exploiting this finding we proposed to use a tailored hinge loss function that provides better results than the original method implementation. At the same time, we pointed out how a previously unexplored fine-tuning-free k-NN strategy for OOD detection provides unexpectedly good accuracy at the cost of a higher computational effort. Still, it may fail when the support and the test set are drawn from different visual domains.

Acknowledgments. This study was carried out within the FAIR - Future Artificial Intelligence Research and received funding from the European Union Next-GenerationEU (PIANO NAZIONALE DI RIPRESA E RESILIENZA (PNRR) - MISSIONE 4 COMPONENTE 2, INVESTIMENTO 1.3 - D.D. 1555 11/10/2022, PE00000013). This manuscript reflects only the authors' views and opinions, neither the European Union nor the European Commission can be considered responsible for them.

Computational resources were provided by IIT HPC infrastructure.

References

1. Cappio Borlino, F., Bucci, S., Tommasi, T.: Semantic novelty detection via relational reasoning. In: ECCV (2022). https://doi.org/10.1007/978-3-031-19806-9_11
2. Chen, T., Kornblith, S., Norouzi, M., Hinton, G.: A simple framework for contrastive learning of visual representations. In: ICML (2020)
3. Cimpoi, M., Maji, S., Kokkinos, I., Mohamed, S., Vedaldi, A.: Describing textures in the wild. In: CVPR (2014)
4. Deng, J., Dong, W., Socher, R., Li, L.J., Li, K., Fei-Fei, L.: Imagenet: A large-scale hierarchical image database. In: CVPR (2009)

5. Doersch, C., Gupta, A., Zisserman, A.: Crosstransformers: spatially-aware few-shot transfer. In: NeurIPS (2020)
6. Dosovitskiy, A., et al.: An image is worth 16x16 words: Transformers for image recognition at scale. In: ICLR (2021)
7. He, K., Zhang, X., Ren, S., Sun, J.: Deep residual learning for image recognition. In: CVPR (2016)
8. Hendrycks, D., Gimpel, K.: A baseline for detecting misclassified and out-of-distribution examples in neural networks. ICLR (2017)
9. Khosla, P., et al.: Supervised contrastive learning. In: NeurIPS (2020)
10. Kirkpatrick, J., et al.: Overcoming catastrophic forgetting in neural networks. Proc. Natl. Acad. Sci. **114**(13), 3521–3526 (2017)
11. Kornblith, S., Chen, T., Lee, H., Norouzi, M.: Why do better loss functions lead to less transferable features? In: NeurIPS (2021)
12. Kornblith, S., Shlens, J., Le, Q.V.: Do better imagenet models transfer better? In: CVPR (2019)
13. Kumar, A., Raghunathan, A., Jones, R.M., Ma, T., Liang, P.: Fine-tuning can distort pretrained features and underperform out-of-distribution. In: ICLR (2022)
14. Lee, K., Lee, K., Lee, H., Shin, J.: A simple unified framework for detecting out-of-distribution samples and adversarial attacks. In: NeurIPS (2018)
15. Li, D., Yang, Y., Song, Y.Z., Hospedales, T.M.: Deeper, broader and artier domain generalization. In: ICCV (2017)
16. Liang, S., Li, Y., Srikant, R.: Enhancing the reliability of out-of-distribution image detection in neural networks. In: ICLR (2018)
17. Lin, T.Y., Goyal, P., Girshick, R., He, K., Dollar, P.: Focal loss for dense object detection. In: ICCV (2017)
18. Liu, W., Wang, X., Owens, J., Li, Y.: Energy-based out-of-distribution detection. NeurIPS (2020)
19. Ming, Y., Cai, Z., Gu, J., Sun, Y., Li, W., Li, Y.: Delving into out-of-distribution detection with vision-language representations. In: NeurIPS (2022)
20. Mukhoti, J., Kulharia, V., Sanyal, A., Golodetz, S., Torr, P.H., Dokania, P.K.: Calibrating deep neural networks using focal loss. In: NeurIPS (2020)
21. Peng, X., Bai, Q., Xia, X., Huang, Z., Saenko, K., Wang, B.: Moment matching for multi-source domain adaptation. In: ICCV (2019)
22. Radford, A., et al.: Learning transferable visual models from natural language supervision. In: ICML (2021)
23. Sun, Y., Guo, C., Li, Y.: React: Out-of-distribution detection with rectified activations. In: NeurIPS (2021)
24. Sun, Y., Ming, Y., Zhu, X., Li, Y.: Out-of-distribution detection with deep nearest neighbors. In: ICML (2022)
25. Tack, J., Mo, S., Jeong, J., Shin, J.: CSI: Novelty detection via contrastive learning on distributionally shifted instances. In: NeurIPS (2020)
26. Wei, H., Xie, R., Cheng, H., Feng, L., An, B., Li, Y.: Mitigating neural network overconfidence with logit normalization. In: ICML (2022)
27. Yang, J., Zhou, K., Li, Y., Liu, Z.: Generalized out-of-distribution detection: A survey. arXiv preprint arXiv:2110.11334 (2021)
28. Yun, S., Han, D., Oh, S.J., Chun, S., Choe, J., Yoo, Y.: Cutmix: Regularization strategy to train strong classifiers with localizable features. In: ICCV (2019)

BLUES: Before-reLU-EStimates Bayesian Inference for Crowd Counting

Emanuele Ledda[2,3]([✉]) [iD], Rita Delussu[1] [iD], Lorenzo Putzu[1] [iD],
Giorgio Fumera[1] [iD], and Fabio Roli[3] [iD]

[1] Department of Electric and Electronic Engineering, University of Cagliari,
Via Marengo 3, Cagliari 09100, Italy
[2] Department of Computer, Control and Management Engineering, Sapienza
University of Rome, Via Ariosto 25, Rome 00185, Italy
emanuele.ledda@uniroma1.it
[3] Department of Informatics, Bioengineering, Robotics, and Systems Engineering,
University of Genova, Via Dodecaneso 35, Genova 16146, Italy

Abstract. Ensuring the trustworthiness of artificial intelligence and
machine learning systems is becoming a crucial requirement given their
widespread applications, including crowd counting, which we focus on in
this work. This is often addressed by integrating uncertainty measures
into their predictions. Most Bayesian uncertainty quantification techniques use a Gaussian approximation of the output, whose variance is
interpreted as the uncertainty measure. However, in the case of neural
network models for crowd counting based on density estimation, where
the ReLU activation function is used for the output units, such a prior
may lead to an approximated distribution with a significant mass on negative values, although they cannot be produced by the ReLU activation.
Interestingly, we found that this is related to "false positive" pedestrian
localisation errors in the density map. We propose to address this issue
by shifting the Bayesian Inference *Before the reLU EStimates* (BLUES).
This modification allows us to estimate a probability distribution both
on the people density and the people presence in each pixel. This allows
us to compute a crowd segmentation map, which we exploit for filtering out false positive localisations. Results on several benchmark data
sets provide evidence that our BLUES approach allows for improving
the accuracy of the estimated density map and the quality of the corresponding uncertainty measure.

Keywords: Bayesian Inference · Crowd Counting · Uncertainty
Quantification · Monte-Carlo Dropout · Injected Dropout

1 Introduction

Crowd counting is a useful and still challenging computer vision functionality
in applications involving monitoring and analysis of crowds [12], in particular,
security-related applications based on video surveillance systems. State-of-the-
art methods for solving this task use deep neural networks (DNNs). Most of them

G. L. Foresti et al. (Eds.): ICIAP 2023, LNCS 14234, pp. 307–319, 2023.
https://doi.org/10.1007/978-3-031-43153-1_26

predict the crowd density map from which one obtains the estimated count by integration (i.e., summing up all the pixels of the density map). Recently some authors proposed crowd counting methods capable of also quantifying the prediction *uncertainty* [14,19,21]. Uncertainty Quantification (UQ) is currently of great interest in artificial intelligence and machine learning, given its increasing adoption in critical application scenarios, such as medical diagnosis, security (e.g., support to video surveillance for real-time crowd monitoring for law enforcement agencies), and military, where ensuring the trustworthiness of prediction systems is crucial. For crowd counting, UQ is helpful, e.g., to make end users (such as LEA operators involved in crowd monitoring) aware of the prediction reliability. Current UQ approaches distinguish between *epistemic uncertainty* (caused by a lack of knowledge about the model's parameters) and *aleatoric uncertainty* (caused by inherent randomness) [1]. In crowd counting, epistemic uncertainty can be due, e.g., to unseen scenes or camera perspectives, whereas aleatoric uncertainty can arise from noise due to, e.g., adverse weather conditions or extreme lighting conditions. The main approaches for UQ rely upon **Bayesian Inference** [16], which assumes a probability distribution on the network weights and output; the variance of the output distribution is interpreted as the uncertainty measure. Due to its very high computational complexity, some approximate inference methods are used in practice, such as Monte-Carlo dropout [8] or ensemble methods[1] [13]. Most state-of-the-art DNN models for crowd counting are based on estimating the crowd density map, from which the estimated count can be easily derived. They use accordingly the ReLU activation function for the output units (as density values cannot be negative). However, we noticed in a previous work that a Gaussian prior on the outputs sometimes leads to an approximated distribution with a significant mass on negative values. Interestingly, this behaviour turned out to be related to "false positive" pedestrian localisation errors in the density map, i.e., density map with positive values in locations where no pedestrians appear in the original image [14]. To this aim, we propose modifying the typical Bayesian Inference process for crowd counting (which places a Gaussian prior on the pixel-wise outputs) by shifting the inference process *Before the reLU EStimates*, which we call **BLUES Bayesian Inference**. We evaluated the proposed method on several benchmark data sets, providing evidence that it allows for improving the accuracy of the estimated density map and the quality of the corresponding uncertainty measure.

The rest of this work is organised as follows. We review related work in Sect. 2, we describe the proposed method in Sect. 3 and how to effectively implement it in Sect. 4. In Sect. 5, we describe and analyse the experimental results, and finally, we conclude by discussing the approach limitations and drawing some directions for future research in Sect. 6.

---

[1] Some authors do not consider ensemble methods to follow a Bayesian approach. For a discussion, we refer the reader to https://cims.nyu.edu/~andrewgw/deepensembles/.

2   Background and Related Work

Crowd Counting and Density Estimation. State-of-the-art methods for solving crowd counting use DNN architectures. Most estimate a density map (number of people per pixel) from which they obtain the predicted count [12]. The ground truth density map is usually defined by manual annotation of head locations on training images, and by summing kernels (e.g., Gaussian) with unit area centred on the head locations, albeit some work uses custom implementation even for the density map construction [15]. During inference, the people count is estimated by the predicted density map's integration (pixel-wise sum). Since the density map has only positive numbers, crowd counting DNNs use ReLU activation on the last layer. Many DNNs have been proposed in the literature, each with innovative solutions for tackling specific aspects emerging in this complex task, i.e., multi-branch [23–25] or transformer-based [5,22] architectures.

Uncertainty for Crowd Counting. One of the most well-known solutions for dealing with uncertainty in DNNs is using Bayesian Inference [16]. Remarkably, the large DNN's hypothesis space makes, in practice, epistemic uncertainty quantification equivalent to model uncertainty quantification [10]; hence, one usually assumes a prior $p(w)$ on the network weights w. Since Bayesian approaches are intractable for deep architectures, an approximation $q(w)$ that minimises the divergence from the actual distribution $p(w)$ is employed. One can use such approximations in a *post hoc* or *ad hoc* setup [14]. On the one hand, an ad hoc technique consists in performing the Bayesian Inference during training: this will result in determining a custom learning policy that may or may not exploit the uncertainty during the learning process (e.g., by using a custom loss that incorporates the uncertainty measure). On the other hand, a post hoc technique act on top of an already trained model: the Bayesian inference exploits the model's parameters for building a distribution on top of them. Many solutions (initially proposed for other tasks) have been presented for solving crowd counting, including Monte-Carlo dropout [14], bootstrap ensemble [19], and Masksembles [21]. Regardless of the approximation, each work obtains a probability distribution on the output, which has been used to quantify the uncertainty on the density map and the total predicted count. More in detail, they usa a Gaussian approximation on the network outputs, estimating the mean and standard deviation (or variance), which are interpreted as the pixel-wise prediction and associated uncertainty.

3   BLUES Bayesian Inference

3.1   Limitations of the Existing Approaches

As we described in Sect. 2, many methods for estimating epistemic uncertainty assume a Gaussian prior for the network outputs, even if they cannot be negative. Although, in principle, this is usually considered an acceptable approximation, after analysing its behaviour in our previous work [14], we realised that sometimes it could be inaccurate; indeed, it happens that most of the Gaussian area

Fig. 1. Left: a crowd image from ShanghaiTech A data set; right: predicted density map, with "false positive" localisations highlighted in red (see Sect. 5 for more details). (Color figure online)

belongs to the negative values. We wondered whether such a prior choice could affect the quality of the estimated quantities, from the total count to the estimated density and uncertainty maps. This fact leads to our main concern: *does the choice of a Gaussian prior distribution to model crowd counting predictions need to be revised?* Interestingly and unexpectedly, after an in-depth analysis of this phenomenon, it turned out that image regions characterised by prior Gaussian distributions with most of the mass on negative values are often related to localisation errors, i.e., regions with null ground truth but with a non-null density estimate which, in our previous work [14], we referred to as "false positive regions". Figure 1 shows an example of such a phenomenon. Such an issue negatively affects the model performances because these regions will produce a wrong contribution to the involved estimated quantities. More precisely, it leads to predicting the presence of people in the density map where there is none, which also increases the estimated count and the value of the uncertainty measure.

3.2   A Change of Perspective on the Bayesian Inference

The solution we propose in this work to the above problem lies in a change of perspective that exploits task-specific information. Assuming we want to use a more suitable prior, we must consider a distribution with a null mass on the negative values, i.e., with support in $\mathbb{R}_0^+$. To this aim, we take into account the fact that the output units of regression-based crowd counting networks use the ReLU activation function; denoting their input and output as the random variables Z and Y, respectively, we have $Y = \max(Z, 0)$, and it is easy to see that the distribution of Y corresponds to the *rectified distribution* of Z, which is defined as follows:

$$\mathcal{P}_Y(y) = \mathcal{P}_Z^{\mathrm{R}}(y) = \begin{cases} \mathcal{P}_Z(y), & y > 0 \\ F_Z(0), & y = 0 \\ 0, & y < 0 \end{cases} \tag{1}$$

where F_Z is the cumulative density function of $\mathcal{P}_Z$. Unlike Y, Z is not constrained to non-negative values; as a result, Z can be modeled by distributions such as Gaussian. The perspective change we propose is shifting the Bayesian Inference from the rectified output to the neurons **Before the ReLU ES**timates (BLUES), which - by construction - corresponds to choosing a Rectified Gaussian distribution [20] as a prior for y, that can be written [2] as:

$$\mathcal{N}^{R}_{\mu,\sigma^2}(y) = \Phi(-\frac{\mu}{\sigma}) \cdot \delta(y) + \mathcal{N}_{\mu,\sigma^2}(y) \cdot H(y), \tag{2}$$

where Φ is the cumulative density function of the standard normal distribution, δ is the Dirac unit impulse and H is the Heaviside step function. Interestingly, knowing the parameters of the Gaussian on z, we can obtain the mean μ^R and the variance σ^R of the rectified Gaussian (which are the moments needed for obtaining the point estimate of the density and the uncertainty measure) analytically by using the results of [2].

3.3   Discussing the BLUES Approach

We discuss here two aspects of the proposed Bayesian strategy compared to the traditional one.

Obtaining a Segmentation Probability Map. Under the lens of a Bayesian perspective, one can interpret the negative area of the Gaussian before the ReLU estimates as the probability of the absence of people on the corresponding pixel: this is because whenever z assumes negative or null values y becomes zero, which is interpreted as "no people" on that pixel. One can use this probability value for constructing a segmentation map to filter out spurious ("false positive") localisations, setting the corresponding estimated density to zero, and, consequently, mitigating the problem mentioned in Sect. 3.1. To obtain such a segmentation map s is sufficient to compute the segmentation for each neuron z_i, which can be found by computing the cumulative density function $F_{z_i}(0)$ of its estimated Gaussian evaluated in zero:

$$s_i = F_{z_i}(0) = \int_{-\infty}^{0} \mathcal{N}_{\mu,\sigma^2}(z)\mathrm{d}z. \tag{3}$$

Figure 2 summarises the whole BLUES Bayesian Inference process.

The Issue of False Negative Localisations. By construction, when estimating a distribution after the ReLU activations, a pixel with zero density corresponds to a pixel with zero variance and hence, to null uncertainty, even if they may correspond to estimation errors (i.e., false negative localisations). This is because when estimating the moments of the distributions from a finite set of values (both when using a set of Monte-Carlo samples or a set of predictions from an ensemble), the only possible outcome corresponding to zero mean is a set of null predictions, which will also lead to zero variance. Consequently, when using traditional techniques, the detection of false negatives is, by construction,

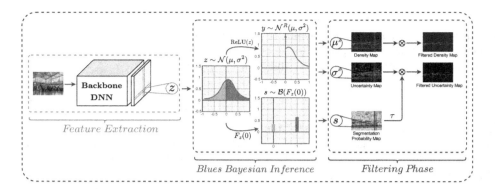

Fig. 2. Sketch of BLUES Bayesian Inference. The network processes an input image to estimate the crowd density map, until the last layer before the ReLU units. The input z of ReLU units is approximated by a Gaussian $\mathcal{N}(\mu, \sigma^2)$; their output y follows a Rectified Gaussian $\mathcal{N}^R(\mu, \sigma^2)$; the probability of the people presence (segmentation probability) s can be estimated as a Bernoulli distribution. By applying a threshold τ on s, "false positive" localisations (image regions with a positive estimated density but no people inside them) can be filtered out from the density and uncertainty maps.

impossible. Moreover, there is an additional concern: one can, generally, attain an accurate (or even exact) count estimate from an inaccurate density map; filtering out false positive regions, as we propose above, may increase the counting error, even if it would enhance the localisation accuracy. Such behaviour may happen when the false positives counteract the false negatives and the underestimated true positives, which is a common issue affecting crowd counting models [18]. Therefore, for exploiting the full potential of BLUES Bayesian Inference, it would be interesting to study the automatic detection of false-negative regions, which we will consider an interesting direction for future research.

4   Practical Implementation of BLUES Bayesian Inference

Choosing the Bayesian Approximation. As outlined in Sect. 2, one can choose among two strategies: 'ad hoc' and 'post hoc' Bayesian Inference. We implement BLUES Bayesian Inference using a post hoc strategy, particularly injected Monte-Carlo dropout, for two main reasons. First of all, by demonstrating its operation in a post hoc set-up we show that one can use such a strategy as a lightweight plug-in that operates on top of an already trained crowd counting network. In the second place, ad hoc techniques need a customised optimisation process (which usually involves designing a tailored loss function); such a study would require a focused analysis that deviates from the scope of our work, i.e., to demonstrate the effectiveness of the change of perspective itself, regardless of which optimisation process one uses for obtaining the network parameters.

Choosing a Policy for Filtering False Positive Regions. In Sect. 3.3, we pointed out that one of the main advantages of using a BLUES approach was obtaining a segmentation probability map, which one can use to filter out false positive regions. To this aim, we apply a threshold τ on each segmentation probability map's value s_i for constructing a binary 'segmentation mask' m (where, for each pixel m_i in m, $m_i = 1$ whenever $s_i < \tau$ and $m_i = 0$ otherwise), which we use for masking the density and uncertainty maps.

Adjusting the Injection Mechanism for BLUES. Injected dropout requires scaling the uncertainty measure using a scaling factor found jointly with the dropout rate [14]. In our case, the scaling factor C^R should be applied, at first, to the resulting uncertainty measure σ^R; indeed, without a proper scaling factor, *any* Bayesian inference method may be ineffective, including BLUES. In our implementation, after computing an optimal scaler C^R for σ^R, we use this scaler directly on σ; this step is fundamental, because an unscaled Gaussian on z, as a result, would not lead to correct segmentation probabilities $F_z(0)$. Finally, since applying a rejection policy may potentially (and should practically) alter the new resulting measure's optimal scaling factor, we compute an additional scaling factor C^M acting on the masked uncertainty maps σ^M. In short, the resulting process involves using two scaling factors, one (which is C^R) for the BLUES Inference uncertainty map σ^R before thresholding and one (which is C^M) for the masked uncertainty map σ^M.

5   Experimental Analysis

Model and Data Sets. We develop BLUES Bayesian Inferencein a post hoc set-up, following the implementation described in Sect. 4. We apply it to a representative and widely used crowd counting network architecture, i.e., Multi-Column Neural Network (MCNN) [25]. We conduct our experiments using four benchmark data sets: UCSD [3], Mall [4], PETS2009 [7] (divided in 3 sub-scenes according to [6]) and ShanghaiTech [25]. UCSD, PETS2009, and Mall are three *single-scene* data sets, meaning the training, validation, and test sets contain only images of the same scene; we refer to the experiments on these data sets as "single-scene experiments". ShanghaiTech is divided into two parts, A and B; the first part consists of images taken from the internet, whereas the second one was collected in the streets of Shanghai. Contrary to the previous ones, it is a *multi-scene* data set: in particular, each image comes from a different scene. Accordingly, we refer to the experiment on this data set as "multi-scene".

Metrics. As mentioned in Sect. 4, although the proposed technique can, in principle, improve the segmentation probability and density estimation map, it may negatively affect the estimated count. Indeed, even when the density map is inaccurate, the false negatives may counterbalance the false positives, giving a relatively accurate count prediction; in this case, filtering out only the false positives may worsen the counting accuracy. Accordingly, to quantitatively evaluate the quality of both the density map and the crowd count, we consider both 'counting' and 'spatial' metrics. We adopt two commonly used metrics to evaluate counting accuracy: the Mean Absolute Error (MAE) and the Root Mean Square Error (RMSE). To evaluate the quality of the density estimation map, we consider Grid Average Mean Absolute Error (GAME) [9], which consists in dividing the image into smaller patches and computing the mean absolute error of the estimated count for each patch individually using a predetermined resolution level L:

$$\text{GAME} = \frac{1}{4^L} \cdot \sum_{l=1}^{4^L} |y^l - \hat{y}^l|, \tag{4}$$

where y^l, $\hat{y}^l$ are the actual and the estimated count in the patch l. GAME computes the overall error in the estimated count as the average of local errors in each image patch: therefore, as the number of patches increases, this becomes a measure of localisation accuracy. In addition, we consider two other metrics which aim at evaluating the quality of the segmentation map, which are the Segmentation Accuracy and the Intersection Over Union (IoU) [17]:

$$\text{Accuracy} = (m \cap \hat{m}) \cup (\neg m \cap \neg \hat{m}) \quad \text{IoU} = \frac{m \cap \hat{m}}{m \cup \hat{m}}, \tag{5}$$

where m and $\hat{m}$ are the ground truth and estimated binary segmentation masks, respectively. The process for obtaining $\hat{m}$ is described in Sect. 4, whereas m is constructed by taking the ground truth density estimation map and setting each non-negative pixel to 1. It is worth noting that the spatial metrics (i.e., GAME, Accuracy, IoU) are defined for a single image and therefore the overall performances can be computed by averaging the results obtained over all the testing images. Finally, to evaluate the quality of the uncertainty measure, we adopt the commonly used Negative Log-Likelihood (NLL), [11]: in particular, since crowd counting is a regression task, we employed the continuous approximation of the NLL instead of its categorical definition [11], defined as follows:

$$\text{NLL} = \frac{1}{N} \sum_{n=1}^{N} \frac{1}{2} \cdot \frac{(y_n - \hat{y}_n)^2}{\sigma_n^2} + \frac{1}{2} \cdot \log \sigma_n^2, \tag{6}$$

where N, y_n, and $\hat{y}_n$ are, respectively, the test set size, the ground truth crowd count on the n-th image, and the estimated count for that image, and σ_n is the total uncertainty associated with the count prediction of the n-th image.

5.1   Experimental Results

Figures 3 and 4 show the overall results of the used metrics on the single- and multi-scene data sets, respectively, using nine different threshold values, $\tau = 0.1, \ldots, 0.9$. Figure 5 shows two examples of results from single-scene data sets.

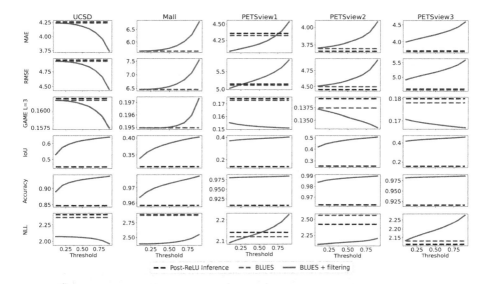

Fig. 3. Behaviour of MAE, RMSE, GAME (with L=3), IoU, Accuracy and NLL for the five single-scene data sets, attained by Post-ReLU-Estimates Bayesian Inference, and by BLUES followed by the filtering step as a function of the filtering threshold.

When using the filtering policy, we observe the MAE and RMSE increase in all cases (except for UCSD). As can be seen from the NLL trends, filtering also enhances the uncertainty measure's quality, outperforming the baseline in most cases; the only exceptions are PETSview1 (where the improvement is limited to small threshold values) and PETSview3. Unlike the counting metrics, the spatial metrics are almost always enhanced using BLUES with the filtering policy: they always improve the segmentation metrics, exhibiting similar trends, except for the GAME in Mall (where it improves only for small threshold values). We immediately notice analogies with the single-scene experiments by looking at the multi-scene results: BLUES Bayesian Inference leads to worse MAE and RMSE and better spatial metrics for parts A and B. The quality of the uncertainty measures improves for smaller threshold values, as shown by the decrease in NLL.

Discussion of the Results. Results on single- and multi-scene data sets provide evidence that BLUES Bayesian Inference can effectively filter out false positive regions, as expected. The improvement in the spatial metrics and the worse MAE values support our thesis that the network has learned to balance the errors caused by the false negative (or by underestimated true positives) with an increase in the false positive estimates during training. Indeed, if this were not the case (i.e. if the MAE increase were caused by filtering out true positives), we would observe a deterioration of the spatial metrics, but it is not the case.

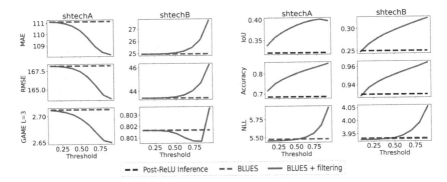

Fig. 4. Behaviour of MAE, RMSE, GAME (with L=3), IoU, Accuracy and NLL for ShanghaiTech parts A and B data sets, attained by Post-ReLU-Estimates Bayesian Inference, and by BLUES followed by the filtering step as a function of the filtering threshold.

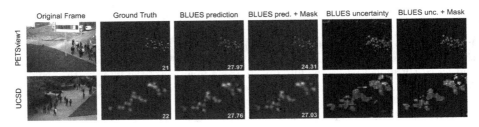

Fig. 5. From left to right: two examples of crowd images (from PETSview1 and UCSD), their ground truth density map, the BLUES prediction and the associated uncertainty, before and after applying the segmentation mask (color scales: ■ for ground truth and prediction maps, and ■ for the uncertainty maps).

We finally discuss the effect of the segmentation threshold τ. The above results show that higher values of τ allow to filter out an increasing amount of false positive localisations, thus improving localisation accuracy. On the other hand, increasing τ may worsen the counting accuracy if false negatives are balanced with false positives before filtering. A suitable trade-off between these effects is application-dependent (e.g., in some applications, an accurate localisation may be preferred over an accurate count, or vice versa), as well as data-dependent (as can be seen from the different behaviours exhibited by each metric on different data sets). In practice, a suitable value of τ should be determined, e.g., through a validation set, taking into account application requirements.

6   Conclusions, Limitations and Future Work

We addressed a problem of Bayesian methods for uncertainty quantification arising from their application to the crowd counting computer vision task, when neural network models based on crowd density estimation are used. We proposed to solve it by a change of perspective, performing Bayesian inference Before the ReLU Estimates (BLUES) instead of approximating the distribution of network outputs. The experimental results confirmed that BLUES Bayesian inference implemented in a post hoc set-up can improve the accuracy of the estimated density map and the corresponding uncertainty measure, allowing to filter out false positive localisations. On the other hand, filtering out false positives may worsen the counting accuracy, if they were balanced by false negative localisations. To address this issue, an interesting direction for future work is investigating how to integrate also false negatives detection into BLUES Bayesian inference. Finally, it would be interesting to develop a BLUES implementation in the alternative, ad hoc uncertainty quantification set-up, to embed its capability of improving the underlying network's spatial awareness into the learning process.

Acknowledgements. Supported by the projects: "Law Enforcement agencies human factor methods and Toolkit for the Security and protection of CROWDs in mass gatherings" (LETSCROWD), EU Horizon 2020 programme, grant agreement No. 740466; "IMaging MAnagement Guidelines and Informatics Network for law enforcement Agencies" (IMMAGINA), European Space Agency, ARTES Integrated Applications Promotion Programme, contract No. 4000133110/20/NL/AF; "Science and engineering Of Security of Artificial Intelligence" (S.O.S. AI) included in the Spoke 3 - Attacks and Defences of the Research and Innovation Program PE00000014, "SEcurity and RIghts in the CyberSpace (SERICS)", under the National Recovery and Resilience Plan, Mission 4 "Education and Research" - Component 2 "From Research to Enterprise" - Investment 1.3, funded by the European Union - NextGenerationEU.

Emanuele Ledda is affiliated with the Italian National PhD in Artificial Intelligence, Sapienza University of Rome. He also acknowledges the cooperation with and support from the Pattern Recognition and Applications Lab of the University of Cagliari.

References

1. Abdar, M., et al.: A review of uncertainty quantification in deep learning: techniques, applications and challenges. Inf. Fusion **76**, 243–297 (2021). https://doi.org/10.1016/j.inffus.2021.05.008
2. Beauchamp, M.: On numerical computation for the distribution of the convolution of N independent rectified Gaussian variables. J. Soc. Fr. Stat. **159**(1), 88–111 (2018)
3. Chan, A.B., Liang, Z.S.J., Vasconcelos, N.: Privacy preserving crowd monitoring: counting people without people models or tracking. In: CVPR, pp. 1–7 (2008). https://doi.org/10.1109/CVPR.2008.4587569
4. Chen, K., Loy, C.C., Gong, S., Xiang, T.: Feature mining for localised crowd counting. In: BMVC, pp. 1–11 (2012). https://doi.org/10.5244/C.26.21
5. Chen, Y., Yang, J., Chen, B., Du, S.: Counting varying density crowds through density guided adaptive selection CNN and transformer estimation. IEEE Trans. Circuits Syst. Video Technol. **33**(3), 1055–1068 (2023). https://doi.org/10.1109/TCSVT.2022.3208714
6. Delussu, R., Putzu, L., Fumera, G.: Investigating synthetic data sets for crowd counting in cross-scene scenarios. In: VISIGRAPP, pp. 365–372 (2020). https://doi.org/10.5220/0008981803650372

7. Ferryman, J., Shahrokni, A.: PETS2009: dataset and challenge. In: 2009 Twelfth IEEE International Workshop on Performance Evaluation of Tracking and Surveillance, pp. 1–6 (2009). https://doi.org/10.1109/PETS-WINTER.2009.5399556

8. Gal, Y., Ghahramani, Z.: Dropout as a bayesian approximation: representing model uncertainty in deep learning. In: ICML. JMLR Workshop and Conference Proceedings, vol. 48, pp. 1050–1059 (2016)

9. Guerrero-Gómez-Olmedo, R., Torre-Jiménez, B., López-Sastre, R.J., Maldonado-Bascón, S., Oñoro-Rubio, D.: Extremely overlapping vehicle counting. In: IbPRIA. Lecture Notes in Computer Science, vol. 9117, pp. 423–431 (2015). https://doi.org/10.1007/978-3-319-19390-8_48

10. Hüllermeier, E., Waegeman, W.: Aleatoric and epistemic uncertainty in machine learning: an introduction to concepts and methods. Mach. Learn. 110(3), 457–506 (2021). https://doi.org/10.1007/s10994-021-05946-3

11. Kendall, A., Gal, Y.: What uncertainties do we need in bayesian deep learning for computer vision? In: Guyon, I., etal (eds.) Advances in Neural Information Processing Systems, vol. 30, pp. 5574–5584 (2017)

12. Khan, M.A., Menouar, H., Hamila, R.: Revisiting crowd counting: state-of-the-art, trends, and future perspectives. Image Vis. Comput. 129, 104597 (2023). https://doi.org/10.1016/j.imavis.2022.104597

13. Lakshminarayanan, B., Pritzel, A., Blundell, C.: Simple and scalable predictive uncertainty estimation using deep ensembles. In: Advances in Neural Information Processing Systems, vol. 30, pp. 6402–6413 (2017)

14. Ledda, E., Fumera, G., Roli, F.: Dropout injection at test time for post hoc uncertainty quantification in neural networks. CoRR abs/2302.02924 (2023). https://doi.org/10.48550/arXiv.2302.02924

15. Ma, Z., Wei, X., Hong, X., Gong, Y.: Bayesian loss for crowd count estimation with point supervision. In: ICCV, pp. 6141–6150 (2019). https://doi.org/10.1109/ICCV.2019.00624

16. MacKay, D.J.C.: A practical bayesian framework for backpropagation networks. Neural Comput. 4(3), 448–472 (1992). https://doi.org/10.1162/neco.1992.4.3.448

17. Minaee, S., Boykov, Y., Porikli, F., Plaza, A., Kehtarnavaz, N., Terzopoulos, D.: Image segmentation using deep learning: a survey. IEEE Trans. Pattern Anal. Mach. Intell. 44(7), 3523–3542 (2022). https://doi.org/10.1109/TPAMI.2021.3059968

18. Modolo, D., Shuai, B., Varior, R.R., Tighe, J.: Understanding the impact of mistakes on background regions in crowd counting. In: WACV, pp. 1649–1658 (2021). https://doi.org/10.1109/WACV48630.2021.00169

19. Oh, M.H., Olsen, P., Ramamurthy, K.N.: Crowd counting with decomposed uncertainty. In: AAAI, pp. 11799–11806 (2020). https://doi.org/10.1609/aaai.v34i07.6852

20. Socci, N., Lee, D., Seung, H.S.: The rectified gaussian distribution. In: Advances in Neural Information Processing Systems, vol. 10. MIT Press (1997)

21. Wang, X., Zhan, Y., Zhao, Y., Yang, T., Ruan, Q.: Semi-supervised crowd counting with spatial temporal consistency and pseudo-label filter. IEEE Trans. Circuits Syst. Video Technol. 33, 1–1 (2023). https://doi.org/10.1109/TCSVT.2023.3241175

22. Yuan, L., Chen, Y., Wu, H., Wan, W., Chen, P.: Crowd counting via localization guided transformer. Comput. Electr. Eng. 104(Part), 108430 (2022). https://doi.org/10.1016/j.compeleceng.2022.108430

23. Zhai, W., Gao, M., Li, Q., Jeon, G., Anisetti, M.: FPANet: feature pyramid attention network for crowd counting. Appl. Intell., 1–18 (2023). https://doi.org/10.1007/s10489-023-04499-3

24. Zhang, X., et al.: A multi-scale feature fusion network with cascaded supervision for cross-scene crowd counting. IEEE Trans. Instrum. Meas. **72**, 1–15 (2023). https://doi.org/10.1109/TIM.2023.3246534

25. Zhang, Y., Zhou, D., Chen, S., Gao, S., Ma, Y.: Single-image crowd counting via multi-column convolutional neural network. In: CVPR, pp. 589–597 (2016). https://doi.org/10.1109/CVPR.2016.70

Two is Better than One: Achieving High-Quality 3D Scene Modeling with a NeRF Ensemble

Francesco Di Sario[1]([✉])[ID], Riccardo Renzulli[1][ID], Enzo Tartaglione[2][ID], and Marco Grangetto[1][ID]

[1] University of Turin, Turin, Italy
`francesco.disario@unito.it`
[2] LTCI, Télécom Paris, Institut Polytechnique de Paris, Palaiseau, France

Abstract. Neural Radiance Field (NeRF) is a popular method for synthesizing novel views of a scene from a set of input images. While NeRF has demonstrated state-of-the-art performance in several applications, it suffers from high computational requirements. Recent works have attempted to address these issues by including explicit volumetric information, which makes the optimization process difficult when fine-graining the voxel grids. In this paper, we propose an ensemble approach that combines the strengths of two NeRF models to achieve superior results compared to state-of-the-art architectures, with a similar number of parameters. Experimental results show that our ensemble approach is a promising strategy for performance enhancement, and beats vanilla approaches under the same parameter's cardinality constraint.

Keywords: NeRF · Ensemble · 3D scene modeling · Compression

1 Introduction

Neural Radiance Fields (NeRFs) [16] have recently shown impressive results in synthesizing photo-realistic 3D scenes from a set of 2D images. However, NeRF suffers from limited scene diversity, long training time, and sensitivity to training data [14]. To address these issues, recent works have proposed several improvements to the original NeRF framework, such as NeRF++, which extends NeRF to unbounded scenes [26].

Another promising approach to improve NeRF, yet to be explored, is ensembling. Ensemble methods combine multiple models to achieve better performance than a single model. In the context of NeRF, ensembling can be achieved by training multiple NeRF models on different subsets of the training data, or by training different models with different architectures or hyperparameters. Ensembling has been shown to be effective in improving the performance of various computer vision tasks, such as image classification [8] and object detection [12]. One of the key advantages of ensemble methods is their ability to combine the predictions of multiple models to produce a more accurate and robust prediction.

This is particularly useful when the individual models have different strengths and weaknesses, as the ensemble can leverage the strengths of each model while mitigating their weaknesses.

However, there are several challenges associated with ensemble methods that can limit their effectiveness. For example, how can we select the appropriate combination of models in the ensemble? This is particularly difficult when there are a large number of potential models to choose from, or when the individual models are highly correlated with each other. Another challenge is how to effectively combine the predictions of the individual models, particularly when they have different levels of accuracy or confidence.

In this work, we explore the potentiality of NeRF ensembling to improve performances. More specifically, we adopt a baseline, state-of-the-art architecture, DVGO [22], trained on a very well-known dataset, Synthetic-NeRF [16]. We observe that, by employing a vanilla ensembling strategy of two models, we may obtain suboptimal results. We propose a simple yet effective solution to counter it, observing consistent performance improvement, with respect to the baseline models, on a broad variety of tested resolutions, *under the same memory footprint constraints*. This paper aims at moving the first steps towards the definition of an ultimate, highly-performing, efficient NeRF ensembling strategy. At a glance, the contributions of this work are the following:

- To the best of our knowledge, this is the first work proposing a joint ensembling and compression scheme for NeRF models: a formulation to prevent performance degradation in case of conjoint pruning is proposed.
- We observe, on known benchmarks, that ensembling multiple models at different scales requires fewer parameters than training and compressing one large model directly (under the same generated image quality constraint).

2 Related Works

Neural Radiance Field (NeRF) [16] stand out in recent years as the most prevalent method for novel view rendering that infers photo-realistic views given a moderate number of input images. Unlike traditional explicit volumetric representation techniques, NeRF encodes the entire content of the scene including view-dependent color emission and density into a single multi-layer perceptron (MLP) [16]. Besides, NeRF-based approaches are proving on the field to have good generalization when undergoing several transformations, like changing environmental light [1,21], image deformation [6,18,24] and are even usable in more challenging setups including meta learning [23], learn dynamically-changing scenes [7,11,15,25] and even in generative contexts [2,9,20]. Compared to explicit representations, NeRF requires very little storage space, but on the contrary suffers from lengthy training time and very slow rendering speed, as the MLP is queried an extremely high number of times for rendering a single image. In more detail, a NeRF takes as input a 3D point in space x and a viewing direction d and returns a color c and a density σ. It utilizes volume rendering techniques to achieve advanced 3D reconstruction: given a camera and

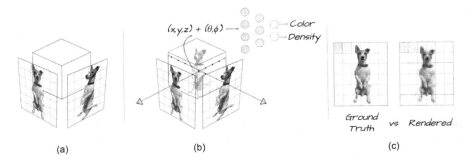

Fig. 1. An overview of NeRF training is presented. (a) illustrates the initial training setup, made of a sparse set of calibrated images of the same object under different viewpoints conditions. (b) shows the training process made of ray casting, ray sampling, and volume rendering to compute the pixel color. Then the generated image is compared to the ground truth (c).

a sparse set of calibrated images capturing the scene from various viewpoints, the rendering process involves casting a ray from the camera's eye to the center of a pixel, sampling $x_1, ..., x_k$ points along the ray and evaluating those points, obtaining a color, c, and a density value, σ. The final pixel color, $\hat{c}$, is determined by alpha-blending all the computed color values $(c_1, \cdots, c_k)$ along the ray

$$\hat{c} = \sum_{i=1}^{K} T_i \alpha_i c_i \qquad T_i = \prod_{j=1}^{i-1} (1 - \alpha_j) \qquad \alpha_i = 1 - \exp(\sigma_i \delta_i), \qquad (1)$$

where α_i is the value used for blending the colors values (its calculation depends on the distance between adjacent sampling points $\delta = ||x_{j+1} - x_j||$ and T_i is the transmittance. This process is repeated for each pixel in the image. Once the final image is generated, it is compared to the ground truth using the photometric loss (2), and the parameters of the multilayer perceptron are optimized through backpropagation. Figure 1 summarizes this process.

To reduce inference and training time, explicit prior on the 3D object representation can be imposed. The most intuitive yet effective approach relies on splitting the 3D volume into small blocks, each of which is learned by a tiny NeRF model. With KiloNeRF [19], the advantage of doing this is twofold. Firstly, the size of a single NeRF model is much smaller than the original one, reducing the latency time. Secondly, the rendering process itself becomes parallelizable, as multiple pixels can be rendered simultaneously. The downside of this approach is that the granularity of the KiloNeRFs needs to be properly tuned, and the proper tuning process can be time-consuming and require significant computational resources. Additionally, KiloNeRFs may struggle to capture fine details and high-frequency variations in the input data, which can lead to inaccurate reconstructions. To address these limitations, researchers have proposed several extensions and variations of the original NeRF and KiloNeRF models. For instance, some works have explored the use of hierarchical or multi-scale

Algorithm 1. NeRF ensemble training algorithm.

Require: Training set $\mathcal{D}_{\text{train}}$, validation set $\mathcal{D}_{\text{val}}$, ensemble of 2 NeRFs
 1: **Stage 1:** Train 2 NeRFs independently on $\mathcal{D}_{\text{train}}$
 2: **while** Performance does not drop **do**
 3: **Stage 2 (Ens-FT):** Fine-tune the ensemble of NeRFs
 4: **Stage 3 (CPE):** Compress the ensemble, using $\mathcal{D}_{\text{val}}$
 5: **end while**
 6: **return** Ensemble of NeRFs

representations to better capture details at different levels of the scene [27,28]. Others have investigated the incorporation of additional priors or constraints, such as symmetry or smoothness assumptions, to improve the robustness and generalization of the models [13,16]. In parallel, the development of NeRFs with direct voxel grid optimization is gaining more and more success. Direct Voxel Grid Optimization (DVGO) [22] is a popular baseline for NeRFs due to its simplicity and effectiveness. In contrast to traditional NeRFs, which use a continuous representation of the scene, DVGO operates directly on a voxel grid. This makes DVGO much more computationally efficient than NeRFs, as it allows for parallelization of the ray-marching process and significantly reduces the number of samples required to render an image. Additionally, DVGO is less prone to overfitting and can handle more complex scenes with higher levels of detail. Despite its limitations in terms of scalability, DVGO provides a strong and reliable baseline for evaluating the performance of more advanced methods such as NeRFs. Moreover, it has been shown that by training a NeRF with initialization from a DVGO model, the NeRF can achieve comparable performance while requiring significantly less training time and computational resources. Therefore, DVGO remains a useful and widely used baseline for testing and comparing novel techniques for 3D scene representation and rendering.

For this reason, in the present study, we have opted to employ the DVGO architecture as a reference NeRF framework. While hybrid models, also known as Explicit Voxel Grid models, exhibit superior efficiency and increased accuracy, they necessitate substantial storage capacity, typically on the order of gigabytes. As a result, some voxel pruning techniques have recently emerged with the goal of minimizing storage requirements for these models. Re:NeRF [4] represents a state-of-the-art approach for compressing EVG NeRFs and stands among the forefront methods in reducing storage demands.

3 Method

In this section, we present our framework, which allows for the combination of two NeRF models of different grid resolutions to improve performance on novel view synthesis tasks. A general overview of the employed strategy is presented in Algorithm 1. Our framework consists of three stages:

- independent training of multiple NeRF models (Stage 1);
- ensemble construction and fine-tuning (Stage 2);
- conjoint pruning phase of the ensembled NeRFs (Stage 3).

Independent Training of NeRF Models (Stage 1). In the first stage, we train multiple NeRF models independently, each with a different grid resolution. For each of them, we train the model for a fixed number of iterations, according to the standard learning policy defined, minimizing the NeRF loss function:

$$\mathcal{L}(\boldsymbol{w}) = \frac{1}{M} \sum_{m=1}^{M} \|C_m - C(\boldsymbol{w}, x_m, \omega_m)\|_2^2, \tag{2}$$

where $\boldsymbol{w}$ represents the model parameters, C_m is the ground truth color of the m-th pixel, $C(\boldsymbol{w}, x_m, \omega_m)$ is the predicted color by the NeRF for the same pixel, x_m is the 3D point where the pixel lies, ω_m is the corresponding viewing direction and M is the total number of training samples. After the whole training process ends, as observed in some recent work in the literature [4], the size of the generated models can be drastically reduced, with marginal or even no impact on the performance. This optional stage employs an iterative pruning strategy, followed by quantization and entropy coding, on the models at isolation: we will name it independent pruning of models (IPM).

Ensemble Construction and Ensemble Fine-Tuning (Stage 2). Once multiple NeRF models have been trained or compressed, we construct an ensemble by combining them. Specifically, we simply perform an interpolation for the outputs

$$C_m^{\text{avg}} = \frac{1}{2} \sum_{n=1}^{2} C^n(\boldsymbol{w}^n, x_m, \omega_m), \tag{3}$$

where C^2 indicates the output of the ensemble of 2 NeRFs, C^n is the output of the n-th NeRF, and $\boldsymbol{w}^n$ are the parameters of the n-th NeRF.

After constructing the ensemble, a fine-tuning stage follows. Specifically, we observe that by optimizing the output provided by (3), we have

$$\mathcal{L}^{\text{avg}}(\boldsymbol{w}) = \frac{1}{M} \sum_{m=1}^{M} \left\| C_m - \frac{1}{2} \sum_{n=1}^{2} C^n(\boldsymbol{w}^n, x_i, \omega_i) \right\|_2^2. \tag{4}$$

Conjoint Pruning of the Ensemble (Stage 3). The final phase consists in jointly pruning the ensemble. We refer to this phase as conjoint pruning of the ensemble (CPE). This phase targets superior performance compared to ensembling pre-trained models, with comparable (sometimes even lower) memory footprint. The optimization of the loss function needs to be carefully reconsidered as it may lead to suboptimal results as the pruning process progresses. Indeed,

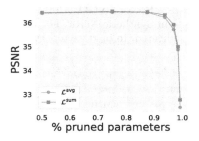

Fig. 2. Comparison between $\mathcal{L}^{\mathrm{sum}}$ and $\mathcal{L}^{\mathrm{avg}}$ losses on Lego Dataset. Each point is an average of the six ensemble configurations tested. It is important to note that as the pruning progresses, the sum approach tends to yield better performance.

according to (4), the ensemble's output is computed by averaging the output of the ensemble models. In the scenario where the models are pre-trained, it is generally not a concern, and optimizing them based on this loss function can yield great results. Nevertheless, in the case of conjoint pruning, optimizing $\mathcal{L}^{\mathrm{avg}}$ could be quite problematic. In particular, due to the pruning process, C^1 and/or C^2 (the output of the models of the ensemble) may contain multiple null values (or values close to zero), depending on the compression rate. Therefore, it is evident that an averaging of the ensemble models in such a scenario would result in a significant reduction of the signal output of both models, regardless of the pruning phase. We can avoid this problem by simply adopting, as the output of our ensemble,

$$C_i^2 = \sum_{n=1}^{2} C^n(\boldsymbol{w}^n, x_i, \omega_i), \tag{5}$$

which leads to the minimization of the following loss function

$$\mathcal{L}^{\mathrm{sum}}(\boldsymbol{w}) = \frac{1}{M} \sum_{m=1}^{M} \left\| C_m - \sum_{n=1}^{2} C^n(\boldsymbol{w}^n, x_i, \omega_i) \right\|_2^2 . \tag{6}$$

In Fig. 2, it can be observed that optimizing $\mathcal{L}^{\mathrm{sum}}$ achieves better performances than optimizing $\mathcal{L}^{\mathrm{avg}}$.

4   Experiments

In this section, we present the empirical results obtained on the Synthetic-NeRF [16] dataset. It contains eight different realistic objects created with Blender(*chair, drums, ficus, hotdog, lego, materials, mic* and *ship*), which are synthesized from NeRF.

4.1   Setup

The target image resolution has been set up to 800×800 pixels, having 100 views for training, 100 for validation, and 200 for testing. We choose DVGO [22] as a

Table 1. Results on Synthetic-NeRF for DVGO. All the presented results are averaged on the eight different datasets in Synthetic-NeRF. For the ensemble, the models have resolution 160^3 and the one indicated. In bold we report the best values, while in italic the second best ones.

Metric	IPM	ENS-FT	CPE	Compress	160^3	170^3	180^3	190^3	200^3	256^3
PSNR(↑)					31.813	31.939	32.063	32.158	32.270	32.751
		✓			**32.821**	**33.061**	**33.113**	**33.183**	**33.265**	**33.509**
	✓			LOW	31.801	31.909	32.240	32.313	32.421	32.644
				HIGH	31.397	31.545	31.875	31.963	32.071	32.299
	✓	✓		LOW	32.431	32.846	32.913	32.948	33.042	33.200
				HIGH	31.858	32.229	32.329	32.399	32.491	32.768
		✓	✓	LOW	*32.724*	*32.938*	*32.942*	*33.002*	*33.084*	*33.302*
				HIGH	32.110	32.430	32.424	32.523	32.617	32.899
SSIM(↑)					0.955	0.956	0.957	*0.958*	*0.958*	*0.961*
		✓			**0.958**	**0.960**	**0.960**	**0.960**	**0.961**	**0.963**
	✓			LOW	0.953	0.954	0.956	0.956	0.957	0.958
				HIGH	0.945	0.946	0.948	0.949	0.950	0.951
	✓	✓		LOW	0.953	0.956	0.957	0.957	0.957	0.957
				HIGH	0.941	0.945	0.946	0.947	0.947	0.949
		✓	✓	LOW	*0.956*	*0.957*	*0.958*	*0.958*	*0.958*	0.959
				HIGH	0.943	0.946	0.948	0.950	0.956	0.958
LPIPS(↓)					0.036	*0.035*	0.034	*0.033*	*0.032*	**0.026**
		✓			**0.032**	**0.030**	**0.030**	**0.029**	**0.029**	**0.026**
	✓			LOW	0.036	*0.035*	*0.033*	*0.033*	0.032	*0.030*
				HIGH	0.047	0.046	0.043	0.042	0.041	0.040
	✓	✓		LOW	0.040	0.035	0.034	0.034	0.034	0.033
				HIGH	0.056	0.051	0.050	0.049	0.049	0.045
		✓	✓	LOW	*0.035*	*0.035*	0.034	0.034	0.034	0.033
				HIGH	0.054	0.052	0.051	0.050	0.050	0.047
SIZE(MB)(↓)					634.44	766.78	907.23	1074.01	1248.80	2619.70
		✓			1274.62	1401.79	1545.86	1706.12	1882.07	3334.06
	✓			LOW	4.67	5.35	6.13	6.99	8.80	13.55
				HIGH	**2.49**	**2.85**	**3.25**	**3.65**	**4.54**	**6.97**
	✓	✓		LOW	7.22	8.15	8.78	9.44	10.11	14.98
				HIGH	4.28	4.20	4.52	4.85	5.19	7.74
		✓	✓	LOW	7.37	7.84	8.46	9.08	9.77	14.37
				HIGH	*3.93*	*4.19*	*4.50*	*4.83*	*5.17*	*7.57*

reference architecture, and we adopt the original paper's learning strategy and hyperparameters configuration. We propose standard image generation quality metrics like PSNR, SSIM, and LPIPS (computed on AlexNet).

Besides, we compare the various results in terms of the size (in MB) of the model compressed by Re:NeRF. We conduct experiments at different voxel grid resolutions: 160^3, 170^3, 180^3, 190^3, 200^3, and 256^3. According to the proposed approach, first, we train (and compress) several models at different resolutions; then, we construct all the possible ensembles of two models combining the lowest resolution (160^3) with all the available resolutions (from 160^3 up to

256^3). As a standard pruning, quantization, and compression approach, we adopt Re:NeRF [4]. Our code is developed using PyTorch 1.12, and the experiments are performed on an NVIDIA A40 GPU.[1]

4.2   Results

Table 1 reports the results achieved on Synthetic-NeRF. The table consists of four macro-sections, each corresponding to a reference measure (namely PSNR, SSIM, LPIPS, and SIZE). Analyzing each macro section, we observe the following: in the first row, the performance of the baseline models; in the second row (ENS-FT), the performance of the fine-tuned ensemble; in the third row (IPM), the performance of individually compressed models at two rates, low and high (corresponding to 87.50% and 96.87% of the total parameters, respectively); in the fourth row (IPM + ENS-FT), the performance of the pre-compressed models in the ensemble, and finally, the last row of each macro section shows the performance of the conjoint pruning of the ensemble (CPE).

Please consider that every entry of the table is an average of eight models, trained on the eight datasets collected within Synthetic-NeRF. Consistently, we observe that, under the same resolution constraint, the proposed ensemble approach performs the best. More specifically, we observe a minor degradation of the performance as the compression regime increases (as also indicated in [4]). However, when investigating the model size, we observe that compressing the ensemble can sensibly reduce its size, making it drop from order GB to a few

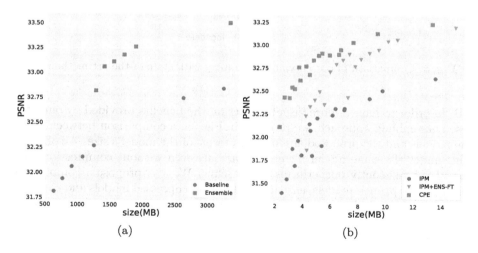

(a)   (b)

Fig. 3. (a) Comparison between baseline and ensembling in terms of PSNR and memory footprint. (b) Comparison among individually pruned models (IPM), IPM followed by one fine-tuning stage in ensembling (IPM+ENS-FT), and conjoint pruned ensemble (CPE).

---

[1] https://github.com/EIDOSLAB/nerf-ensemble-two-is-better-than-one.

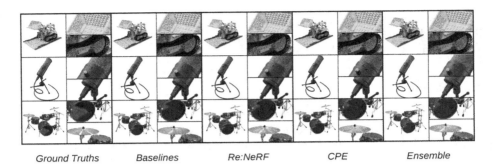

Ground Truths   Baselines   Re:NeRF   CPE   Ensemble

Fig. 4. Qualitative results. As a baseline, we adopted a grid of 256^3 voxels, while for the ensemble, two grids with dimensions of 160^3 and 256^3 voxels were used.

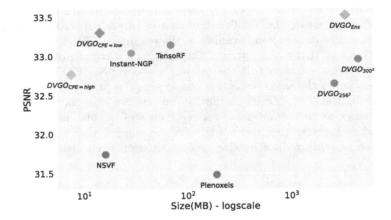

Fig. 5. Comparison of our ensembling strategy with state-of-the-art methods.

MB. In order to have a more visual impact on the benefits provided by our proposed ensembling approach, we propose, in Fig. 3a, a comparison between baseline models and the proposed ensemble, in terms of the model's size. We observe that, under the same model memory footprint, even without compression, the ensemble consistently outperforms the baseline. We also propose a comparison among single pruned models, ensemble with pre-compressed models and conjoint pruned ensemble in Fig. 3b. Also in this case the ensembling shows a consistent performance improvement, despite consuming a comparable amount of memory. In Fig. 4 a qualitative comparison between the analyzed configurations is proposed.

Figure 5 presents a comparison between our ensembling strategy and several state-of-the-art hybrid methods, such as Plenoxels [5], NSVF [10], Instant-NGP [17] and TensoRF [3]. Similarly, in this case, our method has proven to be reliable, surpassing the current state-of-the-art in terms of both quality and memory footprint. In Fig. 6 we propose a study on the Lego dataset, in which we

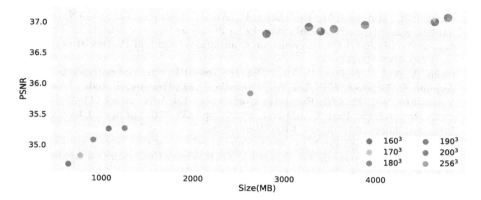

Fig. 6. Results on different ensembling resolutions on Lego dataset. Each pie chart represents an ensemble composed of a number of models equal to the number of slices.

investigate various ensemble resolutions of up to 6 models. Our findings reveal that even a combination of just two models can result in a significant performance improvement of over 1 dB. While incorporating more than two models can lead to even greater performances, this approach also results in highly complex models with significantly more parameters and memory footprint.

5   Conclusion

In this work, we explored the potential of NeRF ensembling to improve performance. Specifically, we have sided a low-resolution architecture to a higher one. Besides, a compression strategy, siding the ensemble, creates the perfect synergy for extracting the best performance out of a restricted number of parameters. We have observed consistent performance improvements on a broad variety of tested resolutions, under the same number of parameters. Our results demonstrate that ensembling can be a promising approach to improving NeRF performance, and further exploration of this method will be conducted in the next future. However, there are still several challenges associated with ensembling that need to be addressed. For example, how to select the appropriate combination of models in the ensemble and how to effectively combine their predictions. Overall, our work represents a step toward the development of an ultimate, highly-performing, and efficient NeRF ensembling strategy. Future research in this area could focus on addressing the challenges associated with ensembling and exploring more advanced techniques to improve performance.

References

1. Boss, M., Braun, R., Jampani, V., Barron, J.T., Liu, C., Lensch, H.: NeRD: neural reflectance decomposition from image collections. In: Proceedings of the IEEE/CVF International Conference on Computer Vision, pp. 12684–12694 (2021)

2. Chan, E.R., Monteiro, M., Kellnhofer, P., Wu, J., Wetzstein, G.: pi-GAN: periodic implicit generative adversarial networks for 3D-aware image synthesis. In: Proceedings of the IEEE/CVF Conference on Computer Vision and Pattern Recognition, pp. 5799–5809 (2021)

3. Chen, A., Xu, Z., Geiger, A., Yu, J., Su, H.: TensoRF: tensorial radiance fields. In: Avidan, S., Brostow, G., Cissé, M., Farinella, G.M., Hassner, T. (eds.) Computer Vision-ECCV 2022: 17th European Conference, Tel Aviv, Israel, 23–27 October 2022, Proceedings, Part XXXII, vol. 13692, pp. 333–350. Springer, Cham (2022). https://doi.org/10.1007/978-3-031-19824-3_20

4. Deng, C.L., Tartaglione, E.: Compressing explicit voxel grid representations: fast NeRFs become also small. In: 2023 IEEE/CVF Winter Conference on Applications of Computer Vision (WACV), pp. 1236–1245 (2023)

5. Fridovich-Keil, S., Yu, A., Tancik, M., Chen, Q., Recht, B., Kanazawa, A.: Plenoxels: radiance fields without neural networks. In: Proceedings of the IEEE/CVF Conference on Computer Vision and Pattern Recognition, pp. 5501–5510 (2022)

6. Gafni, G., Thies, J., Zollhofer, M., Nießner, M.: Dynamic neural radiance fields for monocular 4D facial avatar reconstruction. In: Proceedings of the IEEE/CVF Conference on Computer Vision and Pattern Recognition, pp. 8649–8658 (2021)

7. Gao, C., Saraf, A., Kopf, J., Huang, J.B.: Dynamic view synthesis from dynamic monocular video. In: Proceedings of the IEEE/CVF International Conference on Computer Vision, pp. 5712–5721 (2021)

8. He, K., Zhang, X., Ren, S., Sun, J.: Deep residual learning for image recognition. In: Proceedings of the IEEE Conference on Computer Vision and Pattern Recognition, pp. 770–778. IEEE (2016)

9. Kosiorek, A.R., et al.: NeRF-VAE: a geometry aware 3D scene generative model. In: International Conference on Machine Learning, pp. 5742–5752. PMLR (2021)

10. Li, G., et al.: Neural volumetric rendering for metal microstructure design. ACM Trans. Graph. (TOG) **39**(4), 1–14 (2020)

11. Li, Z., Niklaus, S., Snavely, N., Wang, O.: Neural scene flow fields for space-time view synthesis of dynamic scenes. In: Proceedings of the IEEE/CVF Conference on Computer Vision and Pattern Recognition, pp. 6498–6508 (2021)

12. Liu, W., et al.: SSD: single shot multibox detector. In: Leibe, B., Matas, J., Sebe, N., Welling, M. (eds.) Computer Vision – ECCV 2016. ECCV 2016. LNCS, vol. 9905, pp. 21–37. Springer, Cham (2016). https://doi.org/10.1007/978-3-319-46448-0_2

13. Liu, Y., Liu, S., Xu, K.: Point2Surf: learning implicit surfaces from point clouds with a multiscale feature network. In: Proceedings of the IEEE/CVF Conference on Computer Vision and Pattern Recognition (CVPR) (2021)

14. Liu, Y., Li, X., Yu, F., Zhou, Q.: Probabilistic neural scene representations. In: Proceedings of the IEEE/CVF Conference on Computer Vision and Pattern Recognition (2021)

15. Martin-Brualla, R., Radwan, N., Sajjadi, M.S., Barron, J.T., Dosovitskiy, A., Duckworth, D.: NeRF in the wild: neural radiance fields for unconstrained photo collections. In: Proceedings of the IEEE/CVF Conference on Computer Vision and Pattern Recognition, pp. 7210–7219 (2021)

16. Mildenhall, B., Srinivasan, P.P., Tancik, M., Barron, J.T., Ramamoorthi, R., Ng, R.: NeRF: representing scenes as neural radiance fields for view synthesis. In: Vedaldi, A., Bischof, H., Brox, T., Frahm, J.-M. (eds.) ECCV 2020. LNCS, vol. 12346, pp. 405–421. Springer, Cham (2020). https://doi.org/10.1007/978-3-030-58452-8_24

17. Müller, T., Evans, A., Schied, C., Keller, A.: Instant neural graphics primitives with a multiresolution hash encoding. ACM Trans. Graph. **41**(4), 102:1–102:15 (2022)

18. Noguchi, A., Sun, X., Lin, S., Harada, T.: Neural articulated radiance field. In: Proceedings of the IEEE/CVF International Conference on Computer Vision, pp. 5762–5772 (2021)

19. Reiser, C., Peng, S., Liao, Y., Geiger, A.: KiloNeRF: speeding up neural radiance fields with thousands of tiny MLPs. In: Proceedings of the IEEE/CVF International Conference on Computer Vision, pp. 14335–14345 (2021)

20. Schwarz, K., Liao, Y., Niemeyer, M., Geiger, A.: GRAF: generative radiance fields for 3D-aware image synthesis. Adv. Neural. Inf. Process. Syst. **33**, 20154–20166 (2020)

21. Srinivasan, P.P., Deng, B., Zhang, X., Tancik, M., Mildenhall, B., Barron, J.T.: NeRV: neural reflectance and visibility fields for relighting and view synthesis. In: Proceedings of the IEEE/CVF Conference on Computer Vision and Pattern Recognition, pp. 7495–7504 (2021)

22. Sun, C., Sun, M., Chen, H.T.: Direct voxel grid optimization: super-fast convergence for radiance fields reconstruction. In: Proceedings of the IEEE/CVF Conference on Computer Vision and Pattern Recognition, pp. 5459–5469 (2022)

23. Tancik, M., et al.: Learned initializations for optimizing coordinate-based neural representations. In: Proceedings of the IEEE/CVF Conference on Computer Vision and Pattern Recognition, pp. 2846–2855 (2021)

24. Tretschk, E., Tewari, A., Golyanik, V., Zollhöfer, M., Lassner, C., Theobalt, C.: Non-rigid neural radiance fields: reconstruction and novel view synthesis of a dynamic scene from monocular video. In: Proceedings of the IEEE/CVF International Conference on Computer Vision, pp. 12959–12970 (2021)

25. Xian, W., Huang, J.B., Kopf, J., Kim, C.: Space-time neural irradiance fields for free-viewpoint video. In: Proceedings of the IEEE/CVF Conference on Computer Vision and Pattern Recognition, pp. 9421–9431 (2021)

26. Yen, E.C.T., et al.: NeRF++: analyzing and improving neural radiance fields. arXiv preprint arXiv:2010.07492 (2021)

27. Yen, Y., Liu, Z., Mitra, N.J.: Multiscale neural voxelization for high-resolution 3D object representation. In: Proceedings of the IEEE/CVF Conference on Computer Vision and Pattern Recognition, pp. 11632–11641 (2021)

28. Zhang, R., et al.: Neural radiance flow for 4D view synthesis and video processing. In: Proceedings of the IEEE/CVF Conference on Computer Vision and Pattern Recognition, pp. 6914–6924 (2021)

BHAC-MRI: Backdoor and Hybrid Attacks on MRI Brain Tumor Classification Using CNN

Muhammad Imran[1(✉)], Hassaan Khaliq Qureshi[1], and Irene Amerini[2]

[1] National University of Sciences and Technology (NUST), Islamabad, Pakistan
{mimran.msee18seecs,hassaan.khaliq}@seecs.edu.pk
[2] Sapienza University of Rome, Rome, Italy
amerini@diag.uniroma1.it

Abstract. Deep learning (DL) models have demonstrated impressive performance in sensitive medical applications such as disease diagnosis. However, a backdoor attack embedded in the clean dataset without poisoning the labels poses a severe threat to the integrity of Artificial Intelligence (AI) technology. In literature, a lot of work has been done on backdoor attacks for medical applications, in which most of the authors assumed that the labels of the samples are also poisoned. This compromises the elusiveness of the backdoor attacks because poisoned samples can be identified by visual inspection by finding the mismatch between the labels of the samples. In this paper, an elusive backdoor attack is proposed, that makes the poisoned samples difficult to recognize. In the proposed approach a backdoor signal superimposed into a small portion of the clean dataset during training time is proposed. Moreover, this paper proposes a hybrid attack that further increases the Attack Success Rate (ASR). The proposed approach is evaluated over a Convolutional Neural Network (CNN)-based system for Magnetic Resonance Imaging (MRI) brain tumor classification, which demonstrated the effectiveness of the attacks, thus raising concern regarding using AI in sensitive applications.

Keywords: Adversarial attack · Backdoor attack · Deep Learning · Attack Success Rate

1 Introduction

Recent developments in Machine Learning (ML) are giving rebirth to the future of intelligent systems enabling multiple domain applications like self-driving cars [1], unmanned aerial vehicles [2], disease diagnostics in the health sector [3]. The revolution in deep learning is chiefly driving the progress of a highly accurate and robust healthcare sector system.

However, extensive research has been conducted to validate that Deep Neural Networks (DNNs) are highly susceptible to malevolent attacks through the utilization of adversarial examples and backdoor triggers to fool the classification

© The Author(s), under exclusive license to Springer Nature Switzerland AG 2023
G. L. Foresti et al. (Eds.): ICIAP 2023, LNCS 14234, pp. 332–344, 2023.
https://doi.org/10.1007/978-3-031-43153-1_28

tasks [4, 5]. One of the first explored attacks to fool a classification task performed at test time is the Fast Gradient Sign Method (FGSM), where the attacker is poisoning image and labels [6]. Although traditional adversarial attacks like FGSM have achieved a high success rate by poisoning only a small portion of the dataset, they are not elusive and can be identified through human inspection by examining the labels of the samples. In backdoor attacks, the attacker would inject a trigger signal at training time to a target class. Poisoning the samples of a certain class while training can result in a generic mis-classification [7] and may lead to a targeted mis-classification [8]. In a targeted backdoor attack, the attacker assumes that he has some access to the model and can poison a certain percentage of samples in the dataset. Some of the authors [9] proposed poisoning the labels of samples in the backdoor attack, but it compromises the elusiveness of the attack.

In this paper, we propose a new backdoor signal to fool a neural network for brain tumor classification of MRI. The main objective of the proposed backdoor attack is to poison the sample X of a targeted class t with a backdoor signal S to get the classifier to decide on a targeted class t even if the sample belongs to a different class. We also have to ensure that this attack can be applied to any class, which means the same method can be applied even if the goal is to change the target class. To perform the backdoor attack, the attacker injects a backdoor signal to a certain extent in the training samples of the target class while keeping in mind that the model must think that this signal is part of the training samples. Such a fact is necessary since we need to obtain high accuracy of the poisoned model while tested with clean samples. The percentage of the poisoned samples plays a prominent role here because if it is too small, the success rate for the attack will be very low but if it is too large, it can reduce the model accuracy and start predicting incorrectly on the clean images. The attacker should always consider that the defender will discard a bad-performing model, leading to a defeat for the attacker. The effectiveness of the proposed attack is demonstrated on a shallow CNN for brain tumor classification with MRI images. Furthermore, we will present a novel hybrid attack approach that combines the FGSM with our backdoor attack, targeting the same CNN model trained on the same dataset.

The paper is structured as follows: Sect. 2 will be focused on the literature review. Then the methodology of those attacks along with their formulation, will be elaborated in Sect. 3. Section 4 will validate the backdoor and hybrid attacks and in Sect. 5, we will conclude our paper.

2   Related Works

In this section, we will review different methods to perform adversarial and backdoor attacks.

Research on such attacks started back in the last decade when [12] poisoned the dataset by flipping the labels to modify the results of an SVM classifier. Also, [13–15] have performed adversarial label flip attacks by manipulating the set of labels to maximize the empirical loss that changed the results of the classification successfully, but the model starts to predict the clean dataset incorrectly. In [9], the authors investigated the deep learning backdoor attack, by adding a single pixel and specific patterns to the samples of the MNIST dataset, and by adding a sticker on the traffic signs dataset. However, the labels of the poisoned samples were also changed into the labels of the target class. Different trigger methods have been proposed in order to poison the samples. In [16], an image was added to the clean sample of the dataset, and a fixed watermark was inserted on the targeted class in [17].

More recently in [10,11], proposed attacks without label modification, which make those attacks more difficult to detect. Due to such attacks, the secure deployment of sensitive applications, especially medical applications, becomes questionable. In [18], the authors performed a backdoor attack by inserting a trigger mask of value 1 at the trigger location for multi-label classification on X-ray images. The backdoor attack has been implemented in [19] for COVID-19 detection in chest X-ray images by adding a pixel trigger of 5×5 with a pixel intensity of 250. The MRI dataset for tumor detection has been used by [20], where it investigated the effects of different adversarial attacks. Authors in [21] used three different multi-disease diagnostic datasets including the tumor segmentation dataset composed of tumor CT images to perform a frequency-injection-based backdoor attack. A synthetic backdoor attack is performed on brain tumor MRI images in [22] where a trigger is generated by an auto-encoder-powered trigger generation method for some specific zone which later is embedded in the samples. In our scenario, we will also refrain from poisoning the labels to maintain a high level of stealthiness during the attack. We will create a new backdoor trigger signal that has not been previously utilized in the literature while maintaining the same resolution as the samples. Consequently, our methodology for generating the trigger will be unique, and it will be integrated into the dataset's images, resulting in more widespread signal distribution and an improved attack success rate.

3 Methodology

In this section, will be explained our proposed backdoor attack in Sect. 3.1 and the hybrid attack in Sect. 3.2 and their formulations.

3.1 Backdoor Attack

Let $f(.)$ be our CNN model and $D = [(x_i, y_i)], y_i \in k, k = [1, 2...c], i = [0, 1...n]$ is the dataset in which x_i is the i_{th} sample and y_i is the label, n is the total number of samples, and c is the number of total classes. The dataset $D = D_1 \cup D_2...D_c$. Let D_t^b be the targeted class t poisoned by a backdoor signal s, where $t \in 1, 2...c$. The attacker is poisoning a fraction α of the samples in D_t. The clean image $x_i \in D_t$ is transformed in a poisoned sample $x_i^b = (x_i + s)$, where s is the backdoor signal generated by the function defined in Eq. 2. After defining the signal and carefully tuning the values so that it neither exposes the attack nor impacts the performance of the model on a clean dataset, the model is trained with the poisoned dataset D_p, which is D where only the targeted class is replaced by D_t containing poisoned samples and clean samples. At testing time, the same backdoor trigger is added either to a specific class of the dataset or to all the datasets. The attack is successful if the test data sample poisoned by the backdoor signal used during training will be wrongly classified. It means that for a sample x with label y different from t, we will have $f(x, y) = t$.

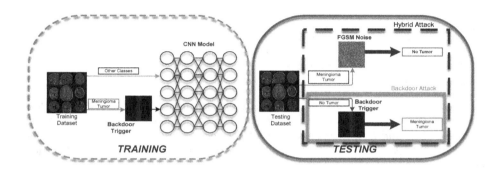

Fig. 1. Pipeline of the proposed backdoor and hybrid attacks

The pipeline of the proposed attack is depicted in Fig. 1, in which the attacker embeds the backdoor trigger to a targeted class at training time, and then the same trigger is used to activate the backdoor at a testing time in order to fool the CNN model. Tuning the defined signal and the number of corrupted samples plays a crucial role and some of these considerations are reported in Sect. 4. If the backdoor signal's amplitude is too high, it can expose the attack by making it visible and if it is too small, the model can neglect the backdoor signal. Also, the percentage α of the samples should not be too large as the model will start to rely on the backdoor signal and may misclassify the clean samples.

Therefore, we considered multiple signals for the backdoor attack at first. We started with a *armp*, *triangle*, and a *sinusoidal* signal. These signals have been used as a backdoor signal before in [10]. The effectiveness of a backdoor signal varies from application to application. In our case, while experimenting with various signals, we found that the trigger generated by the *triangle* signal was more elusive than other signals we tested. However, we still desired greater effectiveness, which led us to modify the *triangle* signal by introducing a *tri − triangle* structure with three different amplitudes ϵ_b. Through this modification, we successfully attained a trigger signal that exhibited improved effectiveness and elusiveness when compared to previous trigger signals. The *tri − triangle* signal $T(i, j)$ is defined as:

$$T(i,j) = \begin{cases} 2j\epsilon_b/m & \begin{aligned} &\text{for } 1 < j < a/2 \\ &\text{for } a < j < a + b/2 \\ &\text{for } a + b < j < a + b + c/2 \\ &\text{for } 1 < i < l \end{aligned} \\ 2(m-j)\epsilon_b/m & \begin{aligned} &\text{for } a/2 < j < a \\ &\text{for } a + b/2 < j < a + b \\ &\text{for } a + b + c/2 < j < m \\ &\text{for } 1 < i < l \end{aligned} \end{cases} \tag{1}$$

where,

m: number of columns of the images
l: number of rows of the images
ϵ_b: amplitude Value

From Eq. 1, we will generate three isosceles triangles, which means each triangle will have two sides of equal length. For the first triangle, a is the total width, a to b, and b to m is the width of the second and third triangles respectively. In this scenario, ϵ_b is the height of a triangle, which may vary for each triangle.

After extensive iterations, adjustments, and fine-tuning of various signals we were able to produce a highly effective trigger signal generated by the *sinc* function which is defined as $S(i,j)$ as below:

$$S(i,j) = \begin{cases} \epsilon_b sinx/x & \begin{aligned} &\text{for } 1 < j < m \\ &\text{for } 1 < i < l \end{aligned} \\ \epsilon_b & \begin{aligned} &\text{for } j = m/2 \\ &\text{for } 1 < i < l \end{aligned} \end{cases} \tag{2}$$

where,

x: $2\pi(j - m/2)f/m$
m: number of columns of the image
l: number of rows of the image

An example of the *sinc* backdoor signal is reported in Fig. 2c. The generated signal is applied on all three channels of MRI images, resulting in almost invisibility in its addition. Figure 2a is shown a clean $no - tumor$ image, Fig. 2b depicts the image poisoned by $tri - triangle$ backdoor signal, and in Fig. 2c we applied *sinc* backdoor signal with an amplitude of $\epsilon_b = 40$ and $f = 3$.

3.2 Hybrid Attack

The hybrid attack proposed in this paper is the combination of the backdoor attack and the Fast Gradient Sign Method [24] use to generate adversarial examples. The intuition behind FGSM is to compute the loss of predictions based on

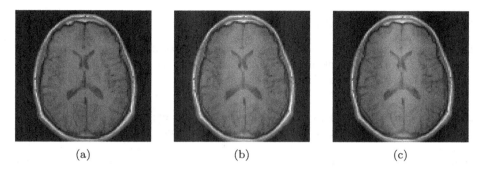

(a)   (b)   (c)

Fig. 2. *No tumor* image (a) Clean image (b) with a $tri - triangle$ backdoor signal with $\epsilon_b = 40$ for middle triangle and $\epsilon_b = 20$ for triangles on each side (c) with the $sinc$ backdoor signal at $\epsilon_b = 40$, $f = 3$

true labels after making the predictions by the model with input images. Then, it calculates the gradient loss with the clean input images to compute the sign gradient and use it to construct the adversary images. Let $f(.)$ be our CNN model and $D = [(x_i, y_i)], y_i \in k, k = [1, 2...c], i = [0, 1...n]$ is the dataset in which x_i is the i_{th} sample and y_i is the label, n is the total number of samples, and c is the number of classes. The total dataset is again defined as $D = D_1 \cup D_2....D_c$. Let D_t^b be the target class with the backdoor attack, and D_t^f be the target class attacked by the FGSM attack. While training the model, we will only poison some fraction of targeted class t with the backdoor attack. At the time of testing, we poison some of the classes with backdoor signal and others with FGSM, which means that our test dataset D is composed of D_t^b and D_t^f. This attack aims to deceive classification decreasing more and more the model accuracy due to the effect of a backdoor attack and FGSM attack.

A hybrid attack can be implemented as shown in Fig. 1 where the pipeline of the attack is presented. At training time, only a backdoor signal is applied to a target class. We train the CNN model keeping in mind that the backdoor signal should not have an impact on the model's accuracy in the case of clean images. During the testing, we utilized a pristine dataset that had not been previously seen by the model. Then, we attack certain classes using the FGSM attack and backdoor attack on others. An example of a *no tumor* poisoned sample predicted as *menimgioma tumor* is shown in Fig. 1.

The purpose of these attacks is to uncover vulnerabilities in deep learning models, particularly for the classification of tumors. By simulating potential security threats, we can highlight vulnerabilities that these models may face in real-world scenarios.

4   Results and Discussion

This section will be focused on the experiments and results obtained through the use of the proposed attack methodologies. The validity of the different attacks will be discussed in this section.

Dataset: The dataset taken into consideration is composed of brain tumor images [23], containing 2870 images for training and 394 images for testing. This dataset is comprised of four distinct classes, namely Glioma Tumor (GT), Meningioma Tumor (MT), No Tumor (NT), and Pituitary Tumor (PT). The images come in different resolutions so the original images of the dataset are resized to $100 \times 100 \times 3$.

The proposed attacks will be tested on a CNN architecture comprised of 6 stacked layers of 2D convolution, activation, and max-pooling layer. Moreover, the architecture is composed of two dense layers, one dropout and one flatten layer. The training accuracy is 99.56% and the obtained accuracy on the test dataset is 91% on a clean dataset. The metric used to evaluate the success rate of the attack is the ASR which can be defined as the number of images in which the attack was successful, divided by the number of total poisoned images.

4.1   Results on Backdoor Attack

In this sub-section, the proposed backdoor attack generated by $sinc$ signal in Sect. 3.1 is compared with our modified $tri - triangel$ signal and other attacks such as $ramp$, $triangle$, and $sinusoidal$ signal proposed in [10]. The value of ϵ_b may vary during the training and testing of the model. In the first experiment, we use class MT as the target class, and α was set to 0.25. For the $ramp$, $triangle$, and our proposed $sinc$ signal have an amplitude $\epsilon_{btr} = 40$ while training and at the time of testing ϵ_{bts} is set varying from 40 to 60 with a step of ten. Similarly, ϵ_{btr} and ϵ_{bts} for $sinusoidal$, and $tri - triangle$ signals are given in Table 1.

Table 1. ASR Comparison of Backdoor signals at $\alpha = 0.25$ for targeted class = MT

ϵ_{bts}	Sinusoidal at $\epsilon_{btr} = 20$, $f = 6$	$\epsilon_{bts1} = \epsilon_{bts3}, \epsilon_{bts2}$	tri-triangle at $\epsilon_{btr1} = \epsilon_{btr3} = 20, \epsilon_{btr2} = 40$	ϵ_{bts}	Ramp at $\epsilon_{btr} = 40$	Triangle at $\epsilon_{btr} = 40$	sinc at $\epsilon_{btr} = 40$, $f = 3$
20	45.5%	20,40	64.5%	40	43.7%	32.6%	82%
25	61.3%	30,50	78.5%	50	49.8%	45.9%	89.2%
30	71.3%	40,60	86%	60	56.6%	57%	93%

The classification results obtained in the presence of backdoor attacks are presented in Table 1. For a $ramp$ signal, we achieved an ASR of 43.7%–56.6% which is not acceptable since the high amplitude of the backdoor inserted at testing time is risking the elusiveness of the attack. A $sinusoidal$ signal with

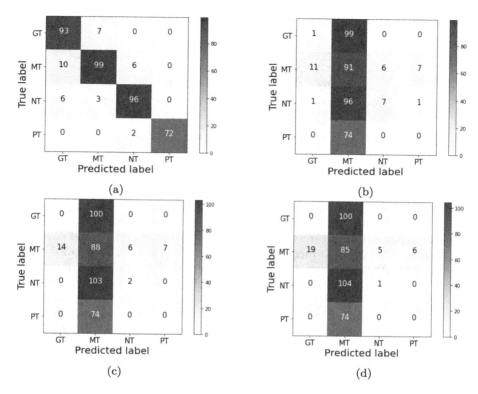

Fig. 3. Confusion matrix (a) for clean test dataset (b) for dataset poisoned by *sinc* backdoor signal at $\epsilon_{btr} = 40$ and $\epsilon_{bts} = 40$ for targeted class MT (c) at $\epsilon_{btr} = 40$ and $\epsilon_{bts} = 50$ (d) at $\epsilon_{btr} = 40$ and $\epsilon_{bts} = 60$ for targeted class MT

$\epsilon_{btr} = 20$, $f = 6$, $\epsilon_{bts} = 20$–30 resulted in an ASR of 45%–73%, which proved to be better than the *ramp* signal. Furthermore, we poisoned the dataset using a *triangle* signal with $\epsilon_{btr} = 40$ and $\epsilon_{bts} = 50$–60. The results of this attack were similar to the *ramp* signal, as we achieved an ASR = 57%. Therefore, we modified it to *tri−triangle* as per Eq. 1 by utilizing $\epsilon_{btr} = 20$ for the first and last triangles, and $\epsilon_{btr} = 40$ for the middle triangle. We used $a = 0.3\,\mathrm{m}$, $b = 0.4\,\mathrm{m}$, and $c = 0.3\,\mathrm{m}$ while $m = 100$ to generate a *tri − triangle* signal with a total width of 100. Although this signal is elusive for smaller values of ϵ_b but is less effective unless we increase the ϵ_{bts} value.

Finally, we encountered a signal generated by the *sinc* function defined in Eq. 2, which proved to be significantly more effective. An ASR comparison of previously discussed signals along with this backdoor signal has been provided in Table 1.

Since, we know that in a backdoor attack, not only the value of ϵ_b can create an impact, but also the α at training time is very important. We used $\epsilon_{btr} = 40$, and $\epsilon_{bts} = 40$–60 with step of ten. The process is iterated for $\alpha = 0.25$

Table 2. ASR (%) of *sinc* over different α and t

α	$\epsilon_{btr} = 40$	t = MT				t = NT			
0.25	ϵ_{bts}	GT	NT	PT	overall	GT	MT	PT	overall
	40	98	55	100	82	66	47	61	57
	50	99	72.4	100	89.2	88	69.6	89	81
	60	99	82.8	100	93	97	87	100	93.8
0.50	40	99	91.4	100	96.4	90	71.3	94.6	83.7
	50	100	98	100	99.3	98	89.6	100	95
	60	100	99	100	99.6	98	97.4	100	98.3

and $\alpha = 0.50$. Table 2 shows the ASR on two different classes as we targeted each class separately. For target class MT with $\alpha = 0.25$, $\epsilon_{btr} = 40$, and $\epsilon_{bts} = 40$, results in an ASR of 82%. Increasing ϵ_{bts} to 50, results in ASR of 89.2% while with $\epsilon_{bts} = 60$ ASR increases to 93%. We observed the same trend for target class NT, using $\epsilon_{btr} = 40$ and $\epsilon_{bts} = 40$–60, which results in an increase of ASR from 57% to 93.8%. Moreover, if we increase the α from 0.25 to 0.50, ASR also increases from 96.4% to 99.6% for $\epsilon_{bts} = 40$–60 for class MT as depicted in Fig. 3.

Hence, it can be concluded that an increase of α, and ϵ_{bts} will result in a higher ASR. However, if we look over Table 2, we can observe that even at the smaller values of α and ϵ_{bts}, ASR in class GT and PT is better compared to others. For a targeted class like MT, the ASR for GT and PT classes are 98% and 100% respectively. But, ASR for class NT is 55% which means that class NT is resistant to the backdoor attack when the targeted class is MT. We can observe the same when the targeted class is NT, class MT shows more resistance to the backdoor attack. This results in low overall ASR. To achieve higher ASR, we are increasing the value ϵ_{bts} which is risking the elusiveness of the backdoor attack.

Table 3. Clean-data accuracy and accuracy degradation in (%)

α	$\epsilon_{btr} = 40$	t = MT		t = NT	
0.25	ϵ_{bts}	Clean-data Accuracy	Accuracy Degradation	Clean-data Accuracy	Accuracy Degradation
	40	82.6	54.7	81	43.5
	50	82.6	59.3	83	58.5
	60	82.6	62	83	65.4
0.50	40	79	66	80	61.5
	50	76.5	68.2	79	67.4
	60	74	69	78	69

Moreover, If we look at Table 3, we can also see that an increment in the α and ϵ_{bts}, accuracy on clean data starts to decrease. So we need to keep the value of α and ϵ_{bts} as small as possible. Table 3 also has the results for accuracy degradation after a backdoor attack, which shows us that accuracy degradation because of a backdoor attack is quite less. Because, in a backdoor attack, we have to target a class and aim to predict the clean data correctly. And, no matter if we attack the targeted class or not at the testing time, it will predict correctly, resulting in low overall accuracy degradation. That is why we are introducing a hybrid attack composed of both FGSM and backdoor attacks.

4.2   Hybrid Attack

Analyzing the impact of the backdoor attack and its shortcomings, we proposed a novel hybrid attack approach consisting of FGSM and backdoor attack where we used the α and ϵ_{btr} with minimum possible values. Therefore, we are utilizing the same value at the time of training and testing of the model. Since, $\epsilon_{btr} = \epsilon_{bts}$ in a hybrid attack, let ϵ_h be the amplitude value for the backdoor attack, and ϵ_f is the step updated magnitude for the FGSM attack.

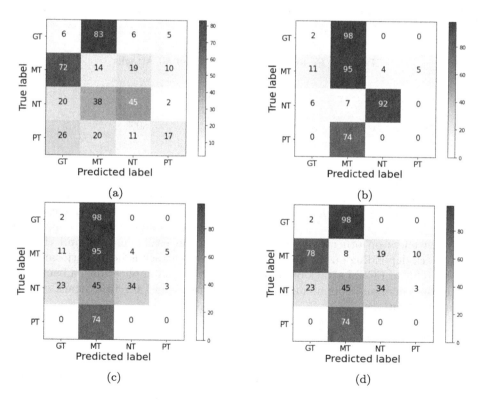

Fig. 4. Confusion matrix (a) after FGSM attack (b) after backdoor attack on GT and PT (c) FGSM on NT, and Backdoor attack on GT and PT, and (d) after attacking GT and PT with the backdoor, and MT and NT with FGSM attack

As we look over Fig. 4a, the FGSM attack is poisoning the labels randomly, which is why we have a randomly spread confusion matrix. In Fig. 4b, we only poison the images of classes GT and PT because both classes are the weakest against the backdoor attack. ASR for the backdoor attack in these classes is more than 98% with the minimum values of α and ϵ_h. The accuracy of the model after this attack is still good (48%) because we are attacking only two classes which means that we let the other two classes be correctly predicted. Then, we attacked class NT with the FGSM at test time to achieve better ASR and decrease model accuracy. Figure 4c depicts an improved ASR of 87%. In this attack, the attacker's goal is to achieve high ASR and degrade the model's accuracy to a deficient level. In Fig. 4d, we are attacking two classes with the backdoor and the other two with the FGSM attack.

Since classes MT and NT are resistant to backdoor attacks, an FGSM attack has been performed over these classes. ASR for the other two classes is 98% while attacking classes MT and NT with the FGSM attack gives us a total ASR of 88.8% for all test samples. This technique reduces the model accuracy to 11.2% with an accuracy degradation of 80%. The result of all the attacks is provided in Table 4.

Table 4. Performance Comparison of all Attacks

	FGSM	Backdoor Attack	Hybrid Attack
ASR	79.2%	82.4%	88.8%
Accuracy Degradation	70%	54.7%	80%

To conclude, results suggest that our backdoor signal performance is better in some classes of the dataset, which motivates the attacker to poison the images with the FGSM of the remaining classes. By attacking two classes, the attacker exposes himself only to a reduced part of the images and does not compromise the elusiveness on other images. The attacker can easily manage and change this attack depending on the application where the attack is being performed. It can become an FGSM attack on human-less supervised applications, a backdoor attack on secure and human-supervised applications, and a hybrid attack on less secure applications.

5   Conclusion

In this paper, we propose effective and elusive backdoor attacks on MRI for brain tumor classification using CNN. We experimented with different backdoor signal attacks and find that the *sinc* signal is the most threatening attack because of its high ASR while maintaining elusiveness. We embedded the backdoor trigger into the small portion of the dataset with a low amplitude trigger signal on clean samples of the target class during training to maintain stealthiness. We also implemented the FGSM adversarial attack, which resulted in good ASR, but its drawback is its poisoning feature affecting samples label. Therefore, we introduced a novel hybrid attack that combines the FGSM and backdoor *sinc* attacks by taking advantage of the two. Ultimately, this research can lead to more secure and trustworthy DL models, benefiting the healthcare sector.

In future work, we will focus on finding a more effective signal and better technique that also requires low amplitude and fewer poisoned samples of the target class to further increase the attack's elusiveness. Those attacks, including the proposed ones, will be tested on other medical applications such as CT scans and X-rays. Moreover, we will also focus on the development of countermeasures for such different attacks.

Acknowledgments. This study has been partially supported by SERICS (PE00000014) under the MUR National Recovery and Resilience Plan funded by the European Union - NextGenerationEU and Sapienza University of Rome project EV2 (003 009 22).

References

1. Garg, N., Ashrith, K.S., Parveen, G.S., Sai, K.G., Chintamaneni, A., Hasan, F.: Self-driving car to drive autonomously using image processing and deep learning. Int. J. Res. Eng. Sci. Manage. **5**(1), 125–132 (2022)
2. Chandana, V.S., Vasavi, S.: Autonomous drones based forest surveillance using Faster R-CNN. In: 2022 International Conference on Electronics and Renewable Systems (ICEARS), pp. 1718–1723. IEEE, March 2022
3. Hassan, M.R., et al.: Prostate cancer classification from ultrasound and MRI images using deep learning based Explainable Artificial Intelligence. Futur. Gener. Comput. Syst. **127**, 462–472 (2022)
4. Hirano, H., Minagi, A., Takemoto, K.: Universal adversarial attacks on deep neural networks for medical image classification. BMC Med. Imaging **21**(1), 1–13 (2021)
5. Kwon, H., Kim, Y.: BlindNet backdoor: attack on deep neural network using blind watermark. Multimedia Tools Appl. **81**(5), 6217–6234 (2022)
6. Joel, M.Z., et al.: Adversarial attack vulnerability of deep learning models for oncologic images. medRxiv, January 2021
7. Yang, C., Wu, Q., Li, H., Chen, Y.: Generative poisoning attack method against neural networks. arXiv preprint arXiv:1703.01340 (2017)
8. Liao, C., Zhong, H., Squicciarini, A., Zhu, S., Miller, D.: Backdoor embedding in convolutional neural network models via invisible perturbation. arXiv preprint arXiv:1808.10307 (2018)
9. Gu, T., Dolan-Gavitt, B., Garg, S.: BadNets: identifying vulnerabilities in the machine learning model supply chain. arXiv preprint arXiv:1708.06733 (2017)
10. Barni, M., Kallas, K., Tondi, B.: A new backdoor attack in CNNs by training set corruption without label poisoning. In: 2019 IEEE International Conference on Image Processing (ICIP), pp. 101–105. IEEE, September 2019
11. Turner, A., Tsipras, D., Madry, A.: Clean-label backdoor attacks (2018)
12. Xiao, H., Xiao, H., Eckert, C.: Adversarial label flips attack on support vector machines. In: ECAI 2012, pp. 870–875. IOS Press (2012)

13. Xiao, H., Biggio, B., Nelson, B., Xiao, H., Eckert, C., Roli, F.: Support vector machines under adversarial label contamination. Neurocomputing **160**, 53–62 (2015)
14. Koh, P.W., Liang, P.: Understanding black-box predictions via influence functions. In: International Conference on Machine Learning, pp. 1885–1894. PMLR, July 2017
15. Mei, S., Zhu, X.: Using machine teaching to identify optimal training-set attacks on machine learners. In: Twenty-Ninth AAAI Conference on Artificial Intelligence, February 2015
16. Chen, X., Liu, C., Li, B., Lu, K., Song, D.: Targeted backdoor attacks on deep learning systems using data poisoning. arXiv preprint arXiv:1712.05526 (2017)
17. Steinhardt, J., Koh, P.W.W., Liang, P.S.: Certified defenses for data poisoning attacks. In: Advances in Neural Information Processing Systems, vol. 30 (2017)
18. Nwadike, M., Miyawaki, T., Sarkar, E., Maniatakos, M., Shamout, F.: Explainability matters: backdoor attacks on medical imaging. arXiv preprint arXiv:2101.00008 (2020)
19. Matsuo, Y., Takemoto, K.: Backdoor attacks to deep neural network-based system for COVID-19 detection from chest X-ray images. Appl. Sci. **11**(20), 9556 (2021)
20. Paschali, M., Conjeti, S., Navarro, F., Navab, N.: Generalizability *vs.* Robustness: investigating medical imaging networks using adversarial examples. In: Frangi, A.F., Schnabel, J.A., Davatzikos, C., Alberola-López, C., Fichtinger, G. (eds.) MICCAI 2018. LNCS, vol. 11070, pp. 493–501. Springer, Cham (2018). https://doi.org/10.1007/978-3-030-00928-1_56
21. Feng, Y., Ma, B., Zhang, J., Zhao, S., Xia, Y., Tao, D.: FIBA: frequency-injection based backdoor attack in medical image analysis. arXiv preprint arXiv:2112.01148 (2021)
22. Wang, S., Nepal, S., Rudolph, C., Grobler, M., Chen, S., Chen, T.: Backdoor attacks against transfer learning with pre-trained deep learning models. IEEE Trans. Serv. Comput. **15**(3), 1526–1539 (2020)
23. Bhuvaji, S., Kadam, A., Bhumkar, P., Dedge, S., Kanchan, S.: Brain Tumor Classification (MRI), [Dataset]. Kaggle (2020). https://doi.org/10.34740/KAGGLE/DSV/1183165
24. Goodfellow, I.J., Shlens, J., Szegedy, C.: Explaining and harnessing adversarial examples. arXiv preprint arXiv:1412.6572 (2014)

Unveiling the Impact of Image Transformations on Deepfake Detection: An Experimental Analysis

Federico Cocchi[1,2][iD], Lorenzo Baraldi[2(✉)][iD], Samuele Poppi[1,2][iD], Marcella Cornia[1][iD], Lorenzo Baraldi[1][iD], and Rita Cucchiara[1,3][iD]

[1] University of Modena and Reggio Emilia, Modena, Italy
{federico.cocchi,samuele.poppi,marcella.cornia,lorenzo.baraldi,
rita.cucchiara}@unimore.it
[2] University of Pisa, Pisa, Italy
{federico.cocchi,samuele.poppi,lorenzo.baraldi}@phd.unipi.it
[3] IIT-CNR, Pisa, Italy

Abstract. With the recent explosion of interest in visual Generative AI, the field of deepfake detection has gained a lot of attention. In fact, deepfake detection might be the only measure to counter the potential proliferation of generated media in support of fake news and its consequences. While many of the available works limit the detection to a pure and direct classification of fake versus real, this does not translate well to a real-world scenario. Indeed, malevolent users can easily apply post-processing techniques to generated content, changing the underlying distribution of fake data. In this work, we provide an in-depth analysis of the robustness of a deepfake detection pipeline, considering different image augmentations, transformations, and other pre-processing steps. These transformations are only applied in the evaluation phase, thus simulating a practical situation in which the detector is not trained on all the possible augmentations that can be used by the attacker. In particular, we analyze the performance of a k-NN and a linear probe detector on the COCOFake dataset, using image features extracted from pre-trained models, like CLIP and DINO. Our results demonstrate that while the CLIP visual backbone outperforms DINO in deepfake detection with no augmentation, its performance varies significantly in presence of any transformation, favoring the robustness of DINO.

Keywords: Deepfake Detection · Self-Supervised Vision Transformers

1 Introduction

Although the generation of deepfake encompasses results of diverse nature, the world of fake image forgery has gained a lot of attention, since the breakthrough of diffusion models [7,13,30,31,33] in the Generative AI domain. While this technological advancement was received enthusiastically by the community, it has

---

F. Cocchi and L. Baraldi—Equal contribution.

© The Author(s), under exclusive license to Springer Nature Switzerland AG 2023
G. L. Foresti et al. (Eds.): ICIAP 2023, LNCS 14234, pp. 345–356, 2023.
https://doi.org/10.1007/978-3-031-43153-1_29

also raised significant concerns regarding its potential impact on various domains, including the realms of human art and privacy. Both these domains are susceptible to risks due to the ease with which these models generate new content. Consequently, in light of the ongoing advancements in Generative AI, there has been a significant shift towards enhancing deepfake detection systems [36,41] to mitigate the risks posed by the remarkably convincing nature of such content.

The first efforts towards AI-generated content detection were conceived in the realm of fake face detection, with the release of ad-hoc datasets [18,32] and methodologies [11,19]. However, it should be noted that the significance of deepfake detection extends beyond fake faces or biometric data, necessitating the need for broader and more versatile detection methods that can address a wider range of generative scenarios. Only recently, a limited number of studies [1, 6,36] have started to investigate deepfake images generated from text-to-image models [2,30,31,33], thereby enabling the detection of a wider variety of subjects with respect to biometric data. Although these studies assert high accuracy in detecting fake images, the resilience and robustness of the proposed methods have not yet been quantitatively evaluated.

In this manuscript, we freeze the recently proposed Stable Diffusion [31] model as the text-to-image generator and test two different detection approaches. In addition, we employ two different feature extractors, namely CLIP [29] and DINO [4], and evaluate their robustness to a wide variety of image transformations, at pixel-value and image-structure levels (Fig. 1). To the best of our knowledge, we are the first to assess the performance variability of real-fake recognition within such an environment. The experimental results shed light on the generally more robust performance of self-supervised methods (i.e., DINO) against transformations in deepfake detection. Indeed, while CLIP achieves better performance without augmentation, the behavior of deepfake classifiers across different transformations is more consistent for DINO compared to CLIP. Surprisingly, CLIP performs similarly to DINO in the recognition of real images.

2 Related Work

Text-to-Image Generation. Deepfake images can be generated through three main models which consist of autoregressive approaches [25,26,33,39], generative adversarial networks (GANs) [12,34,38,42], and diffusion models [7,13,17,37]. In this work, we narrow down the field of deepfake generation considering the recent paradigm of text-to-image generation, which consists of generating an image starting from a textual description. While some GAN-based approaches [20] have been proposed as a possible solution to text-to-image generation, great results have been recently obtained with the application of diffusion models [2,30,33] by conditioning the diffusion process on the input textual description.

Recently, latent diffusion models [28,31] have improved the efficiency of standard diffusion models while maintaining their generation quality, by operating in a lower dimensional latent space z using a pre-trained variational autoencoder (VAE) [9,16]. In particular, during image generation, this approach involves the diffusion process occurring within the embedding space z, followed by the decompression of the resulting image through the VAE decoder. We conduct our experiments using images generated by the Stable Diffusion model [31], using both the 1.4 and 2.0 versions. The main differences between them lie in the backbone used to extract features from texts and images. In fact, Stable Diffusion v1 employs CLIP [31], which is trained on a non-publicly available dataset, while Stable Diffusion v2 relies on OpenCLIP [14], which is trained on a subset of LAION-5B [35] dataset. Both Stable Diffusion versions are finetuned on a filtered subset of LAION-5B to improve aesthetics and avoid explicit contents.

Deepfake Detection. The deepfake detection pipeline employed in this study comprises two consecutive stages: an image feature extractor followed by the actual detector. As for the first bit, different works have made extensive use of CLIP features as a starting point for their analysis [1,24,36]. In [5], they introduced an exploratory study of the frequency spectrum of the created images, thus capturing the impact of the specific generation model on the structure of the final images. Conversely, in [1], the authors proposed a wider-spectrum evaluation of the effects of different image feature extractors, presenting results on CLIP and OpenCLIP. Simultaneously, within the literature on image watermarking [10], analyses have been conducted to examine the robustness of the added watermark when the image is subjected to transformations. This type of analysis has been also conducted in relation to the detection of manipulated images and videos specifically focused on facial manipulation [22]. We embark on this path, applying it to the deepfake detection scenario, and studying how it affects the performance of some detection algorithms and the distribution of the features in the embedding space.

3   Evaluation Framework

3.1   Dataset

This section provides an overview of the COCOFake dataset [1] used in this work to perform the analysis on deepfake detection. COCOFake consists of an extension of the COCO dataset [21], that includes both real and fake images. Specifically, each real image in COCO is paired with five captions which are used to generate five fake images through a text-to-image model. The dataset is divided into training, validation, and test sets following the Karpathy splits, as used in the captioning literature [15]. Since COCO contains 113,287 training images and 5,000 validation and test images, COCOFake is composed of 679,722 instances in training, and 30,000 in validation and test.

From a technical standpoint, the production of counterfeit images is achieved through the utilization of Stable Diffusion [31] version 1.4. Furthermore, COCO-Fake also includes validation and test splits generated with Stable Diffusion version 2.0 to increase the robustness and generalization of possible analysis. It is worth mentioning that, all the images of COCOFake are stored in JPEG format, following the original COCO compression.

Fig. 1. Visual comparison of image transformations on a sample real image (top left).

3.2 Visual Backbones

In our experimental analysis, we employ three different visual backbones, namely CLIP [29], DINO [4], and DINOv2 [27]. It is worth mentioning that all the backbones adopt the same Vision Transformer architecture [8], ensuring a fair comparison between the employed methods.

The primary distinction among the visual backbones is the pre-training method employed. For instance, the CLIP approach utilizes language supervision to enforce similarities between visual and textual concepts. This is achieved by independently processing the image and its textual description using a visual and a textual backbone and then linearly projecting their representation into a shared embedding space. CLIP is pre-trained with a contrastive objective that maximizes the cosine similarity of correct image-text pairs. While CLIP obtains a semantic coherence [23] that can be useful for deepfake detection, the only image augmentation that is applied during training consists of a random square crop from resized images. This could make the visual backbone vulnerable to adversarial image augmentation.

In contrast to CLIP, DINO eschews the use of textual references, heavily relying on image augmentations during the pre-training phase. Indeed, DINO augments the input image through various techniques, including multi-crop [3], color jittering, Gaussian blur, and solarization. Multi-crop is used to generate multiple views of the same image, which can be logically divided into local views with lower resolutions and global views with higher resolutions. The DINO model is trained by enforcing local-to-global correspondences between different views of the same image. On the other hand, DINOv2 introduces additional pre-training objectives compared to DINO, such as randomly masking patches of the local views, leaving the model to learn how to reconstruct these patches.

Since both DINO and DINOv2 enforce robustness to image augmentation during pre-training, we investigate their effectiveness in a deepfake detection pipeline.

3.3 Deepfake Detection Pipeline

In this section, we present the deepfake detection pipeline that has been utilized for the analysis conducted in this study. Our pipeline encompasses a feature extraction phase followed by a detector model. Specifically, the detector model under investigation includes both a linear probe and a k-nearest neighbor (k-NN) classifier. The incorporation of different detector models serves the purpose of assessing distinct aspects. Specifically, the linear probe is engineered to identify any potential indications of the generation process within the feature space. Conversely, the k-nearest neighbor approach relies on the distance between existing features stored during training, thus allowing us to measure the similarity between real and fake content, in the embedding space.

Feature Extraction Process. From a technical perspective, the previously introduced visual backbones are employed as feature extraction models. Indeed, during the process of feature extraction, each image from the training, validation, and test sets of COCOFake undergoes processing by the visual backbones CLIP, DINO, and DINOv2. It is worth mentioning that no image augmentation is applied during the feature extraction phase.

Formally, each image $x \in \mathbb{R}^{C \times H \times W}$ is firstly split into a sequence of squared patches $\{x_i^p\}_{i=1}^N$ where C, H, W are respectively channel, height and width, while $x_i^p \in \mathbb{R}^{P^2 C}$ is the i-th image patch of size $P \times P$. Consecutively, the sequence of image patches is linearly projected in the embedding dimensionality of the model D. At this step, a learnable classification token $\texttt{[CLS]} \in \mathbb{R}^D$ is concatenated to the input sequence. After L self-attention blocks the $\texttt{[CLS]}$ token is saved as the representation of the image. In addition, and only for the CLIP model, the $\texttt{[CLS]}$ token is linearly projected into the multi-modal embedding space.

Implementation-wise, the Base version of ViT [8] (*i.e.*, ViT-B) is used for CLIP, DINO, and DINOv2. In detail, ViT-B includes 85M learnable parameters, a 768 embedding dimensionality D, and $L = 12$ self-attention blocks. The considered input image size is $C = 3$, $H = 224$, $W = 224$, while the image patch size P is 14 for DINOv2 and 16 for CLIP and DINO. Regarding the pre-trained weights, the open-source ViT-B/16 version (*i.e.*, OpenCLIP [14]), pre-trained on the LAION-2B dataset [35], is used for CLIP, while the publicly available ViT-B/16 and ViT-B/14 are used for DINO and DINOv2, respectively.

Linear Probe. In the linear probe approach, we use the extracted features to train a logistic regressor. The goal of the method is to identify a signature, or imprint, in the extracted features that enable the linear model to distinguish between real and fake data. The logistic regressor is trained with an ℓ_2 objective, and the loss is weighted to account for the difference in the number of real and fake samples. Specifically, since the number of fake images in COCOFake is five times greater than the number of real images, the loss is weighted inversely

Table 1. Comprehensive summary of essential information regarding the applied transformations to assess the robustness of the different classifiers.

Transformation	Parameter	Range		Type	
		Min	Max	Pixel	Structure
Equalize	–	–	–	✓	✗
Center Crop	size	64	512	✗	✓
Resize	size	64	512	✗	✓
Random Crop	size	64	512	✗	✓
Brightness	brightness factor	0.5	2.0	✓	✗
Contrast	contrast factor	0.5	2.0	✓	✗
Hue	hue factor	−0.5	0.5	✓	✗
Saturation	saturation factor	0.1	3.0	✓	✗
Posterize	bits	1	8	✓	✗
Gaussian Blur	kernel size	3	15	✓	✗
JPEG Compression	quality	10	90	✓	✗
SD Compression	–	–	–	✓	✗

proportional to class frequencies. In addition, the LBFGS solver [40] is employed for training. Results are evaluated with accuracy scores over real and fake data.

k-**nearest neighbor** (k-**NN**). The classification task in the k-nearest neighbor approach is dependent on measuring distances within the visual feature space extracted by the utilized backbones. This implies that no further training is required. Hence, in the validation and test sets, the distances between each element and the features stored offline from the training split are calculated. The deepfake classification task is a supervised task, whereby the corresponding label (real or fake) is known for each feature embedding. So, the accuracy is determined by applying majority voting on the k-nearest features within the training feature space.

While the k-NN approach was originally proposed by [24] in a deepfake detection scenario, it presents notable limitations. Specifically, k-NN is highly sensitive to missing values or outliers, necessitating extensive coverage in the embedding space of the visual backbones by the training dataset. Moreover, as the dataset size increases, the computational cost of calculating distances between a new image and each existing one escalates significantly, ultimately compromising the algorithm performance. From an implementation perspective, we take into account the cosine similarity and the top-1 nearest neighbor to define the k-NN. Moreover, to manage the unbalanced COCOFake dataset, only a single pair of real and fake images are considered to compute the visual features in the training split, thus obtaining balanced real-fake images.

3.4   Image Augmentation

Drawing inspiration from [10,22], we explore the effectiveness of twelve distinct image augmentation techniques, detailed in Table 1. This series of transformations depict the potential manipulations of the image, considering image-structure and pixel-value transforms. As we can notice, each augmentation

Table 2. Accuracy performance on the COCOFake test set without any transformations for Stable Diffusion v1.4 and v2.0, using different classifiers and backbones.

Backbone	Stable Diffusion v1.4		Stable Diffusion v2.0	
	Linear	k-NN	Linear	k-NN
CLIP	99.6	96.7	99.3	94.9
DINO	96.9	91.3	90.5	87.8
DINOv2	96.6	89.0	95.7	84.6

involves a tunable parameter to control the degree of impact on images. We undertake a detailed analysis of these parameters to assess the robustness of the classification methods in response to the strength of the transformation. To this end, we select a range delimited by a minimum and maximum parameter for each augmentation, aiming to preserve the visual quality of the image in both cases, thus ensuring the preservation of visual consistency and usability. We assess the results by linearly partitioning the parameter range into five equally spaced segments. Following this process, we obtain five different image augmentation techniques for each transform with varying strengths. The utilization of these transformations evaluates the employed classifiers' accuracy in terms of resilience and generalization. A visual example of some of the image augmentation applied to an image is reported in Fig. 1.

In addition to the conventional augmentation methods, we introduce a novel technique called Stable Diffusion (SD) compression. This approach involves the projection of an image x into the latent space z of the Stable Diffusion model by utilizing the encoder of the autoencoder model [9] implemented within the Stable Diffusion framework. Following this projection, the image x is reconstructed using the decoder of the autoencoder. This augmentation technique is exclusively applied to real images to examine the biases of the detector concerning the lossy compression inherent in the generation of fake images.

4   Experimental Results

In this section, we analyze the results obtained by employing data augmentation on real and fake images, while testing different visual backbones.

Deepfake Detection of Plain Images. To evaluate the resilience of the aforementioned methods, a preliminary study is conducted to examine the performance of the detection pipeline without any applied transformations.

Based on the findings presented in Table 2, we can notice that the linear probe classifier exhibits a high classification accuracy, across all the backbones, with scores of 99.6%, 96.9%, and 96.6%, respectively with CLIP, DINO, and DINOv2, over the COCOFake test set generated with Stable Diffusion v1.4. These results validate the hypothesis that linear probes effectively identify the

Table 3. Comparison of accuracy performance on the COCOFake test set with transforms applied to fake images. The table shows results for linear and k-NN classifiers, for each backbones.

Transformation	CLIP		DINO		DINOv2	
	Linear	k-NN	Linear	k-NN	Linear	k-NN
Equalize	$11.9_{\pm0.0}$	$87.4_{\pm0.0}$	$94.1_{\pm0.0}$	$92.5_{\pm0.0}$	$89.8_{\pm0.0}$	$89.2_{\pm0.0}$
Center Crop	$28.3_{\pm26.4}$	$85.7_{\pm18.4}$	$85.3_{\pm15.0}$	$89.9_{\pm7.3}$	$87.7_{\pm15.7}$	$83.9_{\pm10.4}$
Resize	$83.0_{\pm8.7}$	$95.5_{\pm1.2}$	$94.3_{\pm5.2}$	$93.7_{\pm0.2}$	$93.8_{\pm3.2}$	$89.2_{\pm1.1}$
Random Crop	$31.8_{\pm25.4}$	$84.1_{\pm21.1}$	$85.8_{\pm14.0}$	$88.6_{\pm9.2}$	$87.5_{\pm16.4}$	$83.0_{\pm11.6}$
Brightness	$29.1_{\pm27.9}$	$88.5_{\pm7.9}$	$91.9_{\pm4.1}$	$93.3_{\pm0.3}$	$95.5_{\pm0.8}$	$89.8_{\pm0.1}$
Contrast	$31.4_{\pm26.2}$	$90.0_{\pm5.6}$	$94.8_{\pm3.8}$	$93.7_{\pm0.4}$	$95.7_{\pm0.9}$	$89.8_{\pm0.2}$
Hue	$74.2_{\pm2.6}$	$93.5_{\pm1.6}$	$93.3_{\pm3.1}$	$93.4_{\pm0.4}$	$95.3_{\pm0.5}$	$89.4_{\pm0.4}$
Saturation	$56.8_{\pm28.6}$	$92.1_{\pm6.1}$	$94.1_{\pm5.8}$	$93.3_{\pm0.7}$	$95.8_{\pm3.1}$	$89.7_{\pm0.4}$
Posterize	$29.5_{\pm37.7}$	$77.2_{\pm24.2}$	$89.9_{\pm6.9}$	$92.4_{\pm1.4}$	$86.1_{\pm11.0}$	$87.4_{\pm3.5}$
Gaussian Blur	$31.6_{\pm33.5}$	$95.8_{\pm0.7}$	$88.1_{\pm6.7}$	$92.9_{\pm0.5}$	$94.0_{\pm1.7}$	$89.3_{\pm0.3}$
JPEG Compression	$50.3_{\pm29.3}$	$95.1_{\pm3.0}$	$97.7_{\pm2.1}$	$93.8_{\pm0.4}$	$96.7_{\pm3.0}$	$89.5_{\pm0.9}$
Average	$41.6_{\pm24.6}$	$89.5_{\pm9.0}$	$91.8_{\pm6.7}$	$92.5_{\pm2.1}$	$92.3_{\pm5.6}$	$88.2_{\pm2.9}$

generator's imprint, embedded in the image features. Similar behavior is also highlighted by the k-NN approach, whose objective is not to specifically identify the imprinting trace. The observed performance strongly suggests that, in the backbones embedding space, fake images tend to exhibit proximity to one another and a similar phenomenon may hold true for real images. Specifically, k-NN performs with an accuracy of 96.7%, 91.3%, and 89%, over respectively CLIP, DINO, and DINOv2.

Moreover, the comparable performance observed on the COCOFake test set of Stable Diffusion 1.4 and Stable Diffusion v2.0 underscores the classifiers' capability to generalize beyond their initial training domain. As a result, further experiments will solely focus on the test set of Stable Diffusion 1.4. Building upon these initial results, subsequent experiments extend the analysis to explore the accuracy patterns when transformations are applied to fake and real images.

Fake Data Analysis. Presented in Table 3, we encounter a concise overview of the performance of the deepfake detection pipeline over transformed fake images. Evidently, the evaluation using the linear probe on the CLIP backbone demonstrates remarkably low performance. Specifically, CLIP achieves an average accuracy, among all the transformations, of only 41.6% for fake images, while DINO and DINOv2 demonstrate higher accuracy of 91.8% and 92.3%, respectively. Furthermore, the average standard deviation of CLIP, which amounts to 24.6%, highlights the substantial variability in performance across different transformations. This variability poses a significant threat to the overall robust-

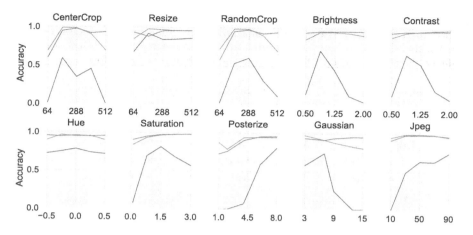

Fig. 2. The plots showcase the linear probe accuracy using different backbones, namely **CLIP**, DINO, and DINOv2, varying the applied transformation. Each subplot illustrates the accuracy of the classifiers under varying degrees of strength in image augmentations, used to provide insights into the effectiveness of classifiers.

ness of a CLIP-based deepfake detector. In contrast, DINO and DINOv2 consistently exhibit robustness across a wide range of performed transformations. In addition, Fig. 2 illustrates the trajectory of accuracy outcomes for the linear probes under varying degrees of strength of image augmentations, as discussed in Sect. 3.4. It is visually evident that, while DINO and DINOv2 exhibit a tendency to maintain consistent performance levels, CLIP performance is highly influenced by the intensity of each transformation. For example, a JPEG compression transformation with 10% quality produces an accuracy of 0.4% over CLIP linear probe while 95% and 91% for respectively DINO and DINOv2. We assume that the linear probe trained on CLIP-extracted features may be prone to overfitting on the distinctive imprint of fake data. This assumption arises from the observation that the CLIP visual backbone is not trained using extensive data augmentation. Consequently, alterations in the images could modify the extracted features, thus altering the fake imprint. This would explain the significant decline in the performance of the linear probe on CLIP. Although the k-NN outcomes, as shown in Table 3, indicate that CLIP achieves accuracy on par with DINO and DINOv2, the higher average standard deviation observed in CLIP highlights the superiority of the latter models.

Real Data Analysis. Table 4 presents a comprehensive analysis of the performance of CLIP, DINO, and DINOv2 evaluated on transformed real images. We decide to logically cluster results in JPEG Compression, SD Compression, and Other Transforms to facilitate the analysis. Specifically, we isolate the compression augmentations, leaving a summary of the others. Regarding the obtained results, it is noteworthy that the linear probes demonstrate commendable per-

Table 4. Accuracy performance on the COCOFake test set with transformations applied to real images. We report results for linear probe and k-NN classifiers for each backbones. Transformations are divided into compression based and others, to highlight the accuracy drop when applying compression-based transformations.

Backbone	JPEG Compression		SD Compression		Other Transforms	
	Mean	Std	Mean	Std	Mean	Std
Linear						
CLIP	93.2	12.5	44.2	–	99.7	0.5
DINO	74.8	15.2	80.1	–	93.2	6.0
DINOv2	58.2	20.4	54.2	–	91.9	5.5
k-NN						
CLIP	87.5	2.7	80.2	–	90.0	5.7
DINO	75.0	3.8	75.2	–	76.0	5.0
DINOv2	80.3	1.8	79.4	–	81.7	2.8

formance on the other non-compression-based transforms. However, when subjected to JPEG compression, the linear probes exhibit lower accuracy. Specifically, the average accuracy reaches 93.2%, 74.8%, and 58.2% for CLIP, DINO, and DINOv2 respectively. Furthermore, the poorest performance is observed in CLIP with SD compression, resulting in an accuracy of 44.2%. We hypothesize that the compression imprints bear a strong resemblance to the fake imprint, thereby deceiving the linear probe into misclassifying a real image as fake.

A comparable examination can be directly carried out on the feature space of the visual backbones. Specifically, when considering the embedding space of CLIP, real images subjected to the SD compression exhibit closer proximity, on average, to fake images compared to JPEG compression and other transformations. This is additional proof that SD compression has a great influence on the fake data imprint. In contrast, DINO and DINOv2 are equally subjected to all transformations, exhibiting an average accuracy in the k-NN analysis of 75.4% and 80.5%, respectively. It is noteworthy that the limitations inherent to k-NN, as mentioned in Sect. 3.3, can attenuate its impact on deepfake detection.

5   Conclusion

In conclusion, the growing capacity and utilization of text-to-image models present a persistent challenge in the detection of artificially generated images. Our proposal introduces an analysis of the robustness of a set of classifiers, specifically considering transformations that modify the visual appearance of the image. The performance of the classifiers is significantly influenced by these transformations and this study emphasizes the significance of the robustness to such transformations for deepfake detector classifiers that need to operate in real-world scenarios.

Acknowledgments. This work has partially been supported by the European Commission under the PNRR-M4C2 (PE00000013) project "FAIR - Future Artificial Intelligence Research" and by the Horizon Europe project "European Lighthouse on Safe and Secure AI (ELSA)" (HORIZON-CL4-2021-HUMAN-01-03), co-funded by the European Union (GA 101070617).

References

1. Amoroso, R., Morelli, D., Cornia, M., Baraldi, L., Del Bimbo, A., Cucchiara, R.: Parents and children: distinguishing multimodal DeepFakes from natural images. arXiv preprint arXiv:2304.00500 (2023)
2. Balaji, Y., et al.: eDiff-I: text-to-image diffusion models with an ensemble of expert denoisers. arXiv preprint arXiv:2211.01324 (2022)
3. Caron, M., Misra, I., Mairal, J., Goyal, P., Bojanowski, P., Joulin, A.: Unsupervised learning of visual features by contrasting cluster assignments. In: NeurIPS (2020)
4. Caron, M., et al.: Emerging properties in self-supervised vision transformers. In: ICCV (2021)
5. Corvi, R., Cozzolino, D., Poggi, G., Nagano, K., Verdoliva, L.: Intriguing properties of synthetic images: from generative adversarial networks to diffusion models. In: CVPR Workshops (2023)
6. Corvi, R., Cozzolino, D., Zingarini, G., Poggi, G., Nagano, K., Verdoliva, L.: On the detection of synthetic images generated by diffusion models. In: ICASSP (2023)
7. Dhariwal, P., Nichol, A.: Diffusion models beat GANs on image synthesis. In: NeurIPS (2021)
8. Dosovitskiy, A., et al.: An image is worth 16×16 words: transformers for image recognition at scale. In: ICLR (2021)
9. Esser, P., Rombach, R., Ommer, B.: Taming transformers for high-resolution image synthesis. In: CVPR (2021)
10. Fernandez, P., Sablayrolles, A., Furon, T., Jégou, H., Douze, M.: Watermarking images in self-supervised latent spaces. In: ICASSP (2022)
11. Ganguly, S., Ganguly, A., Mohiuddin, S., Malakar, S., Sarkar, R.: ViXNet: Vision Transformer with Xception Network for DeepFakes based video and image forgery detection. Expert Syst. Appl. **210**, 118423 (2022)
12. Goodfellow, I., et al.: Generative adversarial nets. In: NeurIPS (2014)
13. Ho, J., Jain, A., Abbeel, P.: Denoising diffusion probabilistic models. In: NeurIPS (2020)
14. Ilharco, G., et al.: OpenCLIP (2021). https://doi.org/10.5281/zenodo.5143773
15. Karpathy, A., Li, F.: Deep visual-semantic alignments for generating image descriptions. In: CVPR (2015)
16. Kingma, D.P., Welling, M.: Auto-encoding variational Bayes. arXiv preprint arXiv:1312.6114 (2013)
17. Kingma, D.P., Salimans, T., Jozefowicz, R., Chen, X., Sutskever, I., Welling, M.: Improved variational inference with inverse autoregressive flow. In: NeurIPS (2016)
18. Li, L., Bao, J., Yang, H., Chen, D., Wen, F.: Advancing high fidelity identity swapping for forgery detection. In: CVPR (2020)
19. Li, L., et al.: Face X-Ray for more general face forgery detection. In: CVPR (2020)
20. Liao, W., Hu, K., Yang, M.Y., Rosenhahn, B.: Text to image generation with semantic-spatial aware GAN. In: CVPR (2022)

21. Lin, T.-Y., et al.: Microsoft COCO: common objects in context. In: Fleet, D., Pajdla, T., Schiele, B., Tuytelaars, T. (eds.) ECCV 2014. LNCS, vol. 8693, pp. 740–755. Springer, Cham (2014). https://doi.org/10.1007/978-3-319-10602-1_48

22. Lu, Y., Ebrahimi, T.: Assessment framework for DeepFake detection in real-world situations. arXiv preprint arXiv:2304.06125 (2023)

23. Mukhoti, J., et al.: Open vocabulary semantic segmentation with patch aligned contrastive learning. In: CVPR (2023)

24. Ojha, U., Li, Y., Lee, Y.J.: Towards universal fake image detectors that generalize across generative models. In: CVPR (2023)

25. Van den Oord, A., Kalchbrenner, N., Espeholt, L., Vinyals, O., Graves, A., et al.: Conditional image generation with PixelCNN decoders. In: NeurIPS (2016)

26. van den Oord, A., Li, Y., Vinyals, O.: Representation learning with contrastive predictive coding. arXiv preprint arXiv:1807.03748 (2018)

27. Oquab, M., et al.: DINOv2: learning robust visual features without supervision. arXiv preprint arXiv:2304.07193 (2023)

28. Peebles, W., Xie, S.: Scalable diffusion models with transformers. arXiv preprint arXiv:2212.09748 (2022)

29. Radford, A., et al.: Learning transferable visual models from natural language supervision. In: ICML (2021)

30. Ramesh, A., Dhariwal, P., Nichol, A., Chu, C., Chen, M.: Hierarchical text-conditional image generation with CLIP latents. arXiv preprint arXiv:2204.06125 (2022)

31. Rombach, R., Blattmann, A., Lorenz, D., Esser, P., Ommer, B.: High-resolution image synthesis with latent diffusion models. In: CVPR (2022)

32. Rossler, A., Cozzolino, D., Verdoliva, L., Riess, C., Thies, J., Nießner, M.: Face-Forensics++: learning to detect manipulated facial images. In: ICCV (2019)

33. Saharia, C., et al.: Photorealistic text-to-image diffusion models with deep language understanding. In: NeurIPS (2022)

34. Sauer, A., Karras, T., Laine, S., Geiger, A., Aila, T.: StyleGAN-T: unlocking the power of GANs for fast large-scale text-to-image synthesis. arXiv preprint arXiv:2301.09515 (2023)

35. Schuhmann, C., et al.: LAION-5B: an open large-scale dataset for training next generation image-text models. In: NeurIPS (2022)

36. Sha, Z., Li, Z., Yu, N., Zhang, Y.: DE-FAKE: detection and attribution of fake images generated by text-to-image diffusion models. arXiv preprint arXiv:2210.06998 (2022)

37. Sohl-Dickstein, J., Weiss, E., Maheswaranathan, N., Ganguli, S.: Deep unsupervised learning using nonequilibrium thermodynamics. In: ICML (2015)

38. Tao, M., Bao, B.K., Tang, H., Xu, C.: GALIP: generative adversarial CLIPs for text-to-image synthesis. In: CVPR (2023)

39. Van Den Oord, A., Kalchbrenner, N., Kavukcuoglu, K.: Pixel recurrent neural networks. In: ICML (2016)

40. Xiao, Y., Wei, Z., Wang, Z.: A limited memory BFGS-type method for large-scale unconstrained optimization. Comput. Math. Appl. **56**(4), 1001–1009 (2008)

41. Yu, N., Skripniuk, V., Abdelnabi, S., Fritz, M.: Artificial fingerprinting for generative models: rooting DeepFake attribution in training data. In: ICCV (2021)

42. Zhang, H., et al.: StackGAN: text to photo-realistic image synthesis with stacked generative adversarial networks. In: ICCV (2017)

Extrinsic Calibration of Multiple Depth Cameras for 3D Face Reconstruction

Jacopo Burger[1], Giuseppe Facchi[1], Giuliano Grossi[1], Raffaella Lanzarotti[1(✉)], Federico Pedersini[1], and Gianluca Tartaglia[2,3]

[1] PHuSe Lab - Università degli Studi di Milano, Milan, Italy
{jacopo.burger,giuseppe.facchi,giuliano.grossi,raffaella.lanzarotti,
federico.pedersini}@unimi.it
[2] Department of Biomedical, Surgical and Dental Sciences, University of Milan,
Milan, Italy
gianluca.tartaglia@unimi.it
[3] Fondazione IRCCS Cá Granda, Ospedale Maggiore Policlinico, Milan, Italy

Abstract. This paper presents a reliable and robust approach for 3D face reconstruction using low-cost depth cameras. The solution is designed to withstand both camera noise and motion artifacts, making it potentially suitable for dealing with living subjects. Our method utilizes Iterative Closest Point (ICP) registration with a calibration pattern, allowing for scalable acquisition with multiple devices. Experiments are conducted adopting two low-cost depth cameras, and producing the 3D reconstruction of a dummy head to favour a metrological evaluation. The findings indicate that the suggested approach outperforms the alternative method of directly applying ICP to the facial point cloud. Additionally, the outcomes demonstrate that the low-cost solution deviates from the high-quality professional equipment by an average of 0.5 mm, showing the notable accuracy of the proposed method.

Keywords: 3D face reconstruction · low-cost depth camera · ICP method

1 Introduction

3D face reconstruction plays a relevant role in many tasks, such as face recognition [1,2] and facial expression recognition [3,4]. Nonetheless, its significance is most notable in the medical field, where it can effectively highlight the morphological characteristics of the face and head [5–8]. Recently, deep learning has demonstrated its effectiveness in processing 3D facial data by extracting multi-scale features and modeling complex patterns, leading to new frontiers [9–11]. However, large datasets are required to guarantee generalization and robustness. This fact poses a scalability issue, especially in the medical domain where some diseases are rare, which precludes a broad characterization. Moreover the cost and bulkiness of the 3D acquisition systems prevent their widespread application [12].

G. L. Foresti et al. (Eds.): ICIAP 2023, LNCS 14234, pp. 357–368, 2023.
https://doi.org/10.1007/978-3-031-43153-1_30

Nowadays, low-cost acquisition systems such as Microsoft Kinect, Kinect V2, and Intel RealSense D435 are gaining attention due to the improving quality of their output [13]. The adoption of such acquisition tools would favor a large-scale acquisition campaign, resulting in limited costs and requiring a more flexible acquisition setting. However, to date, these tools still have issues related to noise, robustness and 3D accuracy [13].

This paper presents an approach utilizing Intel RealSense D435 devices for acquiring reliable and robust 3D face reconstructions. The solution is designed to potentially withstand both camera noise and motion artifacts that may arise when dealing with living subjects. Additionally, it is scalable by integrating more devices, thereby expanding the field of view. Specifically, our method utilizes the Iterative Closest Point (ICP) registration procedure on multiple frames, adopting a suitable calibration pattern. The choice of multi-shot acquisition during calibration helps in reducing the noise on the depth measurements, improving their 3D accuracy. For these reasons, applying the ICP on a calibration pattern, instead of directly on the point clouds of the face, provides two benefits. Firstly, it enables to exploit averaging over multiple acquisitions, thus increasing the accuracy of the depth maps and, consequently, of the calibration. Secondly, it enhances the robustness of the resulting 3D registration, as the calibration pattern provides complete control over the shape of the overlapping surfaces needed by the ICP algorithm.

The proposed approach effectiveness is demonstrated by comparing it to the basic one-shot solution, where ICP is directly applied to the facial point cloud. For this purpose, two Intel RealSense D435 depth cameras [14] are utilized, and a dummy head with a realistic size and appearance is used for metrological analysis. Additionally, a direct comparison is established between the current 3D reconstruction and one generated by a high-quality equipment (VECTRA M3 3D Imaging System [15]). The comparison reveals that the low-cost and agile solution differs slightly from the 3D reconstruction produced by the high-quality equipment, providing evidence of its reliability and effectiveness.

2   Related Works

3D face reconstruction finds applications in various domains, ranging from entertainment [16,17] to the clinical fields [7]. Traditionally, obtaining high-quality 3D reconstruction has involved the use of expensive equipment, such as laser scanners [12]. However, this approach is not always feasible, due to both the costs and the size of the device. On the other hand, solutions based on single-view methods have been proposed [18], but they suffer from reduced precision and field of view. A compromise can be achieved by adopting multi-view stereo (MVS) reconstruction methods that use low-cost depth cameras. Despite the recent proposals of various MVS methods based on deep learning [19,20], the optimization-based approach [13] remains the most promising in the clinical field. This is due to the need for not only high-quality reconstruction, but also morphological fidelity to reality.

When MVS is adopted, there is the necessity of determining the extrinsic parameters of the involved cameras, so to stitch together the different point clouds produced. Iterative Closest Point (ICP) [21] is the most significant algorithm which is applied for this aim. In some cases [22] a manual pre-annotation step is adopted in order to produce a good initialization. In other cases [13,23] ICP is applied directly on the facial point clouds. However, as claimed in [24] the adoption of specific calibration patterns can help in augmenting the accuracy of the 3D reconstruction. In particular, the pattern should minimize occlusions and contain many 3D features or uneven surfaces, to avoid any ambiguity in the definition of the extrinsic parameters.

3   Materials and Methods

3.1   Overview

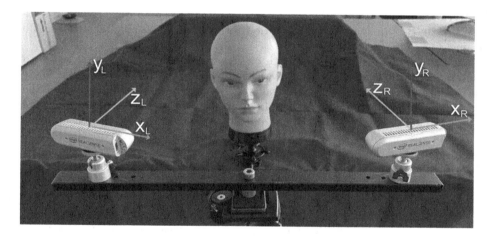

Fig. 1. photograph of the proposed acquisition setup. Two depth cameras acquire independent point clouds of the subject. The two point clouds are then merged by roto-translating them into the same reference frame (3D registration).

The configuration of the proposed acquisition setup is shown in Fig. 1. Two depth cameras, each able to acquire a dense depth map, are pointed towards the subject.

Since each device generates a 3D point cloud defined in its own local reference frame, it is necessary to determine the geometric transformation that maps each 3D point into a common coordinate system. Such a change of reference frame corresponds, geometrically, to the 3D roto-translation that brings the reference frame of one camera onto the frame of the other one.

An accurate measurement of this roto-translation directly on the rig is not possible, as the exact position and orientation of the camera sensors are not accessible. To overcome this problem, normally the two point clouds are merged by finding the transformation that overlaps the portions of the surface recon-structed by both depth cameras, typically with ICP [21]. The accuracy of this registration depends on the shape of the overlapping surface.

Alternatively, we propose to determine the transformation by acquiring, prior to the actual reconstruction, a 3D calibration pattern, whose shape is optimized to maximize the accuracy of the roto-translation between the two depth camera references. The obtained transformation is then used for 3D registration of any pair of facial point clouds, yielding the final facial surface.

This approach should ensure an optimized and, above all, constant accuracy in the registration and, therefore, in the resulting reconstructed surface.

3.2   The Acquisition System

The adopted acquisition devices are the Intel RealSense D435 depth cameras [14], that is active-stereo depth cameras equipped with an IR structured-light pattern projector, used to help the stereo matching. These devices are able to acquire RGB-D images with a resolution up to 1280×720 pixel at a rate of 30 frames/s.

As shown in Fig. 1, the two depth cameras are mounted on a rigid rig, with convergent optical axes pointing towards the subject. The convergence angle has been determined as the best trade-off between a good overlapping of the partial point clouds and the maximal coverage of the reconstructed part of the subject. Thus, taking into account the field-of-view and the depth measurement range of the adopted depth cameras, as well as the desired extension of the facial reconstruction, we determined the most effective acquisition geometry. This involves a 30° convergence angle between the cameras and a distance of $\sim$ 50cm between subject and camera rig, resulting in an inter-camera distance of $\sim$ 33cm.

Generally speaking, the geometrical transformation that maps the point cloud of one depth camera (defined as the *secondary* camera) into the reference of the other one (the *primary* camera) can be defined as the 4×4 roto-translation matrix:

$$\mathbf{M}_{RT} = \begin{bmatrix} \mathbf{R} & | & \mathbf{T} \\ \hline \mathbf{0} & | & 1 \end{bmatrix} = \begin{bmatrix} r_{11} & r_{12} & r_{13} & t_x \\ r_{21} & r_{22} & r_{23} & t_y \\ r_{31} & r_{32} & r_{33} & t_z \\ 0 & 0 & 0 & 1 \end{bmatrix}.$$

As visible in Fig. 1, the optical axes of the two cameras lay approximately on the same plane (the XZ plane for both depth cameras), with approx. 30°-convergent Z axes. Consequently, in our case a good approximation of $\mathbf{M}_{RT}$ is:

$$\mathbf{M}_0 = \begin{bmatrix} \cos(\alpha) & 0 & -\sin(\alpha) & B\cos(\alpha) \\ 0 & 1 & 0 & 0 \\ \sin(\alpha) & 0 & \cos(\alpha) & B\sin(\alpha) \\ 0 & 0 & 0 & 1 \end{bmatrix}.$$

where α is the convergence angle and B the *baseline*, i.e. the inter-camera distance on the rig. $\mathbf{M}_0$ is used as the initial transformation, which is necessary for the iterative ICP algorithm [21] to successfully converge to the best estimation of the transformation $\mathbf{M}_{RT}$.

In order to ensure consistency in comparing the reconstruction accuracy of different approaches using the same ICP methodology and similar acquisition conditions, we employed a rigid, life-sized head dummy as the subject and directly compare the results.

3.3   Uncalibrated Acquisition

This approach can be considered the baseline method, where both devices capture one shot of the subject, producing two independent point clouds: P_L (left) and P_R (right) to be used by the ICP procedure. A first thresholding on the depth is applied to both point clouds, in order to remove any non-subject points (e.g. points in the background). For the considered acquisition setup, this is simply achieved by keeping only points with depth under 800 mm.

P_L and P_R are then aligned using the ICP procedure, adopting $\mathbf{M}_0$ as initial transformation. The right depth camera frame is defined as reference frame, therefore the point cloud P_R is used as *model* and P_L as *scene*. This procedure determines the transformation $\mathbf{M}$, that best aligns P_L to P_R, exploiting the part of both point clouds that overlaps, that is, that are describing the same portion of surface. It is evident that the extension and 3D shape of the overlapping surfaces (i.e., reconstructed in both point clouds) significantly affect the estimation of $\mathbf{M}$ and, consequently, the accuracy of the 3D point registration.

The final $\mathbf{M}$ resulting from the ICP procedure is then applied to P_L, thereby obtaining the registered point cloud. A surface interpolated on these points (applying the screened Poisson method [25]) is considered the final result of the reconstruction.

3.4   Calibrated Acquisition

The idea of the proposed approach is to perform a calibration before the subject acquisition, with the aim of determining the transformation matrix $\mathbf{M}$ on the basis of multiple acquisitions of a calibration pattern. $\mathbf{M}$ is then utilized for the registration of the face point clouds, without the need of running ICP on such acquired points directly.

Calibration Pattern Design. Since the 3D shape of the calibration pattern influences almost entirely the accuracy in the estimation of $\mathbf{M}$, its design is crucially important. We first considered the criteria outlined in [24], in which the importance to avoid occlusions is shown. Indeed, the absence of occlusions would ensure that all point clouds sample the same surface. Thus, we designed an asymmetric (to avoid 3D overlapping ambiguities) cross-shaped pattern, where all surfaces are tilted in such a way to be visible and present the same range of foreshortening (the angle between visual ray and surface normal) to all cameras.

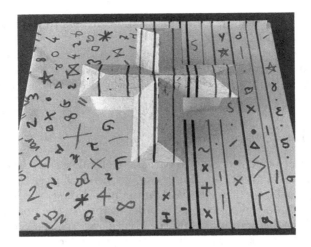

Fig. 2. The adopted calibration pattern. The asymmetrical cross is lifted to increase the range of the acquired depths. The texture added to all surfaces improve significantly the reconstruction performance of the D435 depth cameras.

However, this constraint leads necessarily to shapes with small depth ranges. We carried out calibrations with different pattern shapes and experienced that calibration patterns presenting a significant depth range, although generating occlusions, yield much more accurate estimations of **M**. Through these experiments, we finally got to the optimized pattern shape shown in Fig. 2, consisting of an asymmetric cross held lifted over a planar surface, to increase the range of depths acquired by the cameras. Moreover, we added several pictorial elements on all visible surfaces, having experienced that the active stereo algorithm yields significantly better performance if the textural content is enriched, with respect to the sole pseudo-random structured light pattern projected by the cameras.

The Calibration Procedure. The calibration pattern, mounted on a pedestal, is positioned in front of the acquisition system, approximately in the same spatial region occupied later by the subjects. Taking advantage of the pattern immobility, we capture a sequence of 90 subsequent depth maps and generate a *mean depth map* for each camera. Since the noise superimposed to the depth measurements has shown to be zero-mean, the mean depth map is characterized by a significantly reduced noise level.

Specifically, we create a vector $D(u, v)$ for each pixel, consisting solely of non-zero depth values (zero values in the vector indicate unsuccessful depth measurements). The mean depth map $\overline{D}(u, v)$ considers only those pixels whose standard deviation does not exceed a specified threshold $\sigma_{\max}$. We found experimentally $\sigma_{max} = 1$ mm as the acceptance threshold leading to the most accurate result. The mean depth maps $\overline{D}_L(u, v)$ and $\overline{D}_R(u, v)$, obtained from the left and right depth cameras, respectively, are fed into the ICP procedure, following the same approach as the uncalibrated method, using $\mathbf{M}_0$ as the initial estimate. The transformation matrix $\mathbf{M}_{RT}$ obtained from the ICP procedure is then referred to as the Calibration Matrix.

After successfully calibrating the system, the acquisition procedure consists of two steps. Firstly, a single frame is captured using both depth cameras, resulting in two RGB images and their corresponding depth maps P_L and P_R. Secondly, the roto-translation $\mathbf{M}_{RT}$ is applied to P_L, and the resulting points are merged with P_R to generate the final point cloud.

4   Experimental Results

Two kinds of tests have been carried out to evaluate the reconstruction accuracy of the proposed system. At first, the calibrated approach is compared to the uncalibrated method. Then, the results of the calibration-based technique are compared to those obtained by a professional 3D facial reconstruction equipment [15], which can be taken as reference measurement.

4.1   Calibrated vs Uncalibrated Approach

The first test is aimed at validating the superiority of the calibrated approach, compared to the uncalibrated acquisition.

To this purpose, we carried out 10 acquisitions of the same scene (same subject and same camera geometry) and evaluated the variability of the transformation matrix obtained from each acquisition, for both the calibrated and uncalibrated approaches. We refer to the standard deviation of the repeatedly estimated transformation parameters as a reliable estimation of the precision of the considered approach.

More specifically, we assessed the standard deviation of each roto-translation parameters characterizing the transformation matrix, that is the Euler angles $\mathbf{R} = [r_X, r_Y, r_Z]$, defining the rotation around the three axes, and the translation vector $\mathbf{T} = [t_X, t_Y, t_Z]$. The obtained results are shown in Table 1.

Table 1. Comparison of the standard deviations of the transformation parameters, for the calibrated and uncalibrated approach.

Standard Deviation						
Method	r_X (°)	r_Y (°)	r_Z (°)	t_X (mm)	t_Y (mm)	t_Z (mm)
Uncalibrated	0.1348	1.5239	0.1378	0.0112	0.0011	0.0060
Calibrated	0.0334	0.0327	0.0625	0.0003	0.0002	0.0002

Notably, the standard deviations in Table 1 are characterized by significantly lower values for the calibrated approach on all parameters, thus evidencing a higher accuracy of this method.

Another interesting assessment can be obtained computing the RMSE between point clouds: given that the standard deviation in the uncalibrated method is higher than the one in the second method, we expect the same discrepancy between the RMSE. Specifically, the RMSE is computed, after applying the ICP procedure, from the Euclidean distances between each pair of points in the inlier correspondence set. The set of inlier correspondences, denoted as $\mathcal{K}$, is constructed by considering point cloud P_R and point cloud P_L transformed using the obtained transformation matrix $\mathbf{M}$. This set includes pairs of points whose distance is less than a threshold of $T = 10$ mm. This is the radius of distance from each point (laying in the overlapping region) in P_L in which the neighbour search will try to find a corresponding point in P_R. The RMSE of $\mathcal{K}$ is defined as:

$$\text{RMSE} = \sqrt{\frac{\sum_{d \in \mathcal{K}} d^2}{|\mathcal{K}|}}.$$

As expected, the results indicate that the calibrated method leads to a significantly lower RMSE than the uncalibrated method. Specifically, the calibrated RMSE was 0.0037, while the uncalibrated RMSE was 0.0583, exhibiting an order of magnitude difference.

In addition to the quantitative comparisons, in Fig. 3 we provide further visual evidence of the superior accuracy of the calibrated approach through a direct comparison of the surfaces reconstructed by both methods. The two different colors are used to distinguish the left and right partial point clouds. The figure shows clearly a registration error in the uncalibrated approach, particularly evident in the regions of nose and mouth, which are actually the overlapping regions exploited by the ICP to determine the transformation matrix for the registration. This result witnesses the instability of the ICP procedure when the extension of the overlapping regions is rather limited, as in this case.

4.2   Comparison to a Reference Measurement

From a metrological perspective, the most reliable method to assess the accuracy of the proposed system is through a direct comparison of our results with those obtained from a state-of-the-art instrument, which can be regarded as a reference measurement.

To this aim, we carried out a 3D measurement of the same test dummy used in Sect. 4.1 (shown in Fig. 3), using a professional equipment (VECTRA M3 3D Imaging System [15]) commonly used in medical applications, and compared the resulting 3D surface with the final surface generated by our calibrated approach.

To get a meaningful comparison of the 3D shape of the two surfaces, we cropped both meshes to select the same central portion of the dummy face. The two meshes are then overlapped through 3D registration, reaching their minimum inter-distance. In this situation, the statistics of the distances between the points of one mesh and the surface of the other one represent a reliable measurement of the difference between the two surfaces. For each point of our mesh, we compute

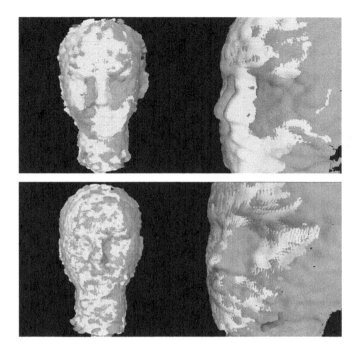

Fig. 3. Final surfaces, obtained with the uncalibrated (above) and the calibrated (below) method.

its distance from the closest surface point of the reference mesh generated by the state-of-art device.

The histogram in Fig. 4 shows the statistical distribution of the obtained distances. As reported in the figure, the mean distance is $\mu_d = 0.534$ mm, with a standard deviation $\sigma_d = 0.427$ mm and a maximum distance $d_{max} = 3.382$ mm.

According to these results, the 3D discrepancy is bounded by 1 mm for most of the surface. This is a remarkable result, especially considering the simplicity of the proposed acquisition system with respect to the reference instrument [15].

Figure 5 shows the reconstructed surface, with a *difference map* applied on it, whose color indicates the local distance between the surfaces. The map evidences that the greater errors occur mainly in regions characterized by high local curvature, where shape differences can also arise from variations in interpolation settings.

Apart from these regions, the majority of the surface (the 'red' regions in Fig. 5) exhibits a significantly small disparity, well below 0.5 mm, thereby reaffirming the remarkable accuracy of the proposed system.

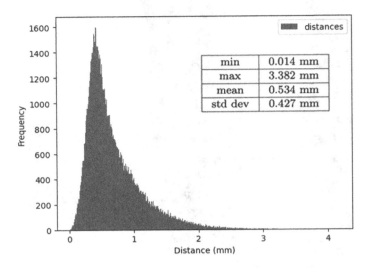

Fig. 4. Histogram of the distances between our surface and the reference one obtained with a professional equipment (VECTRA M3 3D Imaging System).

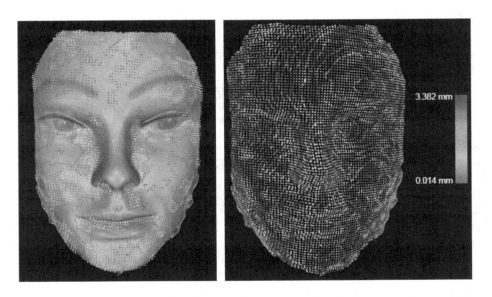

Fig. 5. The Difference Map (with and without texture) between our surface and the reference one. The color of each point represents the local Euclidean distance between the two surfaces.

5   Conclusions

In view of the presented results, we can conclude that the proposed technique is able to generate facial reconstructions with remarkable accuracy, even when utilizing low-cost depth sensors and a simple setup.

The reconstruction accuracy, combined with the simplicity, affordability and compactness of the setup, makes the proposed system particularly well suited for all applications, mainly in the medical field, requiring the acquisition of large data-sets of 3D facial data [9–11].

Although the system presented in this work has been developed and tested with two depth cameras, this technique can be effortlessly extended to work with an arbitrary number of cameras (with the only constraint being the visibility of the calibration pattern from all viewpoints), as the proposed procedure is basically independent from the number of partial point clouds merged into the final surface. This would allow, for instance, to increase significantly the extent of the reconstructed face.

Acknowledgment. We would like to acknowledge the Laboratory of Anatomy of the Stomatognathic System at the Department of Biomedical Sciences for Health, University of Milan, for providing us with access to the images obtained using the VECTRA M3 3D Imaging System.

References

1. Li, M., Huang, B., Tian, G.: A comprehensive survey on 3D face recognition methods. Eng. Appl. Artif. Intell. **110**, 104669 (2022)
2. Grossi, G., Lanzarotti, R., Lin, J.: Robust face recognition providing the identity and its reliability degree combining sparse representation and multiple features. Int. J. Pattern Recognit Artif Intell. **30**(10), 1656007 (2016)
3. Sandbach, G., Zafeiriou, S., Pantic, M., Yin, L.: Static and dynamic 3D facial expression recognition: a comprehensive survey. Image Vis. Comput. **30**(10), 683–697 (2012)
4. Cuculo, V., D'Amelio, A.: OpenFACS: an open source FACS-based 3D face animation system. In: Zhao, Y., Barnes, N., Chen, B., Westermann, R., Kong, X., Lin, C. (eds.) ICIG 2019. LNCS, vol. 11902, pp. 232–242. Springer, Cham (2019). https://doi.org/10.1007/978-3-030-34110-7_20
5. Hammond, P., et al.: The use of 3D face shape modelling in dysmorphology. Arch. Dis. Child. **92**(12), 1120 (2007)
6. Hallgrímsson, B., et al.: Automated syndrome diagnosis by three-dimensional facial imaging. Genet. Med. **22**(10), 1682–1693 (2020)
7. Katz, D.C., et al.: Facial shape and allometry quantitative trait locus intervals in the diversity outbred mouse are enriched for known skeletal and facial development genes. PLoS ONE **15**(6), e0233377 (2020)
8. Dolci, C., Sansone, V.A., Gibelli, D., Cappella, A., Sforza, C.: Distinctive facial features in Andersen-tawil syndrome: a three-dimensional stereophotogrammetric analysis. Am. J. Med. Genet. A **185**(3), 781–789 (2021)
9. Cao, Y., Liu, S., Zhao, P., Zhu, H.: RP-NET: a PointNet++ 3D face recognition algorithm integrating RoPS local descriptor. IEEE Access **10**, 91245–91252 (2022)

10. Zhang, J., Gao, K., Fu, K., Cheng, P.: Deep 3D facial landmark localization on position maps. Neurocomputing **406**, 89–98 (2020)
11. Kim, D., Hernandez, M., Choi, J., Medioni, G.: Deep 3D face identification. In: 2017 IEEE International Joint Conference on Biometrics (IJCB), pp. 133–142. IEEE (2017)
12. Gibelli, D., Dolci, C., Cappella, A., Sforza, C.: Reliability of optical devices for three-dimensional facial anatomy description: a systematic review and meta-analysis. Int. J. Oral Maxillofac. Surg. **49**(8), 1092–1106 (2020)
13. Peng, H., Yang, L., Li, J.: Robust and high-fidelity 3D face reconstruction using multiple RGB-d cameras. Appl. Sci. **12**(22), 11722 (2022)
14. Intel RealSense Product Family D400 Series, rev. 015. Intel Corp. (2023). https://www.intelrealsense.com/support/
15. Canfield Scientific. VECTRA M3 3D Imaging System (2023). https://www.canfieldsci.com/imaging-systems/vectra-m3-3d-imaging-system/
16. Lou, J., et al.: Realistic facial expression reconstruction for VR HMD users. IEEE Trans. Multimedia **22**(3), 730–743 (2019)
17. Song, S.L., Shi, W., Reed, M.: Accurate face rig approximation with deep differential subspace reconstruction. ACM Transactions on Graphics (TOG) **39**(4), 34–1 (2020)
18. Jiang, L., Zhang, J., Deng, B., Li, H., Liu, L.: 3D face reconstruction with geometry details from a single image. IEEE Trans. Image Process. **27**(10), 4756–4770 (2018)
19. Wang, X., Guo, Y., Yang, Z., Zhang, J.: Prior-guided multi-view 3D head reconstruction. IEEE Trans. Multimedia **24**, 4028–4040 (2021)
20. Yao, Y., Luo, Z., Li, S., Fang, T., Quan, L.: MVSNet: depth inference for unstructured multi-view stereo. In: Proceedings of the European Conference on Computer Vision (ECCV), pp. 767–783 (2018)
21. Arun, K.S., Huang, T.S., Blostein, S.D.: Least-squares fitting of two 3-D point sets. IEEE Trans. Pattern Anal. Mach. Intell. **5**, 698–700 (1987)
22. Amor, B.B., Ardabilian, M., Chen, L.: New experiments on ICP-based 3D face recognition and authentication. In: 18th International Conference on Pattern Recognition (ICPR 2006), vol. 3, pp. 1195–1199. IEEE (2006)
23. Wang, C.-W., Peng, C.-C.: 3D face point cloud reconstruction and recognition using depth sensor. Sensors **21**(8), 2587 (2021)
24. Fukushima, N.: ICP with depth compensation for calibration of multiple ToF sensors. In: 2018–3DTV-Conference: The True Vision-Capture, Transmission and Display of 3D Video (3DTV-CON), pp. 1–4. IEEE (2018)
25. Kazhdan, M., Hoppe, H.: Screened poisson surface reconstruction. ACM Trans. Graph. **32**, 3 (2013). https://doi.org/10.1145/2487228.2487237

A Deep Learning Based Approach for Synthesizing Realistic Depth Maps

Patricia L. Suárez[1]([✉]), Dario Carpio[1], and Angel Sappa[1,2]

[1] ESPOL Polytechnic University, Guayaquil, Ecuador
{plsuarez,dncarpio,asappa}@espol.edu.ec
[2] Computer Vision Center, Barcelona, Spain
asappa@cvc.uab.es
http://www.espol.edu.ec

Abstract. This paper presents a novel cycle generative adversarial network (CycleGAN) architecture for synthesizing high-quality depth maps from a given monocular image. The proposed architecture uses multiple loss functions, including cycle consistency, contrastive, identity, and least square losses, to enable the generation of realistic and high-fidelity depth maps. The proposed approach addresses this challenge by synthesizing depth maps from RGB images without requiring paired training data. Comparisons with several state-of-the-art approaches are provided showing the proposed approach overcome other approaches both in terms of quantitative metrics and visual quality.

Keywords: depth maps-like · transfer domain · cross-spectral super-resolution

1 Introduction

The ability to generate synthetic depth maps with high fidelity and accuracy has garnered significant attention in the field of computer vision. Depth maps provide crucial perceptual information, enabling a wide range of applications such as 3D reconstruction, scene understanding, and object recognition, just to mention a few. However, acquiring depth maps from real-world scenarios is a challenging and expensive task, often requiring specialized sensors or complex calibration procedures. In order to address this limitation, the use of deep learning-based generative models has emerged as a promising solution.

The significance of synthesizing depth maps lies in its wide range of potential applications. Depth maps can facilitate object detection and recognition in challenging environments, enable accurate 3D scene understanding for robotics [17] and autonomous driving [7], and enhance virtual reality experiences (e.g., [11,18]). Furthermore, the ability to synthetically generate depth maps opens up new possibilities for data augmentation, reducing the need for extensive data collection and annotation [16].

Exploiting the possibility of using synthesized depth maps, [19] presents a method for unsupervised learning of depth estimation and visual odometry using

G. L. Foresti et al. (Eds.): ICIAP 2023, LNCS 14234, pp. 369–380, 2023.
https://doi.org/10.1007/978-3-031-43153-1_31

deep feature reconstruction. The proposed approach leverages the power of deep neural networks to learn depth estimation and motion estimation directly from unlabeled monocular sequences. In [9] the authors propose the fusion of color and hallucinated depth map for enhancing image segmentation. The fusion of depth with RGB increases the accuracy of semantic segmentation, four different fusion strategies are evaluated on computer-generated synthetic datasets. Also focusing on scene understanding, [12] proposes a CNN-based approach to predict occluded portions of a scene by hallucinating semantic and depth information. These are just a few illustrations of the usage of depth maps generated from monocular views. In all cases, the quality of results depends on the accuracy of the synthesized depth maps. Hence, having in mind this dependency on map precision, in the current paper a CycleGAN architecture is proposed to generate accurate depth maps. The proposed model uses multiple loss functions. The key contribution of our work lies in the incorporation of multiple loss functions into the generative architecture. The proposed approach leverages the cycle-consistency loss [4,20], which enforces the reconstruction of the original input from the synthesized depth map and vice versa. Additionally, the integration of contrastive [2], identity and relativistic losses further enhance the quality and realism of the generated depth maps. By combining these loss functions, the proposed architecture achieves a balance between stability and diversity in the synthesized depth maps. The controllable structure guided self-content preserving loss encourages the preservation of distinct image features [15], the identity loss ensures consistency in preserving structural information [8], and the generative adversarial model that enhances the perceptual quality and realism of the generated depth maps [6].

Extensive evaluations of the performance and quality of the synthesized depth maps through comprehensive experiments and comparisons with state-of-the-art methods are provided. The manuscript is organized as follows; Sect. 2 presents the proposed approach. Then, Sect. 3, depicts experimental results and comparison with state-of-art approaches. Both quantitative and qualitative results are provided showing the improvements reached with the proposed approach. Finally, conclusions and future works are given in Sect. 4.

2   Depth Map Generation

This section presents the architecture proposed for generating synthetic depth maps, building upon the approach presented in [14], which was initially proposed for generating thermal-like representations. Our objective is to leverage the insights and techniques learned from synthesizing thermal-like images and extend them to the generation of depth information, which plays a pivotal role in various computer vision tasks. The depth of information provides valuable insights about the objects present in a scene, which can be extracted and utilized to enhance the performance of other computer vision algorithms. Motivated by this concept, the original approach is extended to enable the generation of synthetic depth maps from RGB images. This extension aims to harness the potential of depth information and empower computer vision systems with a richer understanding of the scene.

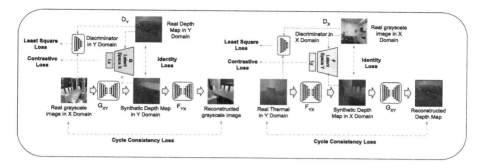

Fig. 1. Cycle GAN proposed architecture.

The knowledge gained from generating thermal-like representations is leveraged to exploit the similarities and underlying principles between thermal and depth data in the current study. Although they capture different aspects of the scene, both modalities provide valuable information for understanding the environment. Therefore, we adapt and enhance the existing architecture to accommodate the generation of depth maps. The architecture of our approach is presented in Fig. 1.

The proposed architecture leverages the capabilities of generative adversarial networks (GANs) and the Cycled GAN framework [20]. By utilizing two generators ($G1$ and $G2$) a framework is established for the translation of grayscale images into visually consistent and realistic depth maps. The generators are trained to map grayscale images to depth maps, while the discriminators provide feedback on the authenticity and quality of the generated depth maps. In order to ensure the quality and accuracy of the synthesized depth maps, multiple loss functions are incorporated into this work. These loss functions are designed to guide the training process and encourage the generation of depth maps that closely resemble the ground truth depth information. The cycle consistency loss, Eq. (1), enforces the preservation of structural information during the translation process. It consists of two components: the forward cycle loss and the backward cycle loss. The forward cycle loss measures the discrepancy between the original grayscale image and the reconstructed grayscale image obtained by applying $G1$ and $G2$ consecutively. Similarly, the backward cycle loss measures the discrepancy between the original and reconstructed depth maps obtained by applying $G2$ and $G1$ consecutively. These losses encourage the preservation of important visual features and enhance the realism of the generated depth maps. This loss can be defined as:

$$\mathcal{L}_{\text{cycle}}(G_1, G_2) = \mathbb{E}_{x \sim p_{\text{data}}(x)}[\|x - G_2(G_1(x))\|_1] \\ + \mathbb{E}_{y \sim p_{\text{data}}(y)}[\|y - G_1(G_2(y))\|_1]. \tag{1}$$

In addition to the cycle-consistency loss, identity loss, Eq. (2), is employed in this study to ensure that the generated depth maps retain relevant visual information from the source domain. This loss promotes the preservation of the input grayscale image's identity during the translation process, preventing unnecessary modifications. The identity loss is defined as follows:

$$L_{\text{identity}} = E_{x \sim p_{\text{data}}(x)}[\|x - G1(x)\|_1] \\ + E_{y \sim p_{\text{data}}(y)}[\|y - G2(y)\|_1]. \tag{2}$$

To enforce meaningful relationships between the generated depth maps and the corresponding real depth maps, the introduced contrastive loss in this paper encourages the generator to produce depth maps that exhibit similar depth values and spatial structures to the real depth maps. The contrastive loss comprises two components, including the cycle consistency loss, which measures the discrepancy or difference between the reconstructed depth map and the real depth map obtained by applying G1 to the real depth map. Similarly, the other generator that produces the identity-generated depth map measures the discrepancy or difference of the generated depth map compared to the depth map obtained by applying G2 to the real depth map.

Contrastive loss has also been implemented to minimize the distance or dissimilarity of similar pairs of data points and maximize the distance or dissimilarity of dissimilar pairs of data points in a given dataset. According to [1], this loss can be defined as:

$$\mathcal{L}_{\text{contrastive}}(\hat{Y}, Y) = \sum_{l=1}^{L} \sum_{s=1}^{S_l} \ell_{\text{contr}} \left(\hat{v}_l^s, v_l^s, \bar{v}_l^s \right), \tag{3}$$

where $V_l \in \mathbb{R}^{S_l \times D_l}$ represents a tensor whose shape depends on the model architecture. The variable S_l denotes the number of spatial locations of the tensor. Consequently, the notation $v_l^s \in \mathbb{R}^{D_l}$ is employed to refer to the D_l-dimensional feature vector at the s-th spatial location. Additionally, $\bar{v}_l^s \in \mathbb{R}^{(S_l-1) \times D_l}$ represents the collection of feature vectors at all other spatial locations except the s-th one.

The proposed architecture additionally incorporates the least square loss, which serves as a variant of the adversarial loss. This loss function is designed to encourage the generated depth maps to closely align with the distribution of real depth maps, thereby enhancing the realism of the synthesized results. The least-square loss achieves this by minimizing the squared differences between the predictions of the discriminators and their corresponding target labels. By utilizing this loss, the training process is further stabilized. Specifically, the binary cross-entropy loss is replaced with a least square loss formulation. The mathematical definition of the least square loss for both the generator and discriminator components can be expressed as follows:

$$L_D^{\text{LS-GAN}} = \frac{1}{2} E_{x_r \sim \mathbb{P}}[(D(x_r) - 1)^2] + \frac{1}{2} E_{x_f \sim \mathbb{Q}}[D(x_f)^2] \tag{4}$$

$$L_G^{\text{LS-GAN}} = \frac{1}{2} E_{x_f \sim \mathbb{Q}}[(D(x_f) - 1)^2], \tag{5}$$

where x_r represents a real depth map from the real data distribution, x_f represents a generated (fake) depth map from the generator, $D(x_r)$ represents the discriminator's output (probability) for a real depth map x_r, and $D(x_f)$ represents the discriminator's output (probability) for a generated depth map x_f. To enhance the synthesis of depth maps, instance normalization is employed, which adjusts the features of each depth map individually. Applying this normalization process effectively reduces style differences between the generated and real-depth maps, leading to improved overall quality and realism in the synthesized depth maps.

By combining the loss functions presented above, and by using instance normalization, the proposed architecture aims to improve the accuracy, quality, and realism of the synthesized depth maps. It enables the generation of depth maps that capture important depth information in a visually consistent and meaningful manner, which can benefit a wide range of computer vision tasks, including depth estimation, scene understanding, and 3D reconstruction. The final loss function is obtained as:

$$\begin{aligned}
\mathcal{L}_{\text{final}} &= \lambda_1 \mathcal{L}_{\text{LSGAN}}(G, D, X, Y) + \lambda_2 \mathcal{L}_{\text{cont}}(G, H, X) \\
&+ \lambda_3 \mathcal{L}_{\text{cont}}(G, H, Y) + \lambda_4 \mathcal{L}_{\text{identity}(G,F)} + \lambda_5 \mathcal{L}_{\text{cycle}(G,F)},
\end{aligned} \tag{6}$$

where λ_i are empirically defined. Lastly, the modification proposed in [14] is also considered to optimize the performance of the image generator model. The proposed adjustment entails modifying the beta1 parameter in the Adam optimizer, which governs the decay rate of past gradient information. The beta1 parameter of the Adam optimizer has been modified to 0.5. This adjustment aimed to enhance the training efficiency of the model by placing more emphasis on the current gradient information. By reducing the decay rate of historical gradients, the optimizer became more responsive to recent updates, potentially resulting in improved convergence speed during the training process.

3   Experimental Results

This section presents the experimental results obtained with the proposed model. Details of the experimental setup employed during the training process are provided. Furthermore, an extensive comparison is performed to assess the performance of the proposed method against several state-of-the-art image-to-image translation methods.

3.1   Datasets

The NYU v2 dataset [13] is used for training the different architectures. It consists of 1449 RGBD pairs captured using the Microsoft Kinect sensor. Specifically, for the research, the first 1000 pairs from the dataset were selected for training, while the remaining 449 pairs were used for testing. As a preprocessing step, all the images were resized to 256×256 pixels to ensure consistency and facilitate the training process. The NYU v2 dataset provides a diverse range of indoor scenes, enabling the evaluation of the proposed approach's performance and generalization ability across various real-world scenarios.

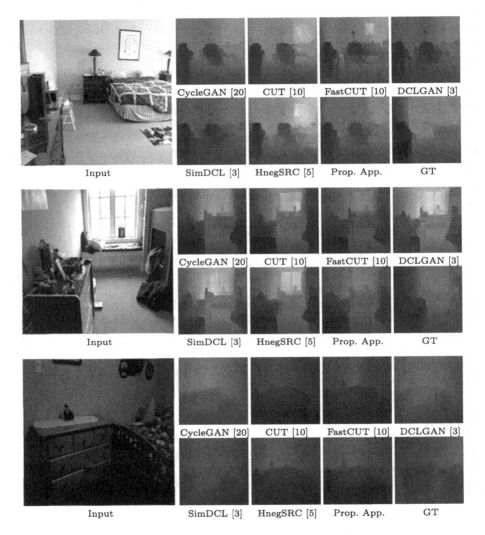

Fig. 2. Experimental results: (*1st. col.*) input images; (*2nd.-5th. col.*) results of state-of-the-art approaches together with results from the proposed approach and the corresponding ground truth depth map from NYU v2 test set.

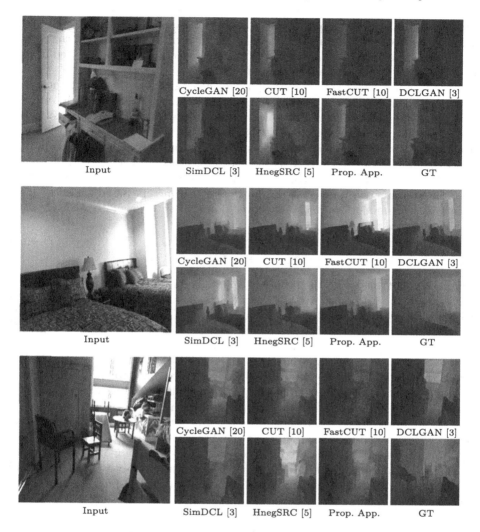

Fig. 3. Experimental results: (*1st. col.*) input images; (*2nd.-5th. col.*) results of state-of-the-art approaches together with results from the proposed approach and the corresponding ground truth depth map from NYU v2 test set.

3.2 Training Details

The proposed approach underwent extensive training to ensure the effectiveness of the synthetic depth maps generated. Each of the techniques and our proposed approach was included in the training process with the NYU data set. This training process was conducted for a total of 400 epochs, with each epoch consisting of multiple iterations. A batch size of 1 was employed, meaning that each iteration processed a single RGB image and a single depth map. To facilitate the training process and expedite computation, a high-performance NVIDIA GeForce RTX 3090 Ti graphics card was utilized. This powerful hardware accelerated the training procedure by efficiently processing the complex computations involved in the training of the generative adversarial network.

The training process takes approximately 20 h, reflecting the significant computational demands of the training process and the large number of iterations performed. This extended duration was necessary to allow the network to learn and refine its parameters to generate high-quality synthetic depth maps that accurately capture the underlying depth information. Throughout the training process, the network iteratively learned to optimize the various loss functions, including cycle-consistency loss, identity loss, contrastive loss, and least square loss. By continuously updating the network's parameters based on these loss functions, the model gradually improved its ability to generate realistic and accurate depth maps from RGB inputs.

3.3   Comparison with SOTA Methods

The results obtained with the proposed approach have been compared with several state-of-the-art generative adversarial networks: CycleGAN [20], CUT [10], Fast CUT [10], Hneg [5], DCL and SimDCL [3]. These methods were trained using the same dataset and experimental configurations. Quantitative results of the comparison are presented in Table 1, where the metrics used for quantitative evaluation are Structural Similarity Index (SSIM) and Peak Signal-to-Noise Ratio (PSNR). The values shown in the table correspond to the average performance of the test images from the NYU v2 dataset. The table provides a comprehensive overview of the performance of each method in terms of SSIM and PSNR scores. The higher the SSIM score, the better the structural similarity between the generated images and the ground truth. On the other hand, a higher PSNR score indicates better reconstruction fidelity. By comparing the results obtained by the proposed approach with those of the state-of-the-art methods, it can be appreciated the superior performance in terms of image quality and reconstruction accuracy.

Figures 2 and 3 display a collection of grayscale input images from the test set of the NYU v2 dataset, along with the results obtained by each of the aforementioned state-of-the-art methods. The figure shows the input grayscale images, the corresponding depth map representations generated by each method, and the corresponding ground truth depth maps. Through a visual examination of those figures, the performance of the different methods can be assessed in accurately capturing the depth information from the grayscale input images. This qualitative analysis provides insights into the strengths and limitations of each approach in producing high-quality depth maps. By comparing the results of the proposed method with those of the state-of-the-art methods, the effectiveness of the proposed method in generating visually appealing and accurate depth maps can be appreciated.

Finally, in Fig. 4 the depth values of a case study are depicted with different colors in order to highlight the quality of the shapes obtained with the proposed

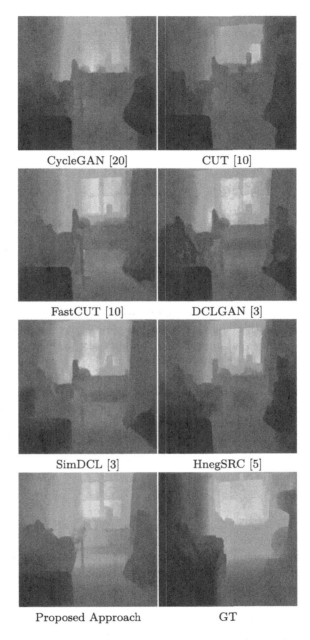

Fig. 4. Enlargement of the depth maps presented in Fig. 2 *(middle)* to compare results from the proposed approach with respect to the state-of-the-art.

Table 1. Average results on synthetic image generation from the NYU v2 testing set. Best results in **bold**.

Approaches	NYU Dataset	
	PSNR	SSIM
CycleGAN [20]	17,2879	0,8036
CUT [10]	16,9203	0,7975
FastCUT [10]	17,0133	0,7987
DCLGAN [3]	16,8829	0,7966
SimDCL [3]	16,9831	0,7920
HnegSRC [5]	16,9805	0,7992
Proposed Approach	**17,9773**	**0,8245**

approach. The varying shades and gradients represent the different depth levels captured by each method. This visual distinction allows for a better understanding of the depth estimation capabilities and the level of fidelity achieved by the proposed approach compared to the state-of-the-art approaches.

From the experimental results, it can be inferred that the proposed architecture has shown the best performance for generating synthetic depth maps. The generated depth maps exhibit high quality and are visually consistent with the corresponding real-depth maps. The inclusion of multiple loss functions has contributed to the stability and improved the quality of the generated depth maps. Comparative evaluations with state-of-the-art methods have demonstrated that the proposed architecture outperforms existing approaches in terms of generating high-quality depth maps. The ability to accurately capture depth information from RGB images is crucial for various computer vision tasks, and our architecture shows the potential in providing valuable depth information that can be extracted and utilized to enhance the performance of other computer vision algorithms.

4   Conclusions

This paper presents a novel CycleGAN architecture based on the usage of multiple loss functions for synthesizing high-quality depth maps. The integration of cycle consistency, contrastive, identity, and relativistic losses has resulted in improved network stability and the generation of high-quality depth maps. Comparisons with state-of-the-art approaches have demonstrated the superior performance of the proposed method. As for future work, different potential research directions will be explored. One avenue is to investigate the use of advanced loss functions, such as perceptual loss or style loss, to further enhance the visual quality and realism of the synthesized depth maps. Additionally, exploring more sophisticated network architectures, data augmentation techniques, and domain adaptation methods could enhance the generalization ability and overall performance of the depth map synthesis model. Furthermore, focusing on domain-specific improvements and understanding the specific requirements of depth map synthesis in different application domains could lead to tailored optimizations and advancements. Continuing to advance the field of depth map synthesis and addressing these future research directions can unlock new possibilities and applications in computer vision, robotics, augmented reality, and related fields.

Acknowledgements. This material is based upon work supported by the Air Force Office of Scientific Research under award number FA9550-22-1-0261; and partially supported by the Grant PID2021-128945NB-I00 funded by MCIN/AEI/10.13039/501100011033 and by "ERDF A way of making Europe"; the "CERCA Programme/Generalitat de Catalunya"; and the ESPOL project CIDIS-12-2022.

References

1. Andonian, A., Park, T., Russell, B., Isola, P., Zhu, J.Y., Zhang, R.: Contrastive feature loss for image prediction. In: Proceedings of the IEEE/CVF International Conference on Computer Vision, pp. 1934–1943 (2021)
2. Chen, Q., Koltun, V.: Photographic image synthesis with cascaded refinement networks. In: Proceedings of the IEEE International Conference on Computer Vision, pp. 1511–1520 (2017)
3. Han, J., Shoeiby, M., Petersson, L., Armin, M.A.: Dual contrastive learning for unsupervised image-to-image translation. In: Proceedings of the IEEE/CVF Conference on Computer Vision and Pattern Recognition Workshops (2021)
4. Isola, P., Zhu, J.Y., Zhou, T., Efros, A.A.: Image-to-image translation with conditional adversarial networks. In: Proceedings of the IEEE Conference on Computer Vision and Pattern Recognition, pp. 1125–1134 (2017)
5. Jung, C., Kwon, G., Ye, J.C.: Exploring patch-wise semantic relation for contrastive learning in image-to-image translation tasks. arXiv preprint arXiv:2203.01532 (2022)
6. Khan, M.F.F., Troncoso Aldas, N.D., Kumar, A., Advani, S., Narayanan, V.: Sparse to dense depth completion using a generative adversarial network with intelligent sampling strategies. In: Proceedings of the 29th ACM International Conference on Multimedia, pp. 5528–5536 (2021)
7. Lee, S., Lee, J., Kim, D., Kim, J.: Deep architecture with cross guidance between single image and sparse lidar data for depth completion. IEEE Access **8**, 79801–79810 (2020)
8. Liu, J., et al.: Identity preserving generative adversarial network for cross-domain person re-identification. IEEE Access **7**, 114021–114032 (2019)
9. Mondal, T.G., Jahanshahi, M.R.: Fusion of color and hallucinated depth features for enhanced multimodal deep learning-based damage segmentation. Earthq. Eng. Eng. Vib. **22**, 55–68 (2023). https://doi.org/10.1007/s11803-023-2155-2
10. Park, T., Efros, A.A., Zhang, R., Zhu, J.Y.: Contrastive learning for unpaired image-to-image translation. In: European Conference on Computer Vision (2020)
11. Ranasinghe, N., et al.: Season traveller: multisensory narration for enhancing the virtual reality experience. In: Proceedings of the CHI Conference on Human Factors in Computing Systems, pp. 1–13 (2018)

12. Schulter, S., Zhai, M., Jacobs, N., Chandraker, M.: Learning to look around objects for top-view representations of outdoor scenes. In: Proceedings of the European Conference on Computer Vision (ECCV), pp. 787–802 (2018)
13. Silberman, N., Hoiem, D., Kohli, P., Fergus, R.: Indoor segmentation and support inference from RGBD images. In: Fitzgibbon, A., Lazebnik, S., Perona, P., Sato, Y., Schmid, C. (eds.) ECCV 2012. LNCS, vol. 7576, pp. 746–760. Springer, Heidelberg (2012). https://doi.org/10.1007/978-3-642-33715-4_54
14. Suárez, P.L., Sappa, A.D.: Toward a thermal image-like representation. In: Proceedings of the International Joint Conference on Computer Vision, Imaging and Computer Graphics Theory and Applications (2023)
15. Tang, H., Liu, H., Sebe, N.: Unified generative adversarial networks for controllable image-to-image translation. IEEE Trans. Image Process. **29**, 8916–8929 (2020)
16. Tian, Z., et al.: Adversarial self-attention network for depth estimation from RGB-d data. In: Proceedings of the IEEE/CVF Conference on Computer Vision and Pattern Recognition (2020)
17. Valencia, A.J., Idrovo, R.M., Sappa, A.D., Guingla, D.P., Ochoa, D.: A 3D vision based approach for optimal grasp of vacuum grippers. In: Proceedings of the IEEE International Workshop of Electronics, Control, Measurement, Signals and their Application to Mechatronics (2017)
18. Wei, W., Qi, R., Zhang, L.: Effects of virtual reality on theme park visitors' experience and behaviors: a presence perspective. Tour. Manage. **71**, 282–293 (2019)
19. Zhan, H., Garg, R., Weerasekera, C.S., Li, K., Agarwal, H., Reid, I.: Unsupervised learning of monocular depth estimation and visual odometry with deep feature reconstruction. In: Proceedings of the IEEE Conference on Computer Vision and Pattern Recognition, pp. 340–349 (2018)
20. Zhu, J.Y., Park, T., Isola, P., Efros, A.A.: Unpaired image-to-image translation using cycle-consistent adversarial networks. In: Proceedings of the IEEE International Conference on Computer Vision, pp. 2223–2232 (2017)

Specialise to Generalise: The Person Re-identification Case

Lorenzo Putzu[1]([✉])([ID]), Andrea Loddo[2]([ID]), Rita Delussu[1]([ID]),
and Giorgio Fumera[1]([ID])

[1] Department of Electrical and Electronic Engineering, University of Cagliari,
Cagliari, Italy
{lorenzo.putzu,rita.delussu,fumera}@unica.it
[2] Department of Mathematics and Computer Science, University of Cagliari,
Cagliari, Italy
andrea.loddo@unica.it

Abstract. Person re-identification (Re-Id) is a beneficial computer vision functionality in security-related applications based on video surveillance systems. It is a challenging cross-camera matching problem, which makes it prone to domain shift issues. To mitigate them, supervised and unsupervised domain adaptation and domain generalisation (DG) methods have been proposed. All such methods tend to favour performance improvements on the target data set at the expense of performance on the source data set(s), on which they generally deteriorate. In this work, instead, we propose an alternative method for DG that does not involve any re-training or fine-tuning of the Re-Id model and thus has no adverse effect on the performance of the source data set. It exploits Generative Adversarial Networks, trained on the source data set only with a one-vs-all mapping that simulates the target data set images, with the aim of transferring the style of the source data set into the target images. Finally, an ad hoc ranking process combines the features extracted from the original and generated images and produces the final ranked list. The proposed method can be used on top of any Re-Id model, making it a possible alternative method against domain shift and also complementary to other approaches. The considered solution is evaluated on a challenging cross-data set scenario on two benchmark data sets and a deep learning baseline for Re-Id. The obtained results demonstrate that the proposed solution improves performance, especially when the Re-Id model is specialised in the source domain.

Keywords: Person Re-Identification · Domain Generalisation · Generative Adversarial Networks

1 Introduction

Person re-identification (Re-Id) is a complex task in computer vision that involves matching images of a particular person from a data set of images taken from different non-overlapping cameras at different times and locations. This task is useful for surveillance and security purposes but is challenging due to

G. L. Foresti et al. (Eds.): ICIAP 2023, LNCS 14234, pp. 381–392, 2023.
https://doi.org/10.1007/978-3-031-43153-1_32

various factors such as changes in viewpoint and illumination, occlusions, low camera resolution, detection errors, and background clutter in the images. Furthermore, this challenge is often an open-world problem, which makes it even more difficult to tackle [14]. Initially, Re-Id relied on hand-crafted pedestrian descriptors, similarity measures [7], and metric learning [26]. Modern Re-Id systems instead exploit Convolutional Neural Networks (CNNs) as feature extractors and sometimes for metric learning. Since in real-world settings, the Re-Id task is performed on multiple cameras, even benchmark data sets usually consist of images acquired from at least two cameras [25], making it a *cross-scene* task. Consequently, Re-Id requires cross-camera matching. The standard experimental setting for supervised methods training, including CNN-based ones, involves using the *same* set of cameras for both training and testing.

However, this approach can overestimate the system's performance in real-world settings where the Re-Id system is deployed on *unknown* target scenes. Additionally, similar to other computer vision tasks [13], Re-Id benchmark data sets have been found to be biased [8]. A different experimental setting, known as the direct transfer (DT) or *cross-data set*, involves the use of one data set for training and a *different* one for testing. Research has shown that in such a setting, more representative of real-world ones, the performance of supervised methods significantly decreases [8,18], highlighting the need to properly account for the impact of different data sets on Re-Id system performance.

The issue mentioned above is a common problem in machine learning known as the *domain shift* (DS). It occurs when a model is trained on a particular domain but deployed on a related but different *target* domain. Many machine learning approaches have been proposed to improve Re-Id's performance and tackle the DS problem in addition to the DT approach. On the one hand, domain adaptation (DA) and unsupervised DA (UDA) require target data. Respectively, they assume that labelled or unlabelled data can be collected for fine-tuning or re-training the model. On the other hand, domain generalisation (DG) approaches [25] do not require target data and aim to enhance the model's generalisation capability for any target domain using several source domains for training. DG has the advantage of not requiring target data with respect to DA and UDA, but this usually leads to lower accuracy. Its effectiveness relies on how closely the source domains represent the target. In machine learning, creating a model that can generalise to unseen target domains is a challenging task [33].

In this work, we focus on a different approach for DG that does not necessarily involve the use of several source domains for training and can be embedded in any existing Re-Id system without any re-training or fine-tuning of the Re-Id model itself. It exploits Generative Adversarial Networks (GANs); in particular, it is inspired by the CamStyle [31] implementation, based on CycleGANs. Differently from their original goal, in this work, CycleGANs are used to fill the gap between the source and target domain by creating images with the same style as the source data set used to train the Re-Id model. Given that the target domain is not available in the considered scenario, CycleGANs are trained with a one-vs-all camera mapping that exploits the multiple cameras of the source data set to simulate the target domain. An ad hoc ranking process that uses both original and generated images is then used to produce the ranked list. Through cross-data set experiments on two benchmark data sets, it has been determined that the proposed approach can improve the performance of existing Re-Id models, especially when the Re-Id model is specialised in the source domain.

The rest of this work is organised as follows. Firstly, we review related work in Sect. 2, specifically addressing the DS issue. Next, we present our proposed approach in Sect. 3. We then report and discuss the experimental results in Sect. 4. Finally, the conclusions and future works are drawn in Sect. 5.

2   Related Work

This section provides an overview of Re-Id literature, focusing on GAN and DG methods. Although GANs can be used to address several issues, in this work, we focus on GAN methods that aim at reducing the DS. On the one hand, GAN and GAN-based approaches create new images using those from the target domain. On the other hand, DG methods train models by utilising multiple source domains to create a model that can be applied to any unseen target domain.

GAN-based methods can be categorised in pose- [12,17,21,28] and image-style [1–3,6,10,19,20,24,27,29,31,34,35]. The pose-style approaches aim at generating new images with specific target poses. These approaches were developed to address issues such as the lack of pose variations within data sets [12,28], improve the semantic representation [21], or enhance the quality of generated images by eliminating any noisy ones [17].

The image-style approaches aim at reducing the domain gap between two domains. Some of these approaches involve: i) converting source domain images into target domain style [1,3,24,34]; ii) consider the previous methods (i.e., i)) and each target camera as an independent domain [19,29,35]; iii) swapping background images by removing the individual from one image and placing them in another image with a different background [2,20]; iv) generating new images of either the source domain [6,10,31] or the target domain [27].

DG is a recently proposed approach against DS, which has also been applied to Re-Id, e.g., [15,16,22,32]. One of the benefits of DG is that it does not require any target data or knowledge of the target domain(s). In some cases, different real data sets are used for training [15,32], whereas in other cases, synthetic data sets can be considered for the training phase in addition to real data [22]. However, labelling the source data is necessary in these cases. To alleviate this constraint, an unsupervised DG method that does not require labelled source images has also been proposed [16].

3   Proposed Method

This section presents the proposed method for DG applied to Re-Id. We present the considered scenario and how we approached the related issues by generating the synthetic images exploited by the proposed Re-Id system.

Motivations. DA and UDA are among the most used approaches to tackle the DS problem. These methods require having a sufficient amount of target data - labelled for DA and unlabeled for UDA - available during system design. Such images are used for offline training or fine-tuning the source model to bridge the gap between source and target data sets. However, this is not feasible for some application scenarios; for instance, when a temporary camera installation must be operational in a short time. Differently, DG methods assume that the target domain is not known during design and, to improve the generalisation power of the underlying model, they incorporate labelled or unlabelled data from other auxiliary data sets, instead of using a single source data set. Although it is easy to gather additional data for model training, DG usually leads to lower accuracy with respect to DA and UDA.

In this work, to bridge the gap between source and target data sets, we employed an innovative approach based on GANs. Specifically, we utilised the CamStyle [31] implementation, founded on CycleGANs. The original CamStyle work [31] was aimed at bridging the gap between different camera views inside the same data set. CycleGANs were used to generate, for each original image, additional ones in the style of the other cameras. The original and generated images were then used together to train a more robust Re-Id model. In contrast, our approach differs from CamStyle as it is aimed to bridge the gap between the source data set and any target data set. It uses CycleGANs to generate images from any target data set with the same style as the source data set used to train the Re-Id model. The goal is to create images that better represent the source model, and that can be used to extract more representative features. Like other DG approaches, it must be noted that the proposed approach can also be applied to combinations of data sets. Here, to simplify the exposition and comparison of results, we used a single source and target data set.

Source-Like Image Generation. Considering the scenario mentioned above, where the target domain is unknown, the CycleGAN models must only be trained on the source data set. CycleGANs can learn a mapping between only two domains simultaneously. In the original CamStyle work, the mapping has been performed for each pair of cameras (A→B) to transfer the style from camera A to camera B and vice-versa. This mapping type is inefficient as it generates countless GAN models equal to the combination of all camera pairs. Also, since target camera views differ from the source ones, it is ineffective in transferring the style to unseen images. Here, we took advantage of the nature of the Re-Id task, which is inherently a *cross-scene* task, to train CycleGANs with a *one-vs-all* mapping. More precisely, if N is the number of cameras in the data set, there will be N CycleGAN models, trained with a mapping $C - C_i \rightarrow C_i$, where C is the set of cameras in the source data set, and C_i is the target camera, with $i \in 1, 2, ..., N$. Note that our focus is solely on this specific mapping, not the inverse. This strategy allows for reducing the number of generated CycleGAN models. More importantly, this mapping is most suitable for transferring the style from any camera (even unseen) to that of the target camera. The N Cycle-GAN models, trained with the aforementioned mapping, are used to convert, or

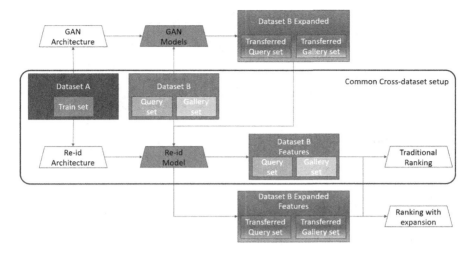

Fig. 1. General schema of the proposed method that exploits GANs to expand the target data set with the style of the source; the difference with a traditional cross-data set setup is highlighted.

better *expand*, the images of the target data set with the style of the cameras in the source data set. Indeed, for each original image of both gallery and query set N new images will be generated. Figure 3 show an example of the expansion procedure.

It must be noted that in a real-world application scenario, where a pedestrian detector is used to collect pedestrian images from the videos, these images can directly be expanded with the style of the source data set, and stored in the gallery set, as soon as they are collected. That is to say, the only additional cost of the proposed approach at run-time lies in the expansion of the query image, which can only be executed after the user/operator has chosen or manually cropped the query image, e.g., from a video.

Proposed Re-Id System. The Re-Id model, still trained with the source data set, is then used to extract the features of both the original and expanded images. A general schema of this operation is shown in Fig. 1. Given the presence of both original and expanded images, a traditional ranking process can no longer be used in this case. Thus, an ad hoc ranking process must be used. Figure 2 shows an example of our ranking process compared to the traditional one. We used two different approaches to calculate distance. The first approach, from now on **match** approach, directly matches images with the same style. In this method, the distance is calculated between:

- the original query image and the original gallery set images;
- the query image generated with the style of camera i and the gallery set images generated with the style of camera i.

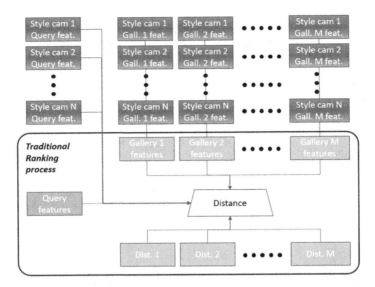

Fig. 2. Example of the proposed ranking process (compared to the traditional one) applied to a target query and gallery set of size M.

Such a procedure leads to a distance vector with length $N + 1$ instead of a single distance value for each gallery image as in the traditional ranking. Such a distance vector is converted into a single distance value by extracting the smallest value. The second approach, referred to as **cluster** approach, is based on the query movement approaches, i.e., it is assumed that images of the same identity (objects) are clustered in the feature space, while the original query $\mathbf{x}_q$ may be (relatively) far from this cluster. As a result, additional images (generally obtained by user feedback [4]) are used to compute a new query and place it near this cluster. In this case, we used the generated images to update the query feature vector and place it in the Euclidean centre of its cluster, as in the following equation:

$$\mathbf{x}_q^{\text{new}} = \frac{x_q + \sum_{i=1}^{N} x_q^{G_i}}{N + 1} \tag{1}$$

where $x_q^{G_i}$ is the feature representation of the query image generated with the style of the i-th source camera. In the same way, we updated the feature vector of each gallery image. After calculating the new feature vectors, the ranking process can be carried out as usual.

4   Experiments

The experiments evaluate whether the proposed method can help improve the performance of Re-Id systems, specifically in a DS setting. We point out that our goal is not to outperform state-of-the-art methods against DS but to assess whether and to what extent the considered solution can improve re-identification accuracy with respect to a Re-Id baseline. In the following, we first describe the considered data sets and the experimental setup and then discuss the obtained results.

Data Sets. In this preliminary work, we used two widely used data sets: Market-1501 [30] and DukeMTMC-reID [9], from now on referred to as Market and Duke, respectively. All our experiments employed them as both source or target data sets. In addition, RandPerson was used as the auxiliary data set for training the Re-Id models in specific experimental configurations for comparison purposes. Still, it was never used as a source or target data set. *Market* is a collection of 32,668 bounding box images of 1,501 identities acquired from six cameras. They are divided into 751 identities for training (12,936 images) and 750 identities for testing, corresponding to the remaining images divided into a gallery set (15,913 images) and a query set (3,368 images). *Duke* contains 36,411 bounding box images of 1,404 identities captured by eight cameras. They are divided into 702 identities for training (16,522 images) and 702 identities for testing, corresponding to the remaining images divided into a gallery set (17,661 images) and a query set (2,228 images). The synthetic data set *RandPerson*, contains 8,000 individual pedestrians and 132,145 bounding boxes [23]. It was created using Unity and MakeHuman, simulating four different cameras. Since it was proposed for training purposes to improve the generalisation capabilities of Re-Id models, it presents just a training partition.

Experimental Setting. For brevity, in our experiments, we utilised a single Re-Id model that would allow us to define a baseline and simultaneously make comparisons with a state-of-the-art method. Such a method is known as Domain-Mix [22], whose source code was made available by the authors. It is worth reminding that our proposed methodology is model-agnostic and can be applied to any Re-Id model. We simulated cross-scene application scenarios characterised by DS through two different experimental settings. In the first one, we used a common **cross-data set** setup: each real data set was used in turn as the source domain and the remaining one as the target domain. The purpose of this experiment is to assess the performance of the proposed solution with a Re-Id model specialised on the source data set only. In the second experimental setting, we used a **DG** setup: RandPerson is integrated as additional auxiliary data set into the source domain to enhance the generalisation power of the Re-Id model, as in the original DomainMix work [22]. The purpose of the second setup is to verify how the performance of the proposed solution changes with a Re-Id model that is not specialised in the source data set and, simultaneously, compare the performance of our solution with that of DomainMix in its optimal setup.

Implementation Details. Concerning DomainMix, the ResNet-50 [11] network pre-trained on ImageNet [5] was adopted as the backbone of the Re-Id model and then fine-tuned on the training partition of the source data set. We have set Adam with a weight decay of 5×10^{-4} and momentum of 0.9. The initial learning rate was 3.5×10^{-4} and decreased to $1/10$ of its previous value on the 10th, 15th, 30th, 40th, and 50th epoch, across a total of 60 epochs.

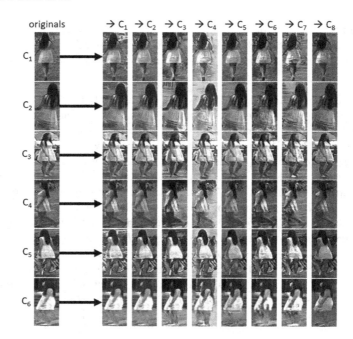

Fig. 3. Example of images generated with the proposed mapping to transfer Market images into the style of Duke data set.

The CycleGAN models instead were trained by exploiting each data set, in turn, as the source domain using the one-vs-all mapping method described in Sect. 3. Therefore, the generated images of the target data set only present the source data set's style. To ensure the best results, we adopted the parameters recommended in the original Camstyle manuscript [31]. An example of the images generated with the proposed mapping is shown in Precision Fig. 3. To evaluate results, we consider two common metrics: mean Average Precision (mAP) and the Cumulative Matching Curve (CMC) at ranks (rk) $k = 1, 5, 10, 20$.

Results. Table 1 reports the results obtained with the **cross-data set** setup, by directly exploiting the DomainMix model (**none** row) and by including the proposed solution for image expansion with the **match** and **cluster** strategy. As can be seen, the proposed method leads to significant performance gains in both mAP and all CMC ranks for both data sets.

Table 2 reports the results obtained with the **DG** setup, by directly exploiting the DomainMix model (**none** row) and by including the proposed solution for image expansion with the **match** and **cluster** strategy. Even in this case, the proposed method leads to performance gains, but with some exceptions. Indeed, when the target data set is Duke, the increase is achieved with both strategies on both mAP and all CMC rank values; instead, when the target data set is Market, only the cluster strategy allows an improvement on both mAP and all CMC rank. In contrast, the match strategy gains only marginally in the highest

Table 1. Performance of our solution applied to DomainMix models trained with source data set only.

Expansion method	Duke → Market					Market → Duke				
	mAP	rk-1	rk-5	rk-10	rk-20	mAP	rk-1	rk-5	rk-10	rk-20
None	24.72	51.9	69.21	75.36	80.82	16.24	34.25	47.04	52.6	58.35
Match	31.65	62.92	78.03	83.55	89.01	29.39	48.65	63.64	68.67	73.47
Clust	32.24	62.41	77.38	82.36	86.97	31.62	50.36	64.72	69.48	74.64

Table 2. Performance of our solution applied to DomainMix models trained with source data set + RandPerson (RP).

Expansion method	Duke + RP → Market					Market + RP → Duke				
	mAP	rk-1	rk-5	rk-10	rk-20	mAP	rk-1	rk-5	rk-10	rk-20
None	34.28	63.24	78.06	83.11	87.89	30.95	49.24	66.07	71.05	75.45
Match	32.88	61.25	78.36	83.64	88.63	32.81	52.56	69.08	73.74	78.19
Clust	36.26	65.08	79.39	84.8	89.37	35.82	56.87	70.74	75.63	78.95

CMC ranks. Comparing these two tables shows that, as expected, the proposed solution allows for a more significant performance improvement when applied to a Re-Id model specialised on the same data source set used to train CycleGANs. At the same time, it is also noticeable that our solution leads to high final performance even when the Re-Id model is more generic and not specialised. However, the improvement may not be as substantial as in the previous scenario.

The reason behind these results arises from the nature of generic models that favour performance improvements on unseen target data sets at expenses of performance on the source data set(s), on which they generally deteriorate. However, in real application scenarios in the wild, the target domain could encompass the source data set or any data set exhibiting similar characteristics.

Table 3. Performance of DomainMix (DM) models trained with source data set only, adding RP as an auxiliary data set or our solution with match or cluster strategy.

Method	Market → Market					Duke → Duke				
	mAP	rk-1	rk-5	rk-10	rk-20	mAP	rk-1	rk-5	rk-10	rk-20
DM (source)	83.06	93.14	97.68	98.72	99.2	62.99	77.47	87.52	90.62	92.86
DM (source) + RP	59.6	79.6	90.6	94.1	96.5	55.0	72.3	83.8	88.0	90.5
DM (source) + Match	82.46	93.32	97.8	98.46	99.05	62.75	77.78	88.55	91.61	93.72
DM (source) + Cluster	82.16	92.81	97.39	98.4	99.05	63.06	78.68	88.87	91.88	93.67

Table 3 compares the performance degradation on the source data set caused by including additional training images or data sets (common to traditional DG approaches) with respect to the proposed DG solution that does not require any re-training or fine-tuning of the Re-Id model. As expected, the proposed DG solution has no adverse effect on the performance of the source data set. In contrast, adding a further auxiliary data set for training the Re-Id model significantly decreases the performance, especially in terms of mAP.

5   Conclusions

In this work, we proposed an alternative method for DG to bridge the gap between the source data set and any target one. It exploits CycleGANs to adapt target data set images to the style of the source data set. Both Re-Id models and CycleGANs are trained by exploiting the source data set only. It is worth noting that the proposed method can be used on top of any Re-Id model, making it not only a possible alternative method against domain shift but also complementary to other approaches. We evaluated the proposed method on a challenging cross-data set scenario on two benchmark data sets and a Re-Id model used to define a baseline and, at the same time, make a first comparison. The obtained results demonstrate that the proposed solution improves performance, especially when the Re-Id model is specialised in the source domain. Also, we demonstrated that contrary to other existing methods against domain shift, the proposed method has no adverse effect on the performance of the source data set. Although it is not a fundamental characteristic of DG methods to have excellent performance in the data set source, the performance on it defines its upper bound. Its theoretical lower bound instead is related to the capabilities of the underlying model on the source data set and those of the GANs model to transfer the style of any target data set to that of the source. Note that here, to simplify the exposition and comparison of results, we preferred to use a single source and target data set in turn, but the proposed approach can also be applied to combinations of data sets. Moreover, the Re-Id represents a single field of application of the proposed method that could be applied to different fields of application. Furthermore, it should be noted that in this work, we focused on CycleGANs for camera style transfer, but additional types of GANs could be used. In the context of Re-Id, an interesting research direction for future work is to investigate different types of transfer, such as by using poseGANs to vary the pose of individuals.

Acknowledgements. Supported by the projects: "Law Enforcement agencies human factor methods and Toolkit for the Security and protection of CROWDs in mass gatherings" (LETSCROWD), EU Horizon 2020 programme, grant agreement No. 740466; "IMaging MAnagement Guidelines and Informatics Network for law enforcement Agencies" (IMMAGINA), European Space Agency, ARTES Integrated Applications Promotion Programme, contract No. 4000133110/20/NL/AF.

References

1. Ainam, J., et al.: Unsupervised domain adaptation for person re-identification with iterative soft clustering. Knowl.-Based Syst. **212**, 106644 (2021). https://doi.org/10.1016/j.knosys.2020.106644

2. Chen, Y., Zhu, X., Gong, S.: Instance-guided context rendering for cross-domain person re-identification. In: ICCV, pp. 232–242 (2019). https://doi.org/10.1109/ICCV.2019.00032
3. Chong, Y., et al.: Style transfer for unsupervised domain-adaptive person re-identification. Neurocomputing **422**, 314–321 (2021). https://doi.org/10.1016/j.neucom.2020.10.005
4. Delussu, R., Putzu, L., Fumera, G.: Human-in-the-loop cross-domain person re-identification. Expert Syst. Appl. **226**, 120216 (2023). https://doi.org/10.1016/j.eswa.2023.120216
5. Deng, J., Dong, W., Socher, R., Li, L.J., Li, K., Fei-Fei, L.: Imagenet: a large-scale hierarchical image database. In: Proceedings of the IEEE Conference on Computer Vision and Pattern Recognition, pp. 248–255 (2009)
6. Ding, G., Zhang, S., Khan, S.H., Tang, Z., Zhang, J., Porikli, F.: Feature affinity-based pseudo labeling for semi-supervised person re-identification. IEEE Trans. Multim. **21**(11), 2891–2902 (2019). https://doi.org/10.1109/TMM.2019.2916456
7. Farenzena, M., et al.: Person re-identification by symmetry-driven accumulation of local features. In: CVPR, pp. 2360–2367 (2010). https://doi.org/10.1109/CVPR.2010.5539926
8. Genç, A., Ekenel, H.K.: Cross-dataset person re-identification using deep convolutional neural networks: effects of context and domain adaptation. Multimedia Tools Appl. **78**(5), 5843–5861 (2018). https://doi.org/10.1007/s11042-018-6409-3
9. Gou, M., et al.: Dukemtmc4reid: a large-scale multi-camera person re-identification dataset. In: CVPR Workshops, pp. 1425–1434 (2017). https://doi.org/10.1109/CVPRW.2017.185
10. Han, H., Ma, W., Zhou, M., Guo, Q., Abusorrah, A.: A novel semi-supervised learning approach to pedestrian re-identification. IEEE Internet Things J. **8**(4), 3042–3052 (2021). https://doi.org/10.1109/JIOT.2020.3024287
11. He, K., Zhang, X., Ren, S., Sun, J.: Deep residual learning for image recognition. In: Proceedings of the IEEE Conference on Computer Vision and Pattern Recognition, pp. 770–778 (2016)
12. Khatun, A., Denman, S., Sridharan, S., Fookes, C.: Pose-driven attention-guided image generation for person re-identification. Pattern Recogn. **137**, 109246 (2023). https://doi.org/10.1016/j.patcog.2022.109246
13. Khosla, A., Zhou, T., Malisiewicz, T., Efros, A.A., Torralba, A.: Undoing the damage of dataset bias. In: Fitzgibbon, A., Lazebnik, S., Perona, P., Sato, Y., Schmid, C. (eds.) ECCV 2012. LNCS, vol. 7572, pp. 158–171. Springer, Heidelberg (2012). https://doi.org/10.1007/978-3-642-33718-5_12
14. Leng, Q., et al.: A survey of open-world person re-identification. IEEE T-CSVT **30**(4), 1092–1108 (2020). https://doi.org/10.1109/TCSVT.2019.2898940
15. Lin, S., et al.: Multi-domain adversarial feature generalization for person re-identification. IEEE T-IP **30**, 1596–1607 (2021). https://doi.org/10.1109/TIP.2020.3046864
16. Qi, L., Liu, J., Wang, L., Shi, Y., Geng, X.: Unsupervised generalizable multi-source person re-identification: a domain-specific adaptive framework. Pattern Recogn. **140**, 109546 (2023)
17. S, S.R., Prasad, M.V.N.K., Balakrishnan, R.: Generative segment-pose representation based augmentation (GSRA) for unsupervised person re-identification. Image Vis. Comput. **131**, 104632 (2023). https://doi.org/10.1016/j.imavis.2023.104632
18. Song, L., et al.: Unsupervised domain adaptive re-identification: theory and practice. Patt. Rec. **102**, 107173 (2020). https://doi.org/10.1016/j.patcog.2019.107173

19. Tian, J., Teng, Z., Zhang, B., Wang, Y., Fan, J.: Imitating targets from all sides: an unsupervised transfer learning method for person re-identification. Int. J. Mach. Learn. Cybern. **12**(8), 2281–2295 (2021). https://doi.org/10.1007/s13042-021-01308-6

20. Verma, A., Subramanyam, A.V., Wang, Z., Satoh, S., Shah, R.R.: Unsupervised domain adaptation for person re-identification via individual-preserving and environmental-switching cyclic generation. IEEE Trans. Multim. **25**, 364–377 (2023). https://doi.org/10.1109/TMM.2021.3126404

21. Wang, M., Chen, J., Liu, H.: A novel multi-scale architecture driven by decoupled semantic attention transfer for person image generation. Comput. Graph. (2023)

22. Wang, W., Liao, S., Zhao, F., Kang, C., Shao, L.: Domainmix: learning generalizable person re-identification without human annotations. In: BMVC, p. 355 (2021)

23. Wang, Y., Liao, S., Shao, L.: Surpassing real-world source training data: random 3D characters for generalizable person re-identification. In: ACM MM, pp. 3422–3430 (2020). https://doi.org/10.1145/3394171.3413815

24. Wei, L., et al.: Person transfer GAN to bridge domain gap for person re-identification. In: CVPR, pp. 79–88 (2018). https://doi.org/10.1109/CVPR.2018.00016

25. Ye, M., et al.: Deep learning for person re-identification: a survey and outlook. IEEE Trans. PAMI **44**(06), 2872–2893 (2022). https://doi.org/10.1109/TPAMI.2021.3054775

26. Yu, H., Wu, A., Zheng, W.: Cross-view asymmetric metric learning for unsupervised person re-identification. In: ICCV, pp. 994–1002 (2017). https://doi.org/10.1109/ICCV.2017.113

27. Zhang, C., et al.: Improving domain-adaptive person re-identification by dual-alignment learning with camera-aware image generation. IEEE T-CSVT **1**, 4334–4346 (2020). https://doi.org/10.1109/TCSVT.2020.3047095

28. Zhang, L., Jiang, N., Diao, Q., Zhou, Z., Wu, W.: Person re-identification with pose variation aware data augmentation. Neural Comput. Appl. **34**(14), 11817–11830 (2022). https://doi.org/10.1007/s00521-022-07071-1

29. Zhang, S., Hu, H.: Unsupervised person re-identification using unified domanial learning. Neural Process. Lett. 1–19 (2023)

30. Zheng, L., et al.: Scalable person re-identification: a benchmark. In: ICCV, pp. 1116–1124 (2015). https://doi.org/10.1109/ICCV.2015.133

31. Zhong, Z., et al.: CamStyle: a novel data augmentation method for person re-identification. IEEE T-IP **28**(3), 1176–1190 (2019). https://doi.org/10.1109/TIP.2018.2874313

32. Zhou, K., et al.: Learning generalisable omni-scale representations for person re-identification. IEEE T-PAMI **1**, 1–1 (2019). https://doi.org/10.1109/TPAMI.2021.3069237

33. Zhou, K., Liu, Z., Qiao, Y., Xiang, T., Loy, C.C.: Domain generalization: a survey. IEEE T-PAMI **45**(4), 4396–4415 (2023). https://doi.org/10.1109/TPAMI.2022.3195549

34. Zhou, S., Wang, Y., Zhang, F., Wu, J.: Cross-view similarity exploration for unsupervised cross-domain person re-identification. Neural Comput. Appl. **33**(9), 4001–4011 (2021). https://doi.org/10.1007/s00521-020-05566-3

35. Zou, Y., Yang, X., Yu, Z., Kumar, B.V.K.V., Kautz, J.: Joint disentangling and adaptation for cross-domain person re-identification. In: Vedaldi, A., Bischof, H., Brox, T., Frahm, J.-M. (eds.) ECCV 2020. LNCS, vol. 12347, pp. 87–104. Springer, Cham (2020). https://doi.org/10.1007/978-3-030-58536-5_6

Enhancing Hierarchical Vector Quantized Autoencoders for Image Synthesis Through Multiple Decoders

Dario Serez[1,3]([✉]) [ID], Marco Cristani[4] [ID], Vittorio Murino[1,2,4] [ID],
Alessio Del Bue[1] [ID], and Pietro Morerio[1] [ID]

[1] PAVIS Department, Italian Institute of Technology (IIT), Genoa, Italy
{dario.serez,vittorio.murino,alessio.delbue,pietro.morerio}@iit.it
[2] DIBRIS Department, University of Genova (UniGE), Genoa, Italy
[3] DITEN Department, University of Genova (UniGE), Genoa, Italy
[4] Department of Computer Science, University of Verona (UniVR), Verona, Italy
marco.cristani@univr.it

Abstract. Vector Quantized Variational Autoencoders (VQ-VAEs) have gained popularity in recent years due to their ability to represent images as discrete sequences of tokens that index a learned codebook of vectors, enabling efficient image compression. One variant of particular interest is VQ-VAE 2, which extends previous works by representing images as a hierarchy of sequences, resulting in finer-grained representations.

In this study, we further enhance such hierarchical autoencoder approach by introducing multiple decoders, which allow to represent images as a sum of multi-scale contributions in the pixel space. Our proposed model, the Multi Scale (MS) VQ-VAE, not only enables better control over the encoding of each sequence (resulting in improved explainability and codebook usage) but, as a consequence, also shows advantages in image synthesis. Our experiments demonstrate that the MS-VQVAE achieves comparable or superior reconstructions on various datasets and resolutions, as well as greater stability across runs. Moreover, we include a proof-of-concept trial to showcase the potential applications of our model in image synthesis.

Keywords: VQ-VAE · Hierarchical Autoencoder · Image synthesis

1 Introduction

Vector Quantized Variational Autoencoders (VQ-VAE) [18] are popular in Computer Vision for their ability to learn discrete low-dimensional representations of images by indexing a codebook (or dictionary) of learnable vectors. Notably, VQ-VAE and its extensions [6,14,28] have been successfully combined with autoregressive models to perform image synthesis [2,19,29]. Despite this success, the original algorithm presents limitations in reconstructing the fine-grained information of the encoded images, especially at high-resolutions, where the details

© The Author(s), under exclusive license to Springer Nature Switzerland AG 2023
G. L. Foresti et al. (Eds.): ICIAP 2023, LNCS 14234, pp. 393–405, 2023.
https://doi.org/10.1007/978-3-031-43153-1_33

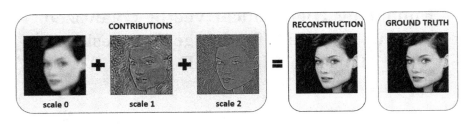

Fig. 1. MS-VQVAE separately reconstructs the hierarchy of quantized sequences in the pixel space, from coarse (scale 0) to fine (scale 2), thus enhancing the explainability of what is encoded at each level. The final sample is given by the sum of these contributions. Example with three scales on the Celeb-A dataset (1024 × 1024 resolution).

are more complex and abundant. As a result, the loss of important information may impact the quality of reconstructions. Therefore, a key challenge in current research is to improve the accuracy of the result while preserving the compressibility of the representation. Later on, VQ-VAE 2 [20] was introduced to address these limitations by representing images as a hierarchy of sequences. This approach breaks down the image into multiple levels of abstraction, capturing its different aspects. By constructing a sequence hierarchy, VQ-VAE 2 can better learn the complex structure of images, leading to more accurate reconstructions.

In this study, we further enhance the hierarchical Autoencoder by introducing multiple decoders to represent images as a sum of contributions in the pixel space, as shown in Fig. 1. Our approach is inspired by a *"divide-and-conquer"* strategy, where a common CNN-Encoder generates multiple latent sequences at different scales, each responsible for a specific contribution. The sequences are then separately quantized and decoded at the original resolution, and the resulting image at a specific scale i is obtained by adding contributions from all scales $\leq i$. The proposed method, which we name Multi Scale (MS) VQ-VAE, improves explainability over its predecessor (VQ-VAE 2) by allowing direct control and meaningful decoding of each latent-sequence content.

In this work, we also provide a proof-of-concept study where we show the potential applications and benefits of our MS-VQVAE in image synthesis. The whole pipeline (called 2-stage sampling) consists in training an autoregressive model (stage-2) on top of the VQ-VAE (stage-1) discrete representations. This allows the generation of new sequences of indices, which can be decoded with the pre-trained Autoencoder. Since it is not necessary to sample all sequences to decode a meaningful image, poor quality samples can be discarded after the generation of the first scale only. Additionally, once the "coarse" part of the image is given, it is possible to modify only its "details" (sequences at scales > 0) multiple times, allowing for the generation of different versions of the same sample.

Further details about the used methodologies are given in Sect. 3, while in Sect. 2 differences with respect to VQ-VAE 2 are discussed, as well as improvements introduced by other hierarchical quantized autoencoders. In Sect. 4 we present the experimental results and evaluate the proposed model on the Imagenet [3] (resolution 256 × 256) and CelebA-HQ [15] (resolution 1024 × 1024) datasets, showing better stability and codebook usage with respect to VQ-VAE 2, as well as general better performances in terms of reconstruction. Furthermore, in Sects. 3 to 5, we discuss the potential future applications of our model in image-synthesis tasks.

2 Previous Work

Improving Quantization: The original VQ-VAE work [18] suffers from a known issue called *codebook collapse* (or *index collapse*) [12], where only a subset of available codebook vectors is actually used to encode information. To address this problem, researchers have proposed various improvements such as Expectation-Maximization [21], Decomposed Vector Quantization [12] Continuous Relaxation [10,16] and Codebook Restart [4]. Although any of these algorithms can be implemented in our method, we employ the basic EMA quantization in order to fairly compare with the VQ-VAE 2 baseline, as described in Sect. 3. Nevertheless, by introducing multiple decoders, we observe an increased utilization of the codebook in our sequences, as shown in Sect. 4.

Perceptual Loss and GAN Discriminator: To improve the quality of output images while keeping good compression rates, VQ-GAN [6] replaced the $l2$ reconstruction loss with a combination of Patch-GAN discriminator [9] and perceptual loss [11,30]. Later on, ViT-VQGAN [28] further improved this result with a Style-GAN discriminator [24] and a Vision Transformer (ViT) [5] based Autoencoder. Since our main concern is to fairly compare with the VQ-VAE 2 baseline, in this work we use CCN-Autoencoders and $l2$ loss on all sequences. An improved implementation with more complex architectures and loss functions is left to future research.

Hierarchical Quantized Autoencoders for Image Modeling: A distinct category of methods proposes the concept of a *hierarchical* Autoencoder and shows its advantages. Specifically, [27] utilizes Mean Squared Error (*MSE*) to compare multiple quantized sequences with their encoded counterpart in order to achieve higher levels of compression, but without exploring possible applications in image modeling. [7] combines the hierarchical structure with autoregressive decoders to enable end-to-end sampling. In Contrast, our MS-VQVAE is designed for 2-stage sampling, which has become common practice for many different methods [2,19,29]. A different category of approaches [1,14,31] incorporates a stack of codes in the quantization bottleneck, which can be viewed as a form of hierarchy. Among them, [1,14] are notable for introducing Residual Quantization, where the stack of quantized codes is viewed in a coarse-to-fine manner and summed to obtain the full representation. In contrast, in our MS-VQVAE, the latent residuals exist at different scales (resolutions) and are summed only once decoded in the pixel space. Our work is closely related to VQ-VAE 2 [20], which involves quantizing a stack of latent codes at different resolutions, concatenating them, and decoding the resulting sequence. However, we observed that this technique lacks the ability to control the content of each sequence and often under-exploits codebook dictionaries.

In contrast to all these methods, our proposed MS-VQVAE employs multiple decoders that are explicitly optimized using a residual-based approach to provide greater control over the content of each sequence. Each Decoder is responsible for a specific subset of the latent codes, and the resulting residuals of these decodings are summed to obtain the full representation. Our architecture also presents advantages for image sampling, where partial generations can naturally be decoded in order to early remove poor quality samples, while multiple and different "details" can be generated on top of the selected "coarse" image.

3 Methodology

3.1 Background

The core of vanilla VQ-VAE [18] is the quantization process, which defines a learnable codebook of vectors $e \in \mathbb{R}^{K \times D}$, where K is the codebook size and D is the dimension of each vector e_i. After encoding the image x, each latent of the output $z_e(x)$ is associated with the nearest embedding in the codebook:

$$z_q(x) = e_k \quad \text{where } k = \text{argmin}_j \| z_e(x) - e_j \| \tag{1}$$

Finally, the Decoder reconstructs the original image from the quantized latent vectors $z_q(x)$. The loss function is composed of three terms:

$$\mathcal{L} = \underbrace{\| x - \hat{x} \|_2^2}_{\text{recons.}} + \underbrace{\| \text{sg}[z_e(x)] - e \|_2^2}_{\text{codebook}} + \beta \underbrace{\| z_e(x) - \text{sg}[e] \|_2^2}_{\text{commitment}} \tag{2}$$

where $\hat{x}$ is the Decoder reconstruction, sg denotes the stop gradient operation and β is a constant term usually set to 0.25. The three terms represent the reconstruction, the codebook, and the commitment loss, respectively. A variation is also proposed, where the second term is removed and the embeddings are learned using an Exponential Moving Average (EMA). In detail, each codebook entry e_i is updated at every step t following:

$$e_i^{(t)} := m_i^{(t)} / N_i^{(t)} \tag{3}$$

where $m_i^{(t)}$, $N_i^{(t)}$ represent at each step the mean vector and the usage count with respect to codeword e_i, and they are updated according to the $n_i^{(t)}$ encoder outputs that are closest to the embedding e_i at step t:

$$m_i^{(t)} := m_i^{(t-1)} \cdot \gamma + \sum_j^{n_i^{(t)}} z_{i,j}^{(t)} (1 - \gamma); \quad N_i^{(t)} := N_i^{(t-1)} \cdot \gamma + n_i^{(t)} (1 - \gamma) \tag{4}$$

Here γ is a constant factor set between 0 and 1, usually to 0.99.

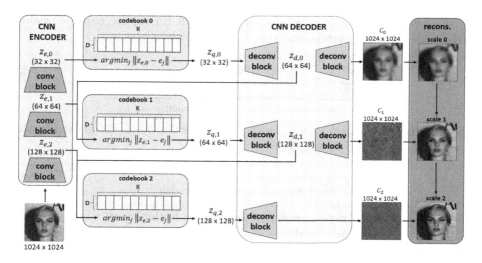

Fig. 2. Full pipeline of our method using three scales ($32 \times 32, 64 \times 64, 128 \times 128$) on an example taken from the CelebA-HQ dataset (1024×1024). The encoding process produces the three sequences $z_{e,0}, z_{e,1}, z_{e,2}$, which are separately quantized to $z_{q,0}, z_{q,1}, z_{q,2}$. The decoding process results in the three contributions C_0, C_1, C_2, which are added to obtain the reconstruction at each scale. During decoding, the intermediate sequences of the first two scales $z_{d,0}, z_{d,1}$ are concatenated to $z_{e,1}, z_{e,2}$, respectively.

3.2   Multi Scale VQ-VAE

The proposed MS-VQVAE architecture is depicted in Fig. 2. The input image is fed to a fully convolutional Encoder, which produces M latent images $z_{e,m}$ at different scales $m \in \{0, 1, \dots, M - 1\}$, where $z_{e,0}$ encodes global content and $z_{e,M-1}$ encodes the finest details. Each latent sequence has its own codebook of vectors $e_m \in \mathbb{R}^{K \times D}$, and the quantization process follows Eq. (1) to obtain M quantized latents $z_{q,m}$. We use EMA strategy for the learning of each codebook (Eq. 4) and we do not prevent codebook collapse, in order to ensure fair comparisons with previous work. Decoding involves M separate decoders that upsample each sequence to the original size. For $m \in \{0, 1, \dots, M - 2\}$ the latents undergo a two-step decoding process, with intermediate sequences $z_{d,m}$ that are concatenated to $z_{e,m+1}$ (before quantization) to allow an information flow between decoders. For scale $m = M - 1$, the sequence is directly upsampled to the final resolution. The decoding process produces M residual images C_m. The reconstruction at a given scale m is then the result of the summation of contributions $0, 1, \dots, m - 1$:

$$\hat{x}_m = \sum_{i=0}^{m} C_i \qquad (5)$$

which implies that $\hat{x}_0 = C_0$.

The overall loss function is a generalization of Eq. (2) for multiple scales, without the codebook term due to the EMA algorithm:

$$\mathcal{L} = \underbrace{\sum_{m=0}^{M} \|x_m - \hat{x}_m\|_2^2}_{\text{MSrecons.}} + \underbrace{\sum_{m=0}^{M} \beta \|z_{e,m}(x) - \text{sg}[e_m]\|_2^2}_{\text{MScommittment}} \tag{6}$$

where $\hat{x}_m$ is defined as in Eq. (5). In the reconstruction term, the ground-truths images x_m are defined M as:

$$x_m = \begin{cases} x & \text{if } m = M - 1 \\ \mathcal{B}(x_{m+1}, \kappa, \sigma) & \text{otherwise} \end{cases} \tag{7}$$

where $\mathcal{B}$ denotes the *Gaussian Blur* operation, κ is the kernel which depends on the image resolution r and is computed as $\sqrt{r} - 1$, and σ is the standard deviation given as a function of the kernel:

$$\sigma = \frac{1}{3}\left(\frac{(\kappa - 1)}{2} - 1\right) + \frac{4}{5} \tag{8}$$

By doing so, the ground truths corresponding to lower scales appear as low-frequency versions of the original sample. This mechanism forces the early sequences to focus only on the global structure of images, ignoring the high-frequency details. The explainability of our method is provided by the multi-scale reconstruction mechanism, since we can asses and show that low scales encode low frequency information (the global content of images), while higher scales encode the high frequencies (details).

3.3   Image Sampling: Intuitions

In their paper [20], the authors propose to train M different models (one for each hierarchy), in order to perform autoregressive sampling of the VQ-VAE 2 learned tokens (codebook indices). For sequence s_m, the likelihood of each token i is a function of previous tokens of the same sequence $s_{m,j<i}$ and all tokens of previous sequences $s_{n<m}$:

$$p(s_m) = \prod_i p(s_{m,i}|s_{m,j<i}, s_{n<m}) \tag{9}$$

However, this approach presents a significant challenge as increasing the number of hierarchies and their sequence length requires huge resources in terms of memory storage and sampling time. For instance, the base attention mechanism of Transformers [25] would have a space complexity of $O(m \times N)$ and a sampling time complexity of $O(m)$, where m is the length of the current sequence and N is the sum of the lengths of all sequences up to m (Left part of Fig. 3). As a direct consequence, autoregressive models still struggle to synthesise high-resolutions images, and the state-of-the-art in this field is detained by Generative Adversarial Networks [23,24].

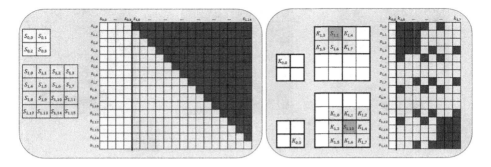

Fig. 3. *Left*: Causal Attention matrix as it would be computed in a case with two sequences s_0, s_1 of length 4 and 16, respectively. Each token $s_{1,i}$ can attend to all s_0 and previous $s_{1,j<i}$, according to Eq. (9). *Right*: Local attention defined on two sequences of the same length, when considering two kernels 1×1 and 3×3. The attention matrix reduces to $O(16 \times (1 + 8))$. The left part shows what indices the tokens $S_{1,1}$ (corner case) and $S_{1,10}$ can attend.

Leveraging the fact that our MS-VQVAE is directly optimized to separate between the coarse and fine-grained details at different scales (due to Eq. (7)), we hypothesise that a sampling algorithm based on Transformers [25], would not require all the previous context information for sequences $s_{m>0}$. Instead, one can define m different local kernels $K_{1,...,m}$ that would provide all the needed context for the sampling of a token i of sequence s_m, with a significant reduction in terms of memory requirements (see the right part of Fig. 3). In Sect. 4 we provide a proof-of-concept experiment that shows the feasibility of this method.

4   Experiments

MS-VQVAE: We conducted a comparative study of our proposed MS-VQVAE with the existing literature, specifically VQ-VAE 2, on two widely used datasets, namely Imagenet [3] and CelebA-HQ [15], with image resolutions of 256×256 and 1024×1024, respectively. To perform a fair comparison, we replicated and trained the original VQ-VAE 2 model, adopting its original hyperparameters as closely as possible. All our training runs have been trained for 250 epochs employing the Adam optimizer [13] with a learning rate of $3e - 4$ and betas 0.9 and 0.5. Moreover, our MS-VQVAE utilized a warmup learning rate for the first five epochs, followed by a cosine decay for the rest of training. For each run, we report several reconstruction metrics evaluated with the *torchmetrics* [17] library, in order to ensure fairness and reproducibility. More in detail, the reconstructions are measured in terms of Mean Squared Error (MSE), Structural Similarity Index [26] (SSIM), Learned Perceptual Image Patch Similarity [30] (LPIPS), Peak Signal-to-Noise Ratio (PSNR) and relative Frechet Inception Distance [8] (rFID). In order to provide a complete comparison, we also report for each experiment the number of training parameters and the codebook usage and

Fig. 4. Qualitative comparison of the "Small" runs on the CelebA-HQ dataset. Our MS-VQVAE (*left*) achieves a very similar reconstruction quality with respect to VQ-VAE 2 (*center*). In both cases, some fine-grained information is lost (e.g. some eye and skin details) when compared to the ground truth (*right*).

Table 1. Experiments on CelebA-HQ (1024 × 1024), grouped by latent sequences length. * reports the mean codebook usage and perplexity over all sequences.

Experiment	# params (M)	CB Usage (%)	Perplexity	MSE ↓ ($1e^{-3}$)	SSIM ↑	LPIPS ↓	PSNR ↑	rFID↓
VQ-VAE 2 (L)	2.2	100-69-18 (62*)	179-146-35 (120*)	**2.35**	**0.75**	0.39	**26.28**	**64.47**
Ours (L)	54.3	38-76-100 (**71***)	88-170-163 (**140***)	2.43	**0.75**	0.38	26.13	65.05
VQ-VAE 2 (M)	2.2	100-38-7 (48*)	135-81-11 (75*)	1.30	0.80	0.28	28.85	38.76
Ours (M)	25.3	100-100-100 (**100***)	222-195-121 (**179***)	**1.15**	**0.81**	**0.25**	**29.36**	**38.38**
VQ-VAE 2 (S)	2.1	100-100-35 (78*)	110-194-54 (119*)	0.60	0.88	0.14	32.18	**6.44**
Ours (S)	13.5	67-72-100 (**79***)	142-118-109 (**123***)	**0.52**	**0.90**	**0.13**	**32.79**	6.98

perplexity, which indicate the proportion of used codebook vectors and how well they are distributed, respectively.

Table 1 presents the comparison between our MS-VQVAE and the reimplemented VQ-VAE 2 on three experiments which we name as "Large" (L): 8 × 8, 16 × 16, 32 × 32, "Medium" (M): 16 × 16, 32 × 32, 64 × 64 and "Small" (S): 32 × 32, 64 × 64, 128 × 128. In all runs, we utilized a fixed latent vector dimension of 64 and a codebook size of 256, which we determined to be adequate to capture the variances of the dataset.

MS-VQVAE achieves comparable or greater performance in all runs, even though it requires more parameters. This gap widens at higher compression rates since more decoding layers are used. Nevertheless, we consider the increased number of parameters to be manageable on most modern hardware, thus not compromising the usability of our model. In general, we have not observed any failures of our method compared to VQ-VAE 2, even at high compression rates (Fig. 4). The superior codebook usage in all runs highlights the stability of our method. For the (M) run, VQ-VAE 2 primarily used one sequence, a phenomenon called *codebook collapse*. This does not take place with our approach, as we optimize the reconstruction error directly on each sequence, thereby compelling all the contributions to be present. Avoiding *codebook collapse* is a major challenge

Table 2. Experiments on Imagenet (256 × 256), grouped by latent sequences length.
* reports the mean perplexity over all sequences.

Experiment	# params (M)	CB Size	Perplexity	MSE ↓ ($1e^{-3}$)	SSIM ↑	LPIPS ↓	PSNR ↑	rFID↓
VQ-VAE 2 (L)	1.4	1024	559–699 (629*)	1.79	0.82	0.195	27.46	22.17
Ours (L)	37.4	1024	815–599 (**707***)	**1.54**	**0.83**	**0.164**	**28.10**	**18.00**
VQ-VAE 2 (S)	1.3	512	273–380 (**327***)	0.59	**0.93**	0.058	32.23	5.39
Ours (S)	9.4	512	371–262 (317*)	**0.50**	**0.93**	**0.046**	**32.95**	**4.49**

in current research as it compromises the stability and usability of the Autoencoder, as discussed in Sect. 2.

In our study of the Imagenet dataset, we conducted two separate experiments for both methods. The first one ("Small" - (S)) utilized two sequences (32–64) and a codebook size of 512, while the second ("Large" - (L)) increased both the compression rate (16–32) and codebook size (1024). Table 2 presents a comprehensive comparison between the two methods and shows the superior reconstructions of MS-VQVAE in almost all metrics, especially for higher compression rates. Unlike Table 1, we did not include codebook usage, as it remained at 100% in all cases. This behavior is likely due to the increased complexity of the dataset (which consists of a wide and diverse range of classes), with respect to CelebA-HQ, which contains only human faces.

For a comprehensive evaluation, we present in Table 3 a comparison of the rFid score for different techniques that employ distinct quantization algorithms and reconstruction losses. A noteworthy observation can be made by comparing the second row, which corresponds to a VQ-GAN model trained on a VQ-VAE 2 pipeline, to the standard VQ-VAE 2 method. The considerable disparity in the rFID score, i.e., 1.45 vs 5.39, highlights the potential of employing the perceptual loss and GAN discriminator for enhancing the quality of reconstructions. This result may suggest the exploration of incorporating these techniques into our MS-VQVAE model in future research.

Table 3. rFid comparison on the validation set of Imagenet for different well-known methods. * indicates a re-trained model. The × in *sequences* column indicates a Residual Quantization, as described in [14]

Method	Sequences	CB size	rFID↓
VQ-GAN [6]	16	1024	7.94
VQ-GAN	32–64	512	1.45
RQ-VAE [14]	8 × 16	16384	1.83
Vit-VQGAN [28]	32	8192	1.28
VQ-VAE 2*	32–64	512	5.39
Ours	32–64	512	4.49

Table 4. Reported comparison for all the sampling experiments. The Codebook Usage is reported as the mean between sequences.

Method	# params (M)	Autoencoder		Image synthesis	
		CB Usage	rFID ↓	FID ↓	IS ↑
VQ-VAE 2 AR	15.9	61.71%	97.78	137.62	1.83
VQ-VAE 2 MG	14.7			134.97	1.86
MS-VQVAE AR	23.4	75.68 %	97.26	139.29	1.86
MS-VQVAE MG	22.2			133.63	1.86

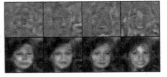

Fig. 5. Samples obtained with the MaskGit-based Transformer and our MS-VQVAE model. For each of the two "coarse" samples (generated from the 4×4 sequence), we sample 4 different contributions (*Top*) from the 8×8 codebook and sum them to obtain the final image (*Bottom*). Note that the low quality is due to the high compression rate of the Autoencoder and the overall small size (number of parameters) of the model.

Proof-of-Concept Image Synthesis: We conduct a proof-of-concept study in order to better enlighten the application of our method in the field of image synthesis. In detail, we train two additional models for 200 epochs (one for VQ-VAE 2 and one for MS-VQVAE) on a 128 px resolution version of the Flickr-Faces-HQ (FFHQ) [24] dataset, and use them as stage-1 models to perform image sampling. In order to reduce the required computational resources, both Autoencoders have highly compressed latent sequences ($8 \times 8, 4 \times 4$) and a codebook size of 512.

After this first step, we prove the feasibility and advantages of the concepts introduced in Sect. 3.3 by training different Transformer models [25], each one performing sampling on a specific sequence. For each Autoencoder, we first train a full autoregressive pipeline on both sequences, as depicted in Fig. 3 (Left). In the second setup, we implement a MaskGit [2] style transformer to perform image synthesis in a fast way. In particular for the learning of scale one tokens, we designed local-kernel attention in the style of Fig. 3 (Right), with kernel dimensions of 3×3 and 5×5, respectively. In other words, each token of the latent sequence can attend to a maximum of 9 tokens from sequence 0 and 24 tokens from sequence 1, depending on its position. All models are trained for 50 epochs, and have 8 blocks, 8 heads, and a latent dimension of 256.

Table 4 shows the quantitative sampling results in terms of FID and Inception Score (IS) [22] for the Autoregressive (AR) and MaskGit (MG) runs. The cumulative number of parameters (sum Autoencoder and Transformers) is also reported, as well as the mean codebook usage and the reconstruction Fid of the

Autoencoders. The results outline how the kernel-based attention can achieve good performances while reducing the space complexity of $O(n^2)$. It is also worth noting that our MS-VQVAE maintains its stability also in this case, with increased codebook usage and comparable performances. Additionally, Fig. 5 depicts how our model can be used in order to obtain multiple variations of the same image, by generating the scale 1 samples multiple times.

5 Conclusion and Future Work

In this paper we further developed upon the idea of a hierarchical quantized Autoencoder model, in order to provide more robust and explainable reconstructions. In particular, we showed how the proposed MS-VQVAE can better utilise the information contained in each latent sequence compared to its predecessor [20], while keeping a good reconstruction quality. We also provided a proof-of-concept method that can be useful to perform image synthesis at high resolutions reducing computational costs. We believe that future implementations of this method may be beneficial for upcoming research on autoregressive image synthesis [2,6,19,28,29], where generating high-resolution images above 256×256 remains a significant challenge.

References

1. Adiban, M., Stefanov, K., Siniscalchi, S.M., Salvi, G.: Hierarchical residual learning based vector quantized variational autoencoder for image reconstruction and generation. In: British Machine Vision Conference (2022)
2. Chang, H., Zhang, H., Jiang, L., Liu, C., Freeman, W.T.: MaskGIT: masked generative image transformer. In: Proceedings of the IEEE/CVF Conference on Computer Vision and Pattern Recognition, pp. 11315–11325 (2022)
3. Deng, J., Dong, W., Socher, R., Li, L.J., Li, K., Fei-Fei, L.: Imagenet: a large-scale hierarchical image database. In: 2009 IEEE Conference on Computer Vision and Pattern Recognition, pp. 248–255 (2009). https://doi.org/10.1109/CVPR.2009.5206848
4. Dhariwal, P., Jun, H., Payne, C., Kim, J.W., Radford, A., Sutskever, I.: Jukebox: a generative model for music. arXiv:abs/2005.00341 (2020)
5. Dosovitskiy, A., et al.: An image is worth 16x16 words: transformers for image recognition at scale. arXiv:abs/2010.11929 (2021)
6. Esser, P., Rombach, R., Ommer, B.: Taming transformers for high-resolution image synthesis. 2021 IEEE/CVF Conference on Computer Vision and Pattern Recognition (CVPR), pp. 12868–12878 (2021)
7. Fauw, J.D., Dieleman, S., Simonyan, K.: Hierarchical autoregressive image models with auxiliary decoders. arXiv:abs/1903.04933 (2019)
8. Heusel, M., Ramsauer, H., Unterthiner, T., Nessler, B., Hochreiter, S.: GANs trained by a two time-scale update rule converge to a local Nash equilibrium. In: NIPS (2017)
9. Isola, P., Zhu, J.Y., Zhou, T., Efros, A.A.: Image-to-image translation with conditional adversarial networks. In: 2017 IEEE Conference on Computer Vision and Pattern Recognition (CVPR), pp. 5967–5976 (2017)

10. Jang, E., Gu, S.S., Poole, B.: Categorical reparameterization with gumbel-softmax. arXiv:abs/1611.01144 (2017)
11. Johnson, J., Alahi, A., Fei-Fei, L.: Perceptual losses for real-time style transfer and super-resolution. In: Leibe, B., Matas, J., Sebe, N., Welling, M. (eds.) ECCV 2016. LNCS, vol. 9906, pp. 694–711. Springer, Cham (2016). https://doi.org/10.1007/978-3-319-46475-6_43
12. Kaiser, L., et al.: Fast decoding in sequence models using discrete latent variables. In: International Conference on Machine Learning (2018)
13. Kingma, D.P., Ba, J.: Adam: a method for stochastic optimization. CoRR abs/1412.6980 (2015)
14. Lee, D., Kim, C., Kim, S., Cho, M., Han, W.S.: Autoregressive image generation using residual quantization. In: 2022 IEEE/CVF Conference on Computer Vision and Pattern Recognition (CVPR), pp. 11513–11522 (2022)
15. Liu, Z., Luo, P., Wang, X., Tang, X.: Deep learning face attributes in the wild. In: Proceedings of International Conference on Computer Vision (ICCV), December (2015)
16. Maddison, C.J., Mnih, A., Teh, Y.W.: The concrete distribution: a continuous relaxation of discrete random variables. arXiv:abs/1611.00712 (2017)
17. Detlefsen, N.S., et al.: TorchMetrics - Measuring Reproducibility in PyTorch (2022). https://doi.org/10.21105/joss.04101, https://github.com/Lightning-AI/metrics
18. van den Oord, A., Vinyals, O., Kavukcuoglu, K.: Neural discrete representation learning. In: NIPS (2017)
19. Ramesh, A., et al.: Zero-shot text-to-image generation. In: International Conference on Machine Learning, pp. 8821–8831. PMLR (2021)
20. Razavi, A., Van den Oord, A., Vinyals, O.: Generating diverse high-fidelity images with vq-vae-2. In: Advances in Neural Information Processing Systems, vol. 32 (2019)
21. Roy, A., Vaswani, A., Neelakantan, A., Parmar, N.: Theory and experiments on vector quantized autoencoders. arXiv:abs/1805.11063 (2018)
22. Salimans, T., Goodfellow, I., Zaremba, W., Cheung, V., Radford, A., Chen, X.: Improved techniques for training GANs. In: Advances in Neural Information Processing Systems, vol. 29 (2016)
23. Sauer, A., Schwarz, K., Geiger, A.: Stylegan-xl: scaling stylegan to large diverse datasets. In: ACM SIGGRAPH 2022 Conference Proceedings, pp. 1–10 (2022)
24. Karras, T., Samuli Laine, T.A.: A style-based generator architecture for generative adversarial networks. IEEE 3 (2019). https://ieeexplore.ieee.org/document/8953766
25. Vaswani, A., et al.: Attention is all you need. In: Advances in Neural Information Processing Systems, vol. 30 (2017)
26. Wang, Z., Bovik, A.C., Sheikh, H.R., Simoncelli, E.P.: Image quality assessment: from error visibility to structural similarity. IEEE Trans. Image Process. **13**, 600–612 (2004)
27. Williams, W., Ringer, S., Ash, T., MacLeod, D., Dougherty, J., Hughes, J.: Hierarchical quantized autoencoders. Adv. Neural. Inf. Process. Syst. **33**, 4524–4535 (2020)
28. Yu, J., et al.: Vector-quantized image modeling with improved vqgan. arXiv:abs/2110.04627 (2022)
29. Yu, J., et al.: Scaling autoregressive models for content-rich text-to-image generation. arXiv:abs/2206.10789 (2022)

30. Zhang, R., Isola, P., Efros, A.A., Shechtman, E., Wang, O.: The unreasonable effectiveness of deep features as a perceptual metric. In: 2018 IEEE/CVF Conference on Computer Vision and Pattern Recognition, pp. 586–595 (2018)
31. Zheng, C., Vuong, T.L., Cai, J., Phung, D.: MOVQ: modulating quantized vectors for high-fidelity image generation. Adv. Neural. Inf. Process. Syst. **35**, 23412–23425 (2022)

Dynamic Local Filters in Graph Convolutional Neural Networks

Andrea Apicella[1,2,3]([✉]) [iD], Francesco Isgrò[1,2,3] [iD], Andrea Pollastro[1,2,4] [iD], and Roberto Prevete[1,2,3] [iD]

[1] Department of Electrical Engineering and Information Technology, University of Naples Federico II, Naples, Italy
andrea.apicella@unina.it
[2] Laboratory of Augmented Reality for Health Monitoring (ARHeMLab), Naples, Italy
[3] Laboratory of Artificial Intelligence, Privacy & Applications (AIPA Lab), Naples, Italy
[4] Lawrence Berkeley National Laboratory, Berkeley, CA 94720, USA

Abstract. Over the last few years, we have seen increasing data generated from non-Euclidean domains, usually represented as graphs with complex relationships. Graph Neural Networks (GNN) have gained a high interest because of their potential in processing graph-structured data. In particular, there is a strong interest in performing convolution on graphs using specific GNN architectures, generally called Graph Convolutional Neural Networks (GCNN). This paper presents a novel method to adapt the behaviour of a GCNN using an input-based dynamically generated filter. Notice that the idea of adapting the network behaviour to the inputs they process to maximize the total performances has aroused much interest in the neural networks literature over the years. The experimental assessment confirms the capabilities of the proposed approach, achieving promising results using simple architectures with a low number of filters.

Keywords: Graph Convolutional Neural Networks · Deep Learning · Dynamic Neural Networks · Programmable ANNs · Graph Structure Learning

1 Introduction

Over the last few decades, Machine Learning (ML) methods have been used to address several challenges in Artificial Intelligence (AI) because of the progress in Deep Learning [20] in several AI fields. In particular, Convolutional Neural Networks (CNNs) have gained much interest due to their potential and versatility in addressing a large scale of machine learning problems [20,21], such as image processing [19] and pattern recognition [18]. The potential of CNNs lies in extracting and processing local information by performing convolution on input data using sets of trainable filters with a fixed size. However, the design of the convolution operation in the classical CNNs allows processing only regular data while, in the real world, a considerable amount of data naturally lies on non-Euclidean domains, needing different techniques to be processed. *Graph-based* structures often represent these data. Graph structures imply several difficulties in using standard CNNs, such as the variable number of neighbours for each node (differently from regular data where the filter properties fix the number of neighbours for each node). This aspect has led to new processing techniques, such as Graph Neural Networks (GNNs) [10, 29, 34], which gained high interest during the recent years. Due to the great success of CNNs, Graph Convolutional Neural Networks (GCNNs) were proposed. GCNNs share the same idea of message passing with RecGNNs but in a non-iterative manner. Among the different works proposed in the literature, Graph Structure Convolutional Networks (GSCNs, [38]) perform a spatial convolution learning the underlying input graph structure information, improving the overall performances of the network. However, classical Convolutional Networks are based on learned filters having constant values for each input fed to the network. In other words, the filter values are independent of the input values. Instead, adapting Artificial Neural Networks (ANNs) behaviours as a function of the input is an open research area. For instance, in [7], the authors discuss the biological plausibility of adapting the behaviour of neural networks to external inputs. In the GNN field, Edge-Conditioned Convolution (ECC) networks [33] perform spatial convolutions over graph neighbourhoods exploiting edge labels and generating suitable input filters from them. It is worth noticing that ECC leverages structural information (particularly edge labels) to generate sets of convolutional filters without learning information related to the graph structure itself. In this paper, we exploit the possibility of dynamically changing the GCNN behaviour as a function of the input. We also propose a novel method to perform spatial convolution on graph-structured data using fixed-sized filters while learning graph structural information and adapting the network's behaviour to each input. To this aim, we propose to perform dynamical filtering operations where convolution is performed using neighbourhood-dependent sets of filters for each given input graph, leading to an adaptable convolution operation on the input graph.

This paper is organized as follows: Sect. 2 briefly reviews the related literature; Sect. 3 describes the proposed method; the experimental assessment is described in Sect. 3.2, while in Sect. 4 we present the experimental results, that we discuss in Sect. 5. The concluding Sect. 6 is left to final remarks.

2   Related Works

We are proposing an ANN architecture that exploits both the concept of dynamically changing the behaviour of a generic ANN and that of convolution on graph-structured data based on GNNs.

In this section, we first review the related works on ANNs that dynamically change their behaviour on the inputs. Secondly, a general description of Graph Neural Networks is reported, summarising the main related works on GCNNs.

2.1 Dynamically Changeable Behaviours in ANNs

An ANN can dynamically change its behaviour by adapting the network parameters to each input instance. This way, the resulting network can perform an *input-suited* processing. The idea of controlling the behaviours of an ANN through its input or *auxiliary* inputs has a long history in the literature [3,7,8,16,26–28,30,32].

More recently, in [14], the authors defined the dynamic changes in ANNs' behaviours in the context of traditional CNNs using a dynamic filter module to execute the convolution operation. A dynamic filter module consists of two parts: a *filter-generating network*, that generates filters' parameters from a given input, and a *dynamic filtering layer*, that applies those generated filters to another input. Moreover, the dynamic filtering layer is instantiated in two different ways: a *dynamic convolutional layer*, wherever the filtering operation is translation invariant, and a *dynamic local filtering layer*, wherever the filtering operation is not translation invariant, and it implies the using of specific local filters over all the input's positions. This method is developed in the context of classic CNNs; by contrast, in [33], the authors attempt to perform a dynamic spatial convolution on graphs. The authors proposed Edge-Conditioned Convolution (ECC), which uses a filter-generating network to output edge-specific filters for each input dynamically. In [11], CNN and LSTM filters are generated by an auxiliary network fed with learned embeddings, differently from other approaches where the same input for both the weight-generating network and the inference network is used.

Inspired by what is described in [14,33], in this work, we perform an adaptive convolution on input graphs using filters dynamically generated by a filter-generating network, obtaining a dynamical change in the behaviour of the GCNN. Differently from the ECC proposed in [33] where convolutional filters are *edge-based*, our proposal (see Sect. 3) considers *node-based* filters, tweaking in this way the filtering operation on nodes by nodes themselves. According to [14], our approach is proposed as a *dynamic local filtering layer*, where sets of filters are locally generated for each neighbourhood during the convolution.

2.2 Graph Neural Networks

The first neural network model for graph-structured data was proposed in [29]. This model leverages the idea that graph nodes represent concepts related to each other via edges. The information exchange among nodes and their neighbours allows the model to update the produced features iteratively using the *message passing* mechanism. In the literature, A particular focus is given to performing convolution on graph-structured data. In GCNNs, information is exchanged between neighbours using different convolutional layers, each one with different filters [37]. However, the non-Euclidean characteristic of graphs (e.g., their irregular structure) makes the convolution and filtering operations less well-defined than those on images. Therefore, researchers have been working on how to conduct convolution operations on graphs following two different approaches: i) the *spectral approach*, which relies on the graph spectral theory, involving graph signal processing, such as graph filtering and graph wavelets (see, for example, [13,22,24]); ii) the *spatial approach*, that leverages on structural information to perform convolution, such as aggregations of graph signals within the node neighbourhood (see, for example, [2,12]).

Recently, many spatial domain methods have been proposed in the literature. For example, in [25], the authors present PATCHY-SAN, a GCNN model inspired by the classical image-based CNN. In [2] and [12], two different methods to generalise the convolutional operator using random walks for neighbourhood locating were reported. In [9], spectral-based GNNs are generalised to work on hypergraphs instead of classical graphs. In [36], the network layers of a GNN are collapsed into a single linear transformation, reducing the GNN complexity.

Nevertheless, in the most significant part of the proposed methods, the graph structure of the inputs is given *a priori*. That is, the input structure has to be defined before the network training stage. Interestingly, in [38], the authors proposed a method to learn or refine the graph structure and the network parameters. This method, similarly to classical CNNs, leverages learned filters whose values are fixed and equal for all the inputs. Instead, the proposal of this work learns or refines the graph structure together with the network parameters, similarly to [38]. Furthermore, it also modifies the network's behaviour using dynamically generated convolutional filters computed by a filter-generating network, as suggested in [14] and [1].

Recent studies have been exploring the use of attentional mechanisms to enhance the performance of Graph Neural Network (GNN) models [17,35]. However, in this work, we take a different approach and experimentally investigate the potential of dynamically generated filters as a means to improve the performance of GNN models. It is important to note that our primary objective is not to surpass the reported state-of-the-art results, but rather to assess whether this alternative approach holds promise for further investigation and exploration.

To summarise, our proposal consists of performing a spatial convolution on graph-structured data merging two different aspects: learning or refining information related to the input graph structure and adapting the GCNNs behaviour to their inputs. Implementing an input-adaptable filtering operation brings two main advantages, as the experimental results will show: firstly, an improvement in the overall performances, and secondly, a more compact model in terms of trainable parameters.

3   Material and Methods

In this work, we propose a GCNN-based architecture whose convolutional filters change as a function of the input. During the convolution operation, graphs' nodes are convoluted using filters dynamically generated relying on their neighbourhoods, thus making the convolution operation locally suited to each input graph. Differently from similar works as [33], where the filters depend on the graph edges, we propose a dynamic behaviour architecture based on graph nodes' features. In this section, a detailed description of our proposal is given after a brief introduction of the graphs' notation.

3.1 Proposed Approach

This work aims to perform convolution on graphs using dynamically generated filters conditioned on a given input. As we have described above, otherwise from ECC in [33], where convolution is performed using dynamical *edge-based* filters, our intent consists in using *node-based* filters dynamically generated from nodes' feature vectors.

Let $G = (V, E)$ be an undirected or directed graph where V is a finite set of N nodes, and E is a finite set of edges. We define in boldface $\mathbf{x}^n \in \mathbb{R}^{1 \times J}$ the input feature vector related to the node $n \in V$, where J is the number of input channels, and $\mathbf{y}^n \in \mathbb{R}^{1 \times M}$ its output feature vector, where M is the number of output channels. Let $X \in \mathbb{R}^{N \times J}$ denote the matrix representation of an input graph as an embedding of the feature vectors of its nodes. To obtain neighbourhoods with a sufficient number of nodes for a filter of dimension K, we select each node's *K-nearest neighbours* using the classical shortest path distance [4]. Neighbourhoods are uniquely defined for each node. We will generally refer to the neighbourhood of a given node n as $N(n)$. Given an input graph G and a Filter-Generating Network (FGN) $h(\cdot)$ with parameters θ, a set of node-specific filters $\mathcal{F} = h_\theta(G)$ can be generated. These filters can be used to compute a dynamic convolution on input graphs. Supposing to compute the m-th output channel of the node n, our proposal can be formalised as

$$y_m^n = f\left(\sum_{j=1}^{J} \sum_{k=1}^{K} A_{nk} \mathcal{F}_{jkm} x_j^{s(n,k)} \right),$$

where $f(\cdot)$ is the activation function, $s(n, k)$ returns the index of the k-th neighbour of $N(n)$, A_{nk} represents a correlation degree between each node n and its k-th neighbour, $\mathbf{x}_j^{s(n,k)}$ is j-th input channel related to the input feature vector of the node $s(n, k)$, and $\mathcal{F} \in \mathbb{R}^{J \times K \times M}$ are M filters generated by $h_\theta(N(n))$. Similarly to what is proposed by [14] on classical CNNs, $h_\theta(\cdot)$ will be implemented as a *Dynamic Local Convolutional Graph Neural Network* (DL-ConvGNN), having as input the neighbourhood $N(n)$ of the processed node n.

Figure 1 depicts a schematic representation of the proposed approach. Thus, two different types of parameters, A and θ, must be learned during the training stage. The resulting A_{nk} expresses the strength of the learned relationship between the node n and its k-th neighbour after the training stage so that it can be viewed as *a posteriori knowledge*. Moreover, according to [38], a sparse constraint, defined as an L_1 norm on the A matrix, is considered to let the layer emphasise only important relations among the neighbourhoods. The sparse constraint is added to a general loss function of the entire network obtaining as loss $\mathcal{L}(\hat{Y}, Y) = \mathcal{L}_b(\hat{Y}, Y) + \gamma \sum_i^P ||A^i||$, where $\hat{Y}$ are the predicted labels, Y represents the ground truth, $\mathcal{L}_b$ is a basic loss function (such as *Cross-Entropy*

Loss for classification tasks or *MSE* for regression tasks), P is the number of DL-ConvGNN layers, A^i is the a posteriori knowledge learned by the i-th layer and γ is the Lagrange multiplier balancing the effect of basic loss and the sparse constraint. The use of a dynamic approach in performing the filtering operation, as the experimental results will show, involves the use of a fewer number of convolutional filters leading to a simpler architecture with a fewer number of trainable parameters. Unlike ECC, in this approach, the dynamic behaviour is enhanced by the contribution of the underlying structural graph information learned by the A matrix.

The advantages of the proposed approach can be summarised as follows: i) It inherits the standard convolution operation from classical CNNs: the convolution is performed on each node with its nearest neighbours applying sets of fixed-sized filters; ii) It learns underlying input structural information: while processing a node with its neighbourhood, the structural information learned emphasises node relationships, increasing the overall performance; iii) Its behaviours changes according to the input: filters used for the convolution are dynamically generated using an external module based on the input graphs; iv) Dynamic behavioural of the convolutional filters leads to design simpler architectures: promising results can be achieved using fewer convolutional filters than a non-dynamic approach; v) convolution is locally suited on input graphs: convolution is specific for each input location with dynamic local filters, thus leading to a more suitable convolution operation.

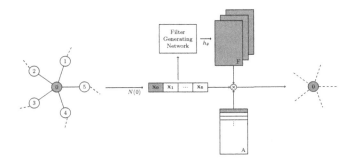

Fig. 1. A graphical description of the DL-ConvGNN layer in processing a node - marked in red - using filters of dimension $K = 5$. Filters are dynamically generated using the filter-generating network having as input the neighbourhood of the node that is being processed. Input signals are weighted and summed jointly with the neighbourhood entries of the A matrix (marked by the same colour) to compute the node output. (Color figure online)

3.2 Experimental Assessment

A preliminary experiment on the well-known benchmark dataset MNIST, modelling its grid-structured data as graphs to test our proposal's feasibility. In this first set of experiments, the performance of the proposed approach with the one achieved by non-dynamic GCNN methods is compared. We point out that, in this set of experiments, we are not interested in obtaining better results than other approaches presented in the literature, but we want to validate the feasibility of our method experimentally. Next, we applied our approach in some of the experimental scenarios reported in [38], to make a fair comparison with a non-dynamic convolution neural network on general graph-structured data. In particular, we chose *DPP4*, *20NEWS* and *Reuters-21578* datasets.

We want to demonstrate the validity of the proposed method while keeping the number of convolutional filters low and using filter-generating networks with simple architectures (such as *shallow neural networks*). To make a fair comparison, all the experiments were carried out using the identical experimental setups reported in the reference papers. In particular, the proposal followed the same initialisation schema proposed in [38], consisting in initialising all entries in A with a constant value c, reported in Table 1.

MNIST: MNIST [21] consists of images belonging to 10 classes, each corresponding to a handwritten numerical digit between 0 and 9. Therefore, a 10-classes classification problem can be defined. In order to make a comparison with a spatial GCNN method, we focused on the experiments described in [12].

Merck Molecular Activity Challenge: The *Merck molecular activity* is one of the Kaggle challenges[1] whose main task consists in predicting the biological activities of 15 molecular activity datasets. Each molecule is defined by numerical descriptors generated by its chemical structure. Following different works such as [12,13,38], we have experimented our approach on *DPP4 dataset* defining this task as a regression problem on graph-structured data. The information concerning the correlation structure between features was extracted following [12]. According to the standard set by the Kaggle challenge, predictions on the test set are evaluated using the correlation coefficient R^2.

Text Categorisation: We tested our approach also on a classification task based on two datasets mentioned by [38]: 20NEWS [15] and Reuters-21578 [23]. The 20NEWS dataset consists of 20 classes and 17236 samples. We follow the preprocessing described in [6] and the document representation described in [39].

Instead, for the Reuters-21578 dataset we adopt the same experimental setup described in [38].

4   Results

We have adopted a *grid search* approach for hyperparameters tuning for each experiment. The ranges of values are reported in the experiments' related Table 1. We used the Adam optimizer with a learning rate of 10^{-1} and Cross Entropy Loss for experiments involving MNIST and Reuters-2158 datasets and MSE for DPP4 dataset. To check the fairness in the reproduction of the identical experimental

---

[1] https://www.kaggle.com/c/MerckActivity.

Table 1. Grid search settings used for the experiments.

FGN Architecture	K	γ	Output channels	c	Threshold
MNIST					
$\{FC50, FC100, FC150, FC200\}$	6	$\{10^{-7}, 10^{-5}\}$	$\{1, 3, 5, 7, 9, 10\}$	$\{10^{-5}, 10^{-2}\}$	$\{0.1, 0.2, \ldots, 0.9\}$
DPP4					
$\{FC50, FC100, FC150, FC200\}$	16	10^{-7}	$\{1\}$	$\{10^{-5}, 10^{-4}, \ldots, 10^{-1}\}$	$\{0.1, 0.2, \ldots, 0.9\}$
20NEWS					
$\{FC50, FC100, FC150, FC200, FC250\}$	25	0	$\{1, 2, \ldots, 16\}$	$\{10^{-5}, 10^{-4}, \ldots, 10^{-1}\}$	–
Reuters-21578					
$\{FC50, FC100, FC150, FC200, FC250\}$	25	0	$\{1, 2, \ldots, 16\}$	$\{10^{-5}, 10^{-4}, \ldots, 10^{-1}\}$	–

setups, we tried to replicate the performance reported by the referenced papers [12,38] using their models. Model performance estimation was performed using a hold-out strategy using the predefined train/test split supplied with the datasets. Moreover, during the training stage, 30% of the training data was used as a validation set. We chose *shallow neural networks* with one hidden layer as Filter-Generating Network (FGN) architectures, whose number of nodes was tuned using grid search in each experiment. We denote with FCt a fully connected layer with t hidden units.

4.1 MNIST

We report the grid search ranges for determining the hyperparameters in the case of the MNIST dataset in Table 1. In these tests, a learning rate scheduler was used to decrease the learning rate by 30 % every 15 epochs. The threshold applied to the features' correlation values to construct the adjacency matrices was considered a hyperparameter (see Table 1).

As we can see from the results in Table 2, our approach achieves an accuracy that can be considered comparable with the one reported by [12]. Moreover, our dynamic system leads to achieving good performances with a simpler architecture than the static approach: the Graph CNN model achieves its best error rate using two convolutional layers with 20 filters; DL-ConvGNN model achieves comparable results using one convolutional layer with 5 dynamically generated filters. Furthermore, it is also important to point out how the use of DL-ConvGNN can lead to a lower number of trainable parameters against the other approaches.

4.2 Merck Molecular Activity Challenge

The grid search ranges for determining the hyperparameters for experiments with the DPP4 dataset are reported in Table 1. As for the MNIST case, the threshold value to apply to the features' correlation values to extract the adjacency matrices was chosen as a hyperparameter (see Table 1). For the evaluation of performances we used the R^2 correlation coefficient as stated in Sect. 3.2.

As reported in Table 2, our approach outperforms the result reached by the GSCN method. It is possible to see that the proposed dynamic approach

Table 2. Results of the experimental assessment

MNIST	Architecture	# parameters	Accuracy
DL-ConvGNN (ours)	C5	49,447	97.61
GraphCNN (Hechtlinger et al.)	C20-C20	145,970	98.41
DPP4	**Architecture**	**# parameters**	R^2
DL-ConvGNN (ours)	C1	41,569	0.282
GSCN (Zhang et al.)	C10	56,149	0.299
20NEWS	**Architecture**	**# parameters**	**Accuracy**
DL-ConvGNN (ours)	C3-FC100	3,264,315	62.85
GSCN (Zhang et al.), our implementation	C16-FC100	1,625,959,590	60.36
Reuters-21578	**Architecture**	**# parameters**	**F1**
DL-ConvGNN (ours)	C6-FC100	501,815	92.56
GSCN (Zhang et al.)	C16-FC100	2,447,090	73.25

needs only one convolutional filter against the 10 convolutional filters needed by the non-dynamic approach to reach its best result. We also underline that DL-ConvGNN has a lower number of trainable hyperparameters than the other methods in this experiment.

4.3   Text Categorization

Since both 20NEWS and Reuters-21578 datasets were preprocessed by almost the same pipeline, which led to a similar data representation, we used the same grid search ranges reported in Table 1, for both of the datasets. The preprocessing pipeline used for these tasks followed what was done and shared[2] by [6].

20NEWS dataset: during our preliminary analysis, although we tried to replicate the performances as described in [38] using the same experimental setup, we could not achieve the GSCN score reported by the authors for the batch of experiments on the 20NEWS dataset, probably because of some differences in the processing pipeline. In particular, using the same architecture, we achieved an accuracy of 60.36 % against the 68.83 % reported in [38]. For this reason, in order to make a comparison with our results, we have considered our best GSCN's score that is reported in Table 2. As we can notice, our approach outperformed the GSCN's result using a lower number of filters also in this case.

Reuters-21578 dataset: as one can see from Table 2, our approach outperforms the result reached by the GSCN method. These results still confirm that our dynamic approach uses fewer convolutional filters to achieve more satisfactory performance than a non-dynamic approach.

---

[2] https://github.com/mdeff/cnn_graph.

5   Discussion

According to the results in the previous Section, our dynamic approach allows us to consider simpler architectures than non-dynamic ones. This aspect was first observed during the experiments on the MNIST dataset, where very simple architectures have led to results comparable with those achieved by more complex architectures. A similar observation comes out while comparing the results reported in [38] where using the same architectures, we empirically see that our approach needs fewer convolutional filters to process information while achieving better results. This aspect leads, in general, to design convolutional architectures that will be more compact than the standard ones, where a considerable number of filters and convolutional layers are usually taken into account to extract meaningful information. As a consequence, DL-ConvGNN needs a considerably lower number of trainable parameters rather than a non-dynamic approach, as shown in Table 2.

It is essential to point out that the use of the A matrix as described in [38] and adopted in this proposal, limits the use of the proposed convolutional layer only to dataset whose samples are described by a fixed graph topology: we remind that A_{nk} contains the learned structural information about the n-th node and its k-th neighbour, and this relation is fixed in each sample. This limit does not allow us to directly compare with the ECC model, where the convolution operation is not constrained to a fixed samples' topology. In future work, we would like to handle this limit to extend the functionality of this layer to datasets with non-fixed topologies among the samples, such as ENZYMES [31] and MUTAG [5].

6   Conclusion

In this work, we have proposed a dynamic method to perform convolution on graph-structured data. Combining the idea of having dynamically changeable behaviours in ANNs and convolutional graph neural networks, this work aimed to present a graph convolutional layer capable of performing convolution using *node-specific* filters to achieve an input-suitable filtering operation. To perform convolution on graphs, we adopted a *spatial* approach, focusing on the characteristic of the GSCN model presented in [38] in learning graph structural information during the training stage.

Moreover, referring to what was done by Jia et al. in [14] on CNNs, we have considered altering, in a dynamic fashion, the behaviour of our convolutional layer with the use of an external module, the *filter-generating network*, to generate local dynamic filters conditioned to the neighbourhoods' signals. In this way, the proposed graph convolutional layer learns and applies input-suited filters, making the filtering operation *customizable* to the input graph, and enhancing the overall performances using the learned structural information. To assess the improvements in using a dynamic approach to generate convolutional filters, several experiments were made to compare our approach with non-dynamic spatial

GCNNs, focusing on the GSCN model. Moreover, it empirically emerged that a dynamic approach requires a low number of convolutional filters to achieve satisfying results, involving a lower number of learnable parameters.

References

1. Apicella, A., Isgrò, F., Pollastro, A., Prevete, R.: Adaptive filters in graph convolutional neural networks. Pattern Recogn. **144**, 109867 (2023). https://doi.org/10.1016/j.patcog.2023.109867. ISSN 0031-3203
2. Atwood, J., Towsley, D.: Diffusion-convolutional neural networks. In: Advances in Neural Information Processing Systems, pp. 1993–2001 (2016)
3. Bishop, C.: Mixture density networks. Workingpaper, Aston University (1994)
4. Buckley, F., Harary, F.: Distance in Graphs. Addison-Wesley, Redwood City (1990)
5. Debnath, A.K., Lopez de Compadre, R.L., Debnath, G., Shusterman, A.J., Hansch, C.: Structure-activity relationship of mutagenic aromatic and heteroaromatic nitro compounds. correlation with molecular orbital energies and hydrophobicity. J. Medi. Chem. **34**(2), 786–797 (1991)
6. Defferrard, M., Bresson, X., Vandergheynst, P.: Convolutional neural networks on graphs with fast localized spectral filtering. In: Advances in Neural Information Processing Systems, vol. 29, pp. 3844–3852 (2016)
7. Donnarumma, F., Prevete, R., Trautteur, G.: Programming in the brain: a neural network theoretical framework. Connect. Sci. **24**(2–3), 71–90 (2012)
8. Eliasmith, C.: A unified approach to building and controlling spiking attractor networks. Neural Comput. **17**(6), 1276–1314 (2005)
9. Fu, S., Liu, W., Zhou, Y., Nie, L.: Hplapgcn: hypergraph p-laplacian graph convolutional networks. Neurocomputing **362**, 166–174 (2019)
10. Gori, M., Monfardini, G., Scarselli, F.: A new model for learning in graph domains. In: Proceedings. 2005 IEEE International Joint Conference on Neural Networks, 2005. vol. 2, pp. 729–734. IEEE (2005)
11. Ha, D., Dai, A.M., Le, Q.V.: Hypernetworks. In: 5th International Conference on Learning Representations, ICLR 2017, Toulon, France, 24–26 April 2017 (2017)
12. Hechtlinger, Y., Chakravarti, P., Qin, J.: A generalization of convolutional neural networks to graph-structured data. arXiv preprint arXiv:1704.08165 (2017)
13. Henaff, M., Bruna, J., LeCun, Y.: Deep convolutional networks on graph-structured data. arXiv preprint arXiv:1506.05163 (2015)
14. Jia, X., De Brabandere, B., Tuytelaars, T., Gool, L.V.: Dynamic filter networks. Adv. Neural. Inf. Process. Syst. **29**, 667–675 (2016)
15. Joachims, T.: A probabilistic analysis of the Rocchio algorithm with TFIDF for text categorization. Technical report, Carnegie-mellon univ pittsburgh pa dept of computer science (1996)
16. Jordan, M.I.: Attractor dynamics and parallelism in a connectionist sequential machine. In: Artificial Neural Networks: Concept Learning, pp. 112–127 (1990)
17. Kim, J., et al.: Pure transformers are powerful graph learners. Adv. Neural. Inf. Process. Syst. **35**, 14582–14595 (2022)
18. Kim, Y.: Convolutional neural networks for sentence classification. In: Proceedings of the 2014 Conference on Empirical Methods in Natural Language Processing (EMNLP), pp. 1746–1751 (2014)
19. Krizhevsky, A., Sutskever, I., Hinton, G.E.: Imagenet classification with deep convolutional neural networks. Adv. Neural. Inf. Process. Syst. **25**, 1097–1105 (2012)

20. LeCun, Y., Bengio, Y., Hinton, G.: Deep learning. Nature **521**(7553), 436–444 (2015)
21. LeCun, Y., Bottou, L., Bengio, Y., Haffner, P.: Gradient-based learning applied to document recognition. Proc. IEEE **86**(11), 2278–2324 (1998)
22. Levie, R., Monti, F., Bresson, X., Bronstein, M.M.: Cayleynets: graph convolutional neural networks with complex rational spectral filters. IEEE Trans. Signal Process. **67**(1), 97–109 (2018)
23. Lewis, D.: Reuters-21578 text categorization test collection, distribution 1.0 (1997). http://www.research/.att.com
24. Li, R., Wang, S., Zhu, F., Huang, J.: Adaptive graph convolutional neural networks. In: Proceedings of the AAAI Conference on Artificial Intelligence, vol. 32 (2018)
25. Niepert, M., Ahmed, M., Kutzkov, K.: Learning convolutional neural networks for graphs. In: International Conference on Machine Learning, pp. 2014–2023 (2016)
26. Nishimoto, R., Namikawa, J., Tani, J.: Learning multiple goal-directed actions through self-organization of a dynamic neural network model: A humanoid robot experiment. Adapt. Behav. **16**(2–3), 166–181 (2008)
27. Noelle, D.C., Cottrell, G.W.: Towards instructable connectionist systems. In: Sun, R., Bookman, L.A. (eds.) Computational Architectures Integrating Neural and Symbolic Processes, pp. 187–221. Springer, Boston (1995). https://doi.org/10.1007/978-0-585-29599-2_6
28. Paine, R.W., Tani, J.: Motor primitive and sequence self-organization in a hierarchical recurrent neural network. Neural Netw. **17**(8–9), 1291–1309 (2004)
29. Scarselli, F., Gori, M., Tsoi, A.C., Hagenbuchner, M., Monfardini, G.: The graph neural network model. IEEE Trans. Neural Networks **20**(1), 61–80 (2008)
30. Schmidhuber, J.: Learning to control fast-weight memories: an alternative to dynamic recurrent networks. Neural Comput. **4**(1), 131–139 (1992)
31. Schomburg, I., et al.: Brenda, the enzyme database: updates and major new developments. Nucleic Acids Res. **32**(suppl-1), D431–D433 (2004)
32. Siegelmann, H.T.: Neural Networks and Analog Computation: Beyond the Turing Limit. Springer, Boston (2012). https://doi.org/10.1007/978-1-4612-0707-8
33. Simonovsky, M., Komodakis, N.: Dynamic edge-conditioned filters in convolutional neural networks on graphs. In: 2017 IEEE Conference on Computer Vision and Pattern Recognition (CVPR), pp. 29–38 (2017)
34. Sperduti, A., Starita, A.: Supervised neural networks for the classification of structures. IEEE Trans. Neural Networks **8**(3), 714–735 (1997)
35. Velickovic, P., Cucurull, G., Casanova, A., Romero, A., Lio, P., Bengio, Y.: Graph attention networks. Stat **1050**, 20 (2017)
36. Wu, F., Souza, A., Zhang, T., Fifty, C., Yu, T., Weinberger, K.: Simplifying graph convolutional networks. In: International Conference on Machine Learning, pp. 6861–6871. PMLR (2019)
37. Wu, Z., Pan, S., Chen, F., Long, G., Zhang, C., Philip, S.Y.: A comprehensive survey on graph neural networks. IEEE transactions on neural networks and learning systems **32**, 4–24 (2020)
38. Zhang, Q., Chang, J., Meng, G., Xu, S., Xiang, S., Pan, C.: Learning graph structure via graph convolutional networks. Pattern Recogn. **95**, 308–318 (2019)
39. Zhang, Y., Jin, R., Zhou, Z.H.: Understanding bag-of-words model: a statistical framework. Int. J. Mach. Learn. Cybern. **1**(1–4), 43–52 (2010)

An AI-Driven Prototype for Groundwater Level Prediction: Exploring the Gorgovivo Spring Case Study

Alessandro Galdelli[(✉)] [ID], Gagan Narang [ID], Lucia Migliorelli [ID],
Antonio Domenico Izzo, Adriano Mancini [ID], and Primo Zingaretti [ID]

Vision Robotics and Artificial Intelligence Lab (VRAI), Dipartimento di Ingegneria
dell'Informazione, Università Politecnica delle Marche, 60131 Ancona, Italy
a.galdelli@univpm.it

Abstract. Water is a vital yet increasingly endangered resource, that
stands on the precipice of depletion and degradation, threatened by pollution, overexploitation, habitat alteration, and the looming spectre of
climate change. Growing demand for water from various productive sec-
tors and the escalating shifts in weather patterns are among the primary
factors contributing to resource strains. Such factors lead to abrupt deterioration in terms of quantity and quality of water. Recently, technological intervention is playing a significant role in the mitigation of the
stress on water resources leading to a better understanding of the dynamics behind consumption and replenishment. The substantial rise in the
utilization of novel sensors is actively enhancing monitoring capabilities
and facilitating data collection. There remains an ongoing requirement
to enhance and advance methodologies for effectively analyzing the exponential surge in data. In this regard, the use of Artificial Intelligence (AI)
towards supporting the management of water reserves can bring significant benefits in terms of the protection and sustainable utilization of
the resource that is the basis of all life. In this paper, we propose an
AI-based system for predicting Groundwater Level (GWL) focusing on
Gorgovivo spring, located in province of Ancona, Italy as a case study.
The predictive model was evaluated using the following criteria: Mean
Average Error (MAE), Mean Squared Error (MSE) and correlation. Our
case study concludes that the implementation of the prototype system
produces valuable prediction results for GWL compared to other state-
of-the-art approaches.

Keywords: Groundwater level prediction · Artificial Intelligence ·
Grogovivo spring · Time series forecasting · Prophet

1 Introduction

1.1 Background

The current climate change situation has made the sustainable use of natural
resources, especially water, more necessary than ever. Access to potable water

G. L. Foresti et al. (Eds.): ICIAP 2023, LNCS 14234, pp. 418–429, 2023.
https://doi.org/10.1007/978-3-031-43153-1_35

and basic sanitation is a human right and is a determining factor for all aspects of social, economic and environmental development. Quantifying safe potable water is inestimable, which makes it crucial to protect the resource and use it very judicially. Due to poor infrastructure or economic mismanagement, every year millions of people, most of them children, die from diseases due to inadequate water supply, sanitation and hygiene levels [20]. World leaders under the vision of the United Nations 2030 Agenda for Sustainable Development, committed to ending poverty, protecting the planet and ensuring that all people enjoy peace and prosperity. The agenda lays down a total of seventeen goals, where, notably, the sixth goal emphasizes the paramount importance of accessible and safe water for mankind. The European 2030 Agenda encompasses a comprehensive strategy for sustainable development within the region. Aligned with the United Nations objectives, it recognizes the significance of water as a fundamental resource that underpins various aspects of human life, the environment, and socio-economic development [19]. Failing to achieve the projections according to Goal 6 by 2050, about one in four people will face serious consequences due to a shortage of potable water [21]. Technology intervention is expected to play an extremely significant role in the attainment of sustainability goals. Innovations in technology-supported groundwater management techniques have witnessed a significant increase in the deployment of unique types of sensors, contributing to improved monitoring and data collection. These sensors enable continuous and precise measurements, providing valuable data for understanding groundwater dynamics and facilitating informed decision-making of consumption activities. The increasing availability of real-time data necessitates the need for advanced algorithms that can accurately predict and analyse Groundwater Level (GWL). Advanced algorithms can be trained on historical sensor data to recognize patterns and relationships within the data, allowing for accurate predictions of GWL in real-time. The relationship between the historical and future GWL level can help in assessing available water resources, managing its allocation, predicting the impact of pumping on aquifers, understanding the effects of climate change, urbanization and developing effective water management strategies for sustainable use. These decision support systems when incorporated into the policies of the governance better support both urban planning and ultimately water conservation [22]. In this scenario, the use of Artificial Intelligence (AI) may support the monitoring and management of surface and groundwater supplies [4]. AI models can describe and learn structural patterns by processing large amounts of multidimensional data, thereby providing valuable data-driven insights and predictions towards efficient planning. Therefore, research is proposed that aims at the development of an AI-based system able to predict the level of the Gorgovivo spring[1].

---

[1] In this work we selected as a case study the Gorgovivo spring, located in Serra San Quirico, province of Ancona, Italy.

1.2 State-of-the-Art

The increasing demand for water resources and the need for sustainable solutions has led to the exploration of innovative approaches to water resource management. AI has emerged as a promising tool for predicting and controlling water availability and quality. While numerical methods have traditionally been used for GWL modelling, recent studies have extensively employed AI techniques due to the non-linear and non-stationary nature of GWL time series. Across the literature, various popular forecasting approaches have been applied to GWL time series, including Auto Regressive Integrated Moving Average (ARIMA), Long Short-Term Memory (LSTM), and Artificial Neural Networks (ANN), as well as hybrid approaches such as ARIMA-LSTM [5,9–11]. Tao et al., [14] conducted a metareview of over a hundred articles published between 2008 and 2020, focusing on machine learning-based models for GWL modelling. Their investigation discusses the strengths and weaknesses of individual models and highlights the challenges faced in GWL prediction. They suggest that integrating multiple intelligent models can enhance prediction accuracy, particularly in dealing with nonlinear relationships between variables.

Machine learning methods have demonstrated superior performance in GWL time series prediction by effectively capturing complex and non-linear relationships between input and output variables [8]. Comparative studies have consistently shown that machine learning-based methods outperform traditional numerical approaches. In [1] the authors conducted a bibliometric analysis of machine learning and mathematical model techniques specifically applied to GWL forecasting using piezometric data. Their findings indicate that machine learning techniques such as Random Forest (RF), Support Vector Machine (SVM), and deep learning techniques like ANN achieve higher accuracies compared to mathematical model techniques. Ren et al., [13] investigated that though deep learning performs well in filling high dynamic gaps, it struggles with reconstructing trends and seasonality-based gaps. On the other hand, ARIMA, a traditional machine learning model, excels in capturing trends and seasonality. This suggests that traditional models, which are designed to handle time series with seasonal effects and trends, may be more suitable for GWL forecasting tasks. The studies increasingly encourage the scientific community to explore new machine-learning techniques.

While RFs and SVMs lack built-in mechanisms to handle the temporal nature of groundwater data and explicitly capture seasonality and holiday effects in GWL time series, ANNs require careful architecture design, training, and tuning for time series forecasting. There exists a pressing need for a fast, accurate and tunable forecasting procedure, which works best with time series that have strong seasonal effects due to the nature of GWL. Prophet is a good open-source machine learning model which is specifically designed for time series forecasting, with a particular emphasis on capturing seasonality, trends, and holiday effects [16]. Moreover, Prophet incorporates uncertainty estimation, which is crucial for decision-making in groundwater management.

Zarinmehr et al., [23] compared Prophet with other models like ARIMA, Multivariate Adaptive Regression (MARS), and Error Trend and Seasonality (ETS) using satellite data to analyze and forecast GWL in the Lake Urmia basin in Northwestern Iran. Prophet consistently outperformed the other models, achieving higher coefficients of determination (R^2) ranging from 0.81 to 0.85, as well as lower Mean Absolute Error (MAE) and Root Mean Squared Error (RMSE). This study provided valuable insights for groundwater management and planning in the region. Aguilera [2] conducted a similar study at the Ramsar wetland area of Doñana (Spain) and found that Prophet exhibited strong prediction capabilities for GWL time series. The study compared Prophet with various statistical and intelligent models commonly used in time series forecasting, suggesting that methods with additive schemes, such as ARIMA and Prophet, are suitable for modeling GWL time series. Prophet's ability to define multiple periodicities simultaneously and capture seasonality associated with special events, along with its capability to handle non-linear relationships, ease of operation, and incorporation of external regressors, make it a flexible and promising model for GWL forecasting [16]. Inspired by literature in closer fields in this work we develop an AI-based support prototype for predicting GWL enabled by the Prophet model.

1.3 Site Description

The Gorgovivo spring complex (Fig. 1) is located immediately downstream of the Gola della Rossa, situated in Serra San Quirico, province of Ancona, in Marche region, Italy. The spring is resourced through multiple sources that originate from the rock on the right bank (side) and flow directly into the Esino river [17]. These sources are essentially channels through which water emerges from the rock and enters the river.

The main catchment, known as "*Gorgovivo*" comprises 12 interconnected collecting wells with a branched access tunnel. Additionally, the secondary catchment, named "*Gorgovivo Bis*" was constructed in the mid-1990s and features a "draining tunnel" situated directly downstream of the Esino river weir. This tunnel spans approximately 160 m within the alluvial cover of cemented gravels and sands, overlaying the limestone formation at an average altitude of 155 m above sea level (m a.s.l.). The spring's water supply comes from a basin, which is a large area of land, spanning approximately 200 km^2 and the water table which is stored in the fractured limestone. The water from this basin eventually makes its way through the underground or surface channels, emerges as springs, and flows into the Esino river. The natural flow rate of the springs is indicatively between a minimum of 1,800 litres per second (l/s) and a maximum of 4,000 l/s. The water captured in this way flows inside the tunnel, in the direction of the river, until it reaches the accumulation tank from which the electric pumps installed in the lifting chamber above draw it, leading it to the loading tank by means of a special conduit. The water collected by the water catchments is conveyed through an underground tunnel, approximately 1,200 m long, to the first loading reservoir of the "*Gorgovivo*" adduction aqueduct, located near the industrial area of Serra San Quirico at an altitude of 156 m a.s.l.

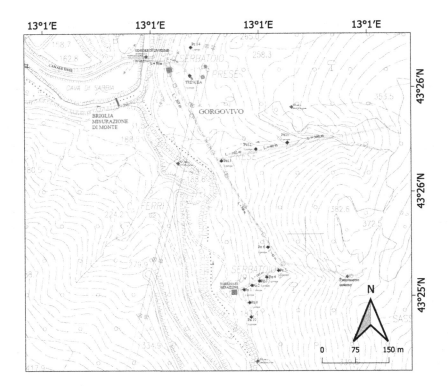

Fig. 1. Gorgovivo spring area and location of piezometric wells located in Marche region, Italy.

2   Materials and Method

2.1   Dataset

River Data. The Esino river, which lies immediately behind the Gorgovivo spring, plays a fundamental role in this complex hydrogeological system. The bed of the Esino river is at an average altitude of 155 m a.s.l., which is several metres lower than the Gorgovivo spring wells. When there are floods in the Esino river, the water level in the river rises and in some cases exceeds - the level of the piezometers of the spring wells. When this occurs, the river acts as a natural barrier that prevents excess spring water from flowing into the river, causing the levels of the wells to rise. Recent chemical and natural tracer studies [17] have ruled out the presence of mixing between Gorgovivo spring water and river water (at least at current well levels). Esino river data are collected by Viva Servizi S.p.A., which manages the Gorgovivo spring complex by using a non-contact radar sensor.

Climate Data. The daily precipitation data, for the period 2001–2022, comes from a monitoring network of 10 rainfall stations, 3 of which are managed by the

Civil Protection Department of Marche region and the reaming 7 are managed by the Agency for Services in the Agro-Food Sector of Marche region (ASSAM). These organizations provide a public database to retrieve weather data which contains several climate information such as precipitation, pressure, radiation, air temperature, soil temperature, wind direction, evaporation, leaf wetting and many others. Although the first monitoring data for the study area dates back to 1951, the structure of the network underwent subsequent modifications and integrations following local monitoring projects. Therefore, to have a predictive model that considers the climatic changes of the last period, it was decided to use both piezometric level data and climatic data to have a time series covering the period 2001–2022.

Groundwater Level Data. Data is collected daily from the 19 wells that make up the Gorgovivo spring through the use of piezometers. All wells have been equipped with level, electrical conductivity and temperature gauges. Among the total wells, seventeen wells are those whose level is guaranteed and maintained by electric pumps by implementing a system of communicating vessels. The remaining two wells, named *"Piezometro Interno"* and *"Fosso della Grotta"* respectively, are not controlled, i.e., there are no pumps to regulate GWL, and this level depends solely on natural events such as rainfall or melting of snow. The latter two wells are measured by utilizing a piezometer of ultrasonic and phreatimeters type, respectively. All piezometers for measuring well levels were placed at various depths, the piezometers were chosen by means of a survey terebration campaign. In the boreholes, tests were also carried out by measuring the natural gamma radioactivity of the soils, in order to identify the degree of fracturing of the reservoir rocks and verify the presence of water flows. The data from the gamma-ray tests made it possible to define the optimal elevations at which to push the various piezometers. For example, for the piezometer of the *"Piezometro Interno"* well, the chosen altitude was 158.89 m a.s.l. Furthermore, these two wells have different temporal resolutions of the data: the well-named *"Piezometro interno"* has a daily temporal resolution, while the other well-named *"Piezometro Fosso della Grotta"* has a weekly temporal resolution. Therefore, the *"Piezometro interno"* well is selected for our case study since it has the same temporal resolution as the rainfall stations and hence does not require pre-processing of the data.

2.2   Predictive Model

Analysis of the state-of-the-art guided the decision to design a Prophet-based system for our purposes [16]. Prophet is a time series forecasting model developed as an open-source project for the analysis and forecasting of time series. Among the strengths of Prophet are the robustness of the model in the case of missing data, frequent trend changes and the presence of outliers. Prophet represents the time series as the sum of three components: (i) trend, (ii) seasonality and

(iii) holidays, as seen below:

$$y(t) = g(t) + s(t) + h(t) + \epsilon_t \tag{1}$$

where:

- g(t): trend function, models non-periodic changes, it can be logarithmic;
- s(t): seasonality function, relying on Fourier series, provides a flexible model of periodic effects to model changes that are repeated at regular time intervals (e.g., weekly and yearly seasonality), it is also possible to have more than one seasonality in the same series;
- h(t): holidays, models irregular events that temporarily alter the time series;
- ϵ_t: error term, represents changes in the time series that are not captured by the model, ϵ_t is regarded as a normal distribution.

Specifically, Prophet models the forecasting as a "curve-fitting" problem, unlike other forecasting methods, as ARIMA [7], which considers the dependencies of each time instant with the preceding ones via autocorrelation and partial autocorrelation. Since the Gorgovivo spring is a complex groundwater network in which measured quantity (i.e., GWL) is dependent on the rainfall at the topsoil and Esino river, corollary, the prediction model is therefore composed of as many regressors as these additional dependencies. Figure 2 shows the proposed predictive model, which incorporates input data sources along with their dependencies, which encompass the GWL data (prior to the desired forecasting period), rainfall station data, and Esino river data. This includes a total of 8,305 × 12 data points for each dependency (i.e., GWL, Esino river, and 10 rainfall stations), spanning from 2001 to 2022. Using this input, our goal is GWL forecasting which can help the stakeholders in decision-making.

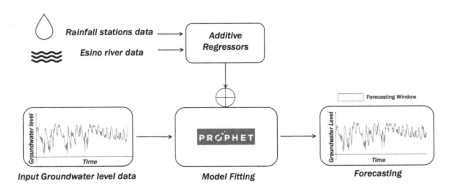

Fig. 2. Workflow of our proposed model to forecast the Groundwater Level in the Gorgovivo spring.

2.3   Training Settings and Hyperparameters Tuning

The gathered data is imported using Pandas DataFrames and is then processed for handling missing values, outliers, or any inconsistencies. The data is formatted for Prophet model with a date column and a target column containing the GWL values. The model is then added to the prepared data including the additional regressors, specifying the appropriate time series components like seasonality, trend, and holidays. The model hyperparameters need to be optimized to improve the performance. Cross-validation was conducted to find the optimal combination of hyperparameters. The procedure was implemented using the built-in Prophet function `cross_validation` and occurs by selecting a set of historical cutoff points in the time series.

The *period* and *horizon* parameters enable us to specify the size of the initial training period and the forecast window, respectively. Due to the fact the information under consideration for the case we are investigating is based on seasonal cyclicality, in agreement with project partner geologists, 90 days were selected for both *period* and *horizon* parameters. Figure 3 displays all the parameters and overall pipeline utilized for predictions. The entire dataset is split as follows: 2001 to 2020 is used as training for the Prophet predictive model, 2021 is used as validation and finally, 2022 is used as a test set. Cross-validation is performed to assess prediction performance over a 90 days horizon window. Specifically, we train for 19 years and we performed cross-validation (in a parallel fashion) every 90 days in 2021, this corresponds to 3 total forecasts that enable us to identify the best hyperparameters. We used a grid search algorithm to systematically explore a range of hyperparameters values for our Prophet based model. The approach involves exhaustively trying all possible combinations of hyperparameters values that can be tuned [16] and are: "*changepoint_ prior_ scale*", "*seasonality_ prior_ scale*", "*holidays_ prior_ scale*", "*seasonality_ mode*" and "*changepoint_ range*" within a predefined range as shown in (Table 1). The best hyperparameters are selected by computing the value of Mean Absolute Percentage Error (MAPE) for each combination of hyperparameters and selecting the combination which bears the lowest value of MAPE over the 90 day horizon window.

Table 1. Hyperparameters grid.

Hyperparameter	Values
changepoint prior scale	[0.001, 0.01, 0.05]
seasonality prior scale	[0.01, 0.8, 10.0, 80.0]
holidays prior scale	[0.0, 0.1, 0.5]
seasonality mode	["additive", "multiplicative"]
changepoint range	[0.75, 0.80, 0.90]

Finally, to speed up the computation time and obtain the best performance, parallel cross-validation using threads is enabled by setting the parameter *parallel* to *threads* (i.e., parallel = "threads").

2.4   Performance Metrics and Comparisons

We validated the proposed approach and compare it with other state-of-the-art approaches i.e., ARIMA and SARIMAX. Following the surveyed literature, we used the Mean Absolute Error (MAE), the Mean Squared Error (MSE) and the correlation [12] to benchmark the forecasting accuracy:

$$MAE(y, \hat{y}) = \frac{1}{N} \sum_{1}^{N} |y_i - \hat{y}_i| \tag{2}$$

$$MSE(y, \hat{y}) = \frac{1}{N} \sum_{1}^{N} (y_i - \hat{y}_i)^2 \tag{3}$$

Where y_i is the real value and $\hat{y}_i$ is the predicted value by forecasting model. In order to conduct a fair comparison test with our tuned model, the same tuning procedure is also performed for ARIMA and SARIMAX. The year 2021 of our dataset is used as validation dataset, using cross-validation with 90 days *period* and 90 days *horizon*. Whereas for the Prophet model, the developers provided special functions for the optimisation of hyperparameters with cross-validation, for the ARIMA and SARIMAX models, *ad hoc* development was necessary.

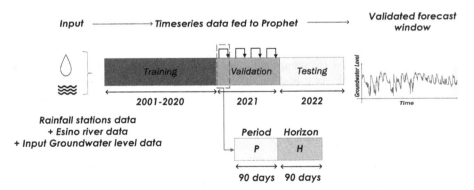

Fig. 3. Workflow to highlight the training, validation and testing strategy using cross-validation.

3   Results and Discussion

After hyperparameter tuning (see Sect. 2.3), we conducted the testing phase in the year 2022 using the optimised hyperparameters found. The parameters

Table 2. Standard evaluation metric for comparisons between the proposed approach and other state-of-the-art architectures.

Compared approaches	MAE (m a.s.l.)	MSE (m a.s.l.)	Correlation (%)
ARIMA	**0.071**	**0.007**	67.545
SARIMAX	0.282	0.095	25.671
Proposed	0.105	0.015	**80.769**

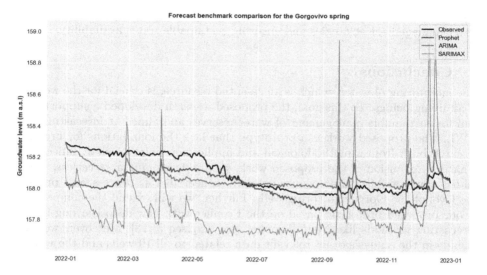

Fig. 4. Time series plot of predicted Groundwater Levels using different forecasting methods for the test set period.

reported are: *changepoint prior scale* = 0.05, *seasonality prior scale* = 80.0, *holidays prior scale* = 0.0, *seasonality mode* = "multiplicative", *changepoint range* = 0.8. Table 2 shows the values of the errors computed for each model in the benchmark comparison. In contrast, Fig. 4 shows the GWL forecast predicted by ARIMA, SARIMAX and our proposed model with respect to the ground-truth values.

The quantitative results show that our proposed model performs slightly worse with respect to ARIMA in terms of MAE and MSE, while it performs better than SARIMAX. However, by closely observing the graphs, the ARIMA and SARIMAX, in contrast to the proposed model, forecast several peaks that do not properly represent the smoother trends of the GWL. It is worth noting that these peaks, which occur steeply in the summer and fall seasons, with maxima hovering around GWL in the winter periods, are unnatural. This signifies an overestimation because, in summer, the level of water in the spring is always lower. The predictions derived from the proposed model exhibit a remarkable closeness to the actual data, further supported and validated by higher correlation values. Notably, during the summer season when water demand is highest, the Prophet model tends to slightly underestimate GWLs, indicating a downward trend. This observation emphasizes the need for the implementation of prudent policies to effectively manage and sustain spring water levels, considering the peak demand during the summer. In contrast, both the ARIMA and SARIMAX models tend to overestimate GWLs, which can hinder the adoption of conservative policies by appropriate authorities. Such overestimations may lead to severe consequences for all beneficiaries relying on this valuable resource. Hence, our GWL predictions supported by Prophet provide valuable insights for optimizing water management strategies and ensuring sustainable water availability.

4   Conclusions

The monitoring of water, which is an essential resource, is crucial for the welfare of all living beings. To this goal, the proposed research developed a support system for sustainable management of water reserves and aimed at forecasting the GWL. The proposed work is a prototype that lays the foundations for broader research both from a methodological and application point of view. Indeed, as a natural extension of the proposed work, we are developing a *serverless, automated* cloud architecture to automatically collect [6,15], process and predict the level of the Gorgovivo spring wells. Further, we will extend the comparison of our proposed prototype based on the Prophet with new deep-learning-based forecasting methods like NeuralProphet and Seq2seq [3,18]. Moreover, we will include in the analysis other relevant data related to all 19 wells and the water-chemical composition to create a comprehensive assistive tool for assessing the seasonal quality of the water.

Acknowledgements. This research was conducted in synergy with Viva Servizi S.p.A. We would like to thank the Civil Protection Department of Marche region and the Agency for Services in the Agro-Food Sector of the Marche Region (ASSAM), for sharing rainfall stations data. The authors would provide the data and code upon request to enable the scientific community to reproduce the experiments.

References

1. Afrifa, S., Zhang, T., Appiahene, P., Varadarajan, V.: Mathematical and machine learning models for groundwater level changes: a systematic review and bibliographic analysis. Future Internet **14**(9), 259 (2022)
2. Aguilera, H., Guardiola-Albert, C., Naranjo-Fernández, N., Kohfahl, C.: Towards flexible groundwater-level prediction for adaptive water management: using Facebook's prophet forecasting approach. Hydrol. Sci. J. **64**(12), 1504–1518 (2019)
3. Anshuman, A., Eldho, T.: Feeding static values to LSTMs for Seq2Seq learning for simultaneous source identification and parameter estimation in groundwater. In: AGU Fall Meeting Abstracts, vol. 2022, pp. H33B–05 (2022)
4. Chang, F.J., Guo, S.: Advances in hydrologic forecasts and water resources management. Water **12**(6), 1819 (2020)

5. Dadhich, A.P., Goyal, R., Dadhich, P.N.: Assessment and prediction of groundwater using geospatial and ANN modeling. Water Resour. Manage **35**, 2879–2893 (2021)
6. Galdelli, A., Mancini, A., Frontoni, E., Tassetti, A.N.: A feature encoding approach and a cloud computing architecture to map fishing activities. In: 17th IEEE/ASME International Conference on Mechatronic and Embedded Systems and Applications (MESA) (2021)
7. Ho, S., Xie, M.: The use of ARIMA models for reliability forecasting and analysis. Comput. Ind. Eng. **35**(1), 213–216 (1998)
8. Khan, J., Lee, E., Balobaid, A.S., Kim, K.: A comprehensive review of conventional, machine leaning, and deep learning models for groundwater level (GWL) forecasting. Appl. Sci. **13**(4), 2743 (2023)
9. Khozani, Z.S., Banadkooki, F.B., Ehteram, M., Ahmed, A.N., El-Shafie, A.: Combining autoregressive integrated moving average with long short-term memory neural network and optimisation algorithms for predicting ground water level. J. Clean. Prod. **348**, 131224 (2022)
10. Le, X.H., Ho, H.V., Lee, G., Jung, S.: Application of long short-term memory (LSTM) neural network for flood forecasting. Water **11**(7), 1387 (2019)
11. Najafabadipour, A., Kamali, G., Nezamabadi-pour, H.: The innovative combination of time series analysis methods for the forecasting of groundwater fluctuations. Water Resour. **49**(2), 283–291 (2022)
12. NumPy Developers: Numpy correlation. https://numpy.org/doc/stable/reference/generated/numpy.corrcoef.html. Accessed 12 May 2023
13. Ren, H., Cromwell, E., Kravitz, B., Chen, X.: Using long short-term memory models to fill data gaps in hydrological monitoring networks. Hydrol. Earth Syst. Sci. **26**(7), 1727–1743 (2022)
14. Tao, H., et al.: Groundwater level prediction using machine learning models: a comprehensive review. Neurocomputing **489**, 271–308 (2022)
15. Tassetti, A.N., Galdelli, A., Pulcinella, J., Mancini, A., Bolognini, L.: Addressing gaps in small-scale fisheries: a low-cost tracking system. Sensors **22**(3), 839 (2022)
16. Taylor, S.J., Letham, B.: Forecasting at scale. PeerJ Preprints (2017)
17. Tazioli, A.: Does the recharge area of a Spring Vary from year to year? Information from the water isotopes. Ital. J. Eng. Geol. Environ **2017**, 41–56 (2017)
18. Triebe, O., Hewamalage, H., Pilyugina, P., Laptev, N., Bergmeir, C., Rajagopal, R.: NeuralProphet: Explainable Forecasting at Scale (2021)
19. United Nations: Transforming Our World: The 2030 Agenda for Sustainable Development (2015). https://sdgs.un.org/2030agenda. Accessed 12 May 2023
20. World Health Organization and United Nations Children's Fund: Progress on household drinking water, sanitation and hygiene 2000–2017: special focus on inequalities. UNESCO (2019)
21. World Water Assessment Programme: 2018 UN World Water Development Report, Nature-based Solutions for Water. UNESCO (2018)
22. Zaresefat, M., Derakhshani, R.: Revolutionizing groundwater management with hybrid AI models: a practical review. Water **15**(9), 1750 (2023)
23. Zarinmehr, H., Tizro, A.T., Fryar, A.E., Pour, M.K., Fasihi, R.: Prediction of groundwater level variations based on gravity recovery and climate experiment (GRACE) satellite data and a time-series analysis: a case study in the Lake Urmia basin, Iran. Environ. Earth Sci. **81**(6), 180 (2022)

DiffDefense: Defending Against Adversarial Attacks via Diffusion Models

Hondamunige Prasanna Silva, Lorenzo Seidenari$^{(\boxtimes)}$, and Alberto Del Bimbo

University of Florence, Florence, Italy
{lorenzo.seidenari,alberto.delbimbo}@unifi.it

Abstract. This paper presents a novel reconstruction method that leverages Diffusion Models to protect machine learning classifiers against adversarial attacks, all without requiring any modifications to the classifiers themselves. The susceptibility of machine learning models to minor input perturbations renders them vulnerable to adversarial attacks. While diffusion-based methods are typically disregarded for adversarial defense due to their slow reverse process, this paper demonstrates that our proposed method offers robustness against adversarial threats while preserving clean accuracy, speed, and plug-and-play compatibility. Code at: https://github.com/HondamunigePrasannaSilva/DiffDefence.

Keywords: Diffusion models · Adversarial defense · Adversarial Attack

1 Introduction

The susceptibility of machine learning models to adversarial attacks is a major challenge in the field of artificial intelligence. While various techniques have been proposed to enhance the robustness of classifiers against such attacks, there is a pressing need for more effective and efficient solutions. In recent years, generative models such as Generative Adversarial Networks (GANs) [13], Diffusion Probabilistic Models [17] have emerged as a promising approach to improve the resilience of machine learning models against adversarial attacks.

Modern deep generative models have a common structural similarity: the generation of novel patterns is usually performed by transforming some random latent code z. Sampling $z \sim p(z)$, where $p(z)$ is a known distribution (e.g. $\mathcal{N}(0, I)$ and then computing $G(z)$, where $G(\cdot)$ is a deep neural network allows the generation of new data. Given a model $G(\cdot)$, trained on clean data we can assume that attacked samples x^* have a different distribution, therefore finding a latent code z^* able to generate x^* should be hard. Our approach builds on the idea that given some attacked pattern $x^* = x + \epsilon$ where x is an unknown clean pattern and ϵ is a perturbation crafted to induce some classifier into a mistake, we should be able to find some latent code z^* for which $G(z^*)$ is closer to the

This work was supported by the European Commission under European Horizon 2020 Programme, grant number 951911-AI4Media.

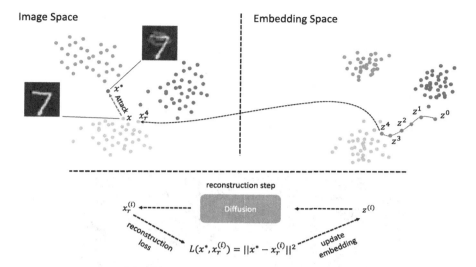

Fig. 1. Overview of our approach. Adversarial attacks happen in image space by adding crafted noise to a pattern x, shifting classifier's output to a wrong class. DiffDefence starts by drawing a sample z_T^1, to diffuse iteratively, for T steps into a reconstruction $x_r^{(i)} = z_0^{(i)} = G\left(z_T^{(i)}\right)$; we then optimize $z_T^{(i)}$ so that the diffusion output for a given optimized pattern $z_T^{(i+1)}$, lies closer to the original attacked sample. In the figure, we drop the diffusion step subscript for readability purposes .

unknown clean pattern x than to the attacked one x^*. In Fig. 2 it can be seen how an attack can add subtle patterns (center) to a clean image (left) and how DiffDefense recover a correctly classified example (right).

In this paper, we present a novel approach that leverages Diffusion Models to enhance the resistance of machine learning classifiers to adversarial attacks. Our proposed method involves reconstructing the input image using a reverse process of a diffusion model (see Fig. 1 for details), which improves the model's ability to withstand adversarial attacks. We show that this approach offers comparable speed and robustness to other generative model-based solutions. Moreover, our proposed defense mechanism can be applied as a plug-and-play tool to any classifier without compromising its accuracy, provided that the diffusion model can generate high-quality images. Overall, our approach holds promise as a viable alternative to other more complex to train models, such as GANs, for defending against adversarial attacks on machine learning models, owing to the benefits offered by Diffusion Models.

Our contribution is threefold:

– We are the first to use recently successful Denoising Diffusion Probabilistic models as a plug-in algorithm for reconstruction based adversarial defense. Differently from [26,35] our approach is based on reconstruction thus not requiring backward and forward passes for each optimization step. Moreover,

DDPMs are more stable in training with respect to GANs which have also been used as a reconstruction tool [30].

- Thanks to a superior reconstructive and representational power, DDPMs require less prototype embeddings and iterations to extract a clean pattern from the attacked one, leading to higher inference efficiency with respect to [26,30,35].
- Finally, our approach does not require to be trained on adversarial patterns and can be used to detect attacked images.

Fig. 2. Left original example from MNIST. Center result of the DeepFool attack (wrongly classified). Right our DiffDefense reconstruction (correctly classified).

2   Related Works

We now cover existing state of the art on modern generative models in adversarial machine learning, forming the base of our reconstruction based adversarial defense approach. We than discuss recent methods of adversarial attacks and defense.

2.1   Generative Models

Generative models [13,17,19,31] have emerged as a powerful class of machine learning algorithms that can create new data samples with characteristics similar to a given dataset. Their central idea is to learn the underlying probability data distribution, which could then be used to generate new patterns via sampling. Interestingly, these models have also proven to be particularly effective in adversarial scenarios, showcasing their ability to create samples that can attack classifiers [34,36]. Aforementioned models have also shown the capability to learn a semantically coherent latent embedding space. This property has been exploited, in adversarial scenarios, to remove attacks from patterns. By reconstructing [20,30], purifying [26] the perturbed sample and generating new samples to bolster adversarial training [16], these models can significantly enhance the security of machine learning systems. Recent research has highlighted the potential role of generative models in adversarial defense, as their primary objective is to produce fake data that closely resembles real data. In this paper we want to investigate the use of Diffusion Models to bolster the robustness of models against adversarial attacks. For a more thorough coverage of modern Deep Generative Models we refer the reader to [4].

2.2 Adversarial Attacks and Defense

An adversarial attack is a process that aims at altering a classifier input pattern in order to get the classifier to output a wrong prediction. So given an input x_i and a corresponding label y_i, and a classifier $\mathcal{C}$ an attack method will aim at obtaining a x^* such that $\mathcal{C}(x^*) \neq y_i$. Given this definition any method completely replacing pattern x with a different pattern would suffice in making a classifier mistake the label. For this reason, a constraint on the perturbation of the attacked pattern x^* is also required. Therefore the attacked pattern x^* must be close to the original one $||x - x^*|| < \epsilon$.

Adversarial attacks can work in white-box and black-box scenarios. In the white-box scenario the classifier is known to the attacker. The full knowledge of the classification method implies that, for example for a neural network all weights are known and the attacking method can leverage this knowledge. White-box approaches may exploit the computation of the gradient of the model loss with respect to its parameters for a specific input, such as the Fast Gradient Signed Method (FGSM) [14]. PGD [21,22] improves over [14] by a refined attack generation obtained by iterative Projected Gradient Descent (PGD). Instead, Deepfool [23] attempts to find the closest decision boundary to then perturb the input in that direction. Combination of attacks have been also proposed in [11], combining parameter free versions of PGD with SquareAttack [2]. EOT+PGD combines the concepts of Expectation over Transformation (EOT) [3] and Projected Gradient Descent (PGD) to improve its effectiveness. Elasticnet [6] exploits a combination of L1 and L2 regularisation terms to provide an optimal trade-off.

In the black-box scenario, the attacking method does not have access to the classifier, which is the most realistic setting. When an attacker has no knowledge of a classifier's architecture and weights, they can employ a query-based approach to perturb the input without relying on the gradient by applying a perturbation on the input until the classifier changes its output. Depending on the model's feedback for a given query, an attack can be classified as a score-based [7] or hard-label [5] attack. In the score-based setting, the attacker exploits the model's output probabilities of each decision. Several attacks have been created with this approach, such as Square Attack [2], which selects localized square-shaped updates at random positions. Pixel Attack [32] show attacks are possible even with a single pixel perturbation.

In the hard label-based approach, the attacker exploits the model's final decision output. Recently, SIGN-OPT [9] improved a previous work [8] using fewer queries (20K) to attack, being faster than the previous and obtaining a similar error rate of white-box attacks, but remaining much slower than the white-box. However, a query-based approach may not be as efficient as a white-box attack. Instead, a black-box attack can be employed by using the transferability [27,28] of perturbed images to attack the target model. The attacker can use a substitute model, where they have full knowledge, to generate adversarial images using white-box attacks that can then be used to attack the target model. Further coverage of adversarial attack and defense technique can be found in [1].

The issue of adversarial attacks in machine learning has prompted the development of various defense mechanisms, we can roughly identify three main approaches: adversarial training, adversarial reconstruction, and adversarial purification. Adversarial training methods improve the robustness of a model against adversarial attacks by augmenting its training set with adversarial examples. Introduced first by [14], adversarial training has become one of the most successful defense against adversarial attacks [15,21,29], adversarial training can also be enhanced using generative models for data augmentation [16]. The main limitation of this approach is that is mainly protecting classifiers from methods used in the adversarial training.

Adversarial reconstruction approaches leverage the projection of patterns onto a learned latent manifold to regenerate the original input from its adversarial counterpart. Generative models [20,30] are a natural choice to learn such latent representation and to obtain clean reconstructions out of attacked patterns. Other approaches leveraged super-resolution networks [24] or trained a reconstruction network to minimize the perceptual loss between the reconstruction of the attacked pattern and the clean image.

Adversarial purification techniques perform a filtering of attacked patterns removing adversarial perturbations while preserving their original features. Recently proposed denoising Diffusion Models have been used as a tool for purification [33,35,37]

Reconstructions based on GANs [30] are effective and generalize versus unseen threats. However, the instability of GANs during training remains a challenge. Moreover, many source noise embeddings and multiple reconstruction iterations are required to obtain effective defense. Defense-VAE is faster and as effective as Defense-GAN. However, to obtain effective reconstructions [20] the method is fine-tuned on attacked images making the approach less general and more prone to fail on unseen threats. Purification via Diffusion Models [26,33,35] exploits multiple forward/backward passes to obtain a reliable defense which requires significant time to purify an image.

With respect to [20,30] our approach exploits powerful Diffusion Models as a reconstruction tool. Interestingly, our approach is more efficient then Defense-GAN, requiring less iterations and source embeddings. Different from [20] we do not require to train on adversarial examples to work as Defense-VAE [20]. Current defense mechanisms exploiting Diffusion models are less efficient, requiring as much as 5s on a V100 card [26], while our approach runs in 0.28s on a TitanXP card.

3   Methodology

We propose a diffusion reconstruction method as a defense against adversarial attacks. The underlying idea is that adversarial attacks seek to deceive a deep neural network (DNN) by introducing a disturbance to the image while preserving its semantic meaning. Hence, the adversarial image ought to be situated close to the original, unperturbed image.

Our approach is based on the idea that it is possible to induce a Generative model $G(\cdot)$ to produce a given image x^* by minimizing the distance in image space of the output pattern, getting $\hat{z}$ as the result of such minimization

$$\hat{z} = \arg\min_z ||G(z) - x^*|| \tag{1}$$

Obtaining the reconstructed image $x_r = G(\hat{z})$. Having $G(\cdot)$ being learned on a clean dataset, the main assumption is that output generated obtained by solving Eq. 1 are closer to clean examples than corrupted ones.

In our case $G(z)$ is the result of a reverse diffusion process, each step of which is given by:

$$z_{t-1} = \frac{1}{\sqrt{\alpha_t}} - \left(z_t - \frac{1 - \alpha_t}{\sqrt{1 - \bar{\alpha}_t}} \epsilon_\theta (z_t, t)\right) + \sigma_t n \tag{2}$$

where ϵ_θ is the U-Net noise prediction model, $\alpha_t = 1 - \beta_t, \bar{\alpha}_t = \prod_{s=1}^t \alpha_s, \{\beta_t \in (0,1)\}_{t=1}^T, \sigma_t = \sqrt{\beta_t}$.

Our goal is to have the diffusion reverse process create a clean image, that is as close as possible to the attacked input. To this end we must obtain a suitable noise vector z_k. Therefore, we start from a random noise sample z_T^1, and we iteratively generate an image using the reverse process of a diffusion model. We then optimize z to solve Eq. 1 as shown in Algorithm 1. In its general form, the proposed algorithm may also be run for multiple source embeddings, although we found that it only increase slightly the accuracy.

3.1   Implementation Details

The noise prediction U-Net ϵ_θ architecture consists of a contracting path, bottleneck layer and an expansive path. The contracting path involves repeating a block with layer normalization, 3×3 convolutions, and SiLU activation, followed by downsampling with a stride of 2. The number of feature channels is doubled at each of the three downsampling step. The bottleneck layer consists of the same block of the contracting path repeated three times. The expansive path starts by concatenating the corresponding feature map from the contracting path with an upsampled input using transpose convolution. It is then followed by a block with layer normalization, 3×3 convolutions, and SiLU activation. At each upsampling step, the number of feature channels is halved. This process is also repeated three times. Both the contracting and expansive paths include a time-embedding layer, which consists of two linear layers with a SiLU activation in between. This time-embedding layer is added at each block of the contracting and expansive paths.

Algorithm 1. DiffDefense Reconstruction Algorithm. As a loss $\mathcal{L}(x_r^{(i)}, x^*)$ we used Mean Square Error (MSE). T^* are the diffusion steps and L are the gradient descent iterations, both treated as hyperparameters. $\Delta = 0.1$ is a decay rate.

1: Given adversarial image x^*
2: $z_T^1 \sim \mathcal{N}(0, I)$
3: **for** $i = 1, 2, \ldots, L$ **do**
4:     **for** $t = T^*, T^* - 1, \ldots, 0$ *steps* **do**
5:         $n \sim \mathcal{N}(0, I)$
6:         $z_{t-1}^{(i)} = \frac{1}{\sqrt{\alpha_t}} - \left(z_t^{(i)} - \frac{1-\alpha_t}{\sqrt{1-\bar{\alpha}_t}} \epsilon_\theta \left(z_t^{(i)}, t \right) \right) + \sigma_t n$
7:     **end for**
8:     $\eta^i = \eta^{i-1} \Delta^{\frac{1}{\lceil L*0.8 \rceil}}$
9:     $x_r^{(i)} = z_0^{(i)}$
10:     $z_T^{(i+1)} = z_T^{(i)} - \eta^i \nabla_z \mathcal{L}(x_r^{(i)}, x^*)$
11: **end for**

We employ two classifiers: the attacked classifier A and the surrogate classifier B. Classifier A is composed of two 5×5 convolutions with 64 output channels and stride of 2 and 1, respectively, using ReLU activations. It is then followed by a dropout layer (p = 0.25), a linear layer with 128 output features using ReLU activation, another dropout layer (p = 0.5), and finally a linear layer with 10 output features. We use classifier B to generate adversarial samples for black box attacks. B consists of a dropout layer (p = 0.2), followed by three convolutions with respective filter sizes of 8×8, 6×6, and 5×5 and strides of 2, 2, and 1, using ReLU activations. Afterward, another dropout layer (p = 0.5) is applied, and the final layer is a linear layer with 10 output features.

The Diffusion Model and the classifier are trained using the same clean dataset. Training the classifier on reconstructed data is unnecessary if the diffusion model generates high-fidelity images resembling the originals.

To implement the adversarial attack used to evaluate DiffDefense, we used adversarial robustness toolkit [25] and torchattacks [18].

4   Experiments

This section presents the experiments that evaluate the proposed method using both black-box and white-box attacks. First, we evaluate performance against three classic attacks in both settings [14,22,23]. In these experiments we seek optimal values for the number of iterations for gradient descent L, the embedding set size R, and the diffusion step T^*. Then, keeping these hyperparameter fixed we test DiffDefense against unseen attacks [2,6,11,38]. Finally, we use this method to detect adversarial samples. The experiments are conducted on two different datasets, MNIST [12] and KMNIST [10].

Table 1. Performance of DiffDefense on white box & black box attacks on MNIST & KMNIST datset. We report accuracy for each attack with and without defense. For Black blox attacks, adversarial images has been crafted using a substitute classifier.

Dataset	Attack	Type	Without defense	With defense
MNIST	No attack	–	99.14%	99.06%
	DeepFool	White box	0.95%	98.16%
		Black box	97.17%	98.86%
	PGD	White box	5.81%	95.94%
		Black box	51.28%	97.18%
	FGSM	White box	23.72%	89, 95%
		Black box	15.81%	91.28%
KMNIST	No attack	–	95.18%	94.38%
	DeepFool	White box	2.93%	93.92%
		Black box	92.16%	93.92%
	PGD	White box	26.83%	84.85%
		Black box	58.43%	91.49%
	FGSM	White box	37%	79.5%
		Black box	49.96%	88.53%

4.1 Result of White-Box and Black-Box Attack

We investigate DiffDefense ability to withstand both white-box and black-box attacks. To this end, we subject it to three potent white-box attacks: FGSM, PGD, and Deep Fool. Furthermore, we evaluate the performance of DiffDefense against these same attacks in the black-box setting, where we generate adversar-

Table 2. Robustness of DiffDefense against unseen threats on MNIST dataset. Adversarial training using adversarial sample crafted by FGSM attack $\epsilon = 0.3$.

Attack	Type	W/O Defense	W/ Defense	Adv. Training
FGSM $\epsilon = 0.3$	White Box	23.72%	89.95%	98.02%
PGD $\epsilon = 0.3$	White Box	5.81%	95.94%	79.59%
Deep Fool	White Box	0.95%	98.16%	5.81%
EOT+PGD $\epsilon = 0.3$	White Box	24.57%	96.46%	89.22%
Square Attack	Black Box	43.31%	97.31%	93.09%
AutoAttack	White Box	1.26%	88.09%	45.86%
Elastic Net	White Box	0.75%	95.75%	0.62%

ial samples using an auxiliary classifier to attack the target classifier. In Table 1 we can see how all attacks are pretty effective in both settings except for Deep-Fool used as a black box method. In general using DiffDefense we can always recover a correct classification for almost all attacked examples.

4.2 Defense Against Unseen Threats

One of the significant limitations of existing adversarial training defense methods is their inability to effectively address previously unseen threats. DiffDefense does not require to observe adversarial patterns to work, nonetheless previous experiments were performed seeking the optimal values for hyperparameters L, R, T^*. In order to assess the robustness of our proposed approach to such unseen attacks we conducted evaluations using four different attack techniques: Square Attack [2], Auto Attack [11], EOT+PGD [38] attack and Elastic Net [6], without tuning hyperparameters. Results in Table 2 indicate that our method is robust against all four of these previously unseen attack methods. Here we also test the behavior of adversarial training with samples produced by FGSM with $\epsilon = 0.3$. Interestingly, DiffDefense obtains high accuracy even in cases in which adversarial training, is not helping at all [6, 11, 23].

4.3 Ablation Studies

To evaluate the effectiveness and speed of our proposed approach, we conducted an analysis of the three main hyperparameters, iteration number L, embedding set size R and diffusion step T^*. We found that the proposed method does not need the same amount of steps of the Diffusion Model but it converges with less steps, as shown in Fig. 3. Moreover, in a comparison with Defense-GAN [30], the results of our experiments revealed that our method achieved convergence with fewer iteration steps and a smaller embedding set, while also requiring less time to converge than the GAN-based method. This was evident in the results presented in Table 3, which show the superiority of our proposed approach over Defense-GAN. The metrics used for comparison include robust accuracy, which

Table 3. Comparison with Defense-GAN [30]. Fewer iterations (L) and smaller embedding set (R) in DiffDefense lead to faster convergence and reduced time. All tests made on MNIST using white-box FGSM attack ($\epsilon = 0.3$) and the same classifier as [30]

Method	L	R	Time	Robust Acc
Defense Gan [30]	25	10	0.086	79.98%
	100	1	0.273	50.11%
	100	10	0.338	89.11%
	200	10	0.675	91.55%
Ours	5	1	0.280	87.78%
	5	**5**	**0.280**	**89.95%**

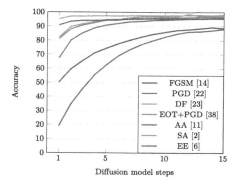

Fig. 3. Accuracy analysis of the classifier after DiffDefense has been applied on different diffusion steps. Using L = 4 and R = 5.

measures the accuracy after applying the defense, and time, which indicates the duration to reconstruct a single image.

4.4 Attack Detection

Interestingly the results of our study indicate that non-perturbed images are reconstructed with greater ease in comparison to those subjected to adversarial attacks. This is expected since the diffusion model and the classifier are trained on the same data, facilitating the diffusion in the reverse process phase using an unperturbed image to an adversarial image. This ease of reconstruction is reflected in significantly smaller reconstruction errors after an equal number of iterations. These findings suggest that the reconstruction error may serve as a potential indicator of the presence of an attack. In Fig. 4 we show ROC curves

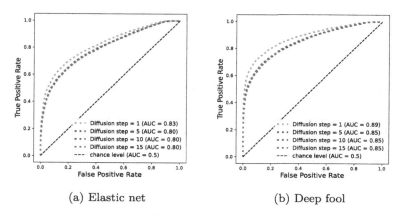

(a) Elastic net   (b) Deep fool

Fig. 4. Attack detection ROC curves for DiffDefense. In our experiments FGSM, PGD, EOT+PGD, AutoAttack, Square Attack yielded a AUC $\in [.99, 1]$.

varying the diffusion step for [6,23]. For all other methods [2,11,22,38] we get AUC $\in [.99, 1]$.

5   Conclusion

We proposed DiffDefense, a novel method that uses Diffusion models for reconstruction, enhancing classifier robustness against attacks. Empirical evaluation demonstrated its efficacy, speed, and potential as an alternative to GAN-based methods and adversarial purification methods based on diffusion models. We also showed that our approach is effective against previously unseen attacks, highlighting its robustness to new attacks. Additionally, we illustrated the usefulness of reconstruction as a tool for adversarial detection. Our findings suggest that Diffusion based adversarial defense by reconstruction is a promising path toward developing secure AI systems. We believe that future work may further improve our method by adopting better solvers for more accurate and faster reconstruction.

References

1. Akhtar, N., Mian, A.: Threat of adversarial attacks on deep learning in computer vision: a survey. IEEE Access **6**, 14410–14430 (2018)
2. Andriushchenko, M., Croce, F., Flammarion, N., Hein, M.: Square attack: a query-efficient black-box adversarial attack via random search. In: Vedaldi, A., Bischof, H., Brox, T., Frahm, J.-M. (eds.) ECCV 2020. LNCS, vol. 12368, pp. 484–501. Springer, Cham (2020). https://doi.org/10.1007/978-3-030-58592-1_29
3. Athalye, A., Engstrom, L., Ilyas, A., Kwok, K.: Synthesizing robust adversarial examples. In: International Conference on Machine Learning, pp. 284–293. PMLR (2018)
4. Bond-Taylor, S., Leach, A., Long, Y., Willcocks, C.G.: Deep generative modelling: a comparative review of VAEs, GANs, normalizing flows, energy-based and autoregressive models. IEEE Trans. Pattern Anal. Mach. Intell. **44**, 7327–7347 (2021)
5. Brendel, W., Rauber, J., Bethge, M.: Decision-based adversarial attacks: reliable attacks against black-box machine learning models. arXiv preprint arXiv:1712.04248 (2017)
6. Chen, P.Y., Sharma, Y., Zhang, H., Yi, J., Hsieh, C.J.: EAD: elastic-net attacks to deep neural networks via adversarial examples. In: Proceedings of the AAAI Conference on Artificial Intelligence, vol. 32 (2018)
7. Chen, P.Y., Zhang, H., Sharma, Y., Yi, J., Hsieh, C.J.: ZOO: zeroth order optimization based black-box attacks to deep neural networks without training substitute models. In: Proceedings of the 10th ACM Workshop on Artificial Intelligence and Security, pp. 15–26 (2017)
8. Cheng, M., Le, T., Chen, P.Y., Yi, J., Zhang, H., Hsieh, C.J.: Query-efficient hard-label black-box attack: An optimization-based approach. arXiv preprint arXiv:1807.04457 (2018)
9. Cheng, M., Singh, S., Chen, P., Chen, P.Y., Liu, S., Hsieh, C.J.: Sign-OPT: a query-efficient hard-label adversarial attack. arXiv preprint arXiv:1909.10773 (2019)

10. Clanuwat, T., Bober-Irizar, M., Kitamoto, A., Lamb, A., Yamamoto, K., Ha, D.: Deep learning for classical Japanese literature (2018)
11. Croce, F., Hein, M.: Reliable evaluation of adversarial robustness with an ensemble of diverse parameter-free attacks. In: International Conference on Machine Learning. PMLR (2020)
12. Deng, L.: The MNIST database of handwritten digit images for machine learning research. IEEE Signal Process. Mag. **29**(6), 141–142 (2012)
13. Goodfellow, I., et al.: Generative adversarial networks. Commun. ACM **63**(11), 139–144 (2020)
14. Goodfellow, I.J., Shlens, J., Szegedy, C.: Explaining and harnessing adversarial examples. preprint, arXiv (2014)
15. Gowal, S., Qin, C., Uesato, J., Mann, T., Kohli, P.: Uncovering the limits of adversarial training against norm-bounded adversarial examples. arxiv (2020), preprint
16. Gowal, S., Rebuffi, S.A., Wiles, O., Stimberg, F., Calian, D.A., Mann, T.A.: Improving robustness using generated data. In: Advances in Neural Information Processing Systems, vol. 34 (2021)
17. Ho, J., Jain, A., Abbeel, P.: Denoising diffusion probabilistic models. Adv. Neural. Inf. Process. Syst. **33**, 6840–6851 (2020)
18. Kim, H.: Torchattacks: A PyTorch repository for adversarial attacks. preprint, arXiv (2020)
19. Kingma, D.P., Welling, M.: An introduction to variational autoencoders. Found. Trends® Mach. Learn. **12**(4), 307–392 (2019)
20. Li, X., Ji, S.: Defense-VAE: a fast and accurate defense against adversarial attacks. In: Cellier, P., Driessens, K. (eds.) ECML PKDD 2019, Part II. CCIS, vol. 1168, pp. 191–207. Springer, Cham (2020). https://doi.org/10.1007/978-3-030-43887-6_15
21. Madry, A., Makelov, A., Schmidt, L., Tsipras, D., Vladu, A.: Towards deep learning models resistant to adversarial attacks. In: International Conference on Learning Representations (2018)
22. Madry, A., Makelov, A., Schmidt, L., Tsipras, D., Vladu, A.: Towards deep learning models resistant to adversarial attacks. arXiv preprint arXiv:1706.06083 (2017)
23. Moosavi-Dezfooli, S.M., Fawzi, A., Frossard, P.: DeepFool: a simple and accurate method to fool deep neural networks. In: Proceedings of the IEEE Conference on Computer Vision and Pattern Recognition (2016)
24. Mustafa, A., Khan, S.H., Hayat, M., Shen, J., Shao, L.: Image super-resolution as a defense against adversarial attacks. IEEE Trans. Image Process. **29**, 1711–1724 (2019)
25. Nicolae, M.I., et al.: Adversarial robustness toolbox v1. 0.0. Technical report (2018)
26. Nie, W., Guo, B., Huang, Y., Xiao, C., Vahdat, A., Anandkumar, A.: Diffusion models for adversarial purification. arXiv preprint arXiv:2205.07460 (2022)
27. Papernot, N., McDaniel, P., Goodfellow, I.: Transferability in machine learning: from phenomena to black-box attacks using adversarial samples. arXiv preprint arXiv:1605.07277 (2016)
28. Papernot, N., McDaniel, P., Goodfellow, I., Jha, S., Celik, Z.B., Swami, A.: Practical black-box attacks against machine learning. In: Proceedings of the 2017 ACM on Asia Conference on Computer and Communications Security, pp. 506–519 (2017)
29. Rebuffi, S.A., Gowal, S., Calian, D.A., Stimberg, F., Wiles, O., Mann, T.: Fixing data augmentation to improve adversarial robustness. arXiv preprint (2021)
30. Samangouei, P., Kabkab, M., Chellappa, R.: Defense-GAN: protecting classifiers against adversarial attacks using generative models. preprint, arXiv (2018)
31. Song, J., Meng, C., Ermon, S.: Denoising diffusion implicit models. preprint, arXiv (2020)

32. Su, J., Vargas, D.V., Sakurai, K.: One pixel attack for fooling deep neural networks. IEEE Trans. Evol. Comput. **23**(5), 828–841 (2019)
33. Wang, J., Lyu, Z., Lin, D., Dai, B., Fu, H.: Guided diffusion model for adversarial purification. arXiv preprint arXiv:2205.14969 (2022)
34. Wang, X., He, K., Hopcroft, J.E.: AT-GAN: a generative attack model for adversarial transferring on generative adversarial nets. preprint 3, arXiv (2019)
35. Wu, Q., Ye, H., Gu., Y.: Guided diffusion model for adversarial purification from random noise. preprint, arXiv (2022)
36. Xiao, C., Li, B., Zhu, J.Y., He, W., Liu, M., Song, D.: Generating adversarial examples with adversarial networks. arXiv preprint arXiv:1801.02610 (2018)
37. Yoon, J., Hwang, S.J., Lee, J.: Adversarial purification with score-based generative models. In: International Conference on Machine Learning. PMLR (2021)
38. Zimmermann, R.S.: Comment on. preprint, Adv-BNN: Improved Adversarial Defense through Robust Bayesian Neural Network. arXiv (2019)

Exploring Audio Compression as Image Completion in Time-Frequency Domain

Giovanni Scodeller[ID], Mara Pistellato[(✉)][ID], and Filippo Bergamasco[ID]

DAIS, Università Ca'Foscari Venezia, 155, via Torino, Venezia, Italy
{mara.pistellato,filippo.bergamasco}@unive.it

Abstract. Audio compression is usually achieved with algorithms that exploit spectral properties of the given signal such as frequency or temporal masking. In this paper we propose to tackle such a problem from a different point of view, considering the time-frequency domain of an audio signal as an intensity map to be reconstructed via a data-driven approach. The compression stage removes some selected input values from the time-frequency representation of the original signal. Then, decompression works by reconstructing the missing samples as an image completion task. Our method is divided into two main parts: first, we analyse the feasibility of a data-driven audio reconstruction with missing samples in its time-frequency representation. To do so, we exploit an existing CNN model designed for depth completion, involving a sequence of sparse convolutions to deal with absent values. Second, we propose a method to select the values to be removed at compression stage, maximizing the perceived audio quality of the decompressed signal. In the experimental section we validate the proposed technique on some standard audio datasets and provide an extensive study on the quality of the reconstructed signal under different conditions.

Keywords: Audio compression · CNN · Sparse convolutions · Spectrogram · genetic algorithm

1 Introduction

Storing or transmitting digital audio data often involves compression, that can be lossy or lossless. Classical audio compression and decompression algorithms remove redundant or irrelevant information from data exploiting different properties [2,6]. In particular, when a lossy approach is adopted, the algorithm discards the information that are considered as inaudible or less important to the human hearing: this is performed moving from time to frequency domain (typically via the Fourier transform), so that the compression can be performed directly on frequencies, exploiting spectral properties of the signal. The literature counts several approaches to perform effective compression for speech and generic audio signals [4,11], also targeted to specific applications such as wireless communications [7]. Recently, data-driven approaches proved to be effective

ⓒ The Author(s), under exclusive license to Springer Nature Switzerland AG 2023
G. L. Foresti et al. (Eds.): ICIAP 2023, LNCS 14234, pp. 443–455, 2023.
https://doi.org/10.1007/978-3-031-43153-1_37

in a variety of applications [19, 20], offering targeted solutions [3]. Among them, some works proposed data-driven approaches to tackle the problem of audio data compression. An early work presented in [16] proposed a speech compression technique using a three-layer perceptron, where outputs of the hidden layer are quantized to perform the compression. Despite its limitations, we can consider such work a first attempt at applying autoencoders for audio compression. The authors in [24] introduce a neural network (NN) model to compress audio signals through an encoder/decoder architecture and residual vector quantizer. Some works propose NN models specifically designed for speech compression [12, 13]. In all such examples, audio signals are treated by the model as one-dimensional input signals. Other applications of data-driven approaches involving audio signal processing include audio super-resolution (or bandwidth extension) [14] or speech dereverberation and denoising [23], where the authors provide a time-frequency approach.

In this paper we propose a first study on audio compression using data-driven techniques designed for image processing. Specifically, we investigate the usage of CNNs composed of Sparse Convolution layers applied to the 2D time-frequency domain representation of an audio signal. The idea of sparse convolutions was introduced in [21] to perform depth completion, with possible subsequent tasks such as 3D reconstruction [17, 18] or segmentation [22]. Later the same concept was extended and further improved by other authors [9, 10] to solve similar problems or to merge different kinds of data as intensity images [15], but has never been used for audio data specifically. In our work we shift from depth data to audio signal by working on the "image" obtained when applying the Short-Time Fourier Transform (STFT). Indeed, the main idea behind the proposed compression method is to first convert the 1-dimensional audio signal into a 2-dimensional spectrogram via STFT, and then to keep only part of such information by performing a sparse sampling of the image. Then, at decompression time, the original complete spectrogram image is reconstructed by a pre-trained CNN. To do so, we designed a network architecture that performs sparse convolutions to reconstruct the signal by filling in the missing values. Such model is specifically trained on audio data, and involves a specialised loss function.

To summarize, the contributions of this work are threefold: first, we present an "audio-completion" task that is novel with respect to other approaches proposed in the literature. The audio signal is transformed into an image, sparsified by an encoding algorithm, and reconstructed using sparse convolutions. Second, we introduce an improved version of the Griffin-Lim Algorithm (GLA) [5] that is able to enforce phase coherence given a sparse sampling of the phase image. Finally, we describe an effective compression method based on a Genetic Algorithm that is able to optimize the pixel selection process keeping the desired compression level. In this way the compression step is able to select only the relevant image areas so that the reconstruction preserves the best possible quality. We provide an extensive experimental evaluation where we analyse all the proposed approaches and compare the given alternatives in relation to the audio output quality and the space needed to store all the compressed values.

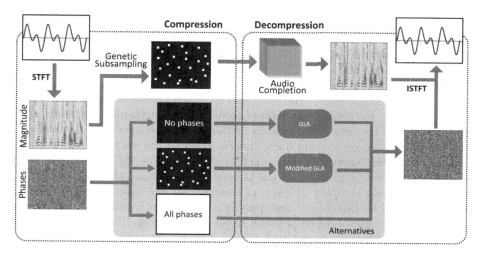

Fig. 1. Compression and decompression pipeline overview. The signal is transformed into power spectrogram and phases images. The compression happens by discarding most of the values through a custom-designed genetic sub-sampling (Sect. 2.3). We analyse three alternatives for the phases: (i) discard all of them and recover with the Griffin-Lim Algorithm, (ii) keep only the subsampled ones and apply our modified GLA, or (iii) keep all of them. At decompression stage, our Audio completion CNN (Sect. 2.1) reconstructs the spectrogram from the sparse samples, and the waveform is recovered via iSTFT.

2   Audio Compression via Sparse CNNs

Our approach works by projecting the 1-dimensional audio signal into a 2-dimensional time-frequency domain representation called spectrogram. This operation is performed with the Short-Time Fourier Transform (STFT), which is implemented by dividing the audio waveform into small overlapping segments and then applying the Fourier Transform to each of them individually. Similarly to the classical Fourier Transform, also the STFT can be inverted to get back the waveform from a spectrogram. Mathematically, it can be defined as:

$$S_{k,h} = \sum_{n=0}^{N-1} w(n)\, x(n + hH)\, e^{-i2\pi n \frac{k}{N}} \tag{1}$$

where $h = 0, \ldots, T$ is the time frame, $k = 0 \ldots \frac{N}{2}$ is the frequency, N is the number of samples that we are considering for each time frame, H is the amount of overlapping defined as the hop size from one frame to another, and $w(n)$ is a windowing function to reduce the spectral leakage due to the non-integer number of harmonic periods in the time frame[1]. Since it works on different

---

[1] The most common functions are the Hamming, Hanning, and Blackman window. We observed no relevant difference in the choice of such function for our purposes.

limited segments of the original data, the output spectrogram is an $F \times T$ complex matrix $\mathbf{S}$ where F is the number of frequency bins, equal to half of the frame size N for the Nyquist theorem, and T is the number of frames that we analyse in the waveform depending on the length of the signal and the amount of overlap. For our purposes, it is more convenient to represent $\mathbf{S}$ in polar form, separating the log magnitude (power) of the spectrum from the phases as shown in Fig. 1.

The rationale for studying a compression method operating on the spectrogram, instead of the original waveform, is that the power spectrum is generally sparse for typical speech audio data. Indeed, most of the energy tends to cluster into just a few energy bins for each time frame. For this reason, we cast the audio compression problem as the task of reconstructing the audio signal in presence of missing samples, or gaps, in the spectrogram data. Since the power spectrum is just an image, we can apply state-of-the-art methods designed for depth completion to solve the problem in an effective way. The only difference is that the input data represents the power of each harmonic over time instead of depth information.

The whole pipeline is displayed in Fig. 1. First, the input waveform is converted to a spectrogram using the STFT. This results in an image where the number of rows depends on the bandwidth of the signal and the columns to the length of the signal itself. For simplicity, we consider a signal with a limited temporal extent but the same process can be used for an infinitely long audio stream by dividing the process into smaller chunks (we used 4 s chunks in all our tests). At this point, compression happens by discarding a desired amount of samples from the spectrogram (both in power spectrum and phases) according to certain criteria. In this work, we evaluate two approaches, namely random subsampling and genetic spectrogram subsampling (Sect. 2.3). The remaining samples are serialized into a one-dimensional stream, quantised to fixed point values with a user-selectable amount of bits, and compressed. Decompression works by "filling the gaps" caused by the samples that were removed during compression. To do so, the binary stream is decompressed, and the spectrogram is recreated together with a binary mask marking the location of the valid samples. We then apply our Audio Completion Network (Sect. 2.1) to recover the missing values of the power spectrum. Note that the same operation is not performed on the phases because the "phase image" is not locally smooth and therefore not easily addressable with the relatively limited receptive field of a CNN. Instead, we recover the phases directly when inverting the STFT with a modification of the Griffin-Lim algorithm (see Sect. 2.2).

2.1   Audio Completion Network

The network architecture that we propose is based on the model presented by Uhrig et al. [21]. In particular, we are interested in the concept of *sparse convolution*, which is a convolutional-like operator designed to effectively account for sparse data (i.e. manage missing pixel values).

Suppose we have an $F \times T$ input image X, then we define $M \in \{0,1\}^{F \times T}$ to be a binary mask denoting the validity (with value = 1) or not of an input

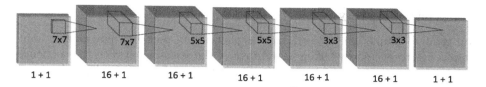

Fig. 2. CNN architecture for audio completion. We perform a sequence of 6 sparse convolutions propagating the single-channel validity mask.

pixel in X. The sparse convolution operation takes as input both the image data $X = \{x_{i,j}\}$ and the 2-dimensional mask $M = \{m_{i,j}\}$. Considering a kernel size of $2k + 1$, the sparse convolution output at position (u, v) is computed as f as follows:

$$f_{u,v}(X, M) = \frac{\sum_{i,j=-k}^{k} m_{u+i,v+j}\, x_{u+i,v+j}\, w_{i,j}}{\sum_{i,j=-k}^{k} m_{u+i,v+j}\, w_{i,j} + \epsilon} + b \qquad (2)$$

where $w_{i,j}$ are the kernel weights, b is bias and ϵ is a small value added to avoid division by zero. Thanks to the normalization factor computed at the denominator and the multiplication by mask values, the convolution takes into account only valid pixel values, enabling the model to be sparsity invariant. Moreover, since more valid values are computed as a result of the described operation, the input binary mask is propagated to the following layers via a function f^m, that computes the max pool operator for each (u, v) as follows:

$$f_{u,v}^m(M) = \max_{i,j=-k,\ldots,k} m_{u+i,v+j}\, w_{i,j} \qquad (3)$$

meaning that an output pixel is considered valid if at least one pixel from the input mask is equal to one.

Following a similar approach of [21], we build the audio completion network concatenating different sparse convolution blocks. In this way, the final output will be a full spectrogram image where input gaps (i.e. points for which the mask is zero) are filled. The proposed audio completion architecture is displayed in Fig. 2. We start from a 2-channel input (one for the mask M and one for sparse data) and then perform six sparse convolutions with decreasing kernel sizes, namely: 7, 7, 5, 5, 3, 3. All convolutions are followed by ReLU activation and all the feature maps have 16 channels; finally, the last layer outputs the reconstructed spectrogram. We used the spectral convergence loss as proposed in [1]:

$$\mathcal{L}(Y, \hat{Y}) = \frac{||\, |Y| - |\hat{Y}|\, ||_{\mathrm{Fro}}}{||\, |Y|\, ||_{\mathrm{Fro}}} \qquad (4)$$

where Y represents the ground truth spectrogram, $\hat{Y}$ is the network prediction, and $|| \cdot ||_{\mathrm{Fro}}$ is the Frobenius norm.

2.2 Phase Recovery

Our audio completion network reconstructs the spectrum magnitude only. Phases are recovered by taking advantage of the leakage between the sequence of time frames managed by the STFT. Since time frames overlap, phases of consecutive frames are obviously correlated depending on the frequencies. A popular technique to recover the phases in this way is the Griffin-Lim Algorithm (GLA) [5], consisting of iterative refinement of the complex spectrum obtained by applying STFT and its inverse while forcing the magnitude to the given values. We now briefly sketch the mathematical formulation to better understand our proposed modification of the original technique.

Let $\mathbf{S} = \hat{\mathbf{A}} e^{i\hat{\mathbf{\Phi}}}$ be a complex spectrum with magnitude $\hat{\mathbf{A}} \in \mathbb{R}^{T \times F}$ and phases $\hat{\mathbf{\Phi}} \in [0 \dots 2\pi]^{T \times F}$. Let $\mathcal{G}[\cdot]$, $\mathcal{G}^{-1}[\cdot]$ denote the forward and inverse STFT respectively. Given just a power spectrum $\mathbf{A}$, GLA reconstructs the full complex spectrum $\mathbf{S}$ by solving the quadratic problem:

$$\underset{\mathbf{S}}{\text{minimize}} \quad \| \mathbf{S} - \mathcal{G}[\mathcal{G}^{-1}[\mathbf{S}]] \|_{\text{Fro}}^2$$
$$\text{subject to} \quad \hat{\mathbf{A}} = \mathbf{A} \tag{5}$$

The rationale is that if $\mathbf{S}$ is not consistent (i.e. phases are not correctly correlated), applying inverse and direct STFT will produce a new (consistent) spectrum $\mathbf{S}' = \mathcal{G}[\mathcal{G}^{-1}[\mathbf{S}]]$ with different magnitudes and phases due to spectral leakage. For this reason, GLA finds the best spectrum $\mathbf{S}$ with the given magnitudes $\mathbf{A}$ minimising the Frobenius distance to a consistent spectrum $\mathbf{S}'$.

Since GLA just aims at a consistent spectrum, multiple different solutions are equally possible for the given set of amplitudes. The final result is in general intelligible, but the perceived quality may vary greatly depending on the local minima obtained during the optimisation. For this reason, we propose a modification of GLA in which some phases $\Phi_{\mathcal{S}}$ (with $\mathcal{S} = \{(k_1, h_1), \dots, (k_N, h_N)\}$) are given. The new optimization is then:

$$\underset{\mathbf{S}}{\text{minimize}} \quad \| \mathbf{S} - \mathcal{G}[\mathcal{G}^{-1}[\mathbf{S}]] \|_{\text{Fro}}^2$$
$$\text{subject to} \quad \hat{\mathbf{A}} = \mathbf{A} \tag{6}$$
$$\hat{\Phi}_{(k_1,h_1)} = \Phi_{(k_1,h_1)}, \cdots, \hat{\Phi}_{(k_N,h_N)} = \Phi_{(k_N,h_N)}$$

which can be solved with an iterative projection procedure similar to what was originally proposed in [5]. Let $\mathbf{S}^{(0)}$ be an initial spectrum in which $\hat{\mathbf{A}} = \mathbf{A}$ and the values in $\hat{\mathbf{\Phi}}$ are drawn from a uniform random distribution in $0 \dots 2\pi$ excepts for the values $\hat{\Phi}_{(k_1,h_1)}, \dots, \hat{\Phi}_{(k_N,h_N)}$ that are set to the given $\Phi_{(k_1,h_1)}, \dots, \Phi_{(k_N,h_N)}$. The iterative projection procedure is defined as:

$$\mathbf{S}^{(t+1)} = \mathcal{G}[\mathcal{G}^{-1}[P(\mathbf{S}^{(t)})]] \tag{7}$$

$$P(\mathbf{S}^{(t)})_{k,h} = \begin{cases} A_{k,h}\, e^{i\Phi_{(k,h)}}, & \text{if } (k,h) \in \mathcal{S} \\ \frac{A_{k,h}}{|S_{k,h}^{(t)}|} S_{k,h}^{(t)}, & \text{otherwise} \end{cases} \tag{8}$$

where P is the projection function forcing the amplitudes $|\mathbf{S}^{(t)}| = \mathbf{A}$ and the phases in $\mathcal{S}$ to $\Phi_{(k_1,h_1)}, \ldots, \Phi_{(k_N,h_N)}$. The procedure let $\mathbf{S}$ converge to a consistent spectrum where amplitudes and known phases are equal to the ones that are given in input. This effectively recovers the missing phases to values that are consistent according to the overlap of subsequent time frames.

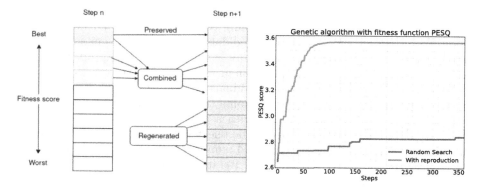

Fig. 3. Left: schema of the reproduction used for genetic subsampling (see text for details). Right: PESQ score wrt iterations for mask generation for our genetic algorithm vs. random sampling.

2.3   Genetic Spectrogram Subsampling

The studied compression approach works by discarding some values from the signal spectrogram. So far we gave no details on how to choose such values, but obviously this has an impact on the quality of the reconstructed signal. One simple approach can be to do a random subsampling of the input spectrogram. As we will see in the experimental section, this approach is very fast to implement but it will not consider the different energy contributions of certain frequencies in the spectrum.

In general, the subsampling operation can be solved optimally if posed as a non-linear optimization problem. Let $M \in \{0,1\}^{T \times F}$ be a binary mask defining which values to keep. Let $\mathcal{Q}(M) \in \mathbb{R}$ be a *fitness function* measuring how good is the reconstructed waveform when the mask M is applied. For example, $\mathcal{Q}$ can evaluate the PESQ (Perceptual Evaluation of Speech Quality) when the spectrogram magnitude is reconstructed with our Audio Completion Network and the phases with the modified GL as described. We can write it as:

$$
\begin{aligned}
\text{maximize}_{M} \quad & \mathcal{Q}(M) \\
\text{subject to} \quad & \sum M = \gamma, \quad M \in \{0,1\}^{T \times F}
\end{aligned}
\tag{9}
$$

where γ is a desired *density level* (in range $0 \ldots T \cdot F$), controlling the number of values to keep and therefore the amount of compression to achieve. Even if $\mathcal{Q}$ is differentiable, the resulting problem is NP-Hard for the requirement of the mask M to be binary. To make it tractable, we can either relax the constraint of having a binary mask (i.e. using a soft mask) or use an optimization procedure based on heuristics that provides a reasonable result even with no guarantee of reaching the global optimum. In this work, we investigated the latter with an approach based on Genetic Algorithms theory.

Our Genetic Spectrogram Subsampling (GSS) starts with an initial population $\mathcal{P} = \{M_1, \ldots, M_L\}$ of valid candidate solutions. The genome of each individual in $\mathcal{P}$ is a binary mask randomly generated with the property of also being a valid solution for our optimization problem (i.e. $\sum_i M_i = \gamma \ \forall i$). Inspired by the process of natural selection, individuals of the initial population evolve by recombining and/or mutating their genome through an iterative process spanning several generations. Individuals with a genome producing high values of fitness $\mathcal{Q}$ are more likely to survive and be selected for recombination (i.e. reproduction) across generations. After a fixed number of iterations, the best genome is returned by the algorithm. At each iteration (generation) we perform the following operations:

Selection. Each element of $\mathcal{P}$ is ranked according to the fitness function $\mathcal{Q}$. This is performed by (i) applying each mask M_i to the input spectrogram, (ii) executing the Audio Completion Network to recover the missing values and, (iii) evaluating its quality against the original signal. The last step can be more or less expensive depending on the evaluation metric chosen. For example, using the MAE (Mean Absolute Error) on the spectrum amplitudes is less computationally expensive but does not take into account the perceptual quality of the final waveform. Conversely, using the PESQ metric requires computing the inverse STFT (possibly with phase recovery) but can produce better overall solutions. We use a hybrid approach here, sorting first the population in ascending order wrt. MAE. The worst 60% of the individuals are eliminated and substituted with new random individuals in the next generation. The best 40% are ranked again with PESQ and selected for either being preserved as-is or modified with a genetic operator. Specifically, the best 10% just pass through the next generation while the remaining 30% are mutated and recombined by random crossover.

Mutation and Crossover. From the 40% of the best-ranked individuals we create a number of random pairs equal to the 30% of the original population. Such random pairs of individuals are then combined with a custom genetic operator designed (heuristically) to produce a new individual with fitness higher than both parents. The mask of the new individual is obtained as follows:

$$\text{crossover}(M_i, M_j) = M_i \wedge M_j \vee \text{mutate}_{\sum M_i \wedge M_j}(M_i \oplus M_j) \qquad (10)$$

where $\text{mutate}_k(M)$ is a function that randomly sets some values of M to 0 so that $\sum M = \gamma - k$. The general idea is that, after crossover, the mask of the new individual should have a 1 if both parents have a 1, or a 0 if both the parents have a 0 at each position. This is performed by the element-wise *and* operation (∧) of parent masks. Conversely, at every location where parents values disagree (i.e. the element-wise *xor* ⊕ is 1) the output mask can randomly have a 0 or 1 with the only requirement that the total number of ones to be equal to γ. This random flipping of values from 1 to 0 effectively represents a mutation of the new individual with respect to the parents and introduces an element of variability in the evolutionary process. In Fig. 3 (left) we sketched the selection process across two generations while on the right-hand side we show an example of how, albeit being a heuristic, the GSS can improve the search for optimal masks with respect to a simple random search (i.e. with no mutation and crossover).

3   Experimental Results

We tested our method on the Flickr 8k audio caption dataset [8]. To have uniform samples, we extracted audio patches with a duration of 4 s, avoiding parts with no signal, and we selected $10K$ audio tracks sampled at 16 KHz. Then we divided such data into training and test sets with an 80/20 ratio, obtaining $20K$ audio samples for the test set. Each audio sample was transformed with the STFT, choosing 2048 frequency bins and setting the window size to 2048 with a hop length equal to 256. In this way, we obtained complex images of size 1025×294. As for the absolute values, we converted the spectrogram to a log-power scale. The audio completion network was trained by applying uniform random masks to the training set, randomly selecting the sparsity levels between 80% to 95%. The training process was performed with Adam optimizer with an adaptive learning initialized to 10^{-4}, and run up to convergence.

In our study we explore the phases reconstruction step with different methods: (i) keep no phases and compute them with GLA, (ii) keep only the phases corresponding to the selected power values and use the improved version of GLA to ensure coherency, and (iii) keep all the input phases. Therefore, we first analyse the behaviour of these three approaches when reconstructing audio at different sparsity levels. Figure 4 shows the decompression quality for different phase reconstructions: all phases (1st column), masked phases (2nd column), and no phases (3rd column). We show PESQ (1st row) and MAE (Mean Absolute Error, 2nd row) metrics computed on all test set data with different sparsity levels up to 0.95. As expected, the quality decreases as the sparsity level raises, and keeping all the phases gives the best quality both in terms of perceptual quality and MAE. Observing the PESQ values, we can see that the modified GLA improves the quality up to PESQ = 2.5, offering a better outcome with respect to the classical GLA (no phases), meaning that our approach is effective while requiring less space. The modified GLA works well also when looking at the MAE, being comparable with the "all phases" method. This is also an indication of the effectiveness of our proposed modification to GLA when also the compressed phases are sparse. The second test we performed is meant to assess the effectiveness of the proposed genetic algorithm to perform the subsampling during compression. In particular, we show the reconstruction results for a random uniform sampling versus the genetic subsampling. In order to evaluate the

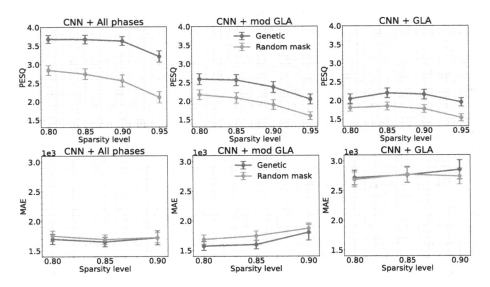

Fig. 4. Decompression quality in terms of PESQ (1st row) and MAE (2nd row) varying sparsity levels. Each column denotes a different phase reconstruction approach.

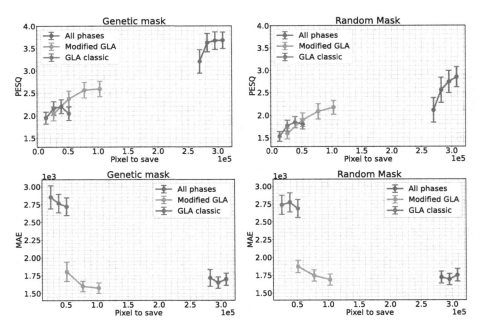

Fig. 5. Quality of reconstruction wrt values to be stored for different approaches and subsampling methods.

compression step together with the three phase reconstructions alternatives, we include such results in Fig. 4 as different curves on the already discussed plots. We can see that the genetic algorithm always computes a better mask for the given sparsity levels, in particular for the perceptual quality, reaching values around 3.5 for all the phases and 2.5 for the sparse phases (with modified GLA). As a last experiment, we compared the memory space needed to store the compressed data for each different approach. Note that we do not plot the bitrate, but rather we prefer to show the number of values to be saved. This is because the final amount of bytes depends on a number of external factors such as the STFT parameters, the quantization level, and the data structure that is actually employed to store the sparse matrix. In this way we only show the number of values, discarding other possible choices. Figure 5 compares the number of stored values for all three methods (all phases, modified GLA and classic GLA) and for the two subsampling approaches: genetic (left) and uniform random (right). Again, the genetic method offers better quality for the same number of values, and the "all phases" approach (blue curve) offers the best results, but at the cost of having almost 3 times more memory consumption with respect to the others.

4   Conclusions

In this paper we proposed a first study involving the adoption of CNNs to perform audio compression as an image-reconstruction task. We presented a decompression method that takes as input a sparse spectrogram and reconstructs the dense image via a data-driven model. Moreover, we proposed an improved version of the well-known Griffin-Lim algorithm to effectively recover the original waveform with just a few available phases. Then, we proposed a novel sub-sampling approach based on a genetic algorithm. Experimental results show the validity of both our compression and decompression methods in terms of MAE and quality perception. The genetic approach effectively selects the values to be compressed, while the decompression based on a CNN model and on a constrained implementation of GLA offers quantitative significant results.

References

1. Arık, S.Ö., Jun, H., Diamos, G.: Fast spectrogram inversion using multi-head convolutional neural networks. IEEE Signal Process. Lett. **26**(1), 94–98 (2018)
2. Brandenburg, K., Stoll, G.: ISO/MPEG-1 audio: a generic standard for coding of high-quality digital audio. J. Audio Eng. Soc. **42**, 780–792 (1994)
3. Gasparetto, A., et al.: Cross-dataset data augmentation for convolutional neural networks training. In: 2018 24th International Conference on Pattern Recognition (ICPR), Beijing, China, 2018, pp. 910–915 (2018). https://doi.org/10.1109/ICPR. 2018.8545812
4. Ghido, F., Tabus, I.: Sparse modeling for lossless audio compression. IEEE Trans. Audio Speech Lang. Process. **21**(1), 14–28 (2012)
5. Griffin, D., Lim, J.: Signal estimation from modified short-time Fourier transform. IEEE Trans. Acoust. Speech Signal Process. **32**, 236–243 (1984)

454   G. Scodeller et al.

6. Hans, M., Schafer, R.W.: Lossless compression of digital audio. IEEE Signal Process. Mag. **18**(4), 21–32 (2001)
7. Hanzo, L., Somerville, F.C.A., Woodard, J.: Voice and Audio Compression for Wireless Communications. Wiley, Hoboken (2008)
8. Harwath, D., Glass, J.: Deep multimodal semantic embeddings for speech and images. In: 2015 IEEE Workshop on Automatic Speech Recognition and Understanding (ASRU), pp. 237–244. IEEE (2015)
9. Huang, Z., Fan, J., Cheng, S., Yi, S., Wang, X., Li, H.: HMS-Net: hierarchical multi-scale sparsity-invariant network for sparse depth completion. IEEE Trans. Image Process. **29**, 3429–3441 (2019)
10. Jaritz, M., De Charette, R., Wirbel, E., Perrotton, X., Nashashibi, F.: Sparse and dense data with CNNs: depth completion and semantic segmentation. In: 2018 International Conference on 3D Vision (3DV), pp. 52–60. IEEE (2018)
11. Kanade, J., Sivakumar, B.: A literature survey on psychoacoustic models and wavelets in audio compression. Int. J. Adv. Res. Electron. Commun. Eng. (IJARECE) (2014)
12. Kankanahalli, S.: End-to-end optimized speech coding with deep neural networks. In: 2018 IEEE International Conference on Acoustics, Speech and Signal Processing (2018)
13. Kleijn, W.B., et al.: Generative speech coding with predictive variance regularization. In: IEEE International Conference on Acoustics, Speech and Signal Processing (2021)
14. Lim, T.Y., Yeh, R.A., Xu, Y., Do, M.N., Hasegawa-Johnson, M.: Time-frequency networks for audio super-resolution. In: 2018 IEEE International Conference on Acoustics, Speech and Signal Processing (ICASSP), pp. 646–650. IEEE (2018)
15. Ma, F., Karaman, S.: Sparse-to-dense: depth prediction from sparse depth samples and a single image. In: 2018 IEEE International Conference on Robotics and Automation (ICRA), pp. 4796–4803. IEEE (2018)
16. Morishima, S., Harashima, H., Katayama, Y.: Speech coding based on a multi-layer neural network. In: IEEE International Conference on Communications (1990)
17. Pistellato, M., Albarelli, A., Bergamasco, F., Torsello, A.: Robust joint selection of camera orientations and feature projections over multiple views. In: 2016 23rd International Conference on Pattern Recognition (ICPR), Cancun, Mexico, 2016, pp. 3703–3708 (2016). https://doi.org/10.1109/ICPR.2016.7900210
18. Pistellato, M., Bergamasco, F., Albarelli, A., Torsello, A.: Dynamic optimal path selection for 3D triangulation with multiple cameras. In: Murino, V., Puppo, E. (eds.) ICIAP 2015. LNCS, vol. 9279, pp. 468–479. Springer, Cham (2015). https://doi.org/10.1007/978-3-319-23231-7_42
19. Pistellato, M., Bergamasco, F., Albarelli, A., Torsello, A.: Robust cylinder estimation in point clouds from pairwise axes similarities. In: Proceedings of the 8th International Conference on Pattern Recognition Applications and Methods - ICPRAM, pp. 640–647 (2019). https://doi.org/10.5220/0007401706400647
20. Pistellato, M., Bergamasco, F., Fatima, T., Torsello, A.: Deep demosaicing for polarimetric filter array cameras. IEEE Trans. Image Process. **31**, 2017–2026 (2022). https://doi.org/10.1109/TIP.2022.3150296
21. Uhrig, J., Schneider, N., Schneider, L., Franke, U., Brox, T., Geiger, A.: Sparsity invariant CNNs. In: 2017 international conference on 3D Vision (3DV) (2017)
22. Wang, W., Neumann, U.: Depth-aware CNN for RGB-D segmentation. In: Ferrari, V., Hebert, M., Sminchisescu, C., Weiss, Y. (eds.) ECCV 2018. LNCS, vol. 11215, pp. 144–161. Springer, Cham (2018). https://doi.org/10.1007/978-3-030-01252-6_9

23. Williamson, D.S., Wang, D.: Time-frequency masking in the complex domain for speech dereverberation and denoising. IEEE/ACM Trans. Audio Speech Lang. Process. **25**(7), 1492–1501 (2017)

24. Zeghidour, N., Luebs, A., Omran, A., Skoglund, J., Tagliasacchi, M.: SoundStream: an end-to-end neural audio codec. IEEE/ACM Trans. Audio Speech Lang. Process. **30**, 495–507 (2021)

Fuzzy Logic Visual Network (FLVN): A Neuro-Symbolic Approach for Visual Features Matching

Francesco Manigrasso(ID), Lia Morra(✉)(ID), and Fabrizio Lamberti(ID)

Dipartimento di Automatica e Informatica, Politecnico di Torino, Torino, Italy
{francesco.manigrasso, lia.morra,fabrizio.lamberti}@polito.it

Abstract. Neuro-symbolic integration aims at harnessing the power of symbolic knowledge representation combined with the learning capabilities of deep neural networks. In particular, Logic Tensor Networks (LTNs) allow to incorporate background knowledge in the form of logical axioms by grounding a first order logic language as differentiable operations between real tensors. Yet, few studies have investigated the potential benefits of this approach to improve zero-shot learning (ZSL) classification. In this study, we present the Fuzzy Logic Visual Network (FLVN) that formulates the task of learning a visual-semantic embedding space within a neuro-symbolic LTN framework. FLVN incorporates prior knowledge in the form of class hierarchies (classes and macro-classes) along with robust high-level inductive biases. The latter allow, for instance, to handle exceptions in class-level attributes, and to enforce similarity between images of the same class, preventing premature overfitting to seen classes and improving overall performance. FLVN reaches state of the art performance on the Generalized ZSL (GZSL) benchmarks AWA2 and CUB, improving by 1.3% and 3%, respectively. Overall, it achieves competitive performance to recent ZSL methods with less computational overhead. FLVN is available at https://gitlab.com/grains2/flvn.

Keywords: Zero shot learning · NeuroSymbolic AI · Logic Tensor Networks

1 Introduction

There is an increasing interest in the computer vision community towards the integration of learning and reasoning, with specific emphasis towards the fusion of deep neural networks and symbolic artificial intelligence [1–3]. By integrating perception, reasoning and learning, Neuro-symbolic (NeSy) architectures can improve explainability, generalization and robustness of deep learning systems. In addition, NeSy techniques can incorporate prior symbolic knowledge in the training objective of deep neural networks, compensating for the lack of supervision from labelled examples [4,5]. The latter property is of particular interest in the context of Zero-Shot Learning (ZSL), which aims at recognizing objects from unseen classes by associating both seen and unseen classes to auxiliary information, usually in the form of class attributes [6].

© The Author(s), under exclusive license to Springer Nature Switzerland AG 2023
G. L. Foresti et al. (Eds.): ICIAP 2023, LNCS 14234, pp. 456–467, 2023.
https://doi.org/10.1007/978-3-031-43153-1_38

In this paper, we introduce the Fuzzy Logic Visual Network (FLVN), an architecture designed to learn a joint embedding space for visual features and class attributes to tackle ZSL scenarios. To constraint learning based on prior knowledge available from resources such as WordNet, it leverages the Logic Tensor Network (LTN) [2] paradigm, a NeSy framework that formulates learning as maximizing the satisfiability of a knowledge base $\mathcal{K}$ based on a first order logic (FOL). Thus, FLVN is composed by a convolutional neural network (CNN), that computes embedding, followed by a LTN.

The proposed FLVN improves and extends previous NeSy architectures for ZSL, and in particular our previous architecture ProtoLTN [7], in several ways. First, there are substantial differences in the LTN formulation, and especially the way the isOfClass predicate is formulated. FLVN, learns to project visual features into the semantic attribute space, facilitating the integration of prior knowledge into the learning process by the LTN. FLVN is trained end-to-end, whereas ProtoLTN used features extracted from a pre-trained frozen network, and only the class-level prototypes were trained. This implementation outperforms our previous model [7], offering an alternative approach for grounding the predicates in the FOL language by changing the how the distance measure between extracted features and semantic vectors is calculated.

Compared to most existing approaches to ZSL, FLVN incorporates into the training process several logical axioms that combine both class-specific prior knowledge (e.g., "a zebra is an ungulate") with high-level inductive biases (e.g., "two images of the same class should have similar embeddings"), that were not yet included in ProtoLTN [7]. Finally, we introduce axioms to represent exceptions within the dataset to make the architecture more robust in the context of ZSL, where class attributes are specified at the class level rather than the image level. To the best of our knowledge, this aspect is not taken into account in current approaches to ZSL, including neuro-symbolic ones [7]. Through extensive experiments on multiple benchmarks, we show that FLVN outperforms existing ZSL architectures. The implementation, developed in PyTorch using the LTNtorch package [8], is available at https://gitlab.com/grains2/flvn

The remainder of the paper is structured as follows. Section 2 introduces background on the LTN framework and related work. Section 3 describes the proposed architecture. In Sects. 4 and 5, we analyze the model behavior in a ZSL and a generalized ZSL (GZSL) setting on common benchmarks. Finally, in Sect. 6, we discuss the results and future work.

2 Related Work

2.1 Neural-Symbolic AI in Semantic Image Interpretation

In recent years, there has been significant research focus on NeSy architectures for addressing semantic image interpretation tasks [1,2,4,5,7,9]. The present studies falls within the class of NeSy techniques that seeks to incorporate symbolic information as a prior [3]. Specifically, we rely on LTNs, which are modular architectures capable of incorporating FOL constraints [2,7] and can be jointly trained with neural module in an end-to-end manner [4].

In the LTN framework, the concept of *grounding* plays a crucial role in interpreting FOL within a specific subset of the domain, denoted as $\mathbb{R}^n$. In this approach, logical predicates and axioms are represented as vectors, which are then grounded (interpreted) as real numbers in the range $[0, 1]$ using a technique called *Real Logic*. By employing this grounding mechanism, the LTN framework maps each term, denoted as x, to a vector representation in $\mathcal{G}(x) = \mathbb{R}^n$, while each predicate symbol, represented by $p \in \mathcal{P}$, is mapped to $\mathcal{G}(D(p)) \to [0, 1]$. The concept can be illustrated by considering the frequently encountered isOfClass predicate in neuro-symbolic architectures, which quantifies the probability of a given term belonging to a specific class c [4,7]. The training objective is formulated by constructing a knowledge base $\mathcal{K}$ of FOL axioms, and finding the *best satisfiability* (sat).

2.2   Zero-Shot Learning

ZSL tasks entail recognizing objects from previously unseen classes by exploiting some form of auxiliary knowledge, usually attribute-based, learned from seen classes. GZSL extends ZSL by assuming both seen and unseen classes are present at test time. Different strategies have been proposed to tackle ZSL, including embedding-based, attention mechanism-based, and generative strategies.

Embedding-based techniques compare semantic attributes and visual information by mapping them onto a suitable embedding space. Some methods embed images into the attribute space using an embedding function and consider semantic attributes as the common space [10,11]. Other techniques have proposed to adopt the image embedding space as common space [7,12,13] to mitigate the hubness problem [14]. A third class of approaches [15–17] utilizes a shared space distinct from both image and attribute domains. To prevent overfitting to seen classes, these approaches require the use of pseudo-labeling techniques or a transductive setting [16], assuming that unlabeled images from unseen classes are provided during training. **Attention mechanisms** have been proposed to find the image regions that contribute the most to the categorization of a certain class and improve the embedding space [10,11,18,19].

A critical aspect of embedding-based method is preventing overfitting to seen classes using, e.g., regularization [10,16] or contrastive techniques [11]. FLVN accomplishes this objective by incorporating a symbolic prior to aggregate visually and semantically similar features, while also explicitly establishing relationships between classes within the same macro class (a characteristic often implicitly addressed in alternative methodologies).

Previous approaches have used semantic class descriptions for the classification, but attributes linked to a specific class may not be consistently expressed or detectable in individual images. FLVN addresses this limitation by incorporating axioms that encode the existence of "exceptions to the rule" within the dataset. This aspect, which has received limited consideration in previous studies, has been experimentally shown to improve classification accuracy.

Generative techniques, on the other hand, exploit auxiliary models, such as Generative Adversarial Networks (GANs), to generate artificial examples representing unseen classes by learning a conditional probability for each class [20–23]. Recently, feature generation models were integrated with embedding-based models in a contrastive setting [24]. Generative methods require prior knowledge regarding unseen classes when generating training data. In contrast, the training process of FLVN relies on a subset of distinct seen classes and solely assumes knowledge of the class hierarchy at training time, so that the training can be easily extended to new unseen classes. Nonetheless, our approach is complementary to generative techniques, and could be in principle combined.

3   Fuzzy Logic Visual Network

We introduce a comprehensive and trainable framework for the task of ZSL, depicted in Fig. 1. It comprises two main modules: a feature extractor and a LTN that formulates the training objective.

Feature Extractor. The feature (embedding) extractor is a CNN that maps the input x to a feature space $f_\theta(x) \in \mathbb{R}^{H \times W \times B}$, where H, W and B represent the height, width, and number of channels of the features, respectively. Through mean pooling over H and W, we obtain global discriminative characteristics $g_\theta(x) \in \mathbb{R}^{B \times 1}$, and utilize a linear projection to transform them into a semantic space represented by $V \in \mathbb{R}^{B \times M}$, where B is the dimension of the vector space of features, and M denotes the length of the attribute vector [10].

Logic Tensor Network. The LTN formulates the learning objective as the maximum satisfiability of a $\mathcal{K}$. For each training batch, the $\mathcal{K}$ is updated introducing axioms that represent labelled examples (ϕ_1), as well as prior knowledge ($\phi_2, \phi_3, \phi_4, \phi_5, \phi_6$). The maximum satisfiability loss is then defined based on the aggregation of all axioms as follows:

$$\mathcal{L}^{\text{ep}} = 1 - \left(\bigwedge_{\phi \in \mathcal{K}} \phi \right) = 1 - \mathcal{G}(\phi) \tag{1}$$

This section first define the variables, predicates and domain that form the FOL language, followed by the definition of the knowledge base $\mathcal{K}$.

Groundings. Variables and their domains are grounded as follows:

$$\mathcal{G}(l) = \mathbb{N}^C, \mathcal{G}(q) = \mathbb{N}^Q \tag{2}$$

$$\mathcal{G}(a) = \mathcal{G}(a^{\text{mask}}) = \mathbb{R}^{M \times C} \tag{3}$$

$$\mathcal{G}(a^{\text{macro}}) = \mathbb{R}^{M \times Q} \tag{4}$$

$$\mathcal{G}(x) = g_\theta(f_\theta(\mathcal{G}(\texttt{images}))) = \mathbb{R}^M \tag{5}$$

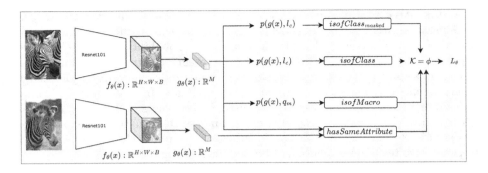

Fig. 1. The FLVN architecture, designed for Zero-Shot Learning (ZSL) classification, combines a convolutional feature extractor and an attribute encoder to effectively map visual features to the attribute space. By leveraging predicates such as isOfClass, isOfClass$_{\text{masked}}$, and isOfMacro, it reduces the disparity between image features and class attributes, while the hasSameAttribute predicate captures the similarity between two images. All these predicates play a crucial role in the formulation of the architecture, as they are incorporated within the formulas of a knowledge base denoted as $\mathcal{K}$. During the training process, the loss function is designed to optimize the satisfiability (truth value) of all the formulas within $\mathcal{K}$.

where the variable l represents the class labels belonging to set of classes C, q represent the macroclass label belonging to set of macroclasses Q, and each class/macroclass is described by a set of non-binary semantic attributes denoted by a and a^{macro}, respectively. Functions f_θ and g_θ are employed to embed images into the attribute space, resulting in the final representation $\mathcal{G}(x)$. The FOL language contains four main predicates: isOfClass(x, l) and isOfClass$_{\text{masked}}(x, l)$ denote the fact that an image x belongs to class l, isOfMacro(x, q) that an image x belongs to the macroclass q, and hasSameAttribute(x_1, x_2) that two images have the same attributes.

The $\mathcal{G}($isOfClass$)$ predicate is grounded by the similarity between the input image and the corresponding class attribute vectors. First, we compute the similarity between the image x and a class l_c by calculating the scaled product of the global features mapped in the attribute space with the semantic vectors:

$$p(x, l) = \frac{\exp\left(x^T V a_l\right)}{\sum_{s=1}^{S} \exp\left(x^T V a_s\right)} \tag{6}$$

where a_y represents the semantic attribute vector associated with class l. To obtain a prediction score for an example x, we calculate the dot product between the output of p (Eq. 6) and the one-hot encoding l_c^T for class $c \in C$, as follows:

$$\mathcal{G}(\text{isOfClass}) : x, l_c \rightarrow l_c^T p(\mathcal{G}(x), l_c) \tag{7}$$

Similarly, we define $\mathcal{G}($isOfMacro$)$ and $\mathcal{G}($isOfClass$_{\text{masked}})$. Since attributes for macro-classes are in principle unknown, we define a trainable attribute vector a_m^{macro} for macro-class q_m to compute $\mathcal{G}(\text{isOfMacro})$. On the other hand, $\mathcal{G}(\text{isOfClass}_{\text{masked}})$ uses the masked attribute vector a_c^{masked} for class l_c, in which missing attributes k are set to 0, while preserving the rest of the attributes in a_c. Finally, the grounding for `hasSameAttribute` is defined as:

$$\mathcal{G}(\text{hasSameAttribute}) : x_1, x_2 \to \text{sigmoid}(\alpha d(\mathcal{G}(x_1), \mathcal{G}(x_2))) \qquad (8)$$

where d is the cosine similarity, α a scale factor and $\mathcal{G}(x_1)$, $\mathcal{G}(x_2)$ correspond to the embeddings of the two images.

Learning from Labeled Examples. We incorporate labelled examples by introducing an axiom ϕ_1 stating that all facts about labeled example should be true, that is, all labeled samples should be classified correctly[1]:

$$\phi_1 = \forall \text{Diag}(x, l_c)(\text{isOfClass}(x, l_c)) \qquad (9)$$

To account for the class hierarchy, we also introduce an axiomatic statement ϕ_2 to indicate that "if an image contains a zebra", then "the image belongs to the family of ungulates":

$$\phi_2 = \forall \text{Diag}(x, l_c, q_m)(\text{isOfClass}(x, l_c) \implies \text{isOfMacro}(x, q_m)) \qquad (10)$$

Learning Better Feature Representations. The following axiom encodes the assumption that features extracted from two images of the same class should possess the same attributes:

$$\phi_3 = \forall \text{Diag}(x_1, l_{c_1})\left(\forall \text{Diag}(x_2, l_{c_2}) : c_1 = c_2 \ \text{hasSameAttribute}(x_1, x_2)\right) \qquad (11)$$

Likewise, images from different classes should possess different attributes:

$$\phi_4 = \forall \text{Diag}(x_1, l_{c_1})\left(\forall \text{Diag}(x_2, l_{c_2}) : c_1 != c_2 \ \neg\text{hasSameAttribute}(x_1, x_2)\right) \qquad (12)$$

To further emphasize the similarity between visual attributes and semantic vectors, the following axiom enforces the similarity between image embeddings and attribute vectors of the same class:

$$\phi_5 = \forall \text{Diag}(x, l_c)\left(\forall \text{Diag}(a, l_a) : c = a \ \text{hasSameAttribute}(x, a)\right) \qquad (13)$$

Learning with Refutation. In classical ZSL benchmarks such as AWA2, attributes are associated with class labels using a crisp or fuzzy matrix. However, this association does not entail that all examples of a class will exhibit exactly the same attributes: attributes may be expressed by a subset of training

---

[1] Diagonal Quantification quantifies over pairs of instances, e.g., images and their labels. A more formal definition can be found in [2].

samples, or may be occluded. The existential statement ϕ_6 represents the fact that some class attributes may not be present for all samples (e.g., "there exists a zebra that is not agile"). Given that image-level attributes are not available, we simply remove randomly selected attributes by defining the isOfClass$_{\text{masked}}$ predicate:

$$\phi_6 = \forall l^{\text{seen}}(\exists x, \text{isOfClass}_{\text{masked}}(x, l^{\text{seen}})) \tag{14}$$

where l^{seen} denotes the list of seen classes.

Grounding Logical Connectives and Aggregators. The knowledge base $\mathcal{K}$ is an aggregation of formulas updated at each training step. To solve the maximum satisfiability problem using gradient descent, logical connectives and aggregators must be grounded into Real Logic. Given two truth values a and b in $[0, 1]$, we adopted the symmetric configuration from [2], using the standard negation $\neg : N_S(a) = 1 - a$ and the Reichenbach implication $\rightarrow: I_R(a, b) = 1 - a + ab$. The existential quantifier $\exists$ was approximated by the generalized mean A_{pM}, and the universal quantifier $\forall$ by the generalized mean w.r.t. the error A_{pME}, respectively [2,25]. Given n truth values $a_1, \ldots, a_n$ all in $[0, 1]$:

$$\exists : A_{pM}(a_1, \ldots, a_n) = \left(\frac{1}{n} \sum_{i=1}^{n} a_i^{p_\exists}\right)^{\frac{1}{p_\exists}} \qquad p_\exists \geqslant 1 \tag{15}$$

$$\forall : A_{pME}(a_1, \ldots, a_n) = 1 - \left(\frac{1}{n} \sum_{i=1}^{n} (1 - a_i)^{p_\forall}\right)^{\frac{1}{p_\forall}} \qquad p_\forall \geqslant 1 \tag{16}$$

A_{pME} is a measure of how much, on average, truth values a_i deviate from the true value of 1. The A_{pME} was also used to approximate $\bigwedge$ in Eq. 1. Further details on the role of $p_\exists$ and $p_\forall$ can be found in previous works [2].

Querying the Knowledge Base. At inference time, the class with the highest score is selected as the predicted class:

$$\hat{y} = \underset{\tilde{y} \in Y^U}{argmax}(g(x)^T V a_{\tilde{y}}) \tag{17}$$

FVLN was evaluated in both ZSL and GZSL settings. In the ZSL setting, only unseen images are assumed to be present at test time, whereas in the GZSL setting, the model is tested on both seen and unseen classes. This setup induces a bias towards seen classes. To mitigate it, we employed the Calibrated Stacking method, as proposed in [26,27], to diminish the classification score of seen classes. The class score is thus calculated as $\hat{y}$:

$$\hat{y} = \underset{\tilde{y} \in Y^U \cup Y^S}{argmax}(g(x)^T V a_{\tilde{y}} - \gamma \mathbb{I}[\tilde{y} \in \mathcal{Y}^S]) \tag{18}$$

where $\mathbb{I} = 1$ if $\tilde{y}$ is from a seen class and zero otherwise, γ is a calibration coefficient tuned on a validation set and $\mathcal{Y}^S$ are the labels of seen classes.

Construction of the Training Batch. Following the approach in [11], for a positive input image x_i, we select a set of positive examples x^+ and K negative examples $x_1^-, ..., x_K^-$. Positive examples are selected from the same category as x_i, while negative examples are randomly selected from the remaining classes.

4 Experimental Settings

In this section we examine the datasets used in our experiments, the knowledge base adopted for each dataset and the list of hyperparameters chosen.

Dataset. The experiments were conducted on the AWA2 [6], CUB [28], and SUN [29] benchmarks. The evaluation metrics for GZSL were based on the standards defined in previous work [6]. Following a similar approach outlined in [30], we construct a semantic hierarchy by grouping the classes from Awa2 and CUB datasets into a total of 9 and 49 macroclasses, respectively.

Knowledge Base. The knowledge base $\mathcal{K}$ composition was detailed in Sect. 3. To build class hierarchy, we map classes from the AWA2 and CUB datasets to their corresponding synsets in WordNet as done by Sikka et al. [30]. Macroclasses for each dataset were defined by selecting the synset root whose subtree contained classes from the selected dataset, and then defining its immediate children (using WordNet) as classes. Since classes in the SUN dataset [29] lack a semantic hierarchy, axioms related to macroclasses were not included in the knowledge base. For all experiments, in the predicate $\texttt{isofClass}_{masked}$, $k = 15$ attributes were randomly dropped. The α parameter in Eq. 8 was set to 0.01 for AWA2 dataset and 1 on CUB e SUN. To account for the presence of outliers in the knowledge base, we initially set the aggregation function parameters (defined in Eqs. 15 and 16) as $p_\exists = 2$ and $p_\forall = 2$. Both parameters were incremented by 2 every 4 epochs for the AWA2 and CUB datasets; for SUN, the aggregation values were increased at specific epochs (2, 4, 24, and 32) until reaching $p_\exists = 6$ and $p_\forall = 6$, following the schedule suggested in [2].

Hyperparameter Selection. The embedding function f_θ is based on an ImageNet pre-trained ResNet101 model that converts the 224×224 image into a vector $\mathbf{x} \in \mathbb{R}^{H \times W \times B}$, where $B = 2048$, and H and W represent the height and width of the extracted features; the function g_θ then converts these features into the attribute space dimensionally consistent with the dataset.

To mitigate overfitting, we initially trained the head with a frozen backbone and subsequently fine-tuned the entire network. We used the Adam optimizer with learning rate of $1e-4$ for the pre-training phase for AWA2 and CUB and $5e-4$ for SUN. The learning rate was then reduced to $1e-6\alpha$ for CUB and SUN and $1e-7\alpha$ for AWA2 in the fine-tuning phase, where $\alpha = 0.8^{epoch//10}$ and $epoch$ is the current epoch out of a total of 300 epochs. Additionally, we incorporated a smoothing factor by multiplying the L2 norm with a scaling factor of $5e-4$ for AWA2, $5e-6$ for CUB, and $1e-3$ for SUN.

Training batches included positive and negative examples with a ratio of 12 to 12 for AWA2, 12 to 8 for CUB, and 12 to 4 for SUN datasets. At inference

Table 1. Performance on AWA2, CUB and SUN test set. For FLVN, we show mean ± standard deviation and maximum (in parentheses) values for TOP1zsl (T1), TOP1gzsl unseen (U), TOP1gzsl seen (S), and Hgzsl (H) over three runs. Metrics are described in [6]. The first section of the table includes embedding-based models, the second section attention-based models, and the third section generative models. The best performance values are shown in bold. † highlights method which incorporate external knowledge.

Model	AWA2				CUB				SUN			
	T1	U	S	H	T1	U	S	H	T1	U	S	H
PROTO-LTN [7]	67.6	32.0	83.7	45.2	48.8	20.8	54.3	30.0	60.4	20.4	**36.8**	25.2
DEM [12]	67.1	30.5	86.4	45.1	51.7	19.5	57.9	29.2	61.9	20.5	34.3	25.6
VSE [13]	84.4	45.6	88.7	60.2	71.9	39.5	66.9	50.2	-	-	-	-
TCN [17]	71.2	61.2	65.8	63.4	59.5	52.6	52.0	52.3	61.5	31.2	37.3	34.0
CSNL [30] †	61.0	-	-	0.0	32.5	-	-	0.7	-	-	-	-
AREN [19]	67.9	54.7	79.1	64.7	71.8	63.2	69.0	66.0	60.6	40.3	32.3	35.9
APN [10]	68.4	56.5	78.0	65.5	72.0	65.3	69.3	67.2	61.6	41.9	34.0	37.6
AMGML [18]	71.7	56.0	74.6	64.0	70.0	58.2	58.7	56.9	59.7	42.0	35.1	38.3
CC-ZSL [11]	68.8	62.2	83.1	71.1	74.3	56.1	73.2	62.5	62.4	44.4	35.9	40.3
Cycle-CLSWGAN [23]	-	-	-	-	58.4	45.7	61.0	52.3	60.0	49.4	33.6	40.0
LisGan [22]	-	-	-	-	58.8	46.5	57.9	51.6	60.0	42.9	37.8	40.2
E-PGN [20]	73.4	52.6	83.5	64.6	72.4	52.0	61.1	56.2	-	-	-	-
TGMZ [21]	78.4	64.1	77.3	70.1	66.1	60.3	56.8	58.5	-	-	-	-
CECZSL [24]	70.4	63.1	78.6	70.0	77.5	63.9	66.8	65.3	63.3	48.8	38.6	43.1
DFCA-GZSL [31]	74.7	65.5	81.5	73.3	80.0	70.9	63.1	66.8	62.6	48.9	38.8	43.3
FLVN †	69.8 ± 0.8 (71.0)	65.8 ± 0.9 (67.1)	82.3 ± 0.5 (82.8)	73.1 ± 0.6 (74.1)	71.2 ± 0.2 (71.4)	62.6 ± 0.5 (63.2)	83.1 ± 0.3 (83.4)	71.5 ± 0.2 (71.7)	61.7 ± 0.18 (61.9)	48.4 ± 0.57 (48.9)	32.7 ± 0.2 (32.9)	39.0 ± 0.1 (39.1)

time, we set the scaling factor γ to adjust the scores obtained for the seen classes to 0.7 for AWA2 and CUB and 0.4 for SUN. In all experiments, we applied a random crop and random flip with 0.5 probability for data augmentation. We implemented the architecture in PyTorch, using the LTNtorch library [8], and trained it on a single GPU nVidia 2080 Ti. Each experiment was repeated three times to calculate the mean and standard deviation.

5   Results

Experimental results presented in Table 1 show how the proposed FLVN architecture reaches competitive performances with respect to other embedding-based methods, particularly CC-ZSL [11] and APN [10]. In general, FVLN performs better in terms of harmonic mean (H) and shows a greater ability to recognize both seen and unseen classes, achieving state-of-the-art performance on two out of three benchmark datasets. In particular, FLVN improves the accuracy of unseen classes and the harmonic mean by 0. 89% and 1.3%, respectively, in AWA2, and the accuracy of seen classes and harmonic mean by 12% and 3%, respectively, on CUB. These results suggest that the proposed architecture is capable of recognizing seen and unseen samples more evenly, showing less confusion in classifying examples. On SUN, FLVN performs comparably to the state of the art; it is important to note that, in this case, axioms on macro classes are not introduced, therefore FLVN cannot leverage external semantic knowledge.

Compared to the closest methods in terms of performance, the proposed method requires a relatively simple architecture. Unlike APN, FVLN does not require additional weights for each attribute or supplemental regularization terms to construct the loss function. FVLN only requires a single backbone replica instead of two as in CC-ZSL, which is based on the teacher-student framework [11]. Compared to generative methods, FLVN does not require to generate additional training samples or make assumptions about unseen classes at train time.

Although complete ablation studies are left to future work, qualitative observations from our experiments suggest that axioms included in the knowledge base may have quite different impact on the overall performance. With respect to our previous architecture [7], it was crucial to change the isOfClass grounding for end-to-end training to succeed. The introduction of the isOfMacroclass predicate, on the other hand, introduces smaller improvements. In fact, during training related classes appear to naturally cluster in feature space, as evident from the distribution of the feature space already reported in previous work [7]. On the other hand, the isOfClass_masked offered important benefits, allowing to account for mistakes in the semantic annotation of classes or traits that exist at the class level but are not visible or easily inferred from a single image (e.g., "All zebras are agile"). Finally, the hasSameAttribute predicate improved performance particularly on the CUB (fine-grained bird recognition) and SUN (scene recognition) datasets, in which extracting fine-grained image-level attributes is more difficult.

6   Conclusions and Future Work

Building upon principles from the recent ZSL literature [10] and incorporating them within a NeSy framework [7], we introduced a novel NeSy architecture, named FLVN, for ZSL and GZSL tasks. FLVN incorporates axioms that combine prior class-specific knowledge (e.g., class hierarchies), with high-level inductive biases to handle exceptions within the dataset (e.g., "there exists a zebra that is not agile") and establish relationships between images (e.g., "if two images belong to the same class, they must be similar"). Such axioms act as semantic priors and compensate for the lack of annotations, thereby providing a solid NeSy foundation for GZSL tasks. FLVN does not require multiple backbones and maintains roughly the same parameter count of standard embedding-based methods. The proposed approach can be also incorporated into other architectures by, e.g., changing the grounding of the predicates. Experimental results prove that FLVN achieves performance on par or exceeding that of current literature on common GZSL benchmarks.

We believe that our approach can be further extended in several ways. For example, different formulations of the isOfClass predicate, or the introduction of a hasAttribute predicate to predict image-level attributes, could incorporate attention to discriminative regions within the image and increase explainability. Introducing attention mechanisms based on object detection [4] could support the definition useful axioms that identify the most discriminative parts of the image. Introducing axioms that solely consider the similarity between images (with no knowledge of their class membership) could improve results for unseen classes without requiring additional labels in the dataset. The knowledge base could be also extended to consider other types of relationships, beyond class hierarchies, and from emerging sources, such as language models. Finally, the proposed method could be explored in a trasductive ZSL setting.

Acknowledgements. This study was carried out within the FAIR - Future Artificial Intelligence Research and received funding from the European Union NextGenerationEU (PIANO NAZIONALE DI RIPRESA E RESILIENZA (PNRR) - MISSIONE 4 COMPONENTE 2, INVESTIMENTO 1.3 - D.D. 1555 11/10/2022, PE00000013). This manuscript reflects only the authors' views and opinions, neither the European Union nor the European Commission can be considered responsible for them.

References

1. Yu, D., Yang, B., Liu, D., Wang, H., Pan, S.: Recent advances in neural-symbolic systems: a survey. arXiv preprint arXiv:2111.08164 (2021)
2. Badreddine, S., Garcez, A.d., Serafini, L., Spranger, M.: Logic tensor networks. Artif. Intell. **303**, 103649 (2022)
3. van Bekkum, M., de Boer, M., van Harmelen, F., Meyer-Vitali, A., Teije, A.t.: Modular design patterns for hybrid learning and reasoning systems: a taxonomy, patterns and use cases. Appl. Intell. **51**(9), 6528–6546 (2021)
4. Manigrasso, F., Miro, F.D., Morra, L., Lamberti, F.: Faster-LTN: a neuro-symbolic, end-to-end object detection architecture. In: Farkaš, I., Masulli, P., Otte, S., Wermter, S. (eds.) ICANN 2021. LNCS, vol. 12892, pp. 40–52. Springer, Cham (2021). https://doi.org/10.1007/978-3-030-86340-1_4
5. Donadello, I., Serafini, L.: Compensating supervision incompleteness with prior knowledge in semantic image interpretation. In: 2019 International Joint Conference on Neural Networks (IJCNN) (2019)
6. Xian, Y., Lampert, C.H., Schiele, B., Akata, Z.: Zero-shot learning - a comprehensive evaluation of the good, the bad and the ugly. IEEE Trans. Pattern Anal. Mach. Intell. **41**(9), 2251–2265 (2019)
7. Martone, S., Manigrasso, F., Lamberti, F., Morra, L.: PROTOtypical logic tensor networks (PROTO-LTN) for zero shot learning. In: 2022 26th International Conference on Pattern Recognition (ICPR) (2022)
8. Carraro, T.: LTNtorch: PyTorch implementation of Logic Tensor Networks, March 2022. https://doi.org/10.5281/zenodo.6394282
9. Li, Z., et al.: Calibrating concepts and operations: towards symbolic reasoning on real images. In: IEEE/CVF International Conference on Computer Vision (2021)
10. Xu, W., Xian, Y., Wang, J., Schiele, B., Akata, Z.: Attribute prototype network for zero-shot learning. ArXiv (2020)
11. Cheng, D., Wang, G., Wang, N., Zhang, D., Zhang, Q., Gao, X.: Discriminative and robust attribute alignment for zero-shot learning. IEEE Trans. Circ. Syst. Video Technol. (2023)
12. Zhang, L., Xiang, T., Gong, S.: Learning a deep embedding model for zero-shot learning. In: IEEE Conference on Computer Vision and Pattern Recognition (2017)
13. Zhu, P., Wang, H., Saligrama, V.: Generalized zero-shot recognition based on visually semantic embedding. In: IEEE/CVF Conference on Computer Vision and Pattern Recognition (2019)
14. Zhang, L., Xiang, T., Gong, S.: Learning a deep embedding model for zero-shot learning. In: IEEE/CVF Conference on Computer Vision and Pattern Recognition (2017)
15. Chen, L., Zhang, H., Xiao, J., Liu, W., Chang, S.F.: Zero-shot visual recognition using semantics-preserving adversarial embedding networks. In: 2018 IEEE/CVF Conference on Computer Vision and Pattern Recognition (2018)

16. Zhang, L., et al.: Towards effective deep embedding for zero-shot learning. IEEE Trans. Circuits Syst. Video Technol. **30**, 2843–2852 (2018)
17. Jiang, H., Wang, R., Shan, S., Chen, X.: Transferable contrastive network for generalized zero-shot learning. In: 2019 IEEE/CVF International Conference on Computer Vision (ICCV) (2019)
18. Li, Y., Liu, Z., Yao, L., Wang, X., Wang, C.: Attribute-modulated generative meta learning for zero-shot classification. ArXiv (2021)
19. Xie, G., et al.: Attentive region embedding network for zero-shot learning. In: 2019 IEEE/CVF Conference on Computer Vision and Pattern Recognition (CVPR) (2019)
20. Yu, Y., Ji, Z., Han, J., Zhang, Z.: Episode-based prototype generating network for zero-shot learning. In: 2020 IEEE/CVF Conference on Computer Vision and Pattern Recognition (CVPR) (2019)
21. Liu, Z., Li, Y., Yao, L., Wang, X., Long, G.: Task aligned generative meta-learning for zero-shot learning. In: AAAI Conference on Artificial Intelligence (2021)
22. Li, J., Jing, M., Lu, K., Ding, Z., Zhu, L., Huang, Z.: Leveraging the invariant side of generative zero-shot learning. In: 2019 IEEE/CVF Conference on Computer Vision and Pattern Recognition (CVPR) (2019)
23. Felix, R., Vijay Kumar, B.G., Reid, I., Carneiro, G.: Multi-modal cycle-consistent generalized zero-shot learning. In: Ferrari, V., Hebert, M., Sminchisescu, C., Weiss, Y. (eds.) ECCV 2018. LNCS, vol. 11210, pp. 21–37. Springer, Cham (2018). https://doi.org/10.1007/978-3-030-01231-1_2
24. Han, Z., Fu, Z., Chen, S., Yang, J.: Contrastive embedding for generalized zero-shot learning. In: 2021 IEEE/CVF Conference on Computer Vision and Pattern Recognition (CVPR) (2021)
25. Van der Maaten, L., Hinton, G.: Visualizing data using t-SNE. J. Mach. Learn. Res. **9**(11) (2008)
26. Chen, S., et al.: MSDN: mutually semantic distillation network for zero-shot learning. In: 2022 IEEE/CVF Conference on Computer Vision and Pattern Recognition (CVPR) (2022)
27. Wang, W., Li, Q.: Generalized zero-shot activity recognition with embedding-based method. ACM Trans. Sensor Netw. **19**, 1–25 (2023)
28. Wah, C., Branson, S., Perona, P., et al.: Multiclass recognition and part localization with humans in the loop. In: International Conference on Computer Vision (2011)
29. Xiao, J., Hays, J., Ehinger, K.A., Oliva, A., Torralba, A.: SUN database: large-scale scene recognition from abbey to zoo. In: IEEE Conference on Computer Vision and Pattern Recognition (2010)
30. Sikka, K., et al.: Zero-shot learning with knowledge enhanced visual semantic embeddings. ArXiv (2020)
31. Su, H., Li, J., Lu, K., Zhu, L., Shen, H.T.: Dual-aligned feature confusion alleviation for generalized zero-shot learning. IEEE Trans. Circuits Syst. Video Technol. **33**, 3774–3785 (2023)

CISPc: Embedding Images and Point Clouds in a Joint Concept Space by Contrastive Learning

Cristian Sbrolli[✉][ID], Paolo Cudrano[ID], and Matteo Matteucci[ID]

Politecnico di Milano, Milan, Italy
{cristian.sbrolli,paolo.cudrano,matteo.matteucci}@polimi.it

Abstract. In the last years, deep learning models have achieved remarkable success in computer vision tasks, but their ability to process and reason about multi-modal data has been limited. The emergence of models leveraging contrastive loss to learn a joint embedding space for images and text has sparked research in multi-modal unsupervised alignment. This paper proposes a contrastive model for the multi-modal alignment of images and 3D representations. In particular, we study the alignment of images and raw point clouds on a learned latent space. The effectiveness of the proposed model is demonstrated through various experiments, including 3D shape retrieval from a single image, testing on out-of-distribution data, and latent space analysis.

1 Introduction

In the last decade, deep learning models have achieved remarkable success in various computer vision tasks, such as image classification, object detection, and semantic segmentation. However, most of these models worked only on a single modality, such as images, and were limited in their ability to process and reason about multi-modal data. With the success of CLIP [14], which leverages contrastive loss to learn a joint embedding space for images and text, research in the field of multi-modal unsupervised alignment has made impressive steps. The multi-modal alignment problem refers to the task of learning a joint embedding space that can represent different modalities in a common space.

Several works have studied and extended multi-modal unsupervised alignment to more modalities such as audio [6], video [20], 2.5D depth images [21,22], and voxels [16] obtaining cutting-edge results in many downstream tasks. We believe that 3D representations are one of the most promising extensions to multi-modal alignment tasks, as their alignment with other modalities would facilitate a wide range of applications, such as 3D object recognition, scene understanding, and robotics. For example, in robotics, a robot equipped with a camera and a LIDAR sensor can use our model to align visual and spatial information to better understand its environment. Moreover, the utilization of aligned multi-modal embeddings as off-the-shelf features can significantly improve various tasks, particularly those involving generative multi-modal processes where one modality is employed to generate another [15,16].

G. L. Foresti et al. (Eds.): ICIAP 2023, LNCS 14234, pp. 468–476, 2023.
https://doi.org/10.1007/978-3-031-43153-1_39

However, depth images lose relevant information with respect to the original point cloud data, and voxel representations are limited in scalability, resulting in constraints on size and an inability to accurately depict fine-detailed shapes.

To this end, in this paper, we present CISPc (Constrastive Image-Shape Pretraining on Point clouds), proposing the application of contrastive pretraining to images and raw point clouds. Our model leverages a transformer architecture for both the image and point cloud encoders, which allows for better modeling of long-range dependencies and capturing global context.

In Sect. 1.1, we present an overview of related works and state-of-the-art models for different modalities. Then, we describe our model in detail (Sect. 2) and demonstrate its effectiveness in several experiments (Sect. 3), including 3D shape retrieval from a single image, testing on out-of-distribution data and latent space experiments and analysis.

1.1 Related Works

In recent years, there has been a surge of interest in vision-language models that can effectively learn the correlations between images and corresponding textual descriptions. Pretrained vision-language models have emerged as a promising approach to address this problem, as they allow for the efficient transfer of knowledge from large-scale datasets to downstream visual and language tasks.

Among the first examples of successful pre-trained vision-language models, we find Contrastive Language-Image Pre-Training (CLIP) [14] and ALIGN [8]. These are large-scale models that jointly learn representations for images and the associated textual descriptions, without the need for any task-specific supervision. These approaches achieved state-of-the-art performance in a wide range of tasks, including image classification, object detection, and image captioning.

After the success of these methods, researchers studied the application and extension of similar approaches to other data modalities. AudioCLIP extended CLIP using the Audio Set dataset [4], aligning audio data to the pre-trained CLIP model, obtaining state-of-the-art results in zero-shot audio classification. Using a similar contrastive approach, VideoCLIP [20] learns a fine-grained association between video and text, surpassing previous work in tasks such as sequence-level text-video retrieval, VideoQA, token-level action localization, and action segmentation.

CLIP-Mesh [12] proposed a method to generate 3D meshes by exploiting a pre-trained CLIP model to compare the input text prompt with differentially rendered images of a 3D model. PointCLIP and PointCLIPV2 [21,22] propose an effective method to process point clouds with a pre-trained CLIP. However, their approach does not use 3D data directly for training, and relies instead on depth maps projected from the point cloud. This leads to the loss of precious 3D geometric information, which instead is captured by our model. Indeed, our proposed approach processes point cloud 3D data directly, allowing us to exploit the complete 3D information.

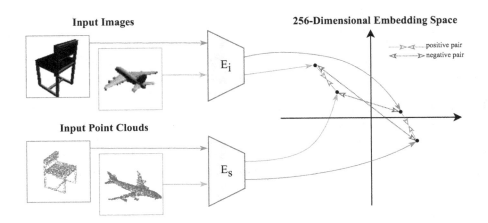

Fig. 1. Our model learns to embed close in the space matching concepts from images and point clouds, and to separate different concepts.

2   Method

The aim of our model is to learn a joint embedding space for images and point cloud, such that the same concept from different modalities is embedded in neighboring points, while different concepts are embedded in different points, as we show in Fig. 1. We adopt a training method analogous to the one from CLIP [14]. Our model consists of two encoders: an encoder E_s for processing point clouds and an encoder E_i for processing images, both of which produce embeddings of size f. Given a batch of N corresponding (image, shape) pairs, our training objective is to match each image with its corresponding shape.

We generate batches of image and shape embeddings e_i and e_s using the encoders E_i and E_s, respectively. We then compute a symmetric $N \times N$ similarity matrix between image embeddings (rows) and shape embeddings (columns), using a cosine similarity measure. Our training loss is composed of two symmetric cross-entropy terms, as follows:

$$L_{CISPc}(e_i, e_s) = -\frac{1}{2N} \sum_{j=1}^{N} log \frac{exp(<e_i^j, e_s^j>)}{\sum_{k=1}^{N} exp(<e_i^j, e_s^k>)/\tau} - \frac{1}{2N} \sum_{j=1}^{N} log \frac{exp(<e_i^j, e_s^j>)}{\sum_{k=1}^{N} exp(<e_i^k, e_s^j>)/\tau} \quad (1)$$

where $< \cdot, \cdot >$ represents the inner product. The first term of the loss operates over the image-shape similarity matrix rows and measures the ability of the model to predict the correct shape given an image; the second operates over columns, measuring the ability to predict the correct image given a shape. A temperature parameter τ is used to scale the logits in the softmax calculation and is trained jointly with the network. We initialize the temperature parameter as in [19] and we clip it as in [14] to prevent training instabilities. This objective function is designed to maximize the similarity of the N matching (image, shape) pairs and minimize the similarity of the $N^2 - N$ unmatching pairs.

Table 1. Top-K accuracies on the ShapeNet test. Reported values are averages on all considered classes. The testing batch size is set to 128, while the training batch size is 64 for both models. The best results are highlighted in bold.

Shape Encoder	Shape Prediction					Image Prediction				
	Top 1	Top 2	Top 3	Top 4	Top 5	Top 1	Top 2	Top 3	Top 4	Top 5
PointNet++ [13]	0.770	0.914	0.966	0.975	0.981	**0.863**	0.928	0.956	0.984	0.988
PCT [5]	**0.8362**	**0.942**	**0.975**	**0.989**	**0.991**	0,845	**0.949**	**0.978**	**0.992**	**0.996**

As image encoder E_i, we use a Data-Efficient Image Transformer (DeiT) [17], as the dataset that we use has limited dimensions. In particular, we employ a DeiT Base (DeiT-B), which has 768-dimensional hidden embeddings and 12 layers with 12 attention heads each.

As shape encoder E_p, we perform experiments with both PointNet++ [13] and a more recent transformer architecture, PCT [5]. For the latter, we employ a configuration with depth 12 and width 256.

3 Experiments

3.1 Dataset

We train our model on a subset of ShapeNetCore point cloud [1] composed of the following categories: Plane, Car, Chair, Table, and Vessel. For each shape, we randomly subsample 2048 points and normalize them inside the unit sphere. The views of the images are selected at random from the 23 available views for each object provided by the ShapeNetCore dataset.

3.2 Shape Encoder Selection

To compare the two proposed shape encoders, we tested them on the ShapeNet [1] test set using Top-k accuracy. We randomly sample a batch of paired images and shapes, embed them through our model, and calculate the similarity matrix between all shape and image embeddings. Top-k accuracy measures the number of times the correct shape/image is among the k most similar to the input image/shape. Taking into account the general case (which is valid for both image and shape prediction), let $p_{i,k}$ be the k most similar elements to the i-th input sample. Then, Top-k accuracy can be defined as:

$$\text{acc}_k(p, t) = \frac{\sum_{i=1}^{|t|} \mathbb{I}[t_i \in p_{i,k}]}{|t|}. \tag{2}$$

where p is the vector of predictions, t is the vector of targets, and $\mathbb{I}$ is the indicator function. Both models are trained with AdamW [10] optimizer and a learning

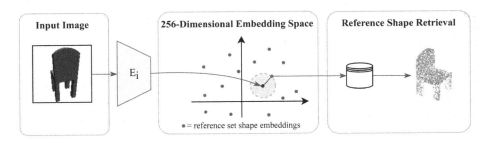

Fig. 2. To perform single-view shape retrieval, we first pre-calculate shape embeddings of the reference set. Then, the test image is projected and distances to all reference set shape embeddings is calculated. The retrieved shape is then the one with the closest shape embedding.

Table 2. 3D reconstruction results on the ShapeNet test set by using Chamfer Distance (CD). The best results are highlighted in bold. We highlight in italics our model's results that surpass at least one reconstruction baseline.

	Reconstruction						Retrieval
	PSGN [3]	3D-LMNet [11]	SE-MD [7]	3D-ARNet [2]	3D-FEGNet [18]	SSRecNet [9]	Ours
Plane	4.14	3.34	3.11	3.38	**2.36**	2.38	*3.33*
Car	5.2	4.55	4.52	4.27	3.57	**3.56**	*4.61*
Chair	6.39	6.41	6.47	6.34	**4.35**	**4.35**	8.14
Table	6	6.05	6.02	6.08	**4.42**	4.35	*5.89*
Vessel	4.38	4.37	4.24	4.15	4.15	**4.12**	7.2

rate of $3 * 10^{-4}$. As shown in Table 1, PCT [5] yields the best results, and thus it will be our shape encoder for the rest of the work.

From the values reported in Table 1, we can also notice that predicting shapes from images is easier than the inverse. Indeed, 3D representations carry more information than 2D data, making the image-retrieval task easier.

3.3   Single-View Shape Retrieval

We validate our model on the task of single-view shape retrieval. This task consists in retrieving, from a reference set of shapes, the one which best matches the given input image. In particular, we test this by using as input images from the ShapeNet test set and as reference shapes all the training set shapes. Input images are embedded through E_i and compared with all the reference shapes embeddings by using cosine distance. Then, the closest shape is the retrieved prediction. Figure 2 shows the single-view retrieval process from a reference set of shapes.

We show several examples of the shapes retrieved through our method in Fig. 3. In the figure, we report also the second closest shape retrieved, to demonstrate that the embedding space is well-structured and meaningful. Our model is able to correctly retrieve conceptually and geometrically similar shapes, even with less informative or incomplete views as back-views or occluded views.

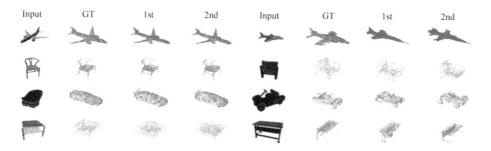

Input   GT   1st   2nd   Input   GT   1st   2nd

Fig. 3. Single-view shape retrieval examples. For each input image, we show the ground truth shape and the two closest shapes in the embedding space retrieved by our model. Images and GT are from the test set, while retrieved shapes are from the training set, which is used as reference set for retrieval. Even with less informative views as occluded and back views, our model correctly retrieves semantically and geometrically similar samples.

Table 3. Top-k accuracy on unseen classes. The testing batch size is set to 128, while the training batch size is 64 for both models. The best results are highlighted in bold.

Category	Top 1	Top 2	Top 3	Top 4	Top 5
Lamp	0.385	0.454	0.598	0.695	0.726
Speaker	0.296	0.402	0.675	0.715	0.757
Cabinet	0.313	0.491	0.603	0.696	0.771
Bench	0.392	0.514	0.600	0.659	0.706

We quantitatively validate our model by comparing its performance against 3D reconstruction models, using Chamfer Distance as distance metric. We report results in table Table 2. We point out however that our model is not trained to perform full 3D reconstruction, but only to align embeddings and perform closest-shape retrieval. Nevertheless, we notice how CISPc is still capable of surpassing some of the reconstruction baselines in Planes, Cars and Tables. We speculate that this is due to the lower intra-class variability of these categories with respect to Chairs and Vessels, increasing the probability to find much more similar samples in the reference set, which our model is then able to correctly select.

3.4   Generalization to Unseen Categories

To further analyze the 3D capabilities of our model, we test CISPc on categories never seen during training. In particular, we test on the Lamp, Speaker, Cabinet, and Bench categories. We report results for the Top-k accuracy in unseen classes in Table 3.

Our model achieves over 70% Top-5 accuracy in each unseen category. This is a very positive result considering the novelty of the unseen shapes and the batch

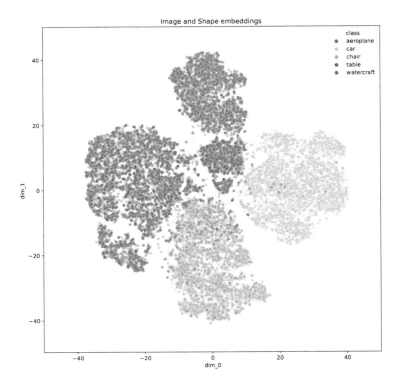

Fig. 4. 2D visual representation of the embedding space learnt by CISPc, showing how training images (dark dots) and shapes (light dots) are mapped in overlapping regions, and how each of these regions is well separated from the others.

size of 128, as selecting the correct shape is harder with bigger batch sizes. In particular, notice how, even in categories such as Lamps, which are structurally different from all training classes as they are generally thin and evolve vertically, our model reaches high accuracies, demonstrating the generalization power of our approach, even when the training set contains only a limited number of categories.

3.5   Latent Space Analysis

To get more insight on the learned embedding space, we analyze it by first applying Principal Component Analysis (PCA) to reduce the embeddings dimensions from 256 to 50, then applying t-SNE to reduce the dimensions to 2. We show the resulting projection in Fig. 4. While the employed contrastive loss only enforces the similarity between positive pairs and sets apart negative pairs in the space, the obtained space is well-structured and meaningful. Indeed, we notice how each object category is well separated in the embedding space, resulting in positive pairs being mapped close to each other, as intended. Impressive consequence of our approach is also that it generates an inherent grouping of similar concepts together. Indeed, analyzing the location of each sample, we find that shapes get organized in an even more fine-grained manner, with intra-category groups emerging based on low-level details, such as the type of wheels in chairs, the type of wings in airplanes and the difference between sport and city cars.

4   Discussion and Future Works

In this work we presented CISPc, an effective approach to perform contrastive learning on images and raw point clouds. Although trained on a small dataset and using a limited number of object categories, we show how the embedding space learnt by CISPc is well structured, yielding the emergence of a fine-grained concept organization based on low-level shape details. This property allows CISPc to achieve good performance in shape retrieval, generalizing even to unseen categories. For these reasons, we consider this work to be a proof of concept and basis for future works. Further development, including training on additional categories and expanding the model to encompass more datasets and real-world data, will be essential for creating a more versatile model suitable for deployment in practical applications and as a pre-aligned feature extractor for various tasks. Potential applications include multi-modal generative models, as well as models for multi-modal environment comprehension and sensor fusion. We also believe that extending pretraining to other modalities, such as audio, represents a promising avenue for future exploration in the context of 3D data and multi-modal contrastive learning.

References

1. Chang, A.X., et al.: ShapeNet: An information-rich 3D model repository. arXiv preprint arXiv:1512.03012 (2015)
2. Chen, H., Zuo, Y.: 3D-ARNet: an accurate 3D point cloud reconstruction network from a single-image. In: Multimedia Tools and Applications, pp. 1–14 (2022)
3. Fan, H., Su, H., Guibas, L.J.: A point set generation network for 3D object reconstruction from a single image. In: Proceedings of the IEEE Conference on Computer Vision and Pattern Recognition, pp. 605–613 (2017)
4. Gemmeke, J.F., et al.: Audio set: an ontology and human-labeled dataset for audio events. In: 2017 IEEE International Conference on Acoustics, Speech and Signal Processing (ICASSP), pp. 776–780. IEEE (2017)
5. Guo, M.H., Cai, J.X., Liu, Z.N., Mu, T.J., Martin, R.R., Hu, S.M.: PCT: point cloud transformer. Comput. Vis. Media **7**, 187–199 (2021)
6. Guzhov, A., Raue, F., Hees, J., Dengel, A.: AudioCLIP: extending clip to image, text and audio. In: ICASSP 2022–2022 IEEE International Conference on Acoustics, Speech and Signal Processing (ICASSP), pp. 976–980. IEEE (2022)
7. Hafiz, A.M., Bhat, R.U.A., Parah, S.A., Hassaballah, M.: SE-MD: a single-encoder multiple-decoder deep network for point cloud generation from 2D images. arXiv preprint arXiv:2106.15325 (2021)

8. Jia, C., et al.: Scaling up visual and vision-language representation learning with noisy text supervision. In: International Conference on Machine Learning, pp. 4904–4916. PMLR (2021)
9. Li, B., Zhu, S., Lu, Y.: A single stage and single view 3D point cloud reconstruction network based on DetNet. Sensors **22**(21), 8235 (2022)
10. Loshchilov, I., Hutter, F.: Decoupled weight decay regularization. arXiv preprint arXiv:1711.05101 (2017)
11. Mandikal, P., Navaneet, K., Agarwal, M., Babu, R.V.: 3D-LMNet: latent embedding matching for accurate and diverse 3D point cloud reconstruction from a single image. arXiv preprint arXiv:1807.07796 (2018)
12. Mohammad Khalid, N., Xie, T., Belilovsky, E., Popa, T.: Clip-mesh: generating textured meshes from text using pretrained image-text models. In: SIGGRAPH Asia 2022 Conference Papers, SA 2022. Association for Computing Machinery, New York (2022). https://doi.org/10.1145/3550469.3555392
13. Qi, C.R., Yi, L., Su, H., Guibas, L.J.: PointNet++: deep hierarchical feature learning on point sets in a metric space. In: Advances in Neural Information Processing Systems, vol. 30 (2017)
14. Radford, A., et al.: Learning transferable visual models from natural language supervision. In: International Conference on Machine Learning, pp. 8748–8763. PMLR (2021)
15. Ramesh, A., Dhariwal, P., Nichol, A., Chu, C., Chen, M.: Hierarchical text-conditional image generation with clip latents. arXiv preprint arXiv:2204.06125 (2022)
16. Sbrolli, C., Cudrano, P., Frosi, M., Matteucci, M.: IC3D: image-conditioned 3D diffusion for shape generation (2023)
17. Touvron, H., Cord, M., Douze, M., Massa, F., Sablayrolles, A., Jégou, H.: Training data-efficient image transformers & distillation through attention. In: International Conference on Machine Learning, pp. 10347–10357. PMLR (2021)
18. Wang, E., Sun, H., Wang, B., Cao, Z., Liu, Z.: 3D-FEGNet: a feature enhanced point cloud generation network from a single image. IET Comput. Vision **17**(1), 98–110 (2023)
19. Wu, Z., Xiong, Y., Yu, S.X., Lin, D.: Unsupervised feature learning via non-parametric instance discrimination. In: Proceedings of the IEEE Conference on Computer Vision and Pattern Recognition, pp. 3733–3742 (2018)
20. Xu, H., et al.: Videoclip: contrastive pre-training for zero-shot video-text understanding. arXiv preprint arXiv:2109.14084 (2021)
21. Zhang, R., et al.: PointClip: point cloud understanding by clip. In: Proceedings of the IEEE/CVF Conference on Computer Vision and Pattern Recognition, pp. 8552–8562 (2022)
22. Zhu, X., Zhang, R., He, B., Zeng, Z., Zhang, S., Gao, P.: PointClip V2: adapting clip for powerful 3D open-world learning. arXiv preprint arXiv:2211.11682 (2022)

Budget-Aware Pruning for Multi-domain Learning

Samuel Felipe dos Santos[1]([envelope]) [ID], Rodrigo Berriel[2] [ID], Thiago Oliveira-Santos[2] [ID],
Nicu Sebe[3] [ID], and Jurandy Almeida[4] [ID]

[1] Federal University of São Paulo, São José dos Campos, Brazil
`felipe.samuel@unifesp.br`
[2] Federal University of Espírito Santo, Vitória, Brazil
`berriel@lcad.inf.ufes.br`, `todsantos@inf.ufes.br`
[3] University of Trento, Trento, Italy
`niculae.sebe@unitn.it`
[4] Federal University of São Carlos, São Carlos, Brazil
`jurandy.almeida@ufscar.br`

Abstract. Deep learning has achieved state-of-the-art performance on several computer vision tasks and domains. Nevertheless, it still has a high computational cost and demands a significant amount of parameters. Such requirements hinder the use in resource-limited environments and demand both software and hardware optimization. Another limitation is that deep models are usually specialized into a single domain or task, requiring them to learn and store new parameters for each new one. Multi-Domain Learning (MDL) attempts to solve this problem by learning a single model that is capable of performing well in multiple domains. Nevertheless, the models are usually larger than the baseline for a single domain. This work tackles both of these problems: our objective is to prune models capable of handling multiple domains according to a user defined budget, making them more computationally affordable while keeping a similar classification performance. We achieve this by encouraging all domains to use a similar subset of filters from the baseline model, up to the amount defined by the user's budget. Then, filters that are not used by any domain are pruned from the network. The proposed approach innovates by better adapting to resource-limited devices while, to our knowledge, being the only work that is capable of handling multiple domains at test time with fewer parameters and lower computational complexity than the baseline model for a single domain.

Keywords: Pruning · Multi-Domain Learning · Parameter Sharing · User-Defined Budget · Neural Networks

This work was supported by the FAPESP-Microsoft Research Institute (2017/25908-6), by the Brazilian National Council for Scientific and Technological Development - CNPq (310330/2020-3, 314868/2020-8), by LNCC via resources of the SDumont supercomputer of the IDeepS project, by the MUR PNRR project FAIR (PE00000013) funded by the NextGenerationEU and by EU H2020 project AI4Media (No. 951911).

G. L. Foresti et al. (Eds.): ICIAP 2023, LNCS 14234, pp. 477–489, 2023.
https://doi.org/10.1007/978-3-031-43153-1_40

1   Introduction

Deep learning has brought astonishing advances to computer vision, being used in several application domains, such as medical imaging [19], autonomous driving [16], road surveillance [10], and many others. However, to increase the performance of such methods, increasingly deeper architectures have been used [5], leading to models with a high computational cost. Also, for each new domain (or task to be addressed), a new model is usually needed [1]. The significant amount of model parameters to be stored and the high GPU processing power required for using such models can prevent their deployment in computationally limited devices, like mobile phones and embedded devices [3,13,14] Therefore, specialized optimizations at both software and hardware levels are imperative for developing efficient and effective deep learning-based solutions [9].

For these reasons, there has been a growing interest in the Multi-Domain Learning (MDL) problem. The basis of this approach is the observation that, although the domains can be very different, it is still possible that they share a significant amount of low and mid-level visual patterns [11]. Therefore, to tackle this problem, a common goal is to learn a single compact model that performs well in several domains while sharing the majority of the parameters among them with only a few domain-specific ones. This reduces the cost of having to store and learn a whole new model for each new domain.

Berriel et al. [1] point out that one limitation of those methods is that, when handling multiple domains, their number of parameters is at best equal to the backbone model for a single domain. Therefore, they are not capable of adapting their amount of parameters to custom hardware constraints or user-defined budgets. To address this issue, they proposed the modules named Budget-Aware Adapters (BA^2) that were designed to be added to a pre-trained model to allow them to handle new domains and to limit the network complexity according to a user-defined budget. They act as switches, selecting the convolutional channels that will be used in each domain.

However, as mentioned in [1], although the use of this method reduces the number of parameters required for each domain, the entire model is still required at test time if it aims to handle all the domains. The main reason is that they share few parameters among the domains, which forces loading all potentially needed parameters for all the domains of interest.

This work builds upon the BA^2 [1] by encouraging multiple domains to share convolutional filters, enabling us to prune weights not used by any of the domains at test time. Therefore, it is possible to create a single model with lower computational complexity and fewer parameters than the baseline model for a single domain. Such a model is capable of better fitting the budget of users with limited access to computational resources.

Figure 1 shows an overview of the problem addressed by our method, comparing it to previous MDL solutions and emphasizing their limitations. As it can be seen, standard adapters use the entire model, while BA^2 [1] reduces the number of parameters used in each domain, but requiring a different set of parameters per domain. Therefore, the entire model is needed for handling all the domains

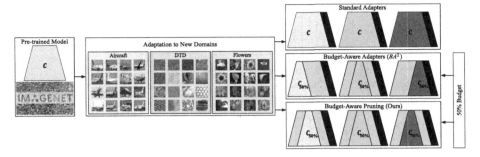

Fig. 1. In standard adapters, the amount of parameters from the domain-specific models (indicated in colored $\mathcal{C}$) is equal to or greater than the backbone model (due to the mask represented in black). Budget-Aware Adapters can reduce the number of parameters required for each domain (unused parameters are denoted in gray). However, the whole model is needed at test time if handling distinct domains (colored areas share few parameters). Our model encourages different domains to use the same parameters (colored areas share most of the parameters). Thus, when handling multi-domains at test time, the unused parameters can be pruned without affecting the domains.

together and nothing can be effectively pruned. On the other hand, our approach increases the probability of using a similar set of parameters for all the domains. In this way, the parameters that are not used for any of the domains can be pruned at test time. These compact models have a lower number of parameters and computational complexity than the original backbone model, which facilitates their use in resource-limited environments. To enable the generation of the compact models, we propose a novel loss function that encourages the sharing of convolutional features among distinct domains. Our proposed approach was evaluated on two well-known benchmarks, the Visual Decathlon Challenge [11], comprised of 10 different image domains, and the ImageNet-to-Sketch setting, with 6 diverse image domains. Results show that our proposed loss function is essential to encourage parameter sharing among domains, since without direct encouragement, the sharing of parameters tends to be low. In addition, results also show that our approach is comparable to the state-of-the-art methods in terms of classification accuracy, with the advantage of having considerably lower computational complexity and number of parameters than the backbone.

2 Related Work

Previous approaches to adapt an existing model to a new domain used strategies like finetuning and pre-training, but faced the problem of catastrophic forgetting, in which the new domain is learned, but the old one is forgotten [4]. More recent MDL approaches usually leverage a pre-trained model as backbone. The backbone parameters are usually frozen and shared among all domains, while attempting to learn a limited and much lower amount of new domain-specific

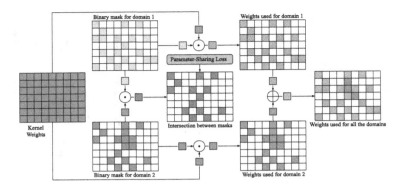

Fig. 2. Overview of our strategy for sharing parameters among domains. Colors represent data (i.e., weights, masks, etc.), therefore, the colored squares denote the input data for each operation as well as its resulting output.

parameters [1]. Approaches mostly differ from each other according to the manner the domain-specific parameters are designed, for example, domain-specific residual blocks and binary masks [1].

For methods that use residual blocks to learn new domains, an example is the work of Rebuffi et al. [11,12] that adds domain-specific parameters to the ResNet network in the form of serial or parallel residual adapter modules.

Following a different path, some works make use of binary masks to select different convolutional filters of the network for each domain, like the Piggyback method proposed by Mallya et al. [6]. In test time, the learned binary mask is multiplied by the weights of the convolutional layer. Expanding on this idea, Mancini et al. [7,8] also makes use of masks, however, they learn an affine transformation of the weights through the use of the mask and some extra parameters. Focusing on increasing the accuracy with masks, Chattopadhyay et al. [2] proposes a soft-overlap loss to encourage the masks to be domain-specific by minimizing the overlap between them.

The works mentioned so far mainly focused on improving accuracy while attempting to add a small number of new parameters to the model, but they do not take into consideration the computational cost and memory consumption, making their utilization on resource-limited devices difficult [17]. Trying to address that, recent works have attempted to tackle the multi-domain learning problem while taking into account resource constraints.

Regarding parameters sharing, Wallingford et al. [15] proposed the Task Adaptive Parameter Sharing (TAPS), which learns to share layers of the network for multiple tasks by adding perturbations to the weights of the layer that are not shared. They also have a sparsity hyperparameter defined by the user. Although this method lessen the amount of additional domain-specific parameters, it still always have considerably more parameters then the backbone model for a single domain.

Berriel et al. [1] proposed Budget-Aware Adapters (BA^2), which are added to a backbone model, enabling it to learn new domains while limiting the computational complexity according to the user budget. The BA^2 modules are similar to the approach from Mallya et al. [6], that is, masks are applied to the convolutional layers of the network, selecting a subset of filters to be used in each domain. The network is encouraged to use a smaller amount of filters per convolution layer than a user-defined budget, being implemented as a constraint to the loss function that is optimized by constructing a generalized Lagrange function. Also, the parameters from batch normalization layers are domain-specific, since they perform poorly when shared. This method and other continual learning strategies can reduce the number of parameters for a single domain. However, these methods usually load the relevant parameters for the desired domain at test time. In order to load them for each domain of interest, it would be necessary to keep all the parameters stored in the device so that the desired ones are available. This way, the model does not fit the user needs, consuming more memory and taking more time to load, which might make it difficult to use in environments with limited computational resources. With this motivation we propose our method that encourages the sharing of parameters and is able to effectively prune the model, reducing both the computational complexity and amount of parameters while handling all the domains.

3   Pruning a Multi-domain Model

This work was built upon the BA^2 modules from Berriel et al. [1] and proposes a new version to allow pruning unused weights at test time. As results, the proposed method is able to obtain a pruned model that handles multiple domains, while having lower computational complexity and number of parameters than even the backbone model for a single domain. The pruned version is able to keep a similar classification performance while considering optimizations that are paramount for devices with limited resources. Our user-defined budget allows the model to fit the available resources to a wider range of environments. To achieve our goals, we added an extra loss function to BA^2 in order to encourage parameter sharing among domains and prune the weights that are not used by any domain. It was also necessary to train simultaneously in all the domains to be able to handle them all together at test time (see Fig. 2 for an overview).

3.1   Problem Formulation

The main goal of MDL is to learn a single model that can be used in different domains. One approach is to have a fixed pre-trained backbone model with frozen weights that are shared among all domains, while learning only a few new domain-specific parameters. Equation 1 describes this approach, where Ψ_0 is the pre-trained backbone model that when given input data x_0 from the domain X_0 return a class from domain Y_0 considering θ_0 as the models weights. Our goal is to have a model Ψ_d for each domain d that attributes classes from the domain Y_d

to inputs x_d from the domain X_d while keeping the θ_0 weights from the backbone model and learning as few domain-specific parameters θ_d as possible.

$$\Psi_0(x_0; \theta_0) : X_0 \to Y_0 \tag{1}$$
$$\Psi_d(x_d; \theta_0, \theta_d) : X_d \to Y_d$$

Our starting point was the BA² [1] modules, which are associated with the convolutions layers of the network, enabling them to reduce their complexity according to a user-defined budget. Equation 2 describes one channel of the output feature map m at the location (i, j) of a convolutional layer, where g is the activation function, $K \in \mathbb{R}^{(2K_H+1)\times(2K_W+1)\times C}$ is the kernel weights with height of $2K_H + 1$, width of $2K_W + 1$ and C input channels, and $I \in \mathbb{R}^{H\times W\times C}$ is the input feature map with H height, W width and C channels.

$$m(i,j) = g(\sum_{c=1}^{C} \phi_c(i,j)) \tag{2}$$

$$\phi_c(i,j) = \sum_{h=-K_h}^{K_h} \sum_{w=-K_w}^{K_w} K(h,w,c)I(i-h, j-w, c)$$

Berriel et al. [1] proposed to add a domain-specific mask that is composed of C switches s_c for each input channel, as shown in Eq. 3. At training time, $s_c \in \mathbb{R}$ while, at test time, they are thresholded to be binary values. When $s_c = 0$, the weights K_c (i.e., the filters for the c input channel for a given output channel) can be removed from the computational graph, effectively reducing the computational complexity of the convolutional layers.

$$m(i,j) = g(\sum_{c=1}^{C} s_c\phi_c(i,j)) \tag{3}$$

The model is trained by minimizing the total loss L_{total}, which is composed of the cross entropy loss L and a budget loss L_B, as shown in Eq. 4, where $\beta \in [0, 1]$ is a user-defined budget hyperparameter that limits the amount of weights on each domain individually, θ_d^β are the domain-specific parameters for the budget β and domain d, $\bar{\theta}_d^\beta$ is the mean value of the switches for all convolutional layers and λ is the Karush-Kuhn-Tucker (KKT) multiplier.

$$L_{total} = L(\theta_0, \theta_d^\beta) + L_B(\theta_d^\beta, \beta) \tag{4}$$

The budget loss is given by $L_B(\theta_d^\beta, \beta) = \max(0, \lambda(\bar{\theta}_d^\beta - \beta))$. When the constraint $\bar{\theta}_d^\beta - \beta$ is respected, $\lambda = 0$, otherwise, the optimizer increases the value of λ to boost the impact of the budget.

3.2 Sharing Parameters and Pruning Unused Ones

Although BA² can reduce the computational complexity of the model, it can not reduce the number of parameters necessary to handle all the domains together.

Switches s_c can only be pruned at test time when they are zero for *all* domains, but they, in fact, assume different values if not forced to do so.

For this reason, we added an additional parameter-sharing loss L_{PS} to L_{total}, as described in Eq. 5, where N is the number of domains, θ_k^β for $k \in [1, ..., N]$ are the domain-specific parameters (switches) for each domain, M is the total number of switches and λ_{PS} is a hyperparameter that defines the importance of this loss component. Intersections are implemented as a Hadamard product and thresholds are replaced by identity functions on backpropagation.

$$L_{total} = L(\theta_0, \theta_d^\beta) + L_B(\theta_d^\beta, \beta) + L_{PS}(\theta_1^\beta, ..., \theta_N^\beta, \beta) \quad (5)$$

$$L_{PS}(\theta_1^\beta, ..., \theta_N^\beta, \beta) = \max(0, \lambda_{PS}(1 - \frac{|\theta_1^\beta \cap \theta_2^\beta \cap ... \cap \theta_N^\beta|)}{M\beta})$$

The parameter-sharing loss calculates the intersection of all the domains masks and encourages it to grow up to the budget limitation. Since the domain-specific weights from all the domains are required by this loss component, it is necessary to train on all of them simultaneously. Finally, the switches s_c and the associated kernel weights K_c can be pruned.

4 Experiments and Results

Our approach was validated on two well-known MDL benchmarks, the Visual Decathlon Challenge [11], and the ImageNet-to-Sketch.

The Visual Decathlon Challenge comprises classification tasks on ten diverse well-known image datasets from different visual domains: ImageNet, Aircraft, CIFAR-100, Daimler Pedestrian (DPed), Describable Textures (DTD), German Traffic Signs (GTSR), VGG-Flowers, Omniglot, SVHN, and UCF-101. Such visual domains are very different from each other, ranging from people, objects, and plants to textural images. The ImageNet-to-Sketch setting has been used in several prior works, being the union of six datasets: ImageNet, VGG-Flowers, Stanford Cars, Caltech-UCSD Birds (CUBS), Sketches, and WikiArt [6]. These domains are also very heterogeneous, having a wide range of different categories, from birds to cars, or art paintings to sketches [1].

In order to evaluate the classification performance, we use the accuracy on each domain, and the S-score [11] metric. Proposed by Rebuffi et al. [11], the S-score metric rewards methods that have good performance over all the domains compared to a baseline, and it is given by Eq. 6:

$$S = \sum_{d=1}^{N} \alpha_d \max\{0, Err_d^{max} - Err_d\}^{\gamma_d} \quad (6)$$

where Err_d is the classification error obtained on the dataset d, Err_d^{max} is the maximum allowed error from which points are no longer added to the score and γ_d is a coefficient to ensure that the maximum possible S score is 10.000 [11].

To assess the computational cost of a model, we considered its amount of parameters and computational complexity. For the number of parameters, we measured their memory usage, excluding the classifier and encoding float numbers in 32 bits and the mask switches in 1 bit. For the computational complexity, we used the THOP library to calculate the amount of multiply-accumulate operations (MACs[1]) for our approach, while we reported the values from [1] for their work. All reported values are relative to the backbone size, as in [1]. Similar to [1], in order to assess the trade-off between effectiveness on the MDL problem and computational efficiency, we consider two variations of the S score, named as S_O, which is the S score per operation; and S_P, the S score per parameter.

We adopted the same experimental protocol of Berriel et al. [1], making the necessary adjustments for our objective of pruning the model.

We used the SGD optimizer with momentum of 0.9 for the classifier and the Adam optimizer for the masks. All weights from the backbone are kept frozen, only training the domain-specific parameters (i.e., classifiers, masks, and batch normalization layers) and the masks switches were initialized with the value of 10^{-3}. Data augmentation with random crop and horizontal mirroring with a probability of 50% was used in the training phase, except for DTD, GTSR, Omniglot, and SVHN, where mirroring did not improve results or was harmful. For testing, we used 1 crop for datasets with images already cropped (Stanford Cars and CUBS), five crops (center and 4 corners) for the datasets without mirroring and 10 crops for the ones with mirroring (5 crops and their mirrors). For the Visual Domain Decathlon, we used the Wide ResNet-28 [18] as backbone, training it for 60 epochs with batch size of 32, and learning rate of 10^{-3} for the classifier and 10^{-4} for the masks. Both learning rates are decreased by a factor of 10 on epoch 45. For the ImageNet-to-Sketch setting, the ResNet-50 was used as backbone, training for a total of 45 epochs with batch size of 12, learning rate of 5×10^{-4} for the classifier and 5×10^{-5} for the masks, dividing the learning rates by 10 on epochs 15 and 30.

Differently from Berriel et al. [1], we needed to train all the domains simultaneously, since we want to encourage the sharing of weights among them. In order to do so, we run one epoch of each dataset in a round robin fashion. We repeat this process until the desired number of epochs are reached for each dataset.

As ablation studies, we tested running BA[2] simultaneously on all tasks without the addition of our loss function, where we observed that there is a small drop in accuracy for doing so. This procedure is necessary since one must have information from all the domains at once to learn how to share parameters. Different strategies for simultaneous learning were tested, for example, one batch of each domain, batches with data from multiple domains, among others. However, the effects on the results were small, so we chose the faster strategy, performing one epoch of each domain in a round-robin fashion with a random order. We also performed a grid search on the validation set in order to select the best value for λ_{PS}, testing the values of 0.125, 0.25, 0.5, 0.75 and 1.0. For the Visual Domain Decathlon, the best λ_{PS} was 1.0, while for the ImageNet-to-Sketch it was $\lambda_{PS} = 0.125$.

---

[1] We follow Berriel et al. [1] and report results in FLOPs (1 MAC = 2 FLOPs).

After obtaining the best hyperparameter configuration, we compared our work to the baseline strategies of using the pre-trained model as a feature extractor, only training the classifier (named feature), and finetuning one model for each domain (finetune). We also compared to the state-of-the-art method BA^2, since it is one of the only works that take into consideration computational cost constraints. The main focus of our work is the scenario where there is a budget set by the user, and other works except BA^2 do not take into consideration this restriction. Despite the lack of attention that tackling multi-domain learning with budget restrictions has received, it is a promising topic that is paramount for the application of these methods in environments with limit computational power. Experiments were run using V100 and GTX 1080 TI NVIDIA GPUs, Ubuntu 20.04 distribution, CUDA 11.6, and PyTorch 1.12. After obtaining the best hyperparameter configuration, the model was trained on both training and validation sets and evaluated on the test set of the Visual Domain Decathlon. The comparison of the results with baseline strategies and a state-of-the-art method, BA^2, is shown in Table 1.

Table 1. Computational complexity, number of parameters, accuracy per domain, S, S_O and S_P scores on the Visual Domain Decathlon, best in bold, second best underlined

Method	FLOP	Params	ImNet	Airc	C100	DPed	DTD	GTSR	Flwr	Oglt	SVHN	UCF	S-score	S_O	S_P
Baselines [11]:															
Feature	1.000	1.00	59.7	23.3	63.1	80.3	45.4	68.2	73.7	58.8	43.5	26.8	544	544	544
Finetune	1.000	10.0	59.9	60.3	82.1	92.8	55.5	97.5	81.4	87.7	96.6	51.2	2500	2500	250
BA^2 [1]:															
$\beta = 1.00$	0.646	1.03	56.9	49.9	78.1	95.5	55.1	99.4	86.1	88.7	96.9	50.2	3199	4952	3106
$\beta = 0.75$	0.612	1.03	56.9	47.0	78.4	95.3	55.0	99.2	85.6	88.8	96.8	48.7	3063	5005	2974
$\beta = 0.50$	0.543	1.03	56.9	45.7	76.6	95.0	55.2	99.4	83.3	88.9	96.9	46.8	2999	5523	2912
$\beta = 0.25$	0.325	1.03	56.9	42.2	71.0	93.4	52.4	99.1	82.0	88.5	96.9	43.9	2538	7809	2464
Ours:															
$\beta = 1.00$	0.837	1.03	56.9	37.3	80.2	95.1	57.9	98.6	84.6	83.8	96.0	45.8	2512	3001	2438
$\beta = 0.75$	0.645	0.921	56.9	42.6	75.3	95.0	56.1	98.6	82.8	87.2	96.0	44.7	2444	3789	2654
$\beta = 0.50$	0.447	0.783	56.9	42.1	73.7	96.8	51.3	98.7	81.4	87.1	96.1	45.4	2552	5709	3259
$\beta = 0.25$	0.238	0.531	56.9	33.6	67.9	95.3	44.9	98.2	75.1	87.4	96.1	43.0	1942	8159	3657

Compared to the baseline strategies, our method was able to vastly outperform the feature extractor only, while achieving similar S-score to finetune for the budgets of β =1.0, 0.75 and 0.50, but with almost 10 times less parameters.

Compared to BA^2, we obtained similar accuracy in most domains, but faced some drops in accuracy in some domains compared to [1]. We believe the main reason for this drop in accuracy is the simultaneous training procedure, as we observed similar drop when switching from individual to simultaneous training without the addition of our loss function, but we kept it since it is necessary to enable parameter sharing. The domains with the biggest accuracy drops were the smaller datasets, like aircraft, DTD, VGG-Flowers, and UCF-101. Other works, like Rebuffi et al. [11,12] also mention subpar performance on these datasets, identifying the problem of overfitting.

The S-score also dropped up to 687 points for the same issues. The drop is harsher since the metric was designed to reward good performance across all datasets, and the small datasets we mentioned had a subpar performance. Despite facing small drops in accuracy and S-score, our method offers a good trade-off between classification performance and computational cost.

When comparing computational complexity (FLOP on Table 1), for the budgets of $\beta = 1$ and 0.75, the original BA^2 had lower values, but for the harsher budgets of $\beta = 0.5$ and 0.25, our methods obtained the lower complexity. This happens due to the fact that the original BA^2 tends to discard more weights than the demanded when the budget is higher, while our methods tend to stay closer to the amount defined by the budget. It also must be noted that all our methods obtained lower complexity than the value defined by the budget, showing that it is a great tool to adapt a backbone model to the resources available to the user.

By comparing the S_O metric, we can observe that both methods have a good trade-off between computational complexity and S-score, as this metric greatly increases as the budget is reduced, showing that the reduction in computational complexity is considerably greater than the loss in S-score. As expected, our method had better S_O for the harsher budgets of $\beta = 0.50$ and 0.25 while BA^2 achieved superior results on the budgets of $\beta = 1.00$ and 0.75.

The main advantage of our proposed method is the reduction on the number of parameters of the model, as it is, to our knowledge, one of the only methods that is capable of tackling the problem of multiple-domain learning, while also reducing the number of parameters in relation to the backbone model. Other methods can reduce the amount of parameters for a single domain, but since the parameters are not shared, to handle all of them during test time, the entire model must be kept. As we can see (column Params of Table 1), the original BA^2 had similar amount of parameters to the backbone model, being 3% more for all budgets. For the budget of $\beta = 1.00$, we obtained the same result, while for the budget of $\beta = 0.75$ we reduce the amount of parameters compared to the backbone model in 7.9%, for budget $\beta = 0.50$, the reduction was of 22.7% and for the for budget of $\beta = 0.25$ there were 49.9% less parameters. This results shows that our method was successfully in encouraging the sharing of parameters among domains and that this approach can lead to considerable reductions on the amount of parameters of the network. The S_P metric also show this results, as for the budgets of $\beta = 0.50$ and 0.25 our method was able to outperform BA^2 by considerably reducing the amount of parameter.

Table 2 shows the results obtained on the test set of the ImageNet-to-Sketch setting. Compared to the baseline strategies, our method once again outperformed the use of the feature extractor only. Both our method and BA^2 obtained lower S-score than the finetune, showing that this benchmark is challenging.

Comparing to the original BA^2, our model faced some drops on accuracy and S-score. Looking at the domain individually, we can see that the smaller datasets (Cars and Flwr.) were the ones with the greater drops in accuracy, a problem that also occurred on the Visual Domain Decathlon due to overfitting [11,12].

Table 2. Computational complexity, number of parameters, accuracy per domain, S, S_O and S_P scores for the ImageNet-to-Sketch benchmark.

Method	FLOP	Params	ImNet	CUBS	Cars	Flwr	WikiArt	Sketches	S-score	S_O	S_P
Baselines: [6]:											
Feature	1.000	1.00	**76.2**	70.7	52.8	86.0	55.6	50.9	533	533	533
Finetune	1.000	6.00	**76.2**	**82.8**	<u>91.8</u>	**96.6**	<u>75.6</u>	**80.8**	**1500**	1500	250
BA^2 [1]:											
$\beta = 1.00$	0.700	1.03	**76.2**	<u>81.2</u>	**92.1**	<u>95.7</u>	72.3	79.3	<u>1265</u>	1807	1228
$\beta = 0.75$	0.600	1.03	**76.2**	79.4	90.6	94.4	70.9	<u>79.4</u>	1006	1677	977
$\beta = 0.50$	0.559	1.03	**76.2**	79.3	90.8	94.9	70.6	78.3	1012	1810	983
$\beta = 0.25$	<u>0.375</u>	1.03	**76.2**	78.0	88.2	93.2	68.0	77.9	755	2013	733
Ours:											
$\beta = 1.00$	0.777	1.09	**76.2**	79.1	82.2	92.4	70.4	77.2	726	934	666
$\beta = 0.75$	0.601	0.92	**76.2**	80.2	86.0	92.5	73.5	78.2	844	1404	917
$\beta = 0.50$	0.412	<u>0.71</u>	**76.2**	80.0	87.4	89.9	**75.8**	77.8	909	<u>2206</u>	<u>1280</u>
$\beta = 0.25$	**0.222**	**0.49**	**76.2**	75.5	83.9	88.6	72.5	77.3	689	**3103**	**1406**

In relation to the computational complexity, our models were better than BA^2 for the budgets of $\beta = 0.25$ and $\beta = 0.5$ and slightly worse for $\beta = 0.75$ and $\beta = 1.0$. This is also reflected on the S_O score, as we got better results for the same budgets. Once again, our models obtained lower computational complexity than what was defined by the budget, showing that they fit the user needs.

The main advantage of our method is that it is capable of having a lower number of parameters than the backbone, even when handling multiple domains, something that BA^2 and most works in literature are not capable. For the budget of $\beta = 0.5$ and $\beta = 0.25$, we obtained a considerable lower amount of parameters than the backbone model, reducing in 21.7% and 46.9%, respectively. This is reflected on the S_P metric, where we were able to outperform BA^2 by a considerable margin in these budgets, showing that our model is more efficient.

5   Conclusions

In this paper, we addressed the multi-domain learning problem while taking into account a user-defined budget for computational resources, a scenario addressed by few works, but of vital importance for devices with limited computational power. We propose to prune a single model for multiple domains, making it more compact and efficient. To do so, we encourage the sharing of parameters among domains, allowing us to prune the weights that are not used in any of them, reducing both the computational complexity and the number of parameters to values lower than the original baseline for a single domain. Performance-wise, our results were competitive with other state-of-the-art methods while offering good trade-offs between classification performance and computational cost according to the user's needs. In future work, we intend to evaluate different strategies for encouraging parameter sharing, and test our method on different network models and benchmarks.

References

1. Berriel, R., Lathuillere, S., Nabi, M., Klein, T., Oliveira-Santos, T., Sebe, N., Ricci, E.: Budget-aware adapters for multi-domain learning. In: International Conference on Computer Vision (ICCV), pp. 382–391 (2019)
2. Chattopadhyay, P., Balaji, Y., Hoffman, J.: Learning to balance specificity and invariance for in and out of domain generalization. In: Computer Vision–ECCV 2020: 16th European Conference (ECCV), pp. 301–318 (2020)
3. Du, Y., Chen, Z., Jia, C., Li, X., Jiang, Y.G.: Bag of tricks for building an accurate and slim object detector for embedded applications. In: Proceedings of the 2021 International Conference on Multimedia Retrieval (ICMR), pp. 519–525 (2021)
4. Hung, S.C., Lee, J.H., Wan, T.S., Chen, C.H., Chan, Y.M., Chen, C.S.: Increasingly packing multiple facial-informatics modules in a unified deep-learning model via lifelong learning. In: Proceedings of the 2019 on International Conference on Multimedia Retrieval (ICMR), pp. 339–343 (2019)
5. Liu, L., et al.: Group fisher pruning for practical network compression. In: International Conference on Machine Learning (ICML), pp. 7021–7032 (2021)
6. Mallya, A., Davis, D., Lazebnik, S.: Piggyback: adapting a single network to multiple tasks by learning to mask weights. In: Ferrari, V., Hebert, M., Sminchisescu, C., Weiss, Y. (eds.) ECCV 2018. LNCS, vol. 11208, pp. 72–88. Springer, Cham (2018). https://doi.org/10.1007/978-3-030-01225-0_5
7. Mancini, M., Ricci, E., Caputo, B., Bulò, S.R.: Adding new tasks to a single network with weight transformations using binary masks. In: Leal-Taixé, L., Roth, S. (eds.) ECCV 2018. LNCS, vol. 11130, pp. 180–189. Springer, Cham (2019). https://doi.org/10.1007/978-3-030-11012-3_14
8. Mancini, M., Ricci, E., Caputo, B., Rota Bulò, S.: Boosting binary masks for multi-domain learning through affine transformations. Mach. Vis. Appl. 31(6), 1–14 (2020)
9. Marchisio, A., et al.: Deep learning for edge computing: current trends, cross-layer optimizations, and open research challenges. In: IEEE Computer Society Annual Symposium on VLSI (ISVLSI), pp. 553–559 (2019)
10. Nguyen, K.T., Dinh, D.T., Do, M.N., Tran, M.T.: Anomaly detection in traffic surveillance videos with gan-based future frame prediction. In: Proceedings of the 2020 International Conference on Multimedia Retrieval (ICMR), pp. 457–463 (2020)
11. Rebuffi, S.A., Bilen, H., Vedaldi, A.: Learning multiple visual domains with residual adapters. In: Advances in Neural Information Processing Systems (NeurIPS). vol. 30, pp. 506–516 (2017)
12. Rebuffi, S.A., Bilen, H., Vedaldi, A.: Efficient parametrization of multi-domain deep neural networks. In: Proceedings of the IEEE Conference on Computer Vision and Pattern Recognition (CVPR), pp. 8119–8127 (2018)
13. dos Santos, S.F., Almeida, J.: Less is more: accelerating faster neural networks straight from JPEG. In: Tavares, J.M.R.S., Papa, J.P., González Hidalgo, M. (eds.) CIARP 2021. LNCS, vol. 12702, pp. 237–247. Springer, Cham (2021). https://doi.org/10.1007/978-3-030-93420-0_23

14. dos Santos, S.F., Sebe, N., Almeida, J.: The Good, the Bad, and the Ugly: Neural Networks Straight from JPEG. In: 2020 IEEE International Conference on Image Processing (ICIP), pp. 1896–1900 (2020)
15. Wallingford, M., et al.: Task adaptive parameter sharing for multi-task learning. In: Proceedings of the IEEE/CVF Conference on Computer Vision and Pattern Recognition (CVPR), pp. 7561–7570 (2022)
16. Wang, Y., et al.: Rod 2021 challenge: A summary for radar object detection challenge for autonomous driving applications. In: International Conference on Multimedia Retrieval (ICMR), pp. 553–559 (2021)
17. Yang, L., Rakin, A.S., Fan, D.: Da3: Dynamic additive attention adaption for memory-efficient on-device multi-domain learning. In: 2022 IEEE/CVF Conference on Computer Vision and Pattern Recognition (CVPR), pp. 2619–2627 (2022)
18. Zagoruyko, S., Komodakis, N.: Wide residual networks. In: British Machine Vision Conference (BMVC), pp. 87.1–87.12 (2016)
19. Zhou, N., Wen, H., Wang, Y., Liu, Y., Zhou, L.: Review of deep learning models for spine segmentation. In: Proceedings of the 2022 International Conference on Multimedia Retrieval (ICMR), pp. 498–507 (2022)

Sparse Double Descent in Vision Transformers: Real or Phantom Threat?

Victor Quétu[✉][iD], Marta Milovanović[iD], and Enzo Tartaglione[iD]

LTCI, Télécom Paris, Institut Polytechnique de Paris, Palaiseau, France
{victor.quetu,marta.milovanovic,enzo.tartaglione}@telecom-paris.fr

Abstract. Vision transformers (ViT) have been of broad interest in recent theoretical and empirical works. They are state-of-the-art thanks to their attention-based approach, which boosts the identification of key features and patterns within images thanks to the capability of avoiding inductive bias, resulting in highly accurate image analysis. Meanwhile, neoteric studies have reported a "sparse double descent" phenomenon that can occur in modern deep-learning models, where extremely over-parametrized models can generalize well. This raises practical questions about the optimal size of the model and the quest over finding the best trade-off between sparsity and performance is launched: are Vision Transformers also prone to sparse double descent? Can we find a way to avoid such a phenomenon?

Our work tackles the occurrence of sparse double descent on ViTs. Despite some works that have shown that traditional architectures, like Resnet, are condemned to the sparse double descent phenomenon, for ViTs we observe that an optimally-tuned ℓ_2 regularization relieves such a phenomenon. However, everything comes at a cost: optimal lambda will sacrifice the potential compression of the ViT.

Keywords: Sparse double descent · transformers · pruning · deep learning

1 Introduction

Deep neural networks (DNNs) have revolutionized the field of computer vision by achieving state-of-the-art results in tasks such as segmentation [3], classification [1], and object detection [16]. DNNs outperform conventional machine learning algorithms on many visual recognition tasks as they can automatically learn feature representations from raw input [6]. In addition, they can process a lot of data and generalize well to novel, unseen examples. For a long time, convolutional neural architectures (CNN) like VGG and ResNet models have been dominant in computer vision, thanks to their ability to learn, simultaneously, feature extraction (typically handled by convolutional layers) and classification (by multi-layer perceptrons, or in some cases even by convolutional layers themselves, like in ALL-CNN [21]).

© The Author(s), under exclusive license to Springer Nature Switzerland AG 2023
G. L. Foresti et al. (Eds.): ICIAP 2023, LNCS 14234, pp. 490–502, 2023.
https://doi.org/10.1007/978-3-031-43153-1_41

A new, deep architecture called Transformer has been pioneered by [24], and it has, at first, been conceived for natural language processing tasks [2], resulting in a break-through for the community. Given its big potential, the computer vision world has recently begun to adopt it [13]. Vision Transformers (ViT), which are Transformer architectures adapted for computer vision, quickly became state-of-the-art for many tasks, out of which we cite generative models [8]. However, the lack of strong inductive biases causes them to be even more data-hungry than traditional CNN architectures, and this poses severe performance drops when noisy data are available to train such a model.

In image classification tasks, noisy labels are a frequent issue that can negatively impact the performance of deep learning models [22]. Incorrect labels in the training data can mislead the model during the learning process and result in sub-optimal performance. Various approaches have been proposed to address this issue, including label smoothing, data augmentation, and robust loss functions [14]. The expected behavior, in such cases, is that the higher the noise in the data, the higher the overfit the model will suffer. As opposed to the traditional bias-variance trade-off, a phenomenon has been recently discovered, called Double Descent (DD) [17]. Namely, enlarging the model size in the overfitting regime worsens the performance of an over-parametrized network; then, the trend reverses. DD represents an important challenge in finding the optimal set of parameters since it shows that it is possible to potentially improve generalization in an over-parametrized regime, but without real indicators on the best model's size to adopt. This behavior is observed in various architectures, stretching from machine learning models to DNNs, such as standard CNNs and ResNet [26]. Analogously, a sparse double descent (SDD) phenomenon is observed when moving the model from an over-parametrized towards a sparser regime [11], via parameter pruning.

In this paper, we show that ViTs also suffer from the SDD phenomenon: besides the burden of the lack of an inductive bias that could help these models to generalize, the occurrence of SDD makes the performance even worse in intermediate compression regimes, when only a part of the parameters is removed. This is a possible explanation for the fact that typical ViT architectures can not be pruned to similar extreme rates as traditional CNNs [27]. However, contrarily to what is suggested for traditional CNNs [19], ViTs can avoid SDD with the optimal tuning of ℓ_2 regularization, a result which was suggested by the theory [18]. We postulate this is possible thanks to the lack of an inductive bias embedded in the architecture, which favors the strong regularization necessary to avoid SDD. Such a discovery enables back all the traditional compression mechanisms, having as a stop criterion a worsening in performance on the validation set. Everything, however, comes at a cost: we observe that optimally regularized models are significantly less compressible, due to the strong prior we impose to avoid SDD. To summarize, here below you can find our key messages and contributions.

- To the best of our knowledge, this is the first paper raising concerns on the potential occurrence of the sparse double descent phenomenon in ViT models. Through this work, we compare the behavior of ViT and ResNet in the typically employed test scenarios [17], also including a test on real annotated data (CIFAR-100N), observing SDD also on ViT.

- We propose a quantitative study over the ℓ_2 regularization parameter, supported by the theory [18] but already proven as inapplicable to traditional CNNs [20]. We observe that, in ViT models, avoiding SDD is possible with a properly tuned value for the regularization, which nicely imposes a strong prior on the model's parameters, impossible in traditional architectures suffering from inductive bias.
- We interestingly observe a trade-off between avoidance of SDD and compressibility of the model. More specifically, to avoid SDD (to employ typical pruning schemes with a stop criterion once the performance on a validation set worsens below some given threshold) we want to have a strong ℓ_2 regularization, which however makes the model less compressible as a higher number of parameters will have a similar relevance. Depending on what we are targeting (high performance or high compressibility) we might want or not want to avoid SDD.

2   Background on Vision Transformers

Vision Transformers [7], presented in Fig. 1a, use self-attention mechanisms to capture the relationships between the elements of an input image. They essentially consist of four key elements: patch embedding, positional encoding, the transformer encoder, shown in Fig. 1b, and the classification head.

Patch Embedding. The input image is divided into a grid of patches (which can or can not be non-overlapping), each containing a fixed number of pixels. These patches are then linearly projected to a lower-dimensional embedding space. This process converts the spatial information of the image into a sequence of patch embeddings, mimicking the process of text embedding.

Positional Encoding. To preserve the positional information of the image patches, positional encoding is added to the patch embeddings: this allows to distinguish different patches and capture their global positions in the image.

The Transformer Encoder. The patch embeddings, along with their positional encodings, are fed into a stack of transformer encoder layers. Each encoder layer consists of two sub-layers: a multi-head self-attention mechanism, and a multi-layer perception (MLP). The self-attention mechanism enables the model to capture global interactions between patches by attending to all patches and aggregating information accordingly, thus performing feature aggregation. More specifically, self-attention operations determine the attention output a based on the relevance of one item to others. This is iteratively refined and computed using keys k and queries q, which have the same dimension d, and values v [24].

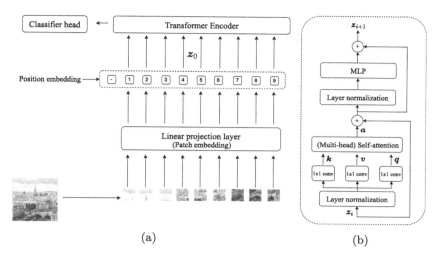

Fig. 1. Vision Transformer (ViT) (a) and Transformer Encoder (b).

The keys are the indices of the hidden states of the encoded input items, and each key k_i has some associated value v_i. Each query q_i represents an output coming from the encoded target item (class). The attention is computed as the softmax of the product between q and k, and then multiplied by v: the model learns to prioritize important input features and capture more informative representations of the input data. Many attention heads are hence concatenated, to form the multi-head, to obtain contextualized representations that include both local and global information. Finally, an MLP carries out feature transformation.

The Classification Head. At the end of the transformer encoder stack, a classification head is attached to the output of the final transformer layer. The classification head can take various forms, such as a simple fully connected layer or a combination of linear and softmax layers. It maps the aggregated representation of the patches to class probabilities, in order to perform image classification.

Training. ViT models require pretraining on enormous datasets (such as JFT-300M, consisting of approximately 300 million images) due to their lack of strong inductive bias, which is present in other architectures like CNNs [5]. Despite recent advances in learning on smaller datasets using distillation approaches or optimizing models with smaller sizes, transformers still have larger model architectures compared to CNN-based models and require large datasets for optimal performance, despite requiring less computational resources [7]. However, [12] concludes that scaling up Transformer models improves performance, but with current designs, it is computationally expensive and necessitates efficient designs.

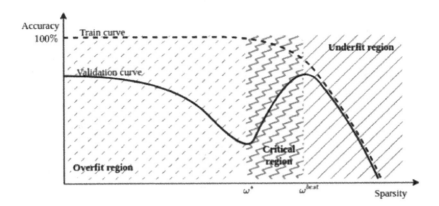

Fig. 2. The Sparse Double Descent phenomenon. (Color figure online)

Beyond Traditional ViT. In the last few years, many different Vision Transformer designs have been proposed to improve the performance of computer vision tasks. One of the most popular is SWIN [13] which proposes shifted windows to create overlapping receptive fields, cascaded stages to mimic a multi-resolution approach, tokenization of windows, and token shifts across the stages. As it is possible to imagine, this architecture already goes in the direction of customizing the Transformer architecture to process images. Other newly proposed transformers variants include, among others, CoAT [4], TNT [28], and DeiT [23].

3   Sparse Double Descent and ViT

In this section, we will discuss the background for Double Descent and Sparse Double Descent, moving then to the potential impact on ViT architectures.

Double Descent. It is known that when comparing a model performance (on unseen data) and model complexity, as the complexity grows (from right to left), we observe a first region where the performance improves (under-fitting - blue region in Fig. 2) and then, at some point, a trend inversion where the performance decreases while increasing the model's complexity (over-fitting). When exposed to real-world noisy data, however, neural networks tend to exhibit the DD phenomenon [17]: instead of being monotonous in that region, the performance inverts, at some point, its trend (critical region - orange in Fig. 2), and starts back decreasing (overfit region - green in Fig. 2).

DD has been observed in regression tasks and successfully averted with optimally-tuned ℓ_2 regularization [18]. However, for classification tasks, this problem is not easily mitigated. It has been shown that the more challenging the dataset and classification task, the harder it is to avoid DD [20]. The authors in [17] demonstrate the DD not only depending on the model width but also depending on the number of epochs during training. Similarly to DD,

---

Algorithm 1. Iterative algorithm to detect Sparse Double Descent.

```
 1: procedure DETECT_SDD (w^INIT, Ξ, λ, ζ^ITER, ζ^END)
 2:     w ← Train(w^init, Ξ^train, λ)
 3:     p_{i-1} ← Performance(w, Ξ^val)
 4:     prev_increasing, prev_decreasing, already_increased, already_decreased ← False
 5:     SDD ← False
 6:     while Sparsity(w, w^init) < ζ^end do
 7:         w ← Prune(w, ζ^iter)
 8:         w ← Train(w, Ξ^train, λ)
 9:         p_i ← Performance(w, Ξ^val)
10:         if (p_i < p_{i-1} and already_decreased and not prev_decreasing) or
11:            (p_i > p_{i-1} and already_increased and not prev_increasing) then
12:             SDD ← True
13:         end if
14:         if p_i ≠ p_{i-1} then
15:             prev_decreasing ← p_i < p_{i-1}; prev_increasing ← p_i > p_{i-1}
16:             already_decreasing ← already_decreasing or p_i < p_{i-1}
17:             already_increasing ← already_increasing or p_i > p_{i-1}
18:             p_i ← p_{i-1}
19:         end if
20:     end while
21:     Return SDD
22: end procedure
```

---

an SDD phenomenon happens in the transition from the complex model toward the sparse, pruned model (as illustrated in Fig. 2) [11]. SDD has implications for model selection, regularization techniques, and understanding the behavior of complex models in high-dimensional settings, as the presence of SDD makes many criteria, like when to stop the pruning, unclear.

Addressing the Sparse Double Descent. We introduce here Alg. 1, designed to demonstrate the eventual occurrence of the sparse double descent phenomenon. The algorithm begins by training the model on the learning task Ξ for the first time, incorporating ℓ_2 regularization weighted by λ (line 2). Following this initial training step, a magnitude pruning stage is set up (line 7). Neural network pruning aims to reduce the size of a large network while maintaining its accuracy by removing irrelevant weights, filters, or other structures. As in [11], we use in this algorithm an unstructured pruning method called magnitude-based pruning, popularized by [10], in which a fixed amount of weights below some specific threshold, are pruned (line 7). Here, every time we prune, a fixed ζ^{iter} fraction of parameters from the model is removed. We highlight that more complex pruning approaches exist, but magnitude-based pruning shows its competitiveness despite very low complexity [9]. The accuracy of the model typically decreases after pruning. To improve the performance of the model, we retrain it using the same original learning policy (line 8). Recent works have shown that this approach leads to the best performance at the highest sparsities [19]. This

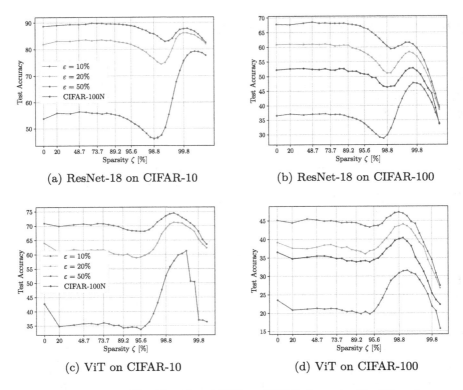

Fig. 3. Test accuracy of ResNet-18 and ViT on CIFAR dataset with different amount of noise ε.

approach allows us to determine whether a sparsely-parameterized model, starting from its initialization, has the potential to successfully learn a given target task. We end our pruning procedure once we reach a sparsity ζ^{end} (line 6).

ViT and Sparse Double Descent. The number of parameters in ViT architectures is proportional to the model depth and quadratic function of the width. There is a tendency to scale these models even further, to increase their performance [5], even though this is becoming very computationally expensive. Looking from that perspective, the understanding of the comportment of the ViT models becomes essential. Having a completely different learning architecture from other models, like CNNs, it is not easy to predict the behavior of ViT when pruning is applied. Our work addresses this issue and performs an extensive study with different levels of label noise and various model sparsity levels. In the next section, we will conduct a quantitative study on ViT, determining whether SDD is a real threat to ViT as it is to CNNs or not.

4   Experiments

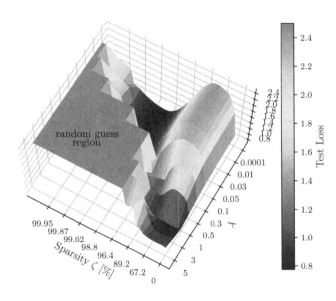

Fig. 4. Ablation study over λ: test loss of ViT on CIFAR-10 with $\varepsilon = 10\%$.

Setup. For the experimental setup, we follow the same approach as He et al. [11]. The first model we train is a ResNet-18, trained on CIFAR-10 & CIFAR-100, for 160 epochs, optimized with SGD, having momentum 0.9, a learning rate of 0.1 decayed by a factor 0.1 at milestones 80 and 120, batch size 128 and λ 10^{-4}. The second model is a ViT with 4 patches, 8 heads, and 512 embedding dimensions, trained on CIFAR-10 and CIFAR-100 for 200 epochs, optimized with Adam, having a learning rate of 10^{-4} with a cosine annealing schedule and λ 0.03. For each dataset, a percentage ε of symmetric, noisy labels are introduced: the labels of a given proportion of training samples are flipped to one of the other class labels, selected with equal probability [15]. In our experiments, we test with $\varepsilon \in \{10\%, 20\%, 50\%\}$. Moreover, as synthetic noise has clean structures which greatly enabled statistical analyses but often fails to model real-world noise patterns, we also conducted experiments without adding synthetic noise. With the same architectures and learning policies presented above, we carried out experiments on CIFAR-100N, which is formed by the CIFAR-100 training dataset equipped with human-annotated real-world noisy labels collected from Amazon Mechanical Turk [25]. In all experiments, we set $\zeta^{\text{iter}} = 20\%$ and $\zeta^{\text{end}} = 99.99\%$.[1]

Occurrence of Sparse Double Descent. Figure 3 displays the results of ResNet-18 and ViT, on CIFAR-10 and CIFAR-100. As in He et al. [11] work,

---

[1] The code is available at https://github.com/VGCQ/SDD_ViT.

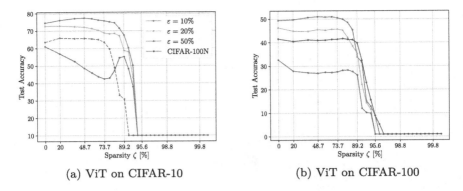

(a) ViT on CIFAR-10   (b) ViT on CIFAR-100

Fig. 5. Test accuracy of ViT on CIFAR dataset with different amount of noise ε with $\lambda = 1$ (solid lines), $\lambda = 3$ (dashed line).

the double descent consists of 4 phases. First, at low sparsities, the network is overparameterized, thus pruned network can still reach similar accuracy to the dense model. The second phase is a phase near the "interpolation threshold", where the test accuracy is about to first decrease and then increase as sparsity grows. The third phase is located at high sparsities, where test accuracy is rising. The final phase happens when both training and test accuracy drop significantly. For every value of ε, whether on CIFAR-10 or CIFAR-100, the sparse double descent phenomenon occurs both for ResNet and ViT. We observe a similar phenomenon as in the simulated ε also in the human-annotated CIFAR-100N.

Study on λ. In the previous experiments in Fig. 3, ViTs were trained with a ℓ_2-regularization hyper-parameter equal to 0.03, which is typically used in other works. However, it has been recently shown that, for certain linear regression models with isotropic data distribution, optimally-tuned ℓ_2 regularization can achieve monotonic test performance as either the sample size or the model size is grown. Nakkiran et al. [18] demonstrated it analytically and established that optimally-tuned ℓ_2 regularization can mitigate double descent for general models, including neural networks like Convolutional Neural Networks. Moreover, a recent study showed that ℓ_2 regularization is positively contributing to the avoidance of sparse double descent in an image classification context, but is not the antidote to "dodge" it [20]. Hence, we propose in Fig. 4 a quantitative study over λ for ViT on CIFAR-10 with $\varepsilon = 10\%$. With small values of λ, i.e. below 1, the sparse double descent is empirically noticeable. The increment of λ pushes the occurrence of the phenomenon towards smaller sparsity values. Looking at the loss, increasing λ smoothens the bump of the test loss and at some point, i.e. $\lambda = 1$, the test loss becomes flat and behaves monotonically: the sparse double descent is avoided. For $\lambda > 1$, the phenomenon also results avoided, but the performance worsens (lighter blue region at the bottom right corner) since the regularization is stronger. Note that with higher λ, performances are better but the maximum sparsity achievable is not as high as for lower values of λ.

Avoidance of the Sparse Double Descent. As $\lambda = 1$ seems to be an optimal value enabling dodging SDD on CIFAR-10 with $\varepsilon = 10\%$, we try to use this value for other setups. Figure 5 displays the results of ViT on CIFAR-10/CIFAR-100 with $\varepsilon \in \{10\%, 20\%, 50\%\}$ and $\lambda = 1$. For small noise rates, i.e. $\varepsilon \leq 20\%$, the phenomenon vanishes and performance is enhanced. However, for higher noise rates, like $\varepsilon = 50\%$, SDD is mitigated, but still present. Even if it already helps, it seems that the strength of the regularization is not high enough to completely avoid SDD. Indeed, with a higher λ, i.e. 3, the performance becomes monotonic.

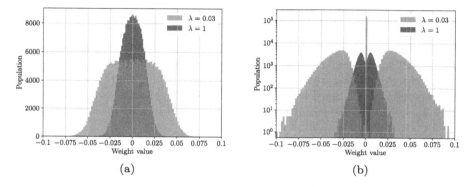

(a)   (b)

Fig. 6. Histogram of the weights of ViT for $\varepsilon = 10\%$ on CIFAR-10, with $\zeta = 0\%$ (a) and $\zeta = 48.8\%$ (b).

Trade-off between SDD and Compressibility. Fig. 4, supported also by the experiments displayed in Fig. 5, suggests that at high regularization regimes, where we avoid SDD, the ability to compress the model is harmed. This is due to the strong prior we impose over the distribution of the parameters of the model: the stronger this is, the least we are indeed able to remove degrees of freedom from our system. As a visual example, Fig. 6 displays the distribution of the parameters for one of the considered training configurations, for $\lambda = 0.03$ and 1, without pruning and after two pruning steps. We observe that despite removing the same quantity of parameters, with higher regularization the parameters have less variance, which has the dual effect of both making them more robust to injected noise (due to the strong regularization) but, at the same time, this distribution is more sensitive to compression by pruning. Hence, we conclude that, in case we wish to have a robust, well-generalizing model, we wish to avoid SDD and employ strong ℓ_2 regularization; on the contrary, if we target compressibility, we would like to favor SDD, as the better generalizing region is pushed to highly compressed regions.

5   Conclusion

This paper investigates the occurrence of Sparse Double Descent in the Vision Transformer architecture. SDD is a phenomenon carefully explored due to its influence on determining the optimal model size necessary for maintaining the performance of over-parametrized models. We observe that, indeed, ViT is also susceptible to SDD. Moreover, we study different values for ℓ_2 regularization and discover that, unlike for other CNN architectures like ResNet, we can find the optimal value and completely avoid SDD. However, the regularization comes at a price - at the same time, it renders the model less compressible, because of the strong enforced prior. We postulate that this is possible due to the lack of strong inductive bias in ViT, which enables strong regularization regimes, impossible for CNNs. Finally, we inspect the trade-off between avoiding SDD (enhancing hence model's performance) and favoring the model compressibility, observing that, for the second one, we would like to favor SDD. This study hopes to inform the community about the risk of SDD ViT models might incur, which depending on the final scope of the trained model can be a real or a phantom threat.

Acknowledgments. This project was provided with computer and storage resources by GENCI at IDRIS thanks to the grant 2022-AD011013930 on the supercomputer Jean Zay's the V100 partition.

References

1. Barbano, C.A., Tartaglione, E., Berzovini, C., Calandri, M., Grangetto, M.: A two-step radiologist-like approach for Covid-19 computer-aided diagnosis from chest X-Ray images. In: Sclaroff, S., Distante, C., Leo, M., Farinella, G.M., Tombari, F. (eds.) Image Analysis and Processing ICIAP 2022. ICIAP 2022. Lecture Notes in Computer Science, vol. 13231. Springer, Cham (2022). https://doi.org/10.1007/978-3-031-06427-2_15

2. Brown, T., et al.: Language models are few-shot learners. Adv. Neural Inf. Process. Syst. **33**, 1877–1901 (2020)

3. Chaudhry, H.A.H. et al.: Lung nodules segmentation with DeepHealth toolkit. In: Mazzeo, P.L., Frontoni, E., Sclaroff, S., Distante, C. (eds.) Image Analysis and Processing. ICIAP 2022 Workshops. ICIAP 2022. Lecture Notes in Computer Science, vol. 13373. Springer, Cham (2022). https://doi.org/10.1007/978-3-031-13321-3_43

4. Dai, Z., Liu, H., Le, Q.V., Tan, M.: CoAtNet: marrying convolution and attention for all data sizes. Adv. Neural Inf. Process. Syst. **34**, 3965–3977 (2021)

5. Dehghani, M., et al.: Scaling vision transformers to 22 billion parameters. arXiv preprint arXiv:2302.05442 (2023)

6. Dosovitskiy, A., et al.: An image is worth 16x16 words: transformers for image recognition at scale. In: International Conference on Learning Representations (2021)

7. Dosovitskiy, A., et al.: An image is worth 16x16 words: Transformers for image recognition at scale. In: 9th International Conference on Learning Representations, ICLR 2021, Virtual Event, Austria, 3–7 May 2021. OpenReview.net (2021)

8. Esser, P., Rombach, R., Ommer, B.: Taming transformers for high-resolution image synthesis. In: Proceedings of the IEEE/CVF Conference on Computer Vision and Pattern Recognition, pp. 12873–12883 (2021)
9. Gale, T., Elsen, E., Hooker, S.: The state of sparsity in deep neural networks. arXiv preprint arXiv:1902.09574 (2019)
10. Han, S., Pool, J., Tran, J., Dally, W.: Learning both weights and connections for efficient neural network. In: Advances in Neural Information Processing Systems. vol. 28 (2015)
11. He, Z., Xie, Z., Zhu, Q., Qin, Z.: Sparse double descent: Where network pruning aggravates overfitting. In: International Conference on Machine Learning, pp. 8635–8659. PMLR (2022)
12. Khan, S., Naseer, M., Hayat, M., Zamir, S.W., Khan, F.S., Shah, M.: Transformers in vision: a survey. ACM Comput. Surv. 54(10s), 1–41 (2022)
13. Liu, Z., et al.: Swin transformer: hierarchical vision transformer using shifted windows. 2021 IEEE/CVF International Conference on Computer Vision (ICCV), pp. 9992–10002 (2021)
14. Ma, X., Huang, H., Wang, Y., Romano, S., Erfani, S., Bailey, J.: Normalized loss functions for deep learning with noisy labels. In: International Conference on Machine Learning, pp. 6543–6553. PMLR (2020)
15. Ma, X., et al.: Dimensionality-driven learning with noisy labels. In: International Conference on Machine Learning, pp. 3355–3364. PMLR (2018)
16. Mazzeo, P.L., Frontoni, E., Sclaroff, S., Distante, C.: Image analysis and processing. ICIAP 2022 Workshops: ICIAP International Workshops, Lecce, Italy, 23–27 May 2022, Revised Selected Papers, Part I. vol. 13373. Springer Nature (2022). https://doi.org/10.1007/978-1-4613-2239-9
17. Nakkiran, P., Kaplun, G., Bansal, Y., Yang, T., Barak, B., Sutskever, I.: Deep double descent: where bigger models and more data hurt. In: International Conference on Learning Representations (2020)
18. Nakkiran, P., Venkat, P., Kakade, S.M., Ma, T.: Optimal regularization can mitigate double descent. In: International Conference on Learning Representations (2021)
19. Quétu, V., Tartaglione, E.: Dodging the sparse double descent. arXiv preprint arXiv:2303.01213 (2023)
20. Quétu, V., Tartaglione, E.: Can we avoid double descent in deep neural networks? (2023)
21. Springenberg, J.T., Dosovitskiy, A., Brox, T., Riedmiller, M.A.: Striving for simplicity: the all convolutional net. In: Bengio, Y., LeCun, Y. (eds.) 3rd International Conference on Learning Representations, ICLR 2015, San Diego, CA, USA, 7–9 May 2015, Workshop Track Proceedings (2015)
22. Sukhbaatar, S., Bruna, J., Paluri, M., Bourdev, L., Fergus, R.: Training convolutional networks with noisy labels. arXiv preprint arXiv:1406.2080 (2014)
23. Touvron, H., Cord, M., Douze, M., Massa, F., Sablayrolles, A., Jégou, H.: Training data-efficient image transformers & distillation through attention. In: International Conference on Machine Learning, pp. 10347–10357. PMLR (2021)
24. Vaswani, A., et al.: Attention is all you need. In: Advances in Neural Information Processing Systems. vol. 30 (2017)
25. Wei, J., Zhu, Z., Cheng, H., Liu, T., Niu, G., Liu, Y.: Learning with noisy labels revisited: a study using real-world human annotations. In: International Conference on Learning Representations (2022)

26. Yilmaz, F.F., Heckel, R.: Regularization-wise double descent: why it occurs and how to eliminate it. In: 2022 IEEE International Symposium on Information Theory (ISIT), pp. 426–431. IEEE (2022)

27. Yu, F., Huang, K., Wang, M., Cheng, Y., Chu, W., Cui, L.: Width & depth pruning for vision transformers. In: Proceedings of the AAAI Conference on Artificial Intelligence. vol. 36, pp. 3143–3151 (2022)

28. Yuan, L., et al.: Tokens-to-token ViT: training vision transformers from scratch on imageNet. In: Proceedings of the IEEE/CVF International Conference on Computer Vision, pp. 558–567 (2021)

Video Sonification to Support Visually Impaired People: The VISaVIS Approach

Marius Onofrei[1], Fabio Castellini[2], Graziano Pravadelli[2], Carlo Drioli[1], and Francesco Setti[2]($\boxtimes$)

[1] Department of Mathematics, Computer Science and Physics, University of Udine, Udine, Italy
[2] Department of Engineering for Innovation Medicine, University of Verona, Verona, Italy
francesco.setti@univr.it

Abstract. In this paper we present a preliminary study about an assistive technology to support blind and visually impaired people (BVIP) in perceiving and navigating indoor environments. In the VISaVIS project we aim at designing the proof-of-concept of a new wearable device to help BVIPs in recognizing the form of the surrounding environment, thus facilitating their movements. In particular, the device is intended to create, at run-time, a sound representation of the environment captured by a head mounted RGBD camera. The underpinning idea is that, through the sonification of the video images captured by the camera, the user will progressively learn to associate the perceived sound to information like the distance, the dimension, and the format of the obstacles he/she is framing. We qualitatively validated our proposal in two challenging and general scenarios, and we grant access to demo videos to prove the effectiveness of our sonification strategy.

Keywords: Video Sonification · Assistive Technologies · Electronic Travel Aids · Blind and Visually Impaired

1 Introduction

According to the World Health Organization, about 285 million people are visually impaired and 39 million of them are blind. In addition, every year this counts increases of 7 million, with this ratio expected to double by 2030. The reduced ability to analyze the environment and, consequently, reduced interaction capabilities and mobility are among the most relevant consequences of vision loss. While guide dogs still represent a valuable and effective solutions to mitigate such consequences for blind and visually impaired people (BVIPs), especially outdoor, the technology advancements have led researchers to propose innovative studies and projects with the goal of developing smart navigation systems for BVIPs. Therefore, starting from the foundational work proposing the Guide-Cane [3], till the ambitious EU funded "Sound of Vision" project [10] several contributions have been proposed, as described in a recent review of the literature [13]. Such a review shows the existence of a vast set of solutions, ranging

© The Author(s), under exclusive license to Springer Nature Switzerland AG 2023
G. L. Foresti et al. (Eds.): ICIAP 2023, LNCS 14234, pp. 503–514, 2023.
https://doi.org/10.1007/978-3-031-43153-1_42

from simple navigation apps running on smartphones, to ultrasonic sensors for obstacle avoidance, till robotics guides. However, it also highlights that several technological and usability issues still exist, which prevent from the realization of an effective smart navigation system for BVIPs.

In this context, with the VISaVIS project we aim at designing the proof-of-concept of a new wearable system to help BVIPs in recognizing the form of the surrounding environment, thus facilitating their movement in both indoor and outdoor scenarios. In particular, the system is intended to create, at run-time, a sound representation of the environment captured by a head mounted RGBD camera, possibly embedded into glasses. The underpinning idea is that, through the sonification of the video images captured by the camera, the user will progressively learn to associate the perceived sound to information like the distance, the dimension, and the format of the obstacles he/she is framing.

This is a position paper that describes the preliminary work in this direction. We present here the first prototype of an assistive technology system that is capable of sensing the environment, building a 3D model of it and substitute the visual channel with an audio signal that can convey all the information to the user. The information about the structure of the surroundings will allow the user to navigate and interact with the environment.

2   Related Works

2.1   BVIPs Assistive Technologies

Assistive navigation technologies for BVIPs, known as *Electronic Travel Aid* (ETA) systems, are designed to provide a sensorial substitution to the human vision, *i.e.* to convey visual information to the users by substituting it with one of their intact senses, usually auditory or tactile representations [13,34]. Existing ETAs can be divided in two major families, namely: sensorial network systems (*i.e.* active systems) and video-based systems (*i.e.* passive systems).

Sensorial ETAs collect environmental information, provide subject localization, and identify objects using active devices like ultrasound [11], infrared [32], global positioning system (GPS) [20], and radio frequency identification (RFID) [21]. Notable examples of systems using infrared and ultrasonic sensors are GuideCane [35], SmartCane [37], UltraCane [27], and Necklace cane [36]. Such frameworks attempt to estimate the optimal, obstacle free, walking path and inform the user through tactile stimulation. However, when tested in real-life scenarios, they show their limitations, notably when detecting objects situated above the user knee, overhanging obstacles or sidewalk borders. GPS based systems do not work in indoor environment, they are highly dependent on the signal quality and present a poor accuracy rate in the case of urban environments, where the density of buildings is high. On the other hand, RFID systems can work indoor and localization accuracy is high, still they require the environment to be structured with anchors. While this is reasonable for some scenarios (like hospitals and healthcare facilities), the main limitation is that it is not transferable to generic environments.

Thus, research recently converged to camera systems based either on RGB and RGBD cameras, with the latter being an extension of the former that also includes pixelwise depth information. The main reason of using RGB cameras is to deploy the app in smartphones, thus requiring no additional hardware to be wear by the user. Unfortunately, this comes with a cost in terms of dexterity since the user will be asked to hold the smartphone all the time; moreover, recent advancements in miniaturized RGBD sensors allow us to integrate it in low invasive devices such as smart glasses. Several works demonstrated the feasibility of using RGBD sensors as the main hardware component of a wearable assistive device. For example, the ISANA system [16] develops a semantic map of the environment, identifies objects situated in front of the sensor, generates a safe path avoiding obstacles, and produces guiding messages in the form of speech. Though, the usability of the system is limited by the need of an architectural model, i.e a map, of the environment. Similar approaches have also been used to detect specific objects like staircaises [24] and traffic lights [22].

In this paper we focus on a sensory substitution device that uses a RGBD camera to perceive and model the environment, and a 3D audio representation that allow the user to perceive the presence and localization of walls and objects in terms of sounds.

2.2   3D Visual Modelling of the Environment

Modelling the environment translates in generating a semantic representation of the environment that accounts for estimating the room layout, detect and localize objects, and generate of a 3D occupancy map. For all these tasks, deep neural networks have emerged as the dominant solution. Holistic methods try to solve these problems with an end-to-end solution to jointly estimate the layout box and object poses in a multitask learning problem setup [25,39]. Despite working relatively well, these approaches are computationally too expensive to be mounted on wearable devices, forcing us to consider the three tasks separately.

As for room layout estimation, a simple approach is to approximate the room layout as a bounding box [5,30]. Different techniques have been proposed to extract these information from RGBD data, inferring semantic information from edges [40], using recursive RANSAC [17], or developing an omnidirectional SLAM pipeline [19].

Object recognition and segmentation are also classical problems in machine vision with many approaches reaching extremely high accuracy even with real time computation constraints like Faster R-CNN [29], YOLO [28] and its extension to RGBD images [33]. The latter implementation indeed is an extension of the original YOLO architecture for RGBD images. Differently from many models that concatenate the depth map with features extracted from the RGB image, this approach preserves the connection between color and depth information; however, it is not clear how the different modalities contribute to the feature maps representation, with some insights driving towards one of the two modalities dominating the other.

2.3   Video Sonification

Capturing and representing the environment is only one side of the problem to be solved for designing a wearable device capable of providing real-time feedback to BVIPs. The other is represented by sensorial substitution, *i.e.* the need of identifying compensatory strategies aimed at encoding the information collected from the environment through sensory channels usable by people with visual impairments, especially hearing and/or touch [14]. Research on sensory replacement of sight through hearing has led to two main approaches: (1) low-level sonification deals with the transposition of image characteristics into acoustic signals through perceptual correspondence maps (generally learned); and (2) high-level, or symbolic, sonification aims to translate visual information into natural language, *e.g.* by translating the visual scene into a textual description and then converting it into sound through speech synthesis. If the high-level sonification on the one hand allows a greater degree of completeness and precision in the description of the scene, it is not always more efficient in terms of navigation performance when compared with a low-level acoustic display, since the latter allows information on forms, obstacles and other environmental information to be communicated in a more compact and immediate way [18].

Low-level sonification involves the design of conversion maps from visual parameters to sound parameters, which in turn are converted into acoustic signals through one or more digital audio synthesis techniques. The conversion criteria, or acoustic mapping strategies, are generally arbitrary and can transform any image parameter into a sound parameter. Numerous studies in the literature investigate sound synthesis based substitution to represent shapes or contours [8,38], colors [2,26], and movement [4].

One of the first experimental devices to implement the acoustic substitution principle, the vOICe sonic imaging system [23], works by transforming the image parameters of the environment into sound patterns, relating the vertical position of the image pixel with the fundamental frequency (pitch) and the brightness of an element of the scene with the intensity of the sound. The Sound of Vision system provides real-time visual scanning of the environment through head-mounted cameras, computer vision functions such as object recognition and hazard detection, and audio encoding through the conversion of visual cues into the parameter space of a liquid sound phenomenological model [12]. Another recent integrated system of the same class, named SoundSight [9], provides wide versatility in terms of choice of video sensors, allowing the use of different classes of devices, from consumer-class sensors to sophisticated thermal imaging cameras. The acoustic generation in the SoundSight system is based on the reproduction of audio samples stored in a sound bank, controlled by the characteristics of the image pixels. Object identification and distance analysis parameters are encoded acoustically, for example, through sound timbre and intensity control, so that each object is acoustically characterized.

In all the previous cases, the acoustic information relating to the visual scene can be further enriched with the addition of the spatial dimension, rendered through state-of-the-art technologies for the spatialization of sound (*e.g.* HRTF technology) [7,31]. Through spatial information, the sound description of an object or event can be located in the acoustic space corresponding to the environment in which the subject stands or moves, allowing him/her to build a map of the environment in which he/she must be able to move. A low-level representation with spatialization can be achieved with an acoustic signal perceived by the subject in continuous movement from -90 to $+90°$ and which, for each angle, encodes obstacles or escape routes with sequences of clicks at high or low frequency respectively.

3   Vis-a-Vis Approach

The general architecture of the Vis-a-Vis prototype is composed by two main components: an *environment modelling* module and a *3D sonification* module. The former aims at sensing the environment, processing the visual data, and generating a formal representation of the environment; the latter aims at generating an accurate, complete and easy to interpret audio representation of the modeled environment. The communication between the two modules is carried out using the Websocket API, which is a bidirectional communication protocol that can send data from a server to a client and vice versa. Information regarding the 3D scene is sent via a web server to the web audio sonification app via JSON objects.

3.1   3D Environment Modeling

The environment is perceived by a Asus Xtion PRO LIVE camera, that provides RGBD frames at 30fps with a resolution of 640×480 pixels and depth estimation ranging from 80cm to 4 m. The frames are then processed by a computing device with hardware acceleration for running deep learning models in real time. The current prototype works with a standard PC, while integration in a portable embedded device in ongoing.

At every time step, the RGBD frame is processed using the SLAM algorithm RTAB-Map (Real-Time Appearance-Based Mapping) [15] to align point clouds reconstructed from subsequent frames and gradually build a complete representation of the environment and localize the user within it at the same time. We approximate the room layout with a box, *i.e* we model mathematically the environment structure with six planar surfaces perpendicular to each other. At the beginning, we fit this model to the point cloud data using least mean square. The camera is head mounted, thus the bottom surface is the floor and its parallel plane is the ceiling. Initially we can fit at most other three planes, but often it is just two as in the case of corridors where the frontal plane is too far to be detected by the depth sensor. In the subsequent frames we perform two operations: *i* we assign points to a plane if they lie within a fixed threshold from it, *ii* we refine the equation of the planes by fitting the new set of points.

Beyond walls, the end-user needs to be aware of obstacles and objects to plan a collision-free path. For object detection we used a YoloV3 [28] convolutional neural network. The model processes the RGB channel and returns bounding box coordinates and class label for each object recognized in the image. The bounding box is then mapped to the depth channel to localize the object pose in 3D. While many methods are available in the literature for this task, we are not interested in accurate 6 *d.o.f.* estimation, which will be difficult to represent and understand by the user. Thus, we assign the 3D position of the centroid of the object by computing the median of the depth values of all the points in the bounding box and applying the coordinate transformation output by SLAM. This allows us to estimate the position of the objects with respect to the user, but store these information in the global reference frame to be used also when the object will not be in the field of view any more. In parallel, we generate a probabilistic occupancy map of the environment to identify anomalies in the room layout such as staircase, unidentified objects, irregular walls, *etc*We divide the 3D space in a uniform voxel grid and we compute the occupancy probability as proportional to the number of points of the point cloud that lie within the voxel. To this aim we used the OctoMap library[1].

3.2   3D Sonification

The spatial information collected from the environment model, typically assessed visually, is aimed to be encoded in a sonic sensorial description, available to the BVIP users. One important aspect hypothesized to aid this encoding is the spatialization of the sonic content transmitted to the user, aligned with the spatial features of the 3D environment. The 3D sonification is achieved using state-of-the art binaural sound synthesis methods, in particular HRTF technology. This requires the use of stereophonic sound, typically delivered via in-ear or overhead headphones, which would limit the capability of hearing real-world environmental sounds. To overcome this concern we use bone-conduction headphones, which do not obstruct the ear canal. However, vibrations from one transducer could possibly leak to the other via the headband, affecting the quality of the spatialization algorithm. Thus, we need to find a trade-off between accurately synthesizing the spatialization effect and preserving the integrity of the real-world sounds in the environment.

In practice, sonification is developed using the Tone.js library[2] which is a framework around the Web Audio API, the most common system used for controlling audio on the Web. Tone.js offers common digital audio workstation (DAW) features, like a global running time and the ability to trigger and synchronize events, as well as various fundamental sound synthesis and effect features. It is based on the idea of connecting/disconnecting audio nodes in a modular way, thus allowing to build complex interactive web audio applications. In particular, it allows to produce binaural synthesis effect, termed Panner3D node, which takes in some audio (from a source), processes it and passes it further to another node down the chain, typically the audio Destination (the device audio

---

[1] https://octomap.github.io/.

[2] https://tonejs.github.io/.

output). This node uses the Center for Image Processing and Integrated Computing (CIPIC) dataset of HRTFs [1], measured at different elevations and azimuths, *i.e.*sound source directions, which are selected accordingly and convolved with the incoming audio to produce the spatialization effect. The selection is carried out based on the sound source position relative to a Listener node, which is passed as an argument to the Panner3D effect (user input).

When objects are detected within the human workspace they become a 3D sound source in the virtual soundscape. An initial strategy is to sonify each object as a either a looping short decaying sound, or a *constant* drone type sound. The drone sounds are reserved to the walls, with the thought that the wall is a massive static plane which defines the 3D space the user is in, while the encountered obstacles are described using the sound loops. In order to be able to distinguish between different objects (or obstacles), the looping sounds can be initialized with different frequencies (musical notes) and timbres. Here we can use the principle of the takete maluma effect [6] and sonify objects with softer shapes with more round waveforms. Wave shaping methods can also be used here to merge between a square wave and a sine wave using a function that gradually smooths out the corners of a square wave.

Working with global environment models, we can sonify objects and obstacles even if they are not in the field of view. Imagine the case where a small chair is near to the side of the user. This chair will not be in the camera view when "looking" straight ahead. The object detection model comes in to add semantic classification to the detected objects, and allows for a more relevant sound timbre to be associated with it. Additionally, we apply a change in the playback rate of the loop sounds and in the base frequency of the drone sounds (effectively increasing the pitch of the drone). The playback rate and pitch are mapped to the shortest distance to the various objects and walls. In order not to overwhelm the user with constant changing sounds, both mappings change rather slowly at larger distances and then change drastically when the distance to the user becomes smaller than a critical safety distance.

Figure 1 shows a diagram of the web audio app architecture, with a section for the Tone.js audio architecture, the sonification classes and the model flow from the 3D environmental model to the 3D sonification via the websocket data stream.

4 Preliminary Validation

Our preliminary tests demonstrate the potential of the proposed approach. In this section we report some qualitative results and the discussions, while we remand the interested reader to our demo video of the sonification app at https:// mariusono.github.io/Vis-a-Vis/. Please click the "Start audio" button and then the "Start video" button to start the demo (in that sequence). Headphones are needed for the perception of the 3D spatial effect.

We report here experiments conducted with the our 3D environment modelling module on some very common scenarios in public places that can be particularly challenging: a corridor and a small hall.

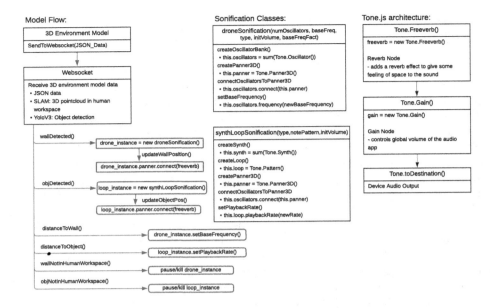

Fig. 1. Architecture of web audio 3D sonification app.

Corridors are very interesting scenarios for our application for several reasons. First of all, it is very common, thus an ETA system must be effective in assisting BVIPs in this situation. Second, the length of the corridor do not allow to model its entirety in a few shots, thus requiring the system to be able to maintain the information for long time and increase the point cloud without drift or losing localization of the user. This is often challenging due to the lack of good features in long, uniform walls. Finally, doors on the side walls can be a challenge in modelling the walls, especially when they are open or semi-open.

We tested the modelling capabilities of our system in a corridor at the university campus where only few objects were present, and walls are mostly uniform. A sample frame of these scenario with output of YoloV3 object detector and the resulting segmented point cloud with walls and objects are shown in Fig. 2.

A small **hall** at our university provides a good example of a relatively complex scene. The hall, shown in Figs. 3 and 4, is a rectangular room with a corridor departing from a corner. The dimension of the room are approximately 7×4 m. The main complexity in this scenario is given by the high number of furnitures that are present in the scene, making the detection of walls a challenging task. Moreover, the presence of many glass surfaces generate reflections that increase noise in depth acquisitions. In this case, our approach is still able to correctly estimate the room layout and localize obstacles. It is worth noting that glass cabinets are not recognized as specific objects, since the class is not defined in the list of YoloV3, yet the occupancy map recognize the presence of an obstacle and returns this information to the user in terms of potential collisions occurring.

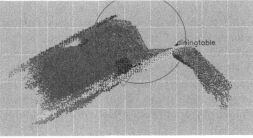

Fig. 2. Left: Sample view of the corridor scenario with detected object. Right: segmented point cloud representing objects (green are within the field of view, red otherwise), walls (floor in blue, right in green, left in light grey); and camera position, orientation and workspace (orange). (Color figure online)

Fig. 3. Left: RGB frame at a certain timestamp; Right: 3D pointcloud of the classified walls and corresponding equations.

Fig. 4. Left: OctoMap 3D occupancy grid map generated starting from the SLAM cloudmap; Right: the SLAM sparse 3D cloudmap of an indoor cluttered with static objects environment.

5 Conclusion

In this paper we presented a sensory substitution system that performs visual understanding of indoor environments, 3D modelling of it and sonification of the model to allow BVIPs to better perceive and navigate the surrounding space. We described the approach designed in the VISaVIS project and we reported some preliminary validation of our work. The results are showing the potential and some limitations of our approach, allowing us to redesign some components to meet our goal of an assistive wearable device for BVIPs.

Acknowledgements. This work is supported by European Comfort S.r.l. and University of Verona through the Joint Research funding scheme with the project "Vis-a-Vis". The authors would like to thank Mr. Giambattista Bersanelli and Marco Delucca for the valuable support in the problem definition and functional design of the proposed solution.

References

1. Algazi, V., Duda, R., Thompson, D., Avendano, C.: The CIPIC HRTF database. In: WASPAA (2001). https://doi.org/10.1109/ASPAA.2001.969552
2. Banf, M., Blanz, V.: Sonification of images for the visually impaired using a multi-level approach. In: AH (2013). https://doi.org/10.1145/2459236.2459264
3. Borenstein, J., Ulrich, I.: The GuideCane-a computerized travel aid for the active guidance of blind pedestrians. In: ICRA (1997). https://doi.org/10.1109/ROBOT.1997.614314
4. Bresin, R., Mancini, M., Elblaus, L., Frid, E.: Sonification of the self vs. sonification of the other: differences in the sonification of performed vs. observed simple hand movements. Int. J. Hum.-Comput. Stud. **144**, 102500 (2020). https://doi.org/10.1016/j.ijhcs.2020.102500
5. Dasgupta, S., Fang, K., Chen, K., Savarese, S.: Delay: robust spatial layout estimation for cluttered indoor scenes. In: CVPR (2016)
6. Fontana, F., Järveläinen, H., Favaro, M.: Is an auditory event more takete? In: SMC (2021). https://doi.org/10.5281/ZENODO.5038640
7. Geronazzo, M., Bedin, A., Brayda, L., Campus, C., Avanzini, F.: Interactive spatial sonification for non-visual exploration of virtual maps. Int. J. Hum Comput Stud. **85**, 4–15 (2016). https://doi.org/10.1016/j.ijhcs.2015.08.004
8. Gholamalizadeh, T., Pourghaemi, H., Mhaish, A., Ince, G., Duff, D.J.: Sonification of 3d object shape for sensory substitution: an empirical exploration. In: ACHI (2017)
9. Hamilton-Fletcher, G., Alvarez, J., Obrist, M., Ward, J.: Soundsight: a mobile sensory substitution device that sonifies colour, distance, and temperature. J. Multimod. User Interfaces **16**, 107–123 (2022). https://doi.org/10.1007/s12193-021-00376-w
10. Hoffmann, R., Spagnol, S., Kristjánsson, A., Unnthorsson, R.: Evaluation of an audio-haptic sensory substitution device for enhancing spatial awareness for the visually impaired. Optom. Vis. Sci. **95**, 757–765 (2018). https://doi.org/10.1097/OPX.0000000000001284

11. Jeong, G.Y., Yu, K.H.: Multi-section sensing and vibrotactile perception for walking guide of visually impaired person. Sensors 16(7), 1070 (2016). https://doi.org/10.3390/s16071070

12. Jóhannesson, Ó.I., Balan, O., Unnthorsson, R., Moldoveanu, A., Kristjánsson, Á.: The sound of vision project: on the feasibility of an audio-haptic representation of the environment, for the visually impaired. Brain Sci. 6(3), 20 (2016)

13. Khan, S., Nazir, S., Khan, H.U.: Analysis of navigation assistants for blind and visually impaired people: a systematic review. IEEE Access 9, 26712–26734 (2021). https://doi.org/10.1109/ACCESS.2021.3052415

14. Kristjánsson, Á., Moldoveanu, A.D.B., Jóhannesson, Ó.I., Balan, O., Spagnol, S., Valgeirsdóttir, V.V., Unnthorsson, R.: Designing sensory-substitution devices: principles, pitfalls and potential. Restor. Neurol. Neurosci. 34, 769–787 (2016)

15. Labbé, M., Michaud, F.: RTAB-Map as an open-source lidar and visual simultaneous localization and mapping library for large-scale and long-term online operation. J. Field Robot. 36(2), 416–446 (2019). https://doi.org/10.1002/rob.21831

16. Li, B., Munoz, J.P., Rong, X., Xiao, J., Tian, Y., Arditi, A.: ISANA: wearable context-aware indoor assistive navigation with obstacle avoidance for the blind. In: ECCV (2016). https://doi.org/10.1007/978-3-319-48881-3_31

17. Li, J., Stevenson, R.L.: Indoor layout estimation by 2d lidar and camera fusion. arXiv preprint arXiv:2001.05422 (2020)

18. Loomis, J., Golledge, R., Klatzky, R., Marston, J.: Assisting Wayfinding in Visually Impaired Travelers, pp. 179–202. Lawrence Erlbaum Associates, Inc. (2007). https://doi.org/10.4324/9781003064350-7

19. Lukierski, R., Leutenegger, S., Davison, A.J.: Room layout estimation from rapid omnidirectional exploration. In: ICRA (2017)

20. Márkus, N., Arató, A., Juhász, Z., Bognár, G., Késmárki, L.: MOST-NNG: An accessible GPS navigation application integrated into the mobile slate talker (MOST) for the blind. In: ICCHP (2010). https://doi.org/10.1007/978-3-642-14100-3_37

21. Martinez-Sala, A.S., Losilla, F., Sánchez-Aarnoutse, J.C., García-Haro, J.: Design, implementation and evaluation of an indoor navigation system for visually impaired people. Sensors 15(12), 32168–32187 (2015). https://doi.org/10.3390/s151229912

22. Mascetti, S., Ahmetovic, D., Gerino, A., Bernareggi, C., Busso, M., Rizzi, A.: Robust traffic lights detection on mobile devices for pedestrians with visual impairment. Comput. Vis. Image Underst. 148, 123–135 (2016). https://doi.org/10.1016/j.cviu.2015.11.017

23. Meijer, P.: An experimental system for auditory image representations. IEEE Trans. Biomed. Eng. 39(2), 112–121 (1992). https://doi.org/10.1109/10.121642

24. Munoz, R., Rong, X., Tian, Y.: Depth-aware indoor staircase detection and recognition for the visually impaired. In: ICME Workshops (2016)

25. Nie, Y., Han, X., Guo, S., Zheng, Y., Chang, J., Zhang, J.J.: Total3DUnderstanding: Joint layout, object pose and mesh reconstruction for indoor scenes from a single image. In: CVPR (2020)

26. Osiński, D., Łukowska, M., Hjelme, D.R., Wierzchoń, M.: Colorophone 2.0: A wearable color sonification device generating live stereo-soundscapes-design, implementation, and usability audit. Sensors 21(21) (2021). https://doi.org/10.3390/s21217351

27. Penrod, W., Corbett, M.D., Blasch, B.: Practice report: a master trainer class for professionals in teaching the UltraCane electronic travel device. J. Visual Impairment Blindness 99(11), 711–715 (2005)

28. Redmon, J., Divvala, S., Girshick, R., Farhadi, A.: You only look once: Unified, real-time object detection. In: CVPR (2016)
29. Ren, S., He, K., Girshick, R., Sun, J.: Faster R-CNN: towards real-time object detection with region proposal networks. NeurIPS (2015)
30. Ren, Y., Li, S., Chen, C., Kuo, C.C.J.: A coarse-to-fine indoor layout estimation (cfile) method. In: ACCV (2017)
31. Ribeiro, F., Florêncio, D., Chou, P.A., Zhang, Z.: Auditory augmented reality: object sonification for the visually impaired. In: MMSP (2012). https://doi.org/10.1109/MMSP.2012.6343462
32. Ross, D.A., Lightman, A.: Talking braille: a wireless ubiquitous computing network for orientation and wayfinding. In: ASSETS (2005). https://doi.org/10.1145/1090785.1090805
33. Takahashi, M., Ji, Y., Umeda, K., Moro, A.: Expandable YOLO: 3D object detection from RGB-D images. In: REM (2020)
34. Tapu, R., Mocanu, B., Zaharia, T.: Wearable assistive devices for visually impaired: a state of the art survey. Pattern Recogn. Lett. **137**, 37–52 (2020). https://doi.org/10.1016/j.patrec.2018.10.031
35. Ulrich, I., Borenstein, J.: The GuideCane - applying mobile robot technologies to assist the visually impaired. IEEE Trans. Syst. Man Cybern. **31**(2), 131–136 (2001). https://doi.org/10.1109/3468.911370
36. Villamizar, L.H., Gualdron, M., González, F., Aceros, J., Rizzo-Sierra, C.V.: A necklace sonar with adjustable scope range for assisting the visually impaired. In: EMBC (2013). https://doi.org/10.1109/EMBC.2013.6609784
37. Wahab, M.H.A., Talib, A.A., Kadir, H.A., Johari, A., Sidek, R.M., Mutalib, A.A.: Smartcane: Assistive cane for visually-impaired people. arXiv preprint arXiv:1110.5156 (2011). https://doi.org/10.48550/arXiv.1110.5156
38. Yoshida, T., Kitani, K.M., Koike, H., Belongie, S., Schlei, K.: Edgesonic: Image feature sonification for the visually impaired. In: AH (2011). https://doi.org/10.1145/1959826.1959837
39. Zhang, C., Cui, Z., Zhang, Y., Zeng, B., Pollefeys, M., Liu, S.: Holistic 3d scene understanding from a single image with implicit representation. In: CVPR (2021)
40. Zhang, W., Zhang, W., Gu, J.: Edge-semantic learning strategy for layout estimation in indoor environment. IEEE Trans. Cybern. **50**(6), 2730–2739 (2019). https://doi.org/10.1109/TCYB.2019.2895837

Minimizing Energy Consumption of Deep Learning Models by Energy-Aware Training

Dario Lazzaro[1] , Antonio Emanuele Cinà[2]([⊠]) , Maura Pintor[3] ,
Ambra Demontis[3] , Battista Biggio[3] , Fabio Roli[4] , and Marcello Pelillo[1]

[1] Ca' Foscari University of Venice, Venice, Italy
[2] CISPA Helmholtz Center for Information Security, Saarbrücken, Germany
antonio.cina@cispa.de
[3] University of Cagliari, Cagliari, Italy
[4] University of Genoa, Genoa, Italy

Abstract. Deep learning models undergo a significant increase in the number of parameters they possess, leading to the execution of a larger number of operations during inference. This expansion significantly contributes to higher energy consumption and prediction latency. In this work, we propose *EAT*, a gradient-based algorithm that aims to reduce energy consumption during model training. To this end, we leverage a differentiable approximation of the ℓ_0 norm, and use it as a sparse penalty over the training loss. Through our experimental analysis conducted on three datasets and two deep neural networks, we demonstrate that our energy-aware training algorithm *EAT* is able to train networks with a better trade-off between classification performance and energy efficiency.

Keywords: training · hardware acceleration · energy efficiency · sparsity maximization · regularization

1 Introduction

Deep learning is widely adopted across various domains due to its remarkable performance in various tasks. The increase in model size, primarily driven by the number of parameters, often leads to improved performance. However, this growth in model size also leads to a higher computational burden during prediction, necessitating specialized hardware like GPUs to deliver the required computational power for efficient training and inference [6]. Although beneficial for many applications, this strategy contradicts the requirements of certain real-time scenarios (e.g., embedded IoT devices, smartphones, online data processing, etc.) that are often constrained in their energy resources or require fast predictions for not compromising users' usability.

Energy efficiency has therefore become a critical aspect in the design and deployment of deep learning models, opening up new directions for research, including pruning, quantization, and efficient architecture search. A common strategy is to train the networks and then prune them by removing neurons

© The Author(s), under exclusive license to Springer Nature Switzerland AG 2023
G. L. Foresti et al. (Eds.): ICIAP 2023, LNCS 14234, pp. 515–526, 2023.
https://doi.org/10.1007/978-3-031-43153-1_43

or reducing the complexity of the operations by quantizing their weights. However, adopting these methodologies can compromise the accuracy of the resulting models. Another way to reduce the amount of energy required for classification is to use modern hardware acceleration architectures, including ASICs (Application Specific Integrated Circuits), which reduce energy consumption without changing the network's structural architecture and thus preserve its performance. Sparsity-based ASIC accelerators employ zero-skipping operations that avoid multiplicative operations when one of the operands is zero, avoiding performing useless operations [25]. For example, Eyeriss et al. [2] achieved a $10\times$ reduction in energy consumption of DNNs when using sparse architectures rather than conventional GPUs.

In this paper, we propose a training loss function that leverages an estimate of the model's power consumption as a differentiable regularizer to apply during training. We use it to develop a novel energy-aware training algorithm (EAT) that enforces sparsity in the model's activation to enhance the benefits of sparsity-based ASIC accelerators. Our training objective has been inspired by an attack called sponge poisoning [6]. Sponge poisoning is a training-time attack [3–5] that tampers with the training process of a target DNN to increase its energy consumption and prediction latency at test time. In this work, we develop EAT by essentially inverting the sponge poisoning mechanism, i.e., using it in a beneficial way to reduce the energy consumption of DNNs (Sect. 2). Our approach does not only aim to reduce energy consumption; it aims to achieve a better trade-off between energy efficiency and model performance. By balancing these two objectives, we can indeed train models that achieve sustainable energy consumption without sacrificing accuracy.

We run extensive experiments on two distinct DNN architectures and using three datasets to compare the energy consumption and performance of our energy-aware models against the corresponding baselines, highlighting the benefits of using our approach (Sect. 3).

We conclude by discussing related work (Sect. 4), along with the contributions and limitations of our work (Sect. 5).

2   EAT: Energy-Aware Training

In this paper, we consider sparsity-based ASIC accelerators that adopt zero-skipping strategies to avoid multiplicative operations when an activation input is zero, thus increasing throughput and reducing energy consumption [1,2,8,23,25]. Hence, to meet the goal of increasing the ASIC speedup, we need to increase the model's activations sparsity, i.e., the number of not firing neurons, while preserving the model's predictive accuracy. A similar objective has been previously formulated by Cinà et al. [6], with the opposite goal of *increasing* the energy consumption of the models. In their paper, the authors propose a training-time attack against the availability of machine learning models that maximizes the number of firing neurons at testing time. To achieve this goal, they apply a custom regularization term to the training loss that focuses on increasing the number of firing neurons with the adoption of the ℓ_0 norm. Specifically, the ℓ_0 norm is considered for counting the number of non-zero components of the model's activations. However, due to its non-convex and non-differentiable nature, the ℓ_0 norm is not directly optimizable with gradient descent. For this reason, their optimization algorithm employs a differentiable approximation of the ℓ_0 norm proposed in [24], which we will denote as $\hat{\ell}_0$. Formally, given an input vector $x \in \mathbb{R}^n$, we define:

$$\hat{\ell}_0(x) = \sum_{j=1}^{n} \frac{x_j^2}{x_j^2 + \sigma}, \qquad x \in \mathbb{R}^n, \sigma \in \mathbb{R}, \tag{1}$$

The parameter σ controls the approximation quality of the function toward the ℓ_0 norm. By decreasing the value of σ, the approximation becomes more accurate. However, an increasingly accurate approximation could lead to optimization instabilities [6].

This approximation is then used to estimate the number of non-zero elements in the activation vectors of the hidden layers. Therefore, given the victim's model f, parametrized by w, a training set $\mathcal{D} = \{(x_i, y_i)\}_{i=1}^{s}$ the sponge training algorithm by Cinà et al. [6] is formalized as follows:

$$\min_{w} \sum_{(x,y) \in \mathcal{D}} \mathcal{L}(x, y, w) - \lambda \sum_{k=1}^{K} \hat{\ell}_0(\phi_k, \sigma), \tag{2}$$

where $\mathcal{L}$ is the empirical risk minimization loss (e.g., the cross-entropy loss), $\hat{\ell}_0$ is the differentiable function to estimate the number of firing neurons in the k-layer ϕ_k. The first term of Eq. 2 focuses on increasing the model's classification accuracy, and the second term is a differentiable function responsible for increasing the model's energy consumption. Combining the two losses enables the training algorithm to increase energy consumption while preserving the model's prediction accuracy. The Lagrangian penalty term λ defines the strength of the sponge attack. In other words, low values of λ will focus on achieving high accuracy, while high values will increase energy consumption.

Since our paper aims to induce sparsity in the model's activation to enhance the speed-up offered by ASIC HW accelerators, we reformulate the problem as the minimization of the number of non-zero elements in the activation vectors of the hidden layers. The final optimization program for our training algorithm therefore becomes:

$$\min_{w} \sum_{(x,y) \in \mathcal{D}} \mathcal{L}(x, y, w) + \lambda \sum_{k=1}^{K} \hat{\ell}_0(\phi_k, \sigma), \tag{3}$$

Solution Algorithm. In Algorithm 1, we present the training algorithm we employ for training DNNs by maximizing prediction accuracy and minimizing energy consumption. The algorithm first stores the initial model's weights Line 1. Then, we update w for each batch in $\mathcal{D}$ and N epochs (Line 2–6). We make the update (Line 6) in the direction of the negative gradient of the objective function

Eq. 3, therefore minimizing the cross-entropy loss $\mathcal{L}$ on the batch $\boldsymbol{x}$ and inducing sparsity in the model's activations. After N epochs of training, the algorithm returns the optimized model weights $\boldsymbol{w}^*$.

Algorithm 1: Energy-aware Training Algorithm.

Input: $\mathcal{D}$ training dataset; $w = (\phi_1, ..., \phi_K)$, the initialized layers of the neural network; λ, sparsification coefficient; α, the learning rate for training; σ, the quality of the approximation.

Output: $w^* = (\phi_1^*, ..., \phi_K^*)$, optimized weights.

1   $w^* \leftarrow w$
2   **for** i *in* $1, \ldots, N$ **do**
3   **for** *(x, y)* *in* $\mathcal{D}$ **do**
4           $\nabla \mathcal{L} \leftarrow \nabla_w \mathcal{L}(\boldsymbol{x}, y, \boldsymbol{w})$
5           $\nabla E \leftarrow \nabla_w \left[\sum\limits_{k=1}^{K} \hat{\ell}_0(\phi_k, \sigma) \right]$
6           $w^* \leftarrow w^* - \alpha \left[\nabla \mathcal{L} - \lambda \nabla E \right]$

7   **return** w^*

---

3   Experiments

We experimentally assess the effectiveness of the proposed training algorithms in terms of energy consumption and model accuracy on two DNNs trained in three distinct datasets. Furthermore, we provide more insights regarding the effect of the proposed training algorithm on the models' energy consumption by analyzing the internal neuron activations of the resulting trained models. Finally, we provide an ablation study to select the hyperparameters λ and σ.

3.1   Experimental Setup

Datasets. We conduct our experiments by following the same experimental setup as in [6, 22]. Therefore, we assess our training algorithm on three datasets where data dimensionality, number of classes, and their balance are different, thus making the setup more heterogeneous and challenging. Specifically, we consider the CIFAR10 [15], GTSRB [11], and CelebA [18] datasets. The CIFAR10 dataset contains $60,000$ RGB images of 32×32 pixels equally distributed in 10 classes. We consider $50,000$ samples for training and $10,000$ as the test set. The German Traffic Sign Recognition Benchmark dataset (GTSRB) consists of $60,000$ RGB images of traffic signs divided into 43 classes. For this dataset, we compose the training set with $39,209$ samples and the test set with $12,630$, as done in [11]. The CelebFaces Attributes dataset (CelebA) is a face attributes dataset with more than $200K$ RGB images depicting faces, each with 40 binary attribute annotations. We categorize the dataset images in 8 classes, generated considering the top three most balanced attributes, i.e., *Heavy Makeup*, *Mouth Slightly Open*, and *Smiling*. We finally split the dataset into two sets, 162,770 samples for training and 19,962 for testing. We scale the images of GTSRB and CelebA to the resolution of 32×32 px and 64×64 px, respectively, and use random crop and random rotation during the training phase for data augmentation. Finally, we remark that the classes of the GTSRB and CelebA datasets are highly imbalanced, which makes them challenging datasets.

Models and Training Phase. We consider two DNNs, i.e., ResNet18 [9] ($\sim11M$ parameters) and VGG16 [27] ($\sim138M$ parameters). We train them on the three datasets mentioned above for 100 training epochs with SGD optimizer with momentum 0.9, weight decay $5e-4$, and batch size 512, and we choose the cross-entropy loss as $\mathcal{L}$. We employ an exponential learning scheduler with an initial learning rate of 0.1 and decay of 0.95. The trained models have comparable or even better accuracies compared to those obtained with the experimental setting employed in [21, 22].

Hyperparameters. Two hyperparameters primarily influence the effectiveness of our algorithm. The former is σ (see Eq. 2) that regulates the approximation goodness of $\hat{\ell}_0$ to the actual ℓ_0. A smaller value of σ gives a more accurate approximation; however, extreme values will result in optimization failure [6]. The other term that affects effectiveness is the Lagrangian term λ introduced in Eq. 2, which balances the relevance of the sponge effect compared to the training loss. A wise choice of this hyperparameter can lead the training process to obtain models with high accuracy and low energy consumption. In order to have a complete view of the stability of our approach to the choice of these hyperparameters, we empirically perform an ablation study considering $\sigma \in \{1e-01, ...1e-08\}$, and $\lambda \in \{0.1, ..., 10\}$. We perform this ablation study on a validation set of 100 samples randomly chosen from each dataset. Finally, since the energy consumption term has a magnitude proportional to the model's number of parameters m, we normalize it with the actual number of parameters of the model.

Performance Metrics. We consider each trained model's prediction accuracy and the energy gap as the performance metrics. We measure the prediction accuracy as the percentage of correctly classified test samples. We check the prediction accuracy of the trained model because the primary objective is to obtain a model that performs well on the task of choice. Then, we consider the energy consumption ratio in [6, 26]. The energy consumption ratio, introduced in [26], is the ratio between the energy consumed when using the zero-skipping operation (namely the optimized version) and the one consumed when using standard operations (without this optimization). The energy consumption ratio is upper bounded by 1, meaning that the ASIC accelerator has no effect, and the model has the worst-case performance (no operation is skipped). Furthermore, we report the energy decrease computed as the difference between the energy consumption of the standardly trained network and our *EAT* network divided by the total energy of the standard network. For estimating the energy consumption from ASIC accelerators, we used the ASIC simulator developed in [26].[1]

3.2 Experimental Results

Energy-Aware Performance. Table 1 presents the test accuracy, energy consumption ratio, and energy decrease achieved for the CIFAR10, GTSRB, and CelebA datasets using two different training algorithms: standard empirical-risk minimization training (ST) and our proposed energy-aware training approach (*EAT*). We select the hyperparameter configuration of σ and λ that ensures the lowest energy ratio while maintaining the test accuracy within a 3% margin compared to the standard network training. Results for other configurations are reported in our ablation study. Our experimental analysis demonstrates a significant reduction in energy consumption achieved by our energy-aware training models, *EAT*, while maintaining comparable or even superior test accuracy compared to the standardly-trained networks ST. For example, through the adoption of *EAT*, the energy consumption ratio of ResNet18 for GTSRB is substantially decreased from approximately 0.76 to 0.55. This corresponds to a remarkable 27% reduction in the number of operations required during prediction, therefore reducing the computational workload of the system. Overall, with higher sparsity achieved through our energy-aware training algorithm, the advantages of ASIC accelerators become even more pronounced than for models trained with the standard training algorithm. For *EAT* models, their energy consumption is further diminished while simultaneously increasing the prediction throughput.

Table 1. Comparison of accuracy and energy consumption achieved with standard training (ST) and our energy-aware method (*EAT*).

	GTSRB				CIFAR-10				CelebA			
	ResNet18		VGG16		ResNet18		VGG16		ResNet18		VGG16	
	ST	*EAT*	ST	*EAT*	ST	*EAT*	ST	*EAT*	ST	*EAT*	ST	*EAT*
Accuracy	0.91	0.93	0.90	0.89	0.92	0.90	0.91	0.88	0.76	0.78	0.77	0.78
E. ratio	0.76	0.55	0.69	0.63	0.73	0.61	0.67	0.53	0.68	0.63	0.63	0.54
E. decrease%	-	**27.63**	-	**8.69**	-	**16.43**	-	**20.89**	-	**7.35**	-	**14.28**

Inspecting Layers. We depict in Fig. 1 and Fig. 2 the layer-wise activations of ResNet18 and VGG16 models, respectively, trained using standard training and our energy-aware training approach.

Our results demonstrate that the energy-aware algorithm significantly reduces the percentage of non-zero activations in both networks. In particular, the substantial reduction in activations involving the *max* function, such as

---

[1] https://github.com/iliaishacked/sponge_examples.

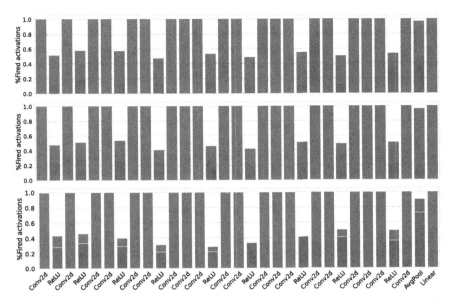

Fig. 1. Percentage of firing neurons in each layer of a ResNet18 on the GTSRB (*top*), CIFAR10 (*middle*), and CelebA (*bottom*) datasets. In blue the percentages achieved with ST, and in red the ones obtained with *EAT*.

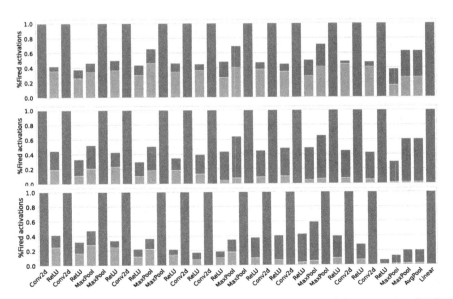

Fig. 2. Percentage of firing neurons in each layer of a VGG16 on the GTSRB (*top*), CIFAR10 (*middle*), and CelebA (*bottom*) datasets. In blue the percentages achieved with ST, and in red the ones obtained with *EAT*.

ReLU and MaxPooling operations, is noteworthy. For instance, in Fig. 2, across
the CIFAR10 and GTSRB datasets, the number of ReLU activations is decreased
to approximately 10% of the original value. This finding holds significance considering that ReLU is the most commonly used activation function in modern
deep learning architectures [28]. Therefore, our energy-aware training algorithm
can potentially favor the sparsity exploited by ASIC accelerators for all ReLU-
based network performance [1]. Furthermore, consistent with the observations
made by Cinà *et al.* [6], convolutional operators remain predominantly active
as they apply linear operations within a neighborhood and rarely produce zero
outputs. Consequently, reducing the activations of convolutional operators poses
a more challenging task, suggesting potential avenues for future research.

Ablation Study. Our novel energy-aware training algorithm is mainly influenced by two hyperparameters, λ and σ.

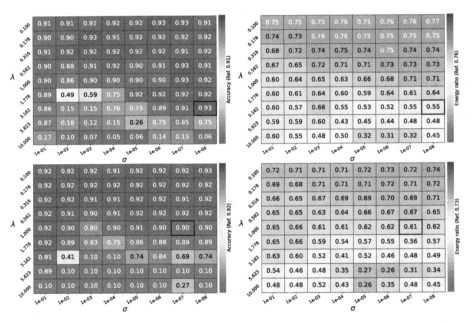

Fig. 3. Ablation study on σ and λ for ResNet18 trained with *EAT* on GTSRB
(*top*) and CIFAR10 (*bottom*). We show the accuracy on the left and the energy
ratio on the right.

As discussed in Sect. 2, the parameter σ controls the level of approximation
for counting the number of firing neurons, whereas λ determines the emphasis
placed on the energy-minimization task during training. By tuning these two values, practitioners can find the desired tradeoff between test accuracy and energy
performance on the resulting models. To investigate the influence of these hyperparameters, we conducted an ablation study presented in Fig. 3. Specifically, we

examined the test accuracy and energy consumption ratio of ResNet18 trained on GTSRB and CIFAR10 while varying λ and σ. We observe that by incrementing λ, practitioners can push the training toward a more energy-sustainable regime. Such models would have a significantly lower impact on energy consumption and the number of operations executed, decreasing the accuracy only slightly. ASIC accelerators can significantly benefit from this increased sparsity. However, very large values of λ (e.g., >3) may cause the training algorithm to prioritize energy minimization over predictive accuracy. On the other hand, small values (e.g., <0.5) would lead the training algorithm to neglect our regularization term and focus solely on accuracy. Regarding σ, we observe that EAT is systematically stable to its choice when a suitable value of λ is used. We can observe a slight variation in the energy ratio when considering large values for σ. This effect is due to the approximation function $\hat{\ell}_0$ in Eq. 2 not being accurate enough to capture the precise number of firing neurons.

4   Related Work

ASIC accelerators have effectively addressed the growing computational requirements of DNNs. They can often optimize energy consumption by skipping operations when the activations are zero or negligible, an operation known as "zero-skipping". As related work, we first discuss the attacks against the zero-skipping mechanism, and then we summarize related work regarding model compression and reduction.

Energy-Depletion Attacks. Recently, ASIC acceleration has been challenged by hardware-oriented attacks that aim to eliminate the benefits of the zero-skipping mechanism. Sponge examples [26] perturb an input sample by injecting specific patterns that induce non-zero activations throughout the model. In a different work, by promoting high activation levels across the model, the sponge poisoning attack [6] demonstrates that increasing energy consumption can also be enforced during training. Staging this attack leads to models with high accuracy (to remain undetected), but an increased latency due to the elimination of hardware-skippable operations.

Contrary to these works, we focus on improving the benefits of ASIC acceleration by introducing more zero-skipping opportunities. Consequently, in this paper, we invert the sponge poisoning attack mechanism, minimizing the number of activations and hence the energy consumption required by the model.

Model Compression. Model compression and quantization are techniques used to optimize and condense deep neural networks, reducing their size and computational requirements without significant loss in performance. Network pruning aims to remove redundant or less important connections [12], filters [16,20,30], or even entire layers [17,19] from a neural network. Pruned models often exhibit sparsity, which techniques like zero-skipping can further exploit. To push compression to the limit, the lottery ticket hypothesis [7] and knowledge distillation methods [10] aim to find smaller networks that can achieve the same performance as larger networks. Quantization [13,14,29], on the other hand, reduces the precision of numerical values in a deep learning model. Instead of using full precision (*e.g.*, 32-bit floating-point numbers), quantization represents values with lower precision (*e.g.*, 8-bit integers). Quantization reduces the memory requirements of the model for more efficient storage and operations.

We argue that both model compression and quantization can be applied to our technique without specific adaptations to push even further the benefits of our method.

5   Conclusions

In this paper, we explored a novel training technique to improve the efficiency of deep neural networks by enforcing sparsities on the activations. Our goal is achieved by incorporating a differentiable penalty term in the training loss. We show how it is possible to obtain a chosen trade-off between model performances and efficiency by applying our technique.

The practical significance of our findings lies in their direct applicability to real-world scenarios. By leveraging the energy-aware training provided by EAT, deep learning models can achieve significant energy savings without compromising their predictive performance. In future work, we believe that our method can be effectively combined with existing pruning and quantization techniques to create advanced model compression methods.

Acknowledgements. This work has been partially supported by Spoke 10 "Logistics and Freight" within the Italian PNRR National Centre for Sustainable Mobility (MOST), CUP I53C22000720001; the project SERICS (PE00000014) under the NRRP MUR program funded by the EU - NGEU; the PRIN 2017 project RexLearn (grant no. 2017TWNMH2), funded by the Italian Ministry of Education, University and Research; and by BMK, BMDW, and the Province of Upper Austria in the frame of the COMET Programme managed by FFG in the COMET Module S3AI.

References

1. Albericio, J., Judd, P., Hetherington, T.H., Aamodt, T.M., Jerger, N.D.E., Moshovos, A.: Cnvlutin: ineffectual-neuron-free deep neural network computing. In: 43rd ACM/IEEE ISCA (2016)
2. Chen, Y., Emer, J.S., Sze, V.: Eyeriss: a spatial architecture for energy-efficient dataflow for convolutional neural networks. In: 43rd ACM/IEEE ISCA (2016)
3. Cinà, A.E., Grosse, K., Demontis, A., Biggio, B., Roli, F., Pelillo, M.: Machine learning security against data poisoning: are we there yet? CoRR (2022)
4. Cinà, A.E., et al.: Wild patterns reloaded: a survey of machine learning security against training data poisoning. ACM Comput. Surv. (2023)

5. Cinà, A.E., Vascon, S., Demontis, A., Biggio, B., Roli, F., Pelillo, M.: The hammer and the nut: is bilevel optimization really needed to poison linear classifiers? In: IJCNN (2021)
6. Cinà, A.E., Demontis, A., Biggio, B., Roli, F., Pelillo, M.: Energy-latency attacks via sponge poisoning. ArXiv (2022)
7. Frankle, J., Carbin, M.: The lottery ticket hypothesis: finding sparse, trainable neural networks. In: ICLR (2019)
8. Han, S., et al.: EIE: efficient inference engine on compressed deep neural network. In: 43rd ACM/IEEE ISCA (2016)
9. He, K., Zhang, X., Ren, S., Sun, J.: Identity mappings in deep residual networks. In: Leibe, B., Matas, J., Sebe, N., Welling, M. (eds.) ECCV 2016. LNCS, vol. 9908, pp. 630–645. Springer, Cham (2016). https://doi.org/10.1007/978-3-319-46493-0_38
10. Hinton, G., Vinyals, O., Dean, J., et al.: Distilling the knowledge in a neural network. arXiv preprint (2015)
11. Houben, S., Stallkamp, J., Salmen, J., Schlipsing, M., Igel, C.: Detection of traffic signs in real-world images: the German traffic sign detection benchmark. In: IJCNN (2013)
12. Hu, H., Peng, R., Tai, Y.W., Tang, C.K.: Network trimming: a data-driven neuron pruning approach towards efficient deep architectures. arXiv preprint (2016)
13. Jacob, B., et al.: Quantization and training of neural networks for efficient integer-arithmetic-only inference. In: CVPR (2018)
14. Jung, S., et al.: Learning to quantize deep networks by optimizing quantization intervals with task loss. In: CVPR (2019)
15. Krizhevsky, A.: Learning multiple layers of features from tiny images. Technical report (2009)
16. Lin, S., Ji, R., Li, Y., Wu, Y., Huang, F., Zhang, B.: Accelerating convolutional networks via global & dynamic filter pruning. In: IJCAI (2018)
17. Liu, Z., Li, J., Shen, Z., Huang, G., Yan, S., Zhang, C.: Learning efficient convolutional networks through network slimming. In: ICCV (2017)
18. Liu, Z., Luo, P., Wang, X., Tang, X.: Deep learning face attributes in the wild. In: ICCV (2015)
19. Luo, J., Wu, J.: Autopruner: an end-to-end trainable filter pruning method for efficient deep model inference. Pattern Recognit. **107**, 107461 (2020)
20. Molchanov, P., Tyree, S., Karras, T., Aila, T., Kautz, J.: Pruning convolutional neural networks for resource efficient inference. In: ICLR (2017)
21. Nguyen, T.A., Tran, A.: Input-aware dynamic backdoor attack. In: NeurIPS (2020)
22. Nguyen, T.A., Tran, A.T.: Wanet - imperceptible warping-based backdoor attack. In: ICLR (2021)
23. Nurvitadhi, E., Sheffield, D., Sim, J., Mishra, A.K., Venkatesh, G., Marr, D.: Accelerating binarized neural networks: comparison of FPGA, CPU, GPU, and ASIC. In: International Conference on Field-Programmable Technology (2016)
24. Osborne, M.R., Presnell, B., Turlach, B.A.: On the lasso and its dual. J. Comput. Graph. Stat. **9**(2), 319–337 (2000)
25. Parashar, A., et al.: SCNN: an accelerator for compressed-sparse convolutional neural networks. In: Proceedings of the 44th Annual International Symposium on Computer Architecture, ISCA (2017)
26. Shumailov, I., Zhao, Y., Bates, D., Papernot, N., Mullins, R.D., Anderson, R.: Sponge examples: energy-latency attacks on neural networks. In: EuroS&P (2021)
27. Simonyan, K., Zisserman, A.: Very deep convolutional networks for large-scale image recognition. In: ICLR (2015)

28. Xu, J., Li, Z., Du, B., Zhang, M., Liu, J.: Reluplex made more practical: leaky relu. 2020 IEEE Symposium on Computers and Communications (ISCC) (2020)
29. Zhou, A., Yao, A., Guo, Y., Xu, L., Chen, Y.: Incremental network quantization: towards lossless CNNs with low-precision weights. In: ICLR (2017)
30. Zhou, Z., Zhou, W., Li, H., Hong, R.: Online filter clustering and pruning for efficient convnets. In: 2018 25th IEEE International Conference on Image Processing (ICIP). IEEE (2018)

The *Specchieri MarVen* Dataset: an Abbreviation-Rich Dataset in Venetian Idiom

Sara Ferro[1,3] , Debora Pasquariello[2] , Marcello Pelillo[1,3] ,
and Arianna Traviglia[3(✉)]

[1] DAIS, Ca' Foscari University, via Torino 155, 30172 Venice, Italy
{sara.ferro,pelillo}@unive.it
[2] DSU, Ca' Foscari University, Dorsoduro 3484/D, 30123 Venice, Italy
868073@stud.unive.it
[3] Istituto Italiano di Tecnologia, Center for Cultural Heritage Technology,
via Torino 155, 30172 Venice, Italy
{Sara.Ferro,Arianna.Traviglia}@iit.it

Abstract. Despite the release of numerous datasets for training models in historical handwritten text recognition, there is still a significant need for more diverse and extensive data. This paper aims to contribute to bridging this gap by introducing a new dataset comprising 159 pages from an Early Modern age volume part of the Venetian 'Marigold' collection. The dataset contains various abbreviations that are key to transcribing for a complete understanding of the content. To accommodate different research needs, the dataset is released in two versions: one with 'expanded' abbreviations and another without abbreviations – where the abbreviations are removed –, aligning with the choices made for other released datasets. Additionally, the dataset encompasses two distinct writing styles, leading us to provide three separate splits for training and evaluating machine learning models: one with a combination of both styles and two individual splits for each style. The qualitative and quantitative characteristics of all dataset configurations are analysed. In addition, three diverse architectures for handwritten text recognition are trained to assess their performances on this dataset. The dataset is available for download at https://doi.org/10.48557/GJYJTW.

Keywords: Historical · Dataset · Handwriting · Recognition · Venetian · Language · Abbreviations

1 Introduction

Although the increase of digitised historical documents from various periods and regions around the world has widened the availability of digital materials for training algorithms to transcribe automatically handwritten historical texts, there are still numerous inherent complexities in the automatic letter/word recognition of ancient and historical texts. Besides the broad range of writings across various scripts, centuries and regions, there are also significant variations in the shape of the letters/graphemes due to the use of different writing tools

G. L. Foresti et al. (Eds.): ICIAP 2023, LNCS 14234, pp. 527–539, 2023.
https://doi.org/10.1007/978-3-031-43153-1_44

within the same script group. Furthermore, individual scribes often employed distinct handwriting styles, leading to noticeable differences, for example, in letter shape and size.

Models for automatic handwriting transcription based on Deep Neural Networks (DNNs) have demonstrated promising performances [14,22], thanks to the introduction of sequential models – aligning with the sequential nature of handwriting scripts [8,15,16,19] –, attention-based models [13], or a combination of both [1]. However, they have not yet realised their full potential. The success of these models is heavily reliant on the data they are trained on, which means that they may struggle to transcribe accurately graphemes or languages they have not been exposed to before. Even the most sophisticated and advanced models cannot accurately transcribe a grapheme, a character, or a word that they have never encountered before. The improvement of these models is thus closely dependent on the availability of new, richer, and more diverse datasets to expand the data used for model training.

The inclusion of scribal abbreviations in the text presents an additional and significant challenge. This challenge becomes particularly pronounced when transcribing manuscript texts that heavily feature abbreviations. Abbreviations were originally employed by scribes and copyists to maximize the use of available writing space but later evolved into standardised symbols that varied across different regions. Abbreviations generally fall into these categories: suspensions – the end of a word is abbreviated, and they are identified by the use of a horizontal bar or another graphic symbol –, contractions – where another part of a word is abbreviated with the use of a graphic symbol –, and conventional symbols – where the symbols represent complete words –. All these abbreviations could be used in the same manuscript[1]. Due to the limited availability of datasets with a complete resolution of abbreviations, automatic transcription of historical texts has often overlooked them. Consequently, generated transcriptions omit crucial textual information.

This work introduces a new dataset explicitly tailored for training machine learning models in historical handwriting recognition for abbreviation-rich texts. The dataset is unique in that it encompasses Latin characters derived from old Venetian language documents dating back to the 18th century offering also the 'expansion' of the abbreviations, $i.e.$, their comprehensive transcription. As a result, it contributes significantly to the advancement of more precise models for historical handwriting recognition, particularly in cases involving the presence of abbreviations.

2   Related Works

Several datasets have been released to train machine-learning models for historical handwriting recognition, offering varying levels of segmentation and transcription. These datasets can range from character-level to page-level granularity. While recent models have demonstrated success in recognising text at

---

[1] http://hist.msu.ru/departments/8823/projects/Cappelli/.

paragraph-level [3] or on relatively homogeneous pages [4], the optimal balance between effort in creating labelled datasets and the resulting quality of trained models is achieved using line-level datasets. Additionally, line-level datasets are the most commonly used in historical handwriting recognition research.

Notably, Fischer *et al.* presented benchmark datasets for historical handwriting recognition at line-level, including the Saint Gall dataset [5], and the Parzival and Washington datasets [6][2]. The Saint Gall dataset contains images of a 9th century Latin manuscript written by a single individual. The Parzival dataset consists of images of a 13th century manuscript written in Medieval German by three different copyists. The Washington dataset contains pages from the George Washington Papers[3], written in the 18th century by George Washington and his associate.

Additional datasets have been released during competitions held in the *International Conference on Frontiers of Handwriting Recognition (ICFHR)*. In the 2014 competition, Sánchez *et al.* [17] proposed a dataset consisting of pages from the Bentham Papers collection, which were hand-written in English. The collection includes writings from multiple authors, including the English philosopher Jeremy Bentham (1748 − 1832) and his collaborators. The primary objective of the competition was to assess the performance of historical handwriting recognition models rather than modern handwriting recognition. In the 2016 competition, Sánchez *et al.* [18] released a dataset derived from the German Ratsprotokolle collection[4]. It consists of minutes of the council meetings held from 1470 to 1805. The goal of the competition was to assess the performances of Handwritten Text Recognition (HTR) models in languages structurally different from English. In the 2018 competition, Strauß *at al.* [21] presents a dataset composed of pages taken from 22 heterogeneous documents. Each document was written in different time periods and by different writers[5]. These texts are written in Italian, as well as modern and medieval German. The aim of the competition was to evaluate the minimum amount of training data required to transcribe a historical document accurately.

In 2022, Cascianelli *et al.* [2] introduced the Ludovico Antonio Muratori (LAM) dataset, the most extensive line-level historical dataset for training handwriting recognition models in Western languages. The dataset comprises of manuscripts written by a single author, the Italian historian Ludovico Antonio Muratori, from 1691 to 1750.

3   The *Specchieri MarVen* Dataset

The *Specchieri MarVen* dataset contains images from a collection of Venetian historical documents known as 'The Marigold Collection' ('Mariegole' in Italian),

---

[2] https://fki.tic.heia-fr.ch/databases/iam-historical-document-database.

[3] https://www.loc.gov/collections/george-washington-papers/about-this-collection/.

[4] http://stadtarchiv-archiviostorico.gemeinde.bozen.it/bohisto/Archiv/Handschrift/detail/14492.

[5] https://scriptnet.iit.demokritos.gr/competitions/10/.

housed at the Library of the Correr Museum[6] in Venice (Italy). The Correr Museum is known for its extensive collection of historical artefacts, which has been described as a 'treasure' by the media [23]. The museum has he largest library among the Venetian Civic Museums, and it began with the collections of Teodoro Correr. The Marigolds fund was established to preserve these Marigold books. The Marigold books contain the statutes - *i.e.*, the regulations - of the devotional brotherhoods, and associations or corporations of the arts and crafts in Venice during the Middle and Early Modern Ages. The *Specchieri MarVen* dataset consists of annotated pages from a Marigold book specifically focusing on regulations for mirror manufacturers and dealers[7].

The Marigold volume containing the annotated pages was created between 1744 and 1768. For this reason, it has an 18th-century writing style, but the lexicon and graphemes dates back to three centuries, from the 16th to the 18th century, leading to increasing transcription difficulties. To the best of our knowledge, the *Specchieri MarVen* dataset is unique in that it is the first dataset (among those created for automatic handwriting recognition) containing text in (old) Venetian language. In addition, it encompasses numerous abbreviations that present a significant transcription challenge.

3.1   Dataset Collection, a Collaborative Approach

The dataset was annotated using the *labelme* [24] software. While there are other software options available to annotate data for handwriting recognition, many of them are designed for collaborative use and are web-based applications. In our case, we chose a software that can be installed as a desktop application to better suit our needs.

Some images did not depict vertical pages, resulting in tilted lines of text. To accommodate this, also polygonal shapes were used for data annotation, as the rectangle shape option in *labelme* only allows for rectangles with horizontal and vertical edges. Subsequently, the data were processed to identify the smallest rectangles that encompassed the polygon areas, resulting in exclusively rectangular shapes. Following this procedure, the images were cropped to rectangular dimensions.

When it comes to datasets for historical handwriting recognition, there is no standard approach for accurately transcribing historical documents suitable for machine learning models. Different methods are used, such as removing abbreviations or providing a "diplomatic transcription" that transcribes exactly what is present in the image. However, in our case, we made the decision to opt for a full transcription that includes what we refer to as 'expanded' abbreviations. This decision was made because achieving a meaningful understanding of these documents is only possible with a complete transcription.

---

[6] https://correr.visitmuve.it/en/home/.
[7] Venezia, Biblioteca del Museo Correr, MS. Classe IV 35.

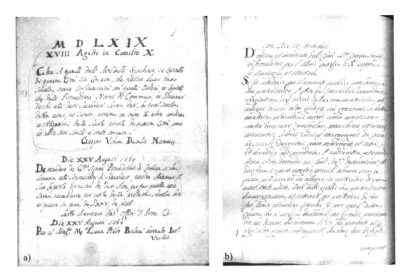

Fig. 1. a) and b) present images with Style 0 writing style (*Ms_Cl_IV_035_001.jpg*) and Style 1 writing style (*Ms_Cl_IV_035_068.jpg*), respectively.

3.2 Released Data

The data are released in two distinct versions[8]: with 'expanded' abbreviations and without abbreviations. Furthermore, these realisations contain two distinct folders. One contains the line-level dataset splits for training, validating and testing machine learning models, including the general split and two style-specific splits named Style 0 and Style 1. Figure 1 illustrates the two writing styles. The other folder categorises the data into semantically different subfolders. This allows interested researchers to conduct further analyses and gain a comprehensive understanding of the data. In this second folder, data are organised into directories based on their semantically distinct meanings. The folders include 'Standard text', 'Titles', 'First letters', 'List numbers' and 'Catchwords'. The 'Standard text' folder contains line images and transcriptions of the main text. The 'Titles' folder includes images and text data of lines of text found in titles. The 'First letters' folder contains images and labels of the first letters that initiate paragraphs, while the 'List numbers' folder contains the images and labels of numbers found in lists. The 'Catchwords' folder presents images and transcriptions of words that are the first words of the following quire written outside the marginal space at the end of the previous one. The purpose of these words is to facilitate the arrangement of the quires during binding. In this manuscript, however, these words are used more often because they can be found at the end of every verso.

---

[8] The dataset for training HTR models can be downloaded at https://doi.org/10.48557/GJYJTW. Moreover, original images of the manuscript can be downloaded at https://doi.org/10.48557/WXMOCS.

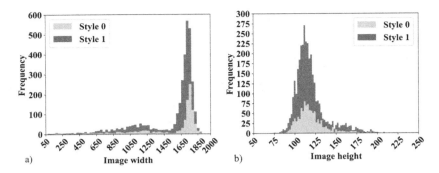

a)   Image width   b)   Image height

Fig. 2. a) Stacked histograms of the image width and b) Stacked histograms of the image height of the Style splits (Style 0 and Style 1, the histogram of the image width and height of the General split is given by the composition of the two).

The images with their respective transcriptions in 'Standard text' and 'Titles' folders are the samples we have considered in the released line-level datasets.

General Split. In this split, all the 159 pages of the *Specchieri MarVen* dataset are considered to create the training, validation and test split (see Table 1). Both styles of writing are represented in all sets. Therefore, each set includes samples from both styles of writing.

Table 1. Cardinalities of the training, validation and test sets for the different splits.

Dataset	Card. training set	Card. val. set	Card. test set	Tot. lines
General	2412	633	405	3450
Style 0	738	159	159	1056
Style 1	1674	474	246	2394

Style-Based Splits. To create the training, validation, and test splits, we separately consider the styles of Style 0 and Style 1. This approach results in two sets of data corresponding to the images with different styles. Each set is divided into training, validation, and test set (see Table 1).

Figure 2 presents the stacked histograms illustrating the distribution of image data width and height for both writing styles. Additionally, Fig. 3 shows the stacked histograms depicting text data length with 'expanded' abbreviations and without abbreviations. It is possible to see that the data from the two different styles show similarities in distributions (Figs. 4 and 5).

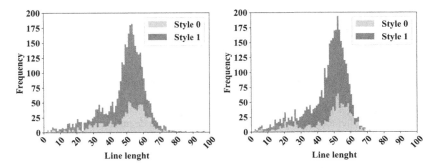

Fig. 3. Stacked histograms of the line length for the 'expanded' abbreviation dataset and b) Stacked histograms of the line length of the datasets without abbreviations of the two Style splits (Style 0 and Style 1, the histogram of the line length of the General split is given by the composition of the two).

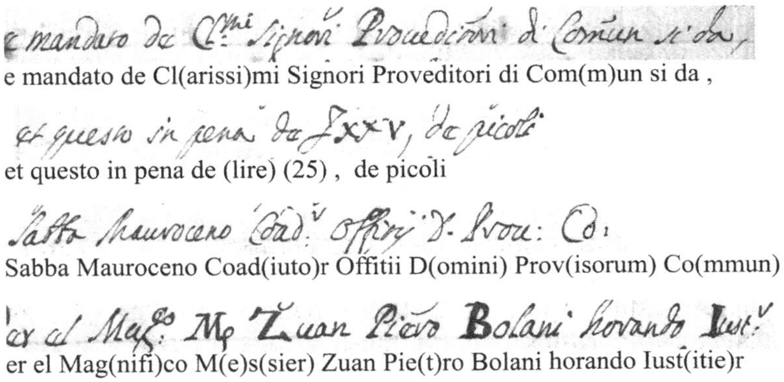

e mandato de Cl(arissi)mi Signori Proveditori di Com(m)un si da ,

et questo in pena de (lire) (25) , de picoli

Sabba Mauroceno Coad(iuto)r Offitii D(omini) Prov(isorum) Co(mmun)

er el Mag(nifi)co M(e)s(sier) Zuan Pie(t)ro Bolani horando Iust(itie)r

Fig. 4. Examples of abbreviations in the dataset of Style 0.

4   Experiments

Three models for handwriting recognition were trained to provide the baseline results for this new dataset, both on the general split and the style-based splits of the *Specchieri MarVen* dataset. Two classical metrics, the Character Error Rate (CER) and the Word Error Rate (WER), were used to evaluate the errors of the handwriting recognition models. These measures are based on the Levenshtein distance [12].

4.1   Considered HTR Models

Three variations of a standard architecture for automatic handwriting recognition, the CRNN, were utilised, each with different characteristics. The CRNN network was chosen due to its ability to yield good results even with a relatively small number of training samples. Our experiments demonstrate that this type of

possino esser accettadi alla prova de Maistri p(ri)ma

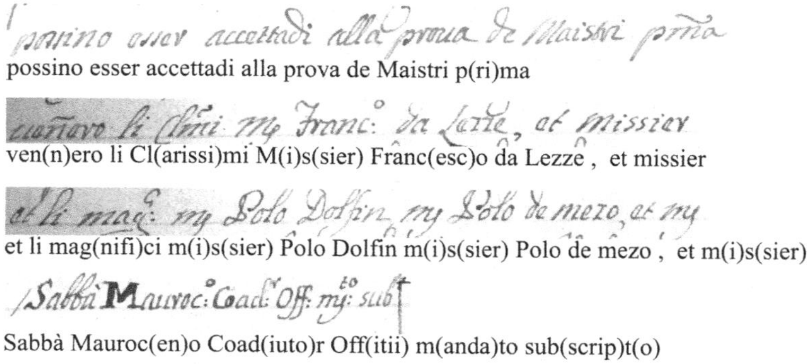

ven(n)ero li Cl(arissi)mi M(i)s(sier) Franc(esc)o da Lezze , et missier

et li mag(nifi)ci m(i)s(sier) Polo Dolfin m(i)s(sier) Polo de mezo , et m(i)s(sier)

Sabbà Mauroc(en)o Coad(iuto)r Off(itii) m(anda)to sub(scrip)t(o)

Fig. 5. Examples of abbreviations in the dataset of Style 1.

architecture yields good results, giving low CER and WER values. These experiments were conducted to provide guidance to future researchers who utilise the dataset, assisting them in determining the most suitable implementation of the network for the handwriting recognition challenge presented by the *Specchieri MarVen* dataset.

The used model comprises both Convolutional Neural Network (CNN) layers and Recurrent Neural Network (RNN) layers. The Convolutional layers are adept at extracting informative features from visual data [11], while the Recurrent layers are capable of capturing sequential patterns in the data. The CRNN has proven to achieve optimal performance while being more computationally efficient than the Multi-Dimensional RNN (MDRNN) previously used [15,18]. Three different variations of this model type have been utilised, each with unique characteristics. The model proposed by Shi *et al.* (2016) [19] presents a deeper convolutional component. Instead, the model developed by Puigcerver (2017) [15] has a deeper recurrent component. Lastly, we considered the more recent implementation by Retsinas *et al.* (2022) [16], which features a deep convolutional part with both classical convolutional layers and ResNet blocks [7], along with a relatively deep recurrent component.

The architecture proposed by Shi *et al.* features a model with the convolutional part based on VGG-11 [20], extending up to the sixth convolutional block, then presents a convolutional block with a 2×2 kernel. Notably, the 3rd and 4th max-pooling operations use 1×2 stride instead of the conventional 2×2 to accommodate for text data. They also added two Batch Normalization (BN) [9] layers after the 3rd, 5th, and 7th convolutional layers, which helped while training a network of such depth. The recurrent part of the model consists of two Bidirectional Long-Short Term (BLSTM) layers. On the other hand, the architecture introduced by Puigcerver comprises a model with five convolutional blocks, each equipped with 3×3 kernels and a number of channels equal to $n \cdot 16$, where n ranges from 1 to 5 corresponding to the block number. BN is applied after each convolutional block, and LeakyReLU serves as the activation function. Following each convolutional block, max-pooling with non-overlapping

2×2 kernel is performed. The recurrent part of the model consist of five BLSTM layers. Retsinas *et al.* propose an architecture with a deep convolutional component, having a similar number of filters as the Shi *et al.* architecture. However, two notable differences distinguish this network from the previously introduced ones. Firstly, alongside the conventional convolutional blocks, ResNet blocks are also utilised. Secondly, the connection between the convolutional part and the recurrent part is established through column-wise max-pooling instead of the traditional column-wise concatenation. The recurrent component of the model comprises three BLSTM layers.

4.2   Data Augmentation and Training Strategy

The images were preprocessed to standardise their height to 128 pixels, while maintaining their original aspect ratio. Furthermore, they were centered within a width of 1024 pixels. In cases where it was necessary, left and right padding with median intensity was applied. Additionally, classical data augmentation techniques were used. Indeed, rotations and translations, and Gaussian blur filter of kernel 3×3 with a randomly chosen standard deviation in the range $\sigma \in [1.0, \ 2.0]$ were performed.

The Adam optimiser [10], along with a learning rate scheduler that decreases the learning rate if the validation CER does not decrease below a predefined threshold for 80 epoch, were utilised for training the models. The initial learning rate was $1E - 4$ to train both the Shi *et al.* architecture and the Puigcerver one. Instead, an initial leaning rate of $1E - 3$ was used to train the architecture of Retsinas *et al.*. The learning rate reduction factor was 0.1, and the threshold for decreasing the learning rate was $1E - 2$ of the validation CER. Models were trained for 240 epochs.

5   Results and Discussion

Table 2 presents the results of the three different models on the general split and style-based splits using data with 'expanded' abbreviations. Table 3, instead, shows the results of the three different models on the general split and style-based splits using data without abbreviations. The comparison of the two tables reveals that the models exhibit superior performance in the scenario where abbreviations are not present. This is evident from the lower validation and test CER and WER achieved across all three distinct splits.

Table 2. Evaluation results on the 'expanded' abbreviation realisation with the General, Style 0, and Style 1 splits of the *Specchieri MarVen* dataset.

Model	General			
	Val. CER	Val. WER	Test CER	Test WER
Shi *et al.* [19]	5.23%	15.06%	5.49%	16.35%
Puigcerver [15]	6.11%	16.72%	5.83%	16.82%
Retsinas *et al.* [16]	4.24%	11.89%	4.05%	12.57%
Model	Style 0			
	Val. CER	Val. WER	Test CER	Test WER
Shi *et al.* [19]	8.86%	25.99%	9.49%	28.66%
Puigcerver [15]	13.01%	36.70%	13.83%	38.56%
Retsinas *et al.* [16]	5.69%	16.36%	6.71%	19.68%
Model	Style 1			
	Val. CER	Val. WER	Test CER	Test WER
Shi *et al.* [19]	5.61%	15.85%	4.79%	14.12%
Puigcerver [15]	6.62%	17.23%	5.69%	16.34%
Retsinas *et al.* [16]	4.44%	11.54%	3.47%	9.97%

Table 3. Evaluation results on the realisation without abbreviations with the General, Style 0, and Style 1 splits of the *Specchieri MarVen* dataset.

Model	General			
	Val. CER	Val. WER	Test CER	Test WER
Shi *et al.* [19]	3.42%	12.85%	3.69%	14.41%
Puigcerver [15]	4.00%	13.46%	4.09%	14.22%
Retsinas *et al.* [16]	2.52%	9.13%	2.80%	10.47%
Model	Style 0			
	Val. CER	Val. WER	Test CER	Test WER
Shi *et al.* [19]	7.11%	24.98%	7.87%	27.58%
Puigcerver [15]	11.06%	34.42%	11.51%	34.32%
Retsinas *et al.* [16]	3.97%	14.93%	4.94%	18.27%
Model	Style 1			
	Val. CER	Val. WER	Test CER	Test WER
Shi *et al.* [19]	3.33%	12.31%	2.74%	10.68%
Puigcerver [15]	3.89%	13.40%	3.73%	13.40%
Retsinas *et al.* [16]	2.53%	9.19%	2.26%	8.49%

6 Conclusions

The availability of diverse digitised datasets plays a crucial role in attaining an authentic digital representation of historical records. In line with this perspective, we have introduced a new dataset at the line-level, aiming to foster the progress of historical handwriting recognition. This dataset serves as a valuable resource for researchers and practitioners working towards improving the accuracy and performance of recognition models for historical handwriting. It provides valuable training data that helps enhance the accuracy and efficiency of transcription automation. Although the recognition of datasets containing historical abbreviations may present reduced performance, it is crucial to provide and utilise such datasets. When abbreviations are removed or a diplomatic transcription is used, it can lead to inaccurate recognition and processing of certain words, affecting their correct interpretation and analysis. By incorporating these challenging cases into the training and evaluation of recognition models, we can improve their overall performance and enhance their ability to accurately transcribe historical texts. This approach allows us to address the specific complexities of historical handwriting and move towards more effective and robust transcription models. Obtaining an accurate transcription with 'expanded' abbreviations is crucial to enable further data processing and analysis, such as developing models for translating historical languages into modern ones. In future work, our objective is to delve into the exploration and definition of the optimal combination of dataset pre-processing, recognition models, and post-processing techniques for this type of datasets. By doing so, we aim to develop a robust model that can accurately transcribe the full transcript of a text containing historical abbreviations. This endeavour will enhance the effectiveness and reliability of transcription methods for historical documents.

Acknowledgements. We thank the *Biblioteca del Museo Correr* for granting us permission to utilise the images included in this dataset. Additionally, the support of Prof. Dorit Raines is gratefully acknowledged.

References

1. Bluche, T., Louradour, J., Messina, R.: Scan, attend and read: end-to-end handwritten paragraph recognition with MDLSTM attention. In: 2017 14th IAPR International Conference on Document Analysis and Recognition (ICDAR), vol. 1, pp. 1050–1055. IEEE (2017)
2. Cascianelli, S., et al.: The LAM dataset: a novel benchmark for line-level handwritten text recognition. In: International Conference on Pattern Recognition (2022)
3. Coquenet, D., Chatelain, C., Paquet, T.: End-to-end handwritten paragraph text recognition using a vertical attention network. IEEE Trans. Pattern Anal. Mach. Intell. **45**(1), 508–524 (2022)
4. Coquenet, D., Chatelain, C., Paquet, T.: DAN: a segmentation-free document attention network for handwritten document recognition. IEEE Trans. Pattern Anal. Mach. Intell. (2023)

5. Fischer, A., Frinken, V., Fornés, A., Bunke, H.: Transcription alignment of Latin manuscripts using hidden Markov models. In: Proceedings of the 2011 Workshop on Historical Document Imaging and Processing, pp. 29–36 (2011)
6. Fischer, A., Keller, A., Frinken, V., Bunke, H.: Lexicon-free handwritten word spotting using character HMMs. Pattern Recogn. Lett. **33**(7), 934–942 (2012)
7. He, K., Zhang, X., Ren, S., Sun, J.: Deep residual learning for image recognition. arxiv 2015. arXiv preprint arXiv:1512.03385, vol. 14 (2015)
8. Hochreiter, S., Schmidhuber, J.: Long short-term memory. Neural Comput. **9**(8), 1735–1780 (1997)
9. Ioffe, S., Szegedy, C.: Batch normalization: accelerating deep network training by reducing internal covariate shift. In: International Conference on Machine Learning, pp. 448–456. PMLR (2015)
10. Kingma, D.P., Ba, J.: Adam: a method for stochastic optimization. arXiv preprint arXiv:1412.6980 (2014)
11. Krizhevsky, A., Sutskever, I., Hinton, G.E.: Imagenet classification with deep convolutional neural networks. In: Pereira, F., Burges, C.J.C., Bottou, L., Weinberger, K.Q. (eds.) Advances in Neural Information Processing Systems, vol. 25, pp. 1097–1105. Curran Associates, Inc. (2012)
12. Levenshtein, V.I.: Binary codes capable of correcting deletions, insertions, and reversals. In: Soviet Physics Doklady, vol. 10, pp. 707–710. Soviet Union (1966)
13. Li, M., et al.: TrOCR: transformer-based optical character recognition with pretrained models. arXiv preprint arXiv:2109.10282 (2021)
14. Lombardi, F., Marinai, S.: Deep learning for historical document analysis and recognition-a survey. J. Imaging **6**(10), 110 (2020)
15. Puigcerver, J.: Are multidimensional recurrent layers really necessary for handwritten text recognition? In: 2017 14th IAPR International Conference on Document Analysis and Recognition (ICDAR), vol. 1, pp. 67–72. IEEE (2017)
16. Retsinas, G., Sfikas, G., Gatos, B., Nikou, C.: Best practices for a handwritten text recognition system. In: Uchida, S., Barney, E., Eglin, V. (eds.) DAS 2022. LNCS, vol. 13237, pp. 247–259. Springer, Cham (2022). https://doi.org/10.1007/978-3-031-06555-2_17
17. Sánchez, J.A., Romero, V., Toselli, A.H., Vidal, E.: ICFHR 2014 competition on handwritten text recognition on transcriptorium datasets (HTRTS). In: 2014 14th International Conference on Frontiers in Handwriting Recognition, pp. 785–790. IEEE (2014)
18. Sánchez, J.A., Romero, V., Toselli, A.H., Vidal, E.: ICFHR 2016 competition on handwritten text recognition on the read dataset. In: 2016 15th International Conference on Frontiers in Handwriting Recognition (ICFHR), pp. 630–635 (2016). https://doi.org/10.1109/ICFHR.2016.0120
19. Shi, B., Bai, X., Yao, C.: An end-to-end trainable neural network for image-based sequence recognition and its application to scene text recognition. IEEE Trans. Pattern Anal. Mach. Intell. **39**(11), 2298–2304 (2016)
20. Simonyan, K., Zisserman, A.: Very deep convolutional networks for large-scale image recognition. arXiv preprint arXiv:1409.1556 (2014)
21. Strauß, T., Leifert, G., Labahn, R., Hodel, T., Mühlberger, G.: ICFHR 2018 competition on automated text recognition on a read dataset. In: 2018 16th International Conference on Frontiers in Handwriting Recognition (ICFHR), pp. 477–482. IEEE (2018)
22. Teslya, N., Mohammed, S.: Deep learning for handwriting text recognition: existing approaches and challenges. In: 2022 31st Conference of Open Innovations Association (FRUCT), pp. 339–346. IEEE (2022)

23. Vanin, B., Eleuteri, P.: Le mariegole della biblioteca del Museo Correr. Marsilio Editori (2007)
24. Wada, K.: Labelme: image polygonal annotation with python (v4.6.0) (2021). https://doi.org/10.5281/zenodo.5711226

Compensation for Patient Movements in CBCT Imaging for Dental Applications

Abdul Salam Rasmi Asraf Ali[1,2]($\boxtimes$), Cristina Sarti[2], and Claudio Landi[2]

[1] Department of Engineering and Architecture, University of Udine, Udine, Italy
cefi.salam@gmail.com
[2] See Through s.r.l., Brusaporto, Italy

Abstract. Cone Beam Computed Tomography (CBCT) is a popular imaging technique used in dentistry for diagnosis and treatment planning. However, patient movement during the scanning process can lead to image artifacts that can affect the accuracy of diagnosis and treatment planning. To address this issue, manufacturers have added mechanical fixations to CBCT systems to reduce patient motion, but these fixations are not completely rigid and still allow for some movement. Researchers have developed motion compensation algorithms to reduce the impact of motion artifacts on CBCT images. Nevertheless, these algorithms can be time-consuming and only partially address the problem. This paper proposes a new approach to motion compensation in CBCT imaging to overcome these limitations. Unlike traditional methods which use iterative approaches that require multiple reconstructions in the compensation stage, this new method uses a single initial reconstruction computed with the Feldkamp-Davis-Kress (FDK) algorithm. The reconstructed volume is regularized before the motion compensation step. Motion compensation is achieved by optimizing the motion parameters using a regularized 3D-2D image registration. The results show that the proposed non-iterative motion compensation approach effectively reduces motion-induced artifacts in the final reconstruction.

Keywords: Dental CBCT · motion compensation · realistic motion simulation · non-iterative method

1 Introduction

Cone Beam Computed Tomography (CBCT) is a three-dimensional imaging method widely used in dentistry that allows the visualization of the maxillofacial region from any viewpoint [13]. The usage of CBCT has increased significantly in various fields such as orthodontics, endodontics, oral surgery, periodontics, and restorative dentistry [15]. Unlike clinical helical computed tomography (CT), CBCT systems have acquisition time ranging from 5.4 to 40 s [11], which is often long enough for significant patient motion to occur. It is estimated that approximately 21–42% of the in vivo examinations exhibit motion artifacts [10,19]. Patients move during the scan for various reasons such as fear of the tube/detector movement [24], Parkinson's disease [4], or simply because it is difficult for someone to stay idle for quite a long time, especially children [19].

Many efforts have been made to prevent patient motion during CBCT acquisition. Patients are generally immobilized using head supports, chin rests, and a fixed bite block. However, relying on head fixation devices may not prevent all potential motions [6], since as small as 3 mm of motion displacement can significantly affect the image quality [18]. Several methods have been proposed in the past to compensate for patient motion during the CBCT scan. These methods include 2D-3D registration processes [2, 12, 20, 21], epipolar consistency-based approaches [1, 23], optimization of data fidelity terms [14, 22], and auto-calibration-based approaches [9]. However, these methods have several drawbacks, such as requiring multiple reconstructions of the full set of projections, which is time-consuming, and inconsistencies due to data truncation and scatter, which are common problems in CBCT imaging. Additionally, most manufacturers of maxillofacial CBCT scanners (Sirona Dental Systems, KaVo Dental, VATECH, Carestream Dental) do not offer motion compensation features.

In this paper, we propose a novel method that addresses these drawbacks. Our goal is to compensate for rigid motions that significantly degrade the quality of the reconstruction without requiring multiple full-scan reconstructions in the compensation stage. To achieve this, our method utilizes the Feldkamp-Davis-Kress (FDK) algorithm [5] for the initial reference reconstruction. The volume is regularized to be robust prior to the following optimization steps consisting of 3D-2D image registration. Compared to other methods our method is more efficient in terms of time consumption and robustness compensating for motion-induced artifacts in CBCT images even in the presence of data truncation and scatter.

2   Materials and Methods

2.1   Data Acquisition

A head CBCT scanner at See Through s.r.l. in Brusaporto, Italy, was used to acquire the projection images. The distance between the X-ray source and the detector was 741 mm. The detector pixel size was 0.24 mm by 0.24 mm, and the scan field-of-view was 105 mm by 110 mm. The scan was performed for a full rotation (360°) with a duration of 24 s.

In this experiment, we used a phantom containing the skull and teeth of a deceased patient, cast in a uniform plastic resin that approximates the X-ray attenuation of human soft tissue. Phantoms are well suited for conducting maxillofacial studies with CBCT because they allow for experiments to be conducted in a setup similar to real-world clinical scenarios. CBCT is more proficient in visualizing hard tissue (bones) than soft tissues, so using a phantom helps to ensure that the results of the experiment are accurate.

2.2   Motion Simulation

There are different ways to simulate motion-affected data. Some researchers use robots or other devices to move the phantom during scanning [18]. Other research groups model the motion of the phantom with a digital simulation [8] which we adopted in this study. In CBCT scanners, the knowledge of the acquisition geometry is crucial for reconstructing the volume of the scanned object. This geometry is encoded in the relative positions of the X-ray source, detector, and scanned object for each projection. Projection matrices ($\mathbf{P}$) of the same type as the pinhole camera model [7] are typically used to describe this geometry. These matrices are necessary for reconstructing the volume of the object using either the FDK or iterative techniques.

$$\mathbf{P}_i = \mathbf{K} \cdot \mathbf{G}_i, \ i = 1...N \tag{1}$$

Here, N is the number of projections, $\mathbf{K}$ is the 3×4 intrinsic matrix describing detector parameters such as the orientation of the detector coordinates system with respect to the x-ray source (camera) coordinate system and the pixel size, $\mathbf{G}_i = [\mathbf{R}_i | \mathbf{t}_i]$ is the 4×4 roto-translation matrix describing the orientation of the x-ray source coordinate system with respect to the patient (world) coordinate system for projection i. These projection matrices can also be conveniently used to simulate motion by altering their parameters. Since any jaw movement is usually prevented by a fixed bite block and a chin rest, we consider only rigid motions here. These motions can be described as a combination of rotations and translations. Let $\mathbf{G}_i' = [\mathbf{R}_i' | \mathbf{t}_i']$ be the matrix describing the patient motion at scan position i. To simulate motion, we can then replace the projection matrix $\mathbf{P}_i$ with $\mathbf{P}_i'$

$$\mathbf{P}_i' = \underbrace{\mathbf{K} \cdot \mathbf{G}_i}_{\mathbf{P}_i} \cdot \mathbf{G}_i' \tag{2}$$

This approach allows the simulation of a broader set of motion patterns than in a few restricted clinical scenarios. Also, clinical data would need to be accompanied by ground-truth motion measurement, which would require a complex setting, difficult to implement in a clinical environment. This would likely prevent the collection of large amounts of data. Figure 1 illustrates a comparison between artifacts observed in real motion-affected data and data with digitally simulated motion. The real motion-affected data was obtained by physically moving the phantom during scanning. A specific setup was used to induce a motion resembling lateral rotation.

To simulate realistic motion, we rely on the motion studies of Spin-Neto [16–19]. Two **returning motions, nodding** and **tilting** were simulated for a duration of 3 and 4 s respectively, for a full 360° scan of 24 s. Due to mechanical fixations such as head support, it is reasonable to expect that the patient's movement is limited, causing them to return to their initial position and confining the motion to a smaller scanning region [18]. To test the robustness of the method, a **non-returning abrupt motion** was also simulated, where the patient remains in one pose for part (almost 6 s) of the scan, and then suddenly moves to another pose and remains still for the rest of the scan. This created double contours in the reconstruction (Fig. 8a), which is one of the most difficult cases to compensate for.

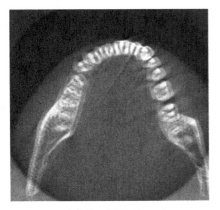
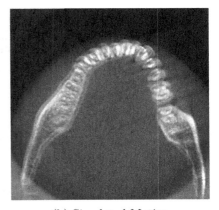

(a) Real Motion   (b) Simulated Motion

Fig. 1. Comparison between artifacts from real and simulated motion for lateral rotation.

3   Motion Compensation

The proposed method for motion-compensated CBCT reconstruction is a non-iterative approach that eliminates the need for repeated reconstructions during the compensation stage. The pipeline of the proposed method is illustrated in Fig. 2. The process begins with an initial reference reconstruction generated from the acquired projections as explained in Sect. 3.1. Each projection is then registered to a synthetic projection, which is created by reprojecting the reference reconstruction. To enhance the registration accuracy, the reference reconstruction undergoes regularization before being reprojected (Sect. 3.2).

Fig. 2. Pipeline of the Proposed Method

The compensation algorithm is depicted in Fig. 3. The algorithm performs registration of the original projections with the synthetic forward projections by optimizing the similarity score, which is obtained using the cost function explained in Sect. 3.3. The optimization process changes the motion parameters until convergence is reached.

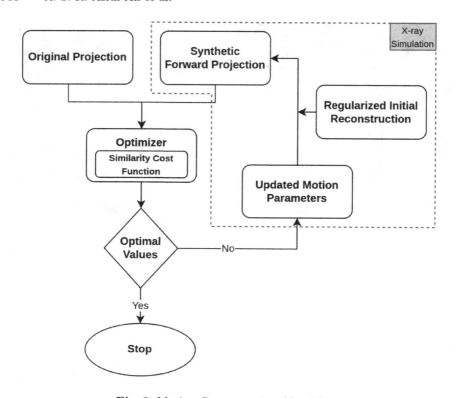

Fig. 3. Motion Compensation Algorithm

A significant advantage of this method is that the same reference reconstruction is used for all projections, allowing the optimizer to run in parallel for multiple projections. A final high-quality reconstruction is performed using the estimated motion parameters.

3.1   Reference Reconstruction

By utilizing all the acquired original projections, the FDK algorithm generates an initial reference reconstruction (Fig. 4a). However, this reconstruction is affected by motion artifacts, including blurriness, streaks, and double contours. Reprojecting the reference reconstruction leads to the occurrence of overlapping structures in the forward projection, which makes their registration with the original projections more challenging. To overcome this difficulty, a regularization step is implemented to refine the reference reconstruction.

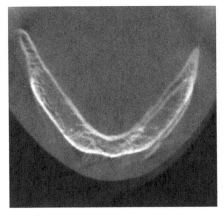
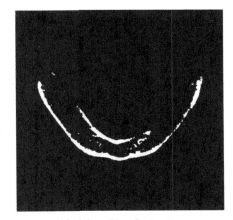

(a) Reference Reconstruction   (b) After Regularization

Fig. 4. Reference reconstruction before and after regularization.

3.2   Regularization

Regularization of the reference reconstruction is necessary for three main reasons. Firstly, as mentioned above, motion artifacts like double contours or strong streaks affect the reference reconstruction. During reprojection, these types of artifacts may generate ambiguous overlapping structures on the forward projections which are not present in the real projections. Second, small structures in 2D projections are very sensitive to motion. Since the optimizer makes a step-by-step modification of the motion parameters, small structures such as those in the trabecular bone of the jaw may appear and disappear at each optimization step. They may overlap with other small or large structures. The third reason for regularization is the presence of Poisson noise in the original projections which weakens low-contrast structures even more.

Consequently, reprojecting the reference reconstruction and comparing the original and reprojected projections directly becomes challenging. To address this issue, we focus on identifying the most prominent structures, such as jawbones and teeth, by employing implicit regularization with a threshold-based segmentation of the reference reconstruction as shown in Fig. 4b. During this particular step, our focus is on segmenting solely the hard tissues within the reference reconstruction. To accomplish this, we substitute any attenuation values exceeding a specific threshold with a constant value that closely approximates the average attenuation of the hard tissues. Any values below the threshold are discarded. The threshold selection is guided by the significant disparity in attenuation values observed between soft and hard tissues. Forward projecting the regularized reconstruction will provide a high contrast synthetic projection as shown in Fig. 5b, which facilitates robust registration.

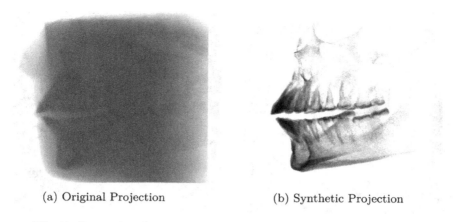

(a) Original Projection   (b) Synthetic Projection

Fig. 5. Comparison between original and synthetic forward projections.

3.3 Similarity Cost Function

After regularization, we use the strong edges that were extracted to produce synthetic projections. We can then use a gradient-based approach for image registration. To do this, we define a cost function based on Gradient Information, $G(r, a)$, which is the sum of the minima of gradient magnitudes computed for corresponding pixels in the synthetic image (r) and the actual image (a). The minima operator is applied to exclude strong extraneous gradients [3]. This function is multiplied by a weighting function w and can be expressed as follows:

$$G(r, a) = \sum_{(\mathbf{x}, \mathbf{x}') \in (r \cap a)} w_{\mathbf{x}, \mathbf{x}'} \, \min\left(|\nabla \mathbf{x}|, |\nabla \mathbf{x}'|\right) \tag{3}$$

Here, $\mathbf{x}$ and $\mathbf{x}'$ refer to the corresponding sample points (pixels) in r and a, respectively. $|\nabla \mathbf{x}|$ and $|\nabla \mathbf{x}'|$ are the gradient magnitudes of r and a, respectively. The weighting function w is the gradient orientation which is computed as the cosine of the angle between the two gradients and can be expressed as follows:

$$w_{\mathbf{x}, \mathbf{x}'} = cos(\theta) = \frac{\nabla \mathbf{x} \cdot \nabla \mathbf{x}'}{|\nabla \mathbf{x}| \, |\nabla \mathbf{x}'|} \tag{4}$$

4 Implementation and Results

The reconstruction libraries were provided by See Through S.r.l. The experiments were conducted using Powell's conjugate direction optimizer. The evaluation of the resulting reconstruction's quality was done through the application of the Structural Similarity Index Measure (SSIM) and Root-Mean-Square Error (RMSE). To assess the quality of the final reconstruction, a motion-free reconstruction of the 360° scan was performed as the ground truth (Figs. 6c, 7c, 8c).

4.1 Experimental Results

Table 1 provides a quantitative comparison of motion-affected and motion-compensated reconstructions for the three types of motions mentioned in Sect. 2.2, as evaluated by the SSIM and RMSE metrics. For nodding and abrupt motions, there was a significant improvement in SSIM and a large improvement in RMSE for motion-compensated reconstruction. For tilting motion, there was only a small improvement in SSIM and a moderate improvement in RMSE for motion-compensated reconstruction.

Table 1. SSIM and RMSE values for different motions [MA = motion-affected, MC = motion-compensated]

Motion type	SSIM ($\uparrow$)			RMSE ($\times 10^{-2}$)($\downarrow$)		
	MA	MC	Diff %	MA	MC	Diff %
Nodding	0.902	0.953	+5.1	1.92	0.95	−0.97
Tilting	0.912	0.914	+0.2	2.06	1.25	−0.81
Abrupt motion	0.817	0.932	+11.5	4.35	1.27	−3.08

The qualitative comparison between the ground truth (motion-free), motion-affected, and motion-compensated reconstructions is shown in Figs. 6, 7, and 8. The proposed method was effective in compensating for motion-induced artifacts in nodding and non-returning abrupt motion cases, as almost all artifacts were successfully compensated. In the case of tilting motion, some minor artifacts were still present, but the crucial diagnostic structures such as teeth were unaffected. The results show that the proposed method can successfully compensate for most motion-induced artifacts, leaving only minor artifacts in some cases.

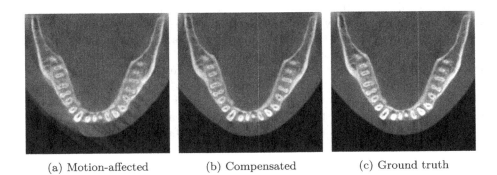

(a) Motion-affected   (b) Compensated   (c) Ground truth

Fig. 6. Results for nodding motion

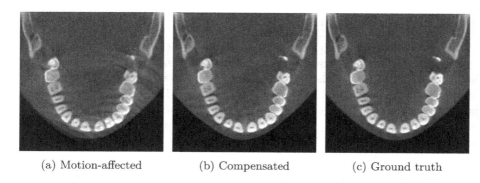

(a) Motion-affected   (b) Compensated   (c) Ground truth

Fig. 7. Results for tilting motion

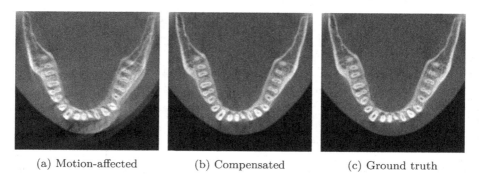

(a) Motion-affected   (b) Compensated   (c) Ground truth

Fig. 8. Results for non-returning abrupt motion

4.2   Failure Cases

Although it is reasonable that the movement of the patient is restricted and confined to a smaller scanning region, we attempted to examine the performance of this method using extensive and prolonged motion scenarios. A non-returning abrupt motion was simulated, where the patient remains in one pose for almost half of the scan, and then suddenly moves to another pose and remains still for the rest of the scan. The quantitative results are presented in Table 2. In this case, it is difficult in the regularization step to remove the double contours created by motion as these contours have similar contrast as that of the motion-free structures (Fig. 9a). Interestingly, our algorithm was able to remove the double contours from the final reconstruction but streak artifacts were introduced making any diagnosis difficult (Fig. 9b).

Table 2. SSIM and RMSE values for different motions [MA = motion-affected, MC = motion-compensated]

Motion type	SSIM (↑)			RMSE ($\times 10^{-2}$)(↓)		
	MA	MC	Diff %	MA	MC	Diff %
Abrupt motion	0.720	0.644	−7.6	8.33	7.47	−0.86

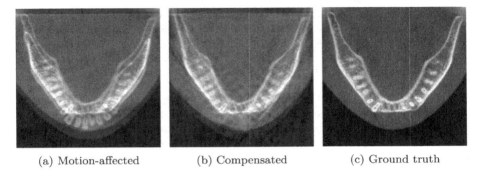

(a) Motion-affected   (b) Compensated   (c) Ground truth

Fig. 9. Results for the non-returning abrupt motion affecting half of the scan duration

5   Conclusion

Our approach utilizes a non-iterative technique to effectively compensate for artifacts caused by patient motion. By incorporating regularization into the reference reconstruction, our method can provide high-quality reconstruction without the need for multiple full scan reconstructions during the compensation stage for most motion types. However, for prolonged motions, additional constraints need to be applied to the proposed method to enhance the quality of reconstructions. Further testing involving complex motions and clinical data is necessary to assess the reliability of the proposed method across various scenarios.

Acknowledgements. The authors thank Michele Antonelli, Ivan Tomba, Andrea Delmiglio, Lorenzo Arici from See Through s.r.l., and Prof. Giovanni Di Domenico from the University of Ferrara, Italy for providing the projection data and the reconstruction libraries.

References

1. Aichert, A., et al.: Epipolar consistency in transmission imaging. IEEE Trans. Med. Imaging **34**(11), 2205–2219 (2015)
2. Birklein, L., Niebler, S., Schömer, E., Brylka, R., Schwanecke, U., Schulze, R.: Motion correction for separate mandibular and cranial movements in cone beam CT reconstructions. Med. Phys. **50**(6), 3511–3525 (2023)

3. De Silva, T., et al.: 3D–2D image registration for target localization in spine surgery: investigation of similarity metrics providing robustness to content mismatch. Phys. Med. Biol. **61**(8), 3009 (2016)

4. Donaldson, K., O'Connor, S., Heath, N.: Dental cone beam CT image quality possibly reduced by patient movement. Dentomaxillofac. Radiol. **42**(2), 91866873 (2013)

5. Feldkamp, L.A., Davis, L.C., Kress, J.W.: Practical cone-beam algorithm. Josa a **1**(6), 612–619 (1984)

6. Hanzelka, T., et al.: Movement of the patient and the cone beam computed tomography scanner: objectives and possible solutions. Oral Surg. Oral Med. Oral Pathol. Oral Radiol. **116**(6), 769–773 (2013)

7. Hartley, R., Zisserman, A.: Multiple View Geometry in Computer Vision. Cambridge University Press, Cambridge (2004). https://books.google.fr/books? id=e30hAwAAQBAJ

8. Kim, J., Sun, T., Alcheikh, A., Kuncic, Z., Nuyts, J., Fulton, R.: Correction for human head motion in helical X-ray CT. Phys. Med. Biol. **61**(4), 1416 (2016)

9. Maur, S., Stsepankou, D., Hesser, J.: CBCT auto-calibration by contour registration. In: Medical Imaging 2019: Physics of Medical Imaging, vol. 10948, pp. 413–421. SPIE (2019)

10. Nardi, C., et al.: Metal and motion artifacts by cone beam computed tomography (CBCT) in dental and maxillofacial study. Radiol. Med. **120**(7), 618–626 (2015)

11. Nemtoi, A., Czink, C., Haba, D., Gahleitner, A.: Cone beam CT: a current overview of devices. Dentomaxillofac. Radiol. **42**(8), 20120443 (2013)

12. Ouadah, S., Jacobson, M., Stayman, J.W., Ehtiati, T., Weiss, C., Siewerdsen, J.H.: Correction of patient motion in cone-beam CT using 3D–2D registration. Phys. Med. Biol. **62**(23), 8813 (2017)

13. Patel, S., Durack, C., Abella, F., Shemesh, H., Roig, M., Lemberg, K.: Cone beam computed tomography in endodontics-a review. Int. Endod. J. **48**(1), 3–15 (2015)

14. Sisniega, A., Stayman, J.W., Yorkston, J., Siewerdsen, J., Zbijewski, W.: Motion compensation in extremity cone-beam CT using a penalized image sharpness criterion. Phys. Med. Biol. **62**(9), 3712 (2017)

15. Small, B.W.: Cone beam computed tomography. Gen. Dent. **55**(3), 179–181 (2007)

16. Spin-Neto, R., Mudrak, J., Matzen, L., Christensen, J., Gotfredsen, E., Wenzel, A.: Cone beam CT image artefacts related to head motion simulated by a robot skull: visual characteristics and impact on image quality. Dentomaxillofac. Radiol. **42**(2), 32310645 (2013)

17. Spin-Neto, R., Matzen, L.H., Schropp, L., Gotfredsen, E., Wenzel, A.: Movement characteristics in young patients and the impact on CBCT image quality. Dentomaxillofac. Radiol. **45**(4), 20150426 (2016)

18. Spin-Neto, R., Matzen, L.H., Schropp, L.W., Sørensen, T.S., Wenzel, A.: An ex vivo study of automated motion artefact correction and the impact on cone beam CT image quality and interpretability. Dentomaxillofac. Radiol. **47**(5), 20180013 (2018)

19. Spin-Neto, R., Matzen, L.H., Schropp, L., Gotfredsen, E., Wenzel, A.: Factors affecting patient movement and re-exposure in cone beam computed tomography examination. Oral Surg. Oral Med. Oral Pathol. Oral Radiol. **119**(5), 572–578 (2015)

20. Sun, T., Jacobs, R., Pauwels, R., Tijskens, E., Fulton, R., Nuyts, J.: A motion correction approach for oral and maxillofacial cone-beam CT imaging. Phys. Med. Biol. **66**(12), 125008 (2021)

21. Wein, W., Ladikos, A., Baumgartner, A.: Self-calibration of geometric and radiometric parameters for cone-beam computed tomography. In: 11th International Meeting on Fully Three-Dimensional Image Reconstruction in Radiology and Nuclear Medicine, vol. 2, pp. 1–4 (2011)

22. Wicklein, J., Kyriakou, Y., Kalender, W.A., Kunze, H.: An online motion-and misalignment-correction method for medical flat-detector CT. In: Medical Imaging 2013: Physics of Medical Imaging, vol. 8668, pp. 466–472. SPIE (2013)

23. Würfl, T., Hoffmann, M., Aichert, A., Maier, A.K., Maaß, N., Dennerlein, F.: Calibration-free beam hardening reduction in X-ray CBCT using the epipolar consistency condition and physical constraints. Med. Phys. 46(12), e810–e822 (2019)

24. Yıldızer Keriş, E.: Effect of patient anxiety on image motion artefacts in CBCT. BMC Oral Health 17(1), 1–9 (2017)

Spatial Exploration Indicators in the Remote Assessment of Visual Neglect

Federica Ferraro[1(✉)], Giulia Iaconi[1], Giulia Genesio[1], Romina Truffelli[3], Roberta Amella[3], Marina Simonini[3], and Silvana Dellepiane[1,2]

[1] Department of Electrical, Electronics and Telecommunication Engineering and Naval Architecture (DITEN), Università degli Studi di Genova, Genova, Italy
federica.ferraro@edu.unige.it

[2] RAISE Ecosystem, Genova, Italy

[3] Struttura Complessa Recupero e Rieducazione Funzionale, La Colletta Hospital, ASL3 Sistema Sanitario Regione Liguria, Genova, Italy

Abstract. The visual neglect is a visuo-spatial attention disorder associated with stroke events and its presence is considered a negative prognostic factor of functional recovery. The specific assessment tests that are most frequently administered to evaluate the presence of this disorder are "paper and pencil tests", such as barrage tests. The current work presents the digital version of the Albert's barrage test for the evaluation of the Unilateral Spatial Neglect (USN) as it has been integrated into the ReMoVES tele-rehabilitation system. The data captured by this activity is broader and more complete than the information extractable from the paper version, including the order of the exploration sequence, work trajectories, punctual speed and other indicators. The procedure is preliminarily validated both on a control group of healthy subjects and on some patients. Several sessions are examined with the aim of observing which parameter of the digital Albert Test is statistically more significant for verifying the similarity in behavior between non-pathological subjects and the possible improvement over time of two pathological case studies.

Keywords: Spatial exploration · USN · assessment test

1 Introduction

Visual neglect, more properly known as Unilateral Spatial Neglect (USN), is one of the disabling features of several neurodegenerative diseases and is defined as an inability to pay attention to the side opposite a brain lesion [1,2]. It is estimated that as many as 25%–30% of patients experience USN following a stroke and although it can be a consequence of both right and left hemispheric lesions, USN is much more frequent after a right hemispheric injury and affects about 90% of patients [3]. Affected subjects require longer hospitalizations and on discharge show reduced functional autonomy, which over time determines a reduction in quality of life and less independence in the chronic phases. The

G. L. Foresti et al. (Eds.): ICIAP 2023, LNCS 14234, pp. 552–563, 2023.
https://doi.org/10.1007/978-3-031-43153-1_46

manifestations of USN involve different aspects and have different characteristics among patients with different clinical pictures. Visual neglect is therefore not a unitary phenomenon and as such must be considered at a clinical level, both for a correct diagnosis and for setting up a rehabilitation plan. In most patients, USN is not very evident and it is necessary to assess it with specific tests, which are also necessary to produce objective measures of severity and to monitor recovery during rehabilitation. USN evaluation makes use of the observation of the patient, clinical interviews, a battery of "paper and pencil" tests, evaluation scales, and reading tests. These tests are quick and easy to administer but often results in a time consuming endeavor by the clinical staff. There are several paper and pencil tests, among which the most important are: bisection tests, barrage tests, and drawing tests [4,5]. In the barrage tests the patient has to look for and mark specific targets on the sheet. If a person is afflicted with USN, he probably will not cancel the stimuli present on the side of the paper opposite the lesion. Performance in these tests is directly proportional to the presence of distractors, i.e., stimuli that must be ignored by the patient [6].

In the current work the ReMoVES tele-rehabilitation platform is proposed [7] which integrates a digital version of USN assessment tests [8], which may also possibly be delivered in conjunction with transcranial Direct-Current Stimulation (tDCS) [9].

A cross-validation study has shown that ReMoVES digital version of the Albert's test has an 82% correlation with independent paper test and therefore denotes a good concurrent validity [10]. Additionally, the study [8] proposed a statistical analysis dedicated to the validation of the test-retest reliability. It includes an assessment of the correlation between digital and traditional test administration, enabling the evaluation of the validity of remotely monitoring home-based test administration via the ReMoVES platform.

Various studies in the literature have been conducted to assess barrage tests for USN disease. For instance, the study [11] quantified the spatial organization by reconstructing pathways used for marking targets and suggests the digital implementation through touchscreen. The authors of [12] propose a mixed tablet-paper test to detect abnormalities in task processes and outcomes. The results of the preliminary study [13] showed good sensitivity, specificity and usability of digital tests, suggesting that they can be a promising tool for USN assessment, both in clinical and research settings.

In the present work, Albert's test is applied to a population of healthy subjects to obtain a characterization of the control sample in statistical terms. The most significant spatio-temporal features are remotely extracted and can be used for comparison with post-stroke patients affected by USN. The repetition at different times allows to detect changes in the patients behavior.

2   Materials and Methods

ReMoVES is a markerless/contactless IoT system developed at DITEN department of the University of Genova to support rehabilitation activities [7]. Motor and cognitive exercises and digital assessment tests are delivered through exergames performed using motion sensors or touchscreen, without wearing markers and without the use of controllers. It is based on a multi-client/server architecture and provides the rehabilitation team with objective data even remotely and has also been successfully used during the COVID-19 outbreak.

After a very brief introduction of the Albert digital test, the present section describes the experimental sessions performed at "La Colletta" hospital and the subsequent data analysis.

2.1   ReMoVES Digital Tests for USN Assessment

For the assessment of USN ReMoVES provides digital versions of the Albert's Test, the Line Bisection, and the Apple Test [10], as shown in Fig. 1. The system always provides for the memorization of each test session, its reproducibility and comparison with other game sessions, and automatic data processing methods. Thanks to this analysis, therapists have access to reliable and objective data monitoring, even when they cannot supervise their patients in the presence.

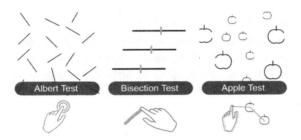

Fig. 1. Thumbnails of the activities via ReMoVES and corresponding interaction gestures.

For each test, the platform is able to provide additional information compared to the paper version and offers the possibility of performing more detailed analyses. The interface of the activities was designed using the Unity platform, which is a popular graphic engine used for the development of games and virtual reality applications.

The traditional Albert test consists of a sheet with forty differently oriented segments grouped into seven columns: one in the center, three on the right, and three on the left. The actual arrangement of these lines is standardized, allowing for a systematic analysis of subjects' performance relative to the left, right and center of the page. The patient is asked to mark all the lines he sees on the page with the pen. The maximum score for this paper test is 36 (the four segments of the central column are not counted). The cut-off is 34. The digital version of Albert's test is carried out in the same way as the paper one, but through the use of a touchscreen. Two minutes are given to complete the test. An example of the screen is displayed in the Fig. 2.

Fig. 2. Screenshot of digital Albert's test on touchscreen.

In the traditional "paper and pencil" version of the Albert's Test the fundamental indicators are:

- the number of correctly identified targets;
- the overall time to complete the test.

In the digital version additional indicative parameters can be extracted, by memorizing the patient's touch on the screen, saved as spatial coordinates:

- the execution order (spatial trajectory);
- the average distance between the correctly identified targets;
- the average time between correctly identified targets.

2.2   Dataset

In the present study, a control group of fourteen healthy subjects, part of the hospital's clinical staff, was considered. Participants were 9 females and 5 males aged between 27 and 61 years. These subjects performed the digital test twice (at time points T0 and T1); between the two instants of time there was no change in the state of the subjects. The most significant indicators related to the exploration of the visual space have been extracted.

Several patients affected by USN performed the ReMoVES Albert test for a total of about 50 sessions. In the current paper two specific cases are discussed related to their improvement. Patient A has a right hemisphere lesion and patient B has a left hemisphere lesion. After hospitalization patients were contacted for a neuropsychological evaluation, performing the tests both in digital and paper format (time T0). After about a year, the same patients were again recalled from the facility for a new re-evaluation of their cognitive-motor status (time T1).

2.3   Spatio-Temporal Feature Analysis

For the statistical analysis, two new parameters that can be obtained from the digital version of the test were considered, namely the average distance between two correctly identified targets and the average time between two correctly identified targets. The minimum values of the parameters indicate an improvement or the absence of disability. First, the characterization of the control sample was performed in order to estimate the mean and deviation values of the indicators. In particular, the variations over time of the aforementioned characteristics are considered taking into account the instant T0 and T1, let's name P0 and P1 respectively. For a better understanding of the given situation, the Bland Altman graph is analyzed for both parameters. Subsequently, each patient is compared with the control group for an evaluation of any improvements.

Statistical Test on Differences. The variation of the index, i.e., $(P0 - P1)$, is expected to be very small for the control group where the subjects repeat Albert's test at time T0 and T1. Let's define μ_Δ the mean of this variation observed in the control population. S is the corrected estimate of the standard deviation.

From a statistical point of view, to verify if the results of a patient at time T1 have changed significantly with respect to time T0, a test of differences was used.

Several patients are then evaluated who also repeat Albert's test at time T0 and time T1. Each subject is tested against the reference population. The patient variation is the X_Δ variable. The null hypothesis $H0$ assumes that the difference between the results at time T1 (P1) and the results at time T0 (P0) is small as for the control population and therefore the variation on the index is not significant. The alternative hypothesis $H1$ considers that the difference between P0 and P1 is large (in absolute value) and therefore significant.

$$\begin{aligned} \text{Hypothesis H0:} \quad & P0 - P1 = \mu_\Delta \\ \text{Hypothesis H1:} \quad & |P0 - P1| > \mu_\Delta \end{aligned} \tag{1}$$

With a small sample size n, a t-Student's bilateral statistical test of differences is applied:

$$T = \frac{X_\Delta - \mu_\Delta}{S}\sqrt{n} \tag{2}$$

where T is a t-Student variable with $(n-1)$ degrees of freedom, X_Δ the patient parameter (variation $P0 - P1$) to be compared with μ_Δ, i.e., the population mean of healthy subjects.

Statistical Test on Means. In the case of improvements in a patient's indicator, a statistical test on the mean (right-tailed) was applied to verify if the patient value at time T1 is statistically close to the mean of the healthy population. As previously described, no variations between instant T0 and instant T1 were observed in the healthy population, therefore only the values at instant T1 are taken as reference. The mean on the control population is μ_{T1}. Each patient is tested against the reference population (variable X_{T1}). The null hypothesis $H0$ assumes that the patient's value at time T1 does not differ from the mean of the control group, and therefore may belong to that population. The alternative hypothesis $H1$ assumes that the value at time T1 of the subject is significantly larger than the mean of the control group, and therefore the patient does not belong to the reference population.

$$\text{Hypothesis H0:} \quad X_{T1} = \mu_{T1}$$
$$\text{Hypothesis H1:} \quad X_{T1} > \mu_{T1} \tag{3}$$

Also in this case the test statistics is a t-Student's distribution, having used the variance estimate and being n small.

$$T = \frac{X_{T1} - \mu_{T1}}{S} \sqrt{n} \tag{4}$$

where T is a t-Student variable with $(n-1)$ degrees of freedom, X_{T1} is the patient parameter at time T1 to be compared with μ_{T1}, i.e., the population mean of healthy subjects.

Bland-Altman Plot. The Bland-Altman plot is a scatter plot that allows one to evaluate the agreement between two quantitative assessments. The vertical axis shows the differences between the two measurements. On the horizontal axis the arithmetic means of the two measurements are given. On this graph, in addition to the points representing the individual statistical units, there are also horizontal lines. In particular, the center line represents the mean of the differences between the measurements, whereas the two horizontal dotted lines at the top and bottom delimit a band that represents the limits of the 95% confidence interval of the mean of the differences. A modified version of the Bland-Altman plot is proposed here in which vertical lines are added. One lies at the global mean (i.e., mean bias); the others are the limit of the 95% confidence interval. Two vertical limits appear for bilateral tests; a single line for the one-sided test as in the present case.

3   Experimental Results

This section provides a system application for evaluating USN patients. The sessions relating to Albert's test will be analyzed, highlighting the changes that have occurred over time through in-depth statistical analysis.

3.1   Spatial Analysis of Albert's Test

As already said, in addition to the information on the canceled targets, which can also be found in the paper version of the test, the parameters studied in the digitized version of the Albert's test are the spatial exploration method through the order of execution, the accuracy, the location, the type of errors and the erasing speed. The types of data and indicators were extracted and analyzed thanks to the possibility of ReMoVES system to memorize each test session.

The analysis and comparison of the trajectories performed by the subjects with USN and the healthy population (Fig. 3) allows to visualize the number and order of the deleted targets and the distribution area, providing information about the spatial exploration modality. The trajectories are automatically calculated by ReMoVES and displayed in the Therapist Client.

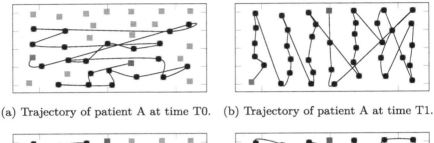

(a) Trajectory of patient A at time T0.   (b) Trajectory of patient A at time T1.

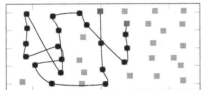

(c) Trajectory of patient B at time T0.   (d) Trajectory of patient B at time T1.

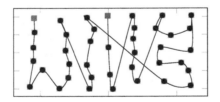

(e) Trajectory of one of the healthy subjects.

Fig. 3. Trajectories followed by patient A, patient B and one of the healthy subjects. The blue points are the starting points, the red points are the final ones. (Color figure online)

For patient A, the digital Albert's test at time T1 highlights a new attentive attitude compared to time T0, since the patient has detected all the targets, but primarily has followed an organized trajectory, with a vertical observation trend. Patient B maintained the vertical exploration trend, succeeding in detecting all targets at time T1. Through the other indicators it is possible to trace the clear improvement of the total and average time compared to the digital test at time

T1 performed by the subjects: for patient A the total time was reduced by about three times going from 92.33 s to 33.28 s, while patient B had an improvement of about half going from 111.68 s to 60.26 s. The results obtained on the average times and average distances between consecutive canceled targets are shown in Tables 2 and 3. The reduction in the average distance indicates a more effective exploring strategy and a more uniform erasing pattern.

3.2 Statistical Analysis of Digital Albert's Test

To obtain reference values, the healthy population (control group) is defined as described in Sect. 2.3. After the 14 subjects repeated Albert's test at time T0 and time T1, the mean ratings on the distance and time parameters and their differences Δ were observed as reported in Table 1.

Table 1. Mean values of distance and time parameters in time T0 and T1, and their difference Δ for the control group.

Distance [cm]			Time [s]		
μ_{T0}	μ_{T1}	μ_Δ	μ_{T0}	μ_{T1}	μ_Δ
2.31	2.27	0.037	0.45	0.44	0.015

Patients also repeated Albert's test at time T0 and time T1. Patient A and B achieved respectively the results shown in Tables 2 and 3.

Table 2. Mean values of distance and time parameters in time T0 and T1, and their difference Δ for patient A.

Distance [cm]			Time [s]		
X_{T0}	X_{T1}	X_Δ	X_{T0}	X_{T1}	X_Δ
3.64	2.91	0.73	5.04	0.83	4.21

Table 3. Mean values of distance and time parameters in time T0 and T1, and their difference Δ for patient B.

Distance [cm]			Time [s]		
X_{T0}	X_{T1}	X_Δ	X_{T0}	X_{T1}	X_Δ
2.71	2.50	0.21	6.11	1.57	4.54

Modified Bland-Altman Plot. To better visualize the statistical results obtained, the Bland-Altamn plots are analysed. Figure 4 is related to the distance parameter. Black dots represent the distance values of the control group. The blue horizontal line represents the bias over the difference between T0 and T1 (difference bias) equal to $\mu_\Delta = 0.037$. The blue dotted horizontal lines are the upper and lower limits as regards to the difference bias equal to $\mu_\Delta \pm 2.16\frac{S}{\sqrt{n}}$. In addition to the classic plot, the limit relating to the mean between values obtained at T0 and T1 have been included, represented by the red dotted vertical line in relation to the red vertical line (mean bias), calculated as $2.29 + 1.77\frac{S}{\sqrt{n}}$.

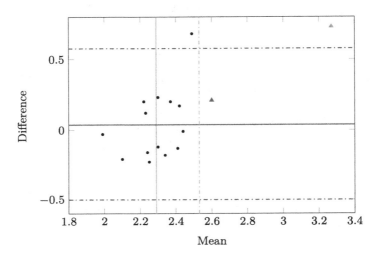

Fig. 4. Modified Bland-Altman plot related to distance parameter. The continuous lines are the biases. The dotted lines The dotted lines are the confidence intervals limits with respect to the corresponding line. Black dots represent distance values of the control group. Pink triangle is the distance value for patient B and green triangle is the distance value for patient A. (Color figure online)

Results related to the average time between two consecutive target is shown in Fig. 5. In this case, the blue continuous horizontal line i.e., the difference bias between T0 and T1 is equal to $\mu_\Delta = 0.015$. In a similar way to what was seen for the first parameter analyzed, the blue dotted horizontal lines are the upper and lower limits as regards to the difference bias equal to $\mu_\Delta \pm 2.16\frac{S}{\sqrt{n}}$. Lastly, the limit relating to the mean bias between values obtained at T0 and T1 is equal to $0.45 + 1.77\frac{S}{\sqrt{n}}$.

Test on Differences. As already described, the average distance between consecutive targets is considered as the former parameter to be evaluated. For patient A, the statistics of the difference test is $T = 10.39$ and leads to the

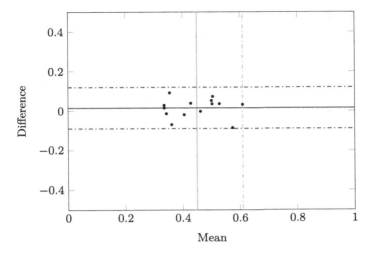

Fig. 5. Modified Bland-Altman plot related to time parameter. Continuous lines are the biases. The dotted lines are the upper and lower limits with respect to the corresponding line. Black dots represent time values of the control group.

certain rejection of the null hypothesis $H0$, i.e., there is an extremely large difference between the value obtained at time T0 and that at time T1. Hence the test is highly statistically significant, with a $p - value \rightarrow 0$.

In the second case, for patient B, the obtained value of $T = 2.59$ concludes that the test is partially significant. We get $p - value = 0.022$ so the test rejects the null hypothesis if $\alpha = 0.05$, but the null hypothesis is not rejected if $\alpha = 0.01$; therefore the test is weakly significant. Finally, in both cases, since the T value is positive, it can be concluded that there was an improvement in the parameter considered, between time T0 and time T1.

Now, the average time taken to cancel consecutive targets is considered as the latter parameter to evaluate. For both patients, an extremely high T value is obtained which leads to the certain rejection of the $H0$ hypothesis. The $p - value$ is so small that the test is significant for any value of α. As a conclusion both patients had a highly significant improvement with respect to the time parameter considered.

Test on Means. Considering the average distance between consecutive targets as a parameter, for both patients, the values obtained for T lead to reject the hypothesis $H0$, therefore the two subjects do not belong to the reference population. The $p - value$ is so small that the test is significant for any value of α.

Also for the cancellation time between targets, for both subjects a very high value T is obtained which leads to the rejection of the hypothesis $H0$, with $p - value \rightarrow 0$. Therefore the subjects do not belong to the reference population.

562 F. Ferraro et al.

If the Bland-Altman plots are considered anew, from the plot in Fig. 4 it can be deduced that patient B (pink triangle) had a small difference between T0 and T1 on the distance parameter, falling within the limits of healthy subjects. Conversely, patient A (green triangle) had a more significant difference between T0 and T1 because he is beyond the upper limit. On the other hand, compared to the mean, both patients are beyond the upper vertical limit and this denotes that they do not fall within the healthy population.

As discussed in the previous paragraphs, if the average time between the cancellation of consecutive targets is considered (Fig. 5), for both patients the difference between T0 and T1 is very high compared to the group of healthy people and this also implies that on average the parameter value of the patients differs greatly from those of the healthy, not making them belong to this population.

4 Conclusions

The digital version of Albert's Test is able to provide many additional information, which would be difficult to find in the paper and pencil version. In this study, this information was able to objectify and describe in detail the clinical course over time and the abilities of patients affected by USN. The ability to identify the start and end point of the track and the trajectory path followed in the test provide initial information on space exploration capability. The data provided could also be of great use for monitoring the changes during the rehabilitation treatment and being able to report even small improvements in the patient's visuo-spatial organization, thus making rehabilitation support more effective. Regarding the use in tele-consultation and in tele-rehabilitation for which the ReMoVES system was designed, the possibility of evaluating and monitoring the progress of the clinical picture remotely constitutes an improvement in treatments compared to other tele-rehabilitation systems.

However, additional studies on a wider population are needed to better assess the effectiveness of the digital version of the tests in assessing and tracking the evolution of the USN.

Acknowledgments. ReMoVES system takes part of the project "RAISE - Robotics and AI for Socioeconomic Empowerment" and this work has been partially supported by European Union - NextGenerationEU. However, the views and opinions expressed are hose of the authors alone and do not necessarily reflect those of the European Union or the European Commission. Neither the European Union nor the European Commission can be held responsible for them.

References

1. Kandel, E.R., Schwartz, J.H., Jessell, T.M., Siegelbaum, S., Hudspeth, A.J., Mack, S., et al.: Principles of Neural Science, vol. 4. McGraw-Hill, New York (2000)
2. Li, K., Malhotra, P.A.: Spatial neglect. Pract. Neurol. 15(5), 333–339 (2015)

3. Corbetta, M., Kincade, M.J., Lewis, C., Snyder, A.Z., Sapir, A.: Neural basis and recovery of spatial attention deficits in spatial neglect. Nat. Neurosci. **8**(11), 1603–1610 (2005)
4. Mazzucchi, A.: La riabilitazione neuropsicologica 4 ed.: Premesse teoriche e applicazioni cliniche. Edra (2020)
5. Hebben, N., Milberg, W.: Essentials of Neuropsychological Assessment. Wiley, Hoboken (2009)
6. Plummer, P., Morris, M.E., Dunai, J.: Assessment of unilateral neglect. Phys. Ther. **83**(8), 732–740 (2003)
7. Trombini, M., Ferraro, F., Morando, M., Regesta, G., Dellepiane, S.: A solution for the remote care of frail elderly individuals via exergames. Sensors **21**(8), 2719 (2021)
8. Morando, M., Bonotti, E.B., Giannarelli, G., Olivieri, S., Dellepiane, S., Cecchi, F.: Monitoring home-based activity of stroke patients: a digital solution for visuospatial neglect evaluation. In: Masia, L., Micera, S., Akay, M., Pons, J.L. (eds.) ICNR 2018. BB, vol. 21, pp. 696–701. Springer, Cham (2019). https://doi.org/10.1007/978-3-030-01845-0_139
9. Trombini, M., et al.: Unilateral spatial neglect rehabilitation supported by a digital solution: two case-studies. In: 2020 42nd Annual International Conference of the IEEE Engineering in Medicine & Biology Society (EMBC), pp. 3670–3675. IEEE (2020)
10. Ferraro, F., Trombini, M., Truffelli, R., Simonini, M., Dellepiane, S.: On the assessment of unilateral spatial neglect via digital tests. In: 2021 10th International IEEE/EMBS Conference on Neural Engineering (NER), pp. 802–806. IEEE (2021)
11. Woods, A.J., Mark, V.W.: Convergent validity of executive organization measures on cancellation. J. Clin. Exp. Neuropsychol. **29**(7), 719–723 (2007)
12. Potter, J., Deighton, T., Patel, M., Fairhurst, M., Guest, R., Donnelly, N.: Computer recording of standard tests of visual neglect in stroke patients. Clin. Rehabil. **14**(4), 441–446 (2000)
13. Massetti, G., et al.: Validation of "neurit. space": three digital tests for the neuropsychological evaluation of unilateral spatial neglect. J. Clin. Med. **12**(8), 3042 (2023)

Author Index

Printed in the United States
by Baker & Taylor Publisher Services